BASIC CALCULUS WITH APPLICATIONS

DONALD R. WILLIAMS

North Texas State University

THOMAS J. WOODS

Central Connecticut State College

WADSWORTH PUBLISHING COMPANY

Belmont, California *A division of Wadsworth, Inc.*

Quantitative Methods Editor: Jon Thompson
Production Editor: Greg Hubit Bookworks
Cover Designer: Henry Breuer
Technical Illustrator: Scientific Illustrators, Inc.

To Mary and Connie

Printed in the United States of America

3 4 5 6 7 8 9 10—83 82

Library of Congress Cataloging in Publication Data

Williams, Donald R.
Basic calculus with applications.

Includes index.
1. Calculus. I. Woods, Thomas J., joint author.
II. Title.
QA303.W495 515 79-925
ISBN 0-534-00685-X

PREFACE

This textbook is written to assist students in developing calculus as a tool for use in the areas of management, biology, behavioral science, and social science. A sufficient number of examples and exercises are included so that the material can be tailored to emphasize applications in any one of these areas. The book is designed for a one- or two-semester course, depending on the background of the students and the goals of the instructor. The only prerequisite is two years of high school algebra or the equivalent. Following are some noteworthy features of this book:

1. *Style and Approach:* This text is written to be very readable from the students' viewpoint. The style of writing is "let us take this path together" rather than a "lecturing" presentation. The approach is intuitive, using the students' background and experience to introduce new ideas and concepts. Properties and rules are established on the basis of geometric considerations, observation of data, and induction. Generalizations are clearly stated, and arguments for their validity are given. We have avoided rigorous proofs, although the arguments used can be expanded if desired. The arguments selected are both convincing to the student and mathematically sound. Based on experience with our own students, we find that this approach prevents the material from becoming a "cookbook," and at the same time eliminates the possibility of having "rigor mortis" set in. For example, limits are introduced in Chapter 2 as an outgrowth of the consideration of horizontal and vertical asymptotes of rational functions, and then limits are used in the discussion of continuous growth and decay in Chapter 3. Further limit properties are

established intuitively in Chapter 4, and these properties are stated as rules that can be proven to be true. After these rules are applied, they are used in conjunction with the definition of derivative to arrive at the various differentiation rules.

2. *Applications:* An important feature of this textbook is the large number of applications chosen from the areas of economics, business, biology, ecology, medicine, psychology, and sociology. Many of these applications are used to introduce new ideas and concepts, while others are discussed as examples in the text or presented as exercises. The applications that have been selected have a minimum number of specialized technical terms. When these terms are introduced, they are clearly defined and thoroughly explained, so that the reader need not be a specialist in that particular field to understand the problem. An indexed sampling of the many applications is given on the page following *A Note to the Student*.

3. *Examples:* Many examples are given to illustrate each new concept that is introduced. These range from direct application of a rule or formula to extended applied problems. The solutions of these examples are worked out in detail, enabling the student to follow the algebraic manipulations as well as the underlying logic. Where appropriate, examples are accompanied by carefully drawn illustrations. In some cases there are several drawings to show how the final solution has evolved. For example, in Chapter 5 in the discussion of curve sketching, two—and sometimes three—sketches are used to develop a solution, rather than just presenting the student with the completed sketch.

4. *Tables and Summaries:* There are more than the usual number of tables summarizing the suggested steps to carry out important procedures. For example, there are tabulated lists to consider for help in constructing mathematical models of real-world problems, for solving extrema problems, and for curve sketching. At the end of each chapter, important terms, concepts, rules and formulas are summarized for the student. These organized lists reinforce the most important topics covered in the chapter and aid the student in reviewing for exams.

5. *Calculator Problems:* Although this text is not designed to be used exclusively with a hand-held calculator, the student will find it advantageous to use one when working many of the exercises. For example, several of the applied problems involving exponential growth and decay (Chapters 3 and 6) lend themselves to solution with a calculator. For this reason, in *A Note to the Student*, we suggest the purchase of a calculator. A simple calculator should be sufficient, since we provide tables of exponential and logarithmic functions in the appendix. However, the purchase of a calculator with a power key should be beneficial as the increased versatility justifies the minimal additional cost.

6. *Organization of Content:* More material is included in the nine chapters of the text than can be expected to be covered in a one-semester or a two-quarters course. Choice of topics, examples, and exercises allows the

individual instructor to design a course to meet his or her students' needs. In the first three chapters, new ideas and concepts, such as limit, mathematical modeling, and continuous growth, are interwoven with a careful review and extension of the concept of function. These chapters lay the necessary groundwork for the calculus, and the pace at which this material is covered will vary. Chapters 4 through 7 form the bulk of a one-semester or a two-quarters course. Several sections in these chapters, marked by an asterisk in the table of contents, may be omitted without affecting the continuity of the text. Omitting these sections and emphasizing topics of interest in these chapters will provide time to include either or both Chapters 8 and 9. If a slower pace and more thorough coverage are desired, the nine chapters provide sufficient material for a three-quarters or two-semester course.

Answers to selected exercises are supplied at the back of the book.

Acknowledgments: Many individuals have assisted us in the preparation of this book, and we gratefully acknowledge their contributions. We express our appreciation to the following professors who reviewed the manuscript at various stages of its development: Charles Armstrong, University of Rhode Island; J. R. Foote, University of New Orleans; Thomas M. Green, Contra Costa Community College; Robert C. Juola, Boise State University; Joel Kagan, University of Hartford; Mel Mitchell, Clarion State College; Carla B. Oviatt, Montgomery College; Gordon Shilling, University of Texas at Arlington; Dale E. Walston, University of Texas at Austin.

We wish to thank Professor Elmer Delventhal of Central Connecticut State College for his many helpful suggestions during the preparation of the manuscript and for checking the galley proofs. A special note of gratitude is given to Rex Smith, our graduate assistant, who worked closely with us on the development of the manuscript, providing us with input from the student's viewpoint. We are deeply indebted to Joanne Bower, our typist, for her competence and cheerfulness in preparing the entire manuscript. Finally, we thank the staff of Wadsworth Publishing Company, especially Jon Thompson, for their encouragement and patient editorial assistance.

A NOTE TO THE STUDENT

In the preparation and design of this textbook we have kept several goals in mind. Our primary goal was to write a text that is both readable and interesting to you. We believe that the language, style, and presentation of material in this book achieve this goal. The necessary groundwork is laid for understanding and applying each new concept. Familiar examples from the real world are presented *before* abstract rules or unmotivated definitions are stated. Once a property or rule is sufficiently motivated, we formally state it and give several examples to illustrate its use. At the end of each chapter we summarize the important terms, concepts, rules, and formulas introduced for ease of reference.

Another goal was to relate the mathematics you are learning to the real world. In this text you will see how calculus can be used to solve problems from the areas of economics, business, biology, ecology, medicine, psychology, and sociology. A sampling of these applications is given on the following page. They are provided to encourage you to explore and appreciate the variety of uses of calculus prior to beginning the course.

Finally, we hope that you will become familiar with the use of hand-held electronic calculators. Although the text is not designed to be used exclusively with a calculator, you will save considerable time and energy by using this tool. A simple calculator (one that adds, subtracts, multiplies, and divides) will suffice, since we have provided necessary tables in the appendix. However, we recommend the purchase of a calculator with a power (y^x) key. This type is more versatile, and the additional cost is minimal.

Applications

The following list represents approximately one-fourth of the applied examples and exercises in the text. They are provided here to encourage you to explore and appreciate the variety of uses of the calculus prior to beginning the course.

Business and Economics

Biology and Medicine

Ecology

Psychology

Sociology

General Interest

CONTENTS

*Optional section which may be omitted without loss of continuity.

1 RELATIONS AND FUNCTIONS

1.1 INTRODUCTION

Every form of the news media reminds us daily of the challenges that face us collectively and as individuals. Pollution, energy, decaying cities, unemployment, and a troubled economy are just a few of the major issues. These are complex problems that do not lend themselves to simple analyses and easy solutions. However, many problems of the real world that evolve from or are part of these major issues are within the scope of analysis and solution by individuals or groups of individuals. Over the years various procedures have been developed to analyze and solve such problems. As these problems become more complex, old procedures are modified and new ones developed. For years engineers and architects have used physical models to study bridge construction, design of high-rise buildings, city redesign, etc. For example, a physical model of a car may be used by an automotive engineer to provide information on how bumper redesign will affect the other components of the car. In the same manner, mathematical models of real-world problems can be created to provide us with information about the various components of the problem we are studying.

A striking example of the use of such a model occurred in the fall of 1977. Just a few months earlier, New York City was three months from financial bankruptcy. To determine what had caused this near catastrophe, a number of large companies joined Pace University and the Wharton School of Finance to develop a mathematical model of New York City that would help answer some questions about the Big Apple's future. The team was able to design a mathematical model by identifying the economic, financial and social variables and constants. After collecting relevant data, developing computer programs, and using the computer, they were able to make conclusions on the effects of changes in tax laws, business conditions, social patterns, and other variables that make up an area's economy.

In this text we will consider problems of much less magnitude; however, our paths will be similar; that is, to first create mathematical models of problems and then analyze these models to predict the behavior of the components of the problem. One tool of analysis that we find most helpful is calculus. Although founded and developed in the 17th century by the English mathematical physicist Isaac Newton (1642–1727) and the German mathematician Gottfried Wilhelm Leibnitz (1646–1716), calculus was primarily used in the physical sciences until the last century. Today, applications of calculus are found in such diverse fields as business, biology, ecology, social sciences, and medicine. The goal of this text is to introduce you to calculus and make it available to you as a mathematical tool.

1.2 RELATIONS AND FUNCTIONS

One of the most important ideas in the study of real-world problems is the concept of a *relation*. Frequently we are interested in how one quantity is related to another. For example, a businessperson would be interested in

1. the dollar amount of new sales in relation to the dollars spent on a new advertising campaign.
2. the profit made in relation to the number of units manufactured and sold.

An ecologist would be interested in

1. the amount of pollutant in a river in relation to the distance from the source of pollution.
2. the growth rate of a planting of tree seedlings in relation to the density of the planting.

A sociologist would be interested in

1. the percentage of families on welfare in relation to population density.
2. the population growth rate in urban areas in relation to the median income of families in the area.

A biologist would be interested in

1. the number of bacteria present in a culture in relation to the number of hours it has been contaminated.
2. the light intensity in water in relation to the depth at which the light is being measured.

... And you are interested in the grade you receive in a course in relation to the amount of effort you expend in that course.

The relationship between two quantities, or variables, of concern can often be expressed by a rule or formula. For example, it can be shown that if you deposit $1,000 in a bank and let it accumulate interest at the rate of 8% compounded annually, you will have an amount of A dollars in t years given by the formula

$$A = 1{,}000(1.08)^t \tag{1.1}$$

This equation is written to imply that A, the amount of money accumulated, depends upon t, the number of years the $1,000 has been left on deposit. For this reason A is called the *dependent variable* and t the *independent variable*.

There are situations, on the other hand, where it is difficult or even impossible to find a simple rule or formula to describe the relationship between the two quantities of interest. In such cases we may devise a rule or formula that gives a good approximation of the relationship, or we may be satisfied with a listing or tabulation of the data. For example, suppose we are interested in describing the relationship between the time of day in New York City and the Dow Jones industrial stock average. Obviously no rule or formula exists that will predict this relationship for any given day in the future. In such situations we can study historical data. The data for any given day may be presented in tabular form such as

Time of day	10	11	12	1	2	3	4
D.J.I.	840	842.5	843	844	843	843	842

or in set notation as

$$S = \{(10, 840), (11, 842.5), (12, 843), (1, 844), (2, 843), (3, 843), (4, 842)\}$$

In the latter case we have listed or rostered the data or members. Each member of our set S is a *pair of numbers* that is always written in the same order; that is, the *first component* is the time and the *second component* is the Dow average. Each of these elements is called an *ordered pair*. The following is a more formal definition of an ordered pair.

Definition 1.1 An ordered pair is an element of the form (x,y) where the variable x is called the first component and the variable y is called the second component. The ordered pairs (x,y) and (s,t) are equal if and only if $x=s$ and $y=t$.

Sets of ordered pairs can be constructed by combining the elements from two sets and forming the *cartesian product* or cross-product of the two sets. If

$X = \{2,4,6\}$ and $Y = \{1,2,3,4\}$, their cartesian product is

$$X \times Y = \{(2,1), (2,2), (2,3), (2,4), (4,1), (4,2), (4,3), (4,4), (6,1), (6,2), (6,3), (6,4)\}$$

The cartesian product $X \times Y$ (read "X cross Y") is formed by applying the following definition.

Definition 1.2 The cartesian product of the two sets X and Y, denoted by $X \times Y$, is the set of all ordered pairs (x,y) such that the first element x is from the set X and the second element y is from the set Y.

If a formula or rule is known that relates two variables, we can express this relationship using set builder notation. The relationship between A and t given by Eq. (1.1) expressed in set builder notation is

$$S = \left\{(t,A)\text{: } A = 1{,}000(1.08)^t, \quad t \text{ is a whole number}\right\}$$

This is read, "the set S equals the set of all ordered pairs (t,A) such that $A = 1{,}000(1.08)^t$ and t is a whole number." The rule or formula $A = 1{,}000(1.08)^t$ is called the *set selector* or *set builder* since it selects the ordered pairs that belong to the set S.

Example 1.1 If $S = \{(t,A)\text{: } A = 1{,}000(1.08)^t, t \in W\}$, determine five elements of the set S.

Solution Since t is any whole number, we arbitrarily select t values of 0, 1, 2, 3, and 4 and calculate the values of A by using the formula $A = 1{,}000(1.08)^t$ as shown in Table 1.1.

Table 1.1 $A = 1{,}000(1.08)^t$

t	$1{,}000(1.08)^t$	=			$A(\$)$
0	$1{,}000(1.08)^0$	=	$1{,}000(1)$	=	1,000
1	$1{,}000(1.08)^1$	=	$1{,}000(1.08)$	=	1,080
2	$1{,}000(1.08)^2$	=	$1{,}000(1.1664)$	=	1,166.40
3	$1{,}000(1.08)^3$	=	$1{,}000(1.25971)$	=	1,259.71
4	$1{,}000(1.08)^4$	=	$1{,}000(1.36049)$	=	1,360.49

We have used the words "relation," "relationship," "is related to," etc. several times in this section and have indicated that a relation is a basic concept of what is to follow. But what exactly is meant by a relation in a mathematical sense? Is it the verbal description that describes the relationship between the two variables of interest, or is it a set of ordered pairs, or the rule or formula that selects or generates the ordered pairs in the set? Since in all of the real-world problems that we will consider either a formula or rule will be given or we will be able to derive such a rule or formula to generate the relationship between the variables of concern, we define a relation as follows.

Definition 1.3 A relation is a rule or formula that associates an element from one set with an element from another set.

An element, x_1, from the first set and its associated element, y_1, from the second set form an ordered pair (x_1,y_1) that belongs to the relation. The set of first components of all ordered pairs belonging to the relation is called the *domain*, and the set of second components of all ordered pairs belonging to the relation is called the *range*. [It is easy to keep this straight by recalling that alphabetically D (domain) comes before R (range) as does first component before second component.] We now realize that the formula $A=1{,}000(1.08)^t$, Eq. (1.1), is a relation. The domain of this relation is the set of whole numbers, W. The range of this relation is the set of all the second components of the ordered pairs (t,A). In general, given any relation and its domain, we can establish a set of ordered pairs that belong to the relation. Five such ordered pairs belonging to the relation $A=1{,}000(1.08)^t$ are listed in Table 1.1.

Relations are mathematical models of real-world problems as they describe the behavior of the variables of interest. As previously noted, in many cases we have to determine the mathematical model from the verbal description of the problem. Let us consider a simple illustration in the following example.

Example 1.2 Part-Time Wages Assume you are earning \$3 per hour working as a part-time cashier in the college bookstore. Determine the relation, formula, or rule that associates x, the number of hours you work per week, with W, your weekly wages. Using this relation as the set selector, describe the set of ordered pairs that belong to this relation.

Solution Since you earn \$3/hr, in 2 hr you earn $3\cdot 2$ or \$6; in 3 hr, $3\cdot 3$ or \$9; and in x hr, $3\cdot x$ or \$$(3x)$. Therefore, the relation is

$$W = 3x$$

where $x=$hours worked per week and $W=$weekly wages in dollars. Using this rule as the set selector, the set of ordered pairs that belong to the relation is written as

$$S = \{(x, W)\colon\ W=3x,\quad x=\text{number of hours worked per week}\}$$

which is read "S equals the set of all ordered pairs (x, W) such that W equals $3x$ and x equals the number of hours worked per week."

The relation $W=3x$ in Example 1.2 is written to imply that W, the second component of the ordered pair, depends upon x, the first component. To emphasize this dependency, we write the formula

$$W(x) = 3x$$

which is read "W of x equals $3x$." $W(x)$ serves the same purpose as the variable W; that is, it is the second component of the ordered pair. Using this notation, $W(2)$ represents the wages earned for working 2 hr, $W(3)$ the wages earned for working 3 hr, etc.; therefore,

$$W(2) = 3\cdot 2 = \$6$$
$$W(3) = 3\cdot 3 = \$9$$
$$W(4) = 3\cdot 4 = \$12$$

This notation has one obvious advantage over the $W=3x$ notation. An equation such as $W(6)=\$18$ gives us both components of the ordered pair; that is, $x=6$ and $W=\$18$, whereas $W=\$18$ gives us only the second component.

The domain of a relation that models a real-world problem is usually implied in the description of the problem. Suppose in Example 1.2 the manager of the bookstore has limited you to a maximum of 20 hr per week and has requested that you work an integral number of hours each week. Given these restrictions the domain of the relation $W(x)=3x$ is

$$D = \{0, 1, 2, 3, \ldots, 20\}$$

and the range of this relation determined by this domain and the rule $W(x)=3x$ is

$$R = \{0, 3, 6, 9, \ldots, 60\}$$

We can picture the relation $W(x)=3x$ as a *mapping* of elements from the domain into elements in the range as illustrated in Fig. 1.1.

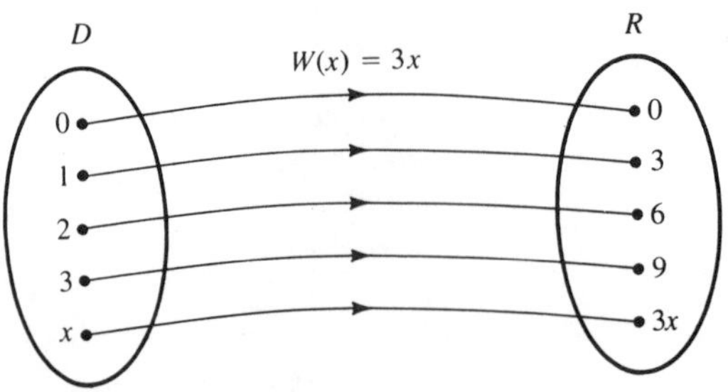

FIGURE 1.1

If no restriction is placed on the domain of the relation, it is assumed that the first component of the ordered pair can be any real number as long as its associated second component given by the relation is also a real number.

Example 1.3 Determine the domain and range of the relation $S(x)=5+\sqrt{x}$.

Solution The equation $S(x)=5+\sqrt{x}$ is written to imply that $S(x)$ depends on the choice of x; hence, x is the independent variable and an element of the domain. Since no restrictions are placed on x, we assume that x is any real number as long as its associated second component $S(x)$ is also a real number. The second component $S(x)=5+\sqrt{x}$ is a real number only if $x \geqslant 0$. (The square root of a negative number is not a real number since no real number squared is negative.) Therefore, the domain, D, of the relation $S(x)=5+\sqrt{x}$ is

$$D = \{x: x \geqslant 0\}$$

The range of the relation is determined by examining the values of $S(x)$ as x assumes various values in the domain D. Since $x \geqslant 0$ implies $\sqrt{x} \geqslant 0$, the range, R, of the relation $S(x)=5+\sqrt{x}$ is

$$R = \{S(x): S(x) \geqslant 5\}$$

The range can also be represented by the use of the single variable y if desired; that is,

$$R = \{y: y \geqslant 5\}$$

Relations are classified according to the way the elements in the domain are matched or associated with the elements in the range. Figure 1.2 illustrates the four ways of classifying relations.

Figure 1.2*a* is a *one-to-one relation*. Each element in the domain is matched with exactly one element in the range and no two elements in the domain are matched with the same element in the range. Figure 1.2*b* is a *many-to-one relation*. At least two elements in the domain are associated with the same element in the range. Figure 1.2*c* is a *one-to-many relation*. At least one element in the domain is associated with two or more elements in the range. Figure 1.2*d* is a *many-to-many relation*. At least two elements in the domain are matched to one element in the range and at least one element in the domain is matched to two or more elements in the range. Relations that are either one-to-one (Fig. 1.2*a*) or many-to-one (Fig. 1.2*b*) are called *functions*. Almost without exception real-world relations are functions. Knowing this, we will concentrate on functions. We can best describe a function with the following definition.

Definition 1.4 A function is a relation that associates or assigns each element of the domain with exactly one element of the range.

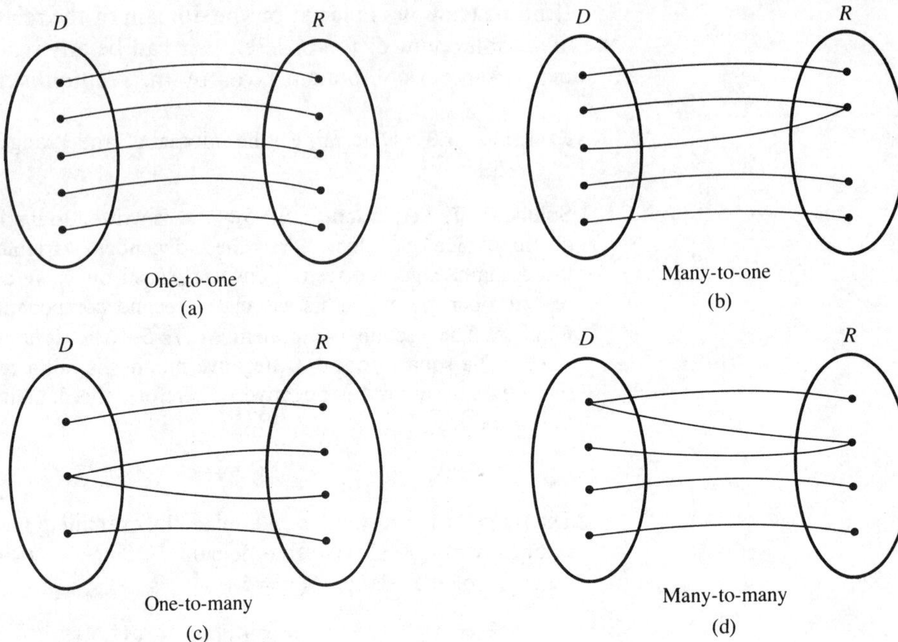

FIGURE 1.2

This definition implies that no two different ordered pairs that belong to the function will have the same first component; for if they did, the same element from the domain would be matched with two different elements of the range. (*A word of caution:* Definition 1.4 does not imply that two different ordered pairs belonging to the function must have different second components.)

Example 1.4 Show that the relations given in Examples 1.1 and 1.2 are functions.

Solution The relation given in Example 1.1 is

$$A = 1{,}000(1.08)^t \quad \text{or} \quad A(t) = 1{,}000(1.08)^t$$

where t is an element of the domain and A an element of the range. For any element t_1 of the domain, the formula $A = 1{,}000(1.08)^t$ assigns exactly one value in the range $A = 1{,}000(1.08)^{t_1}$; hence, this relation is a function. For example, if $t_1 = 2$, then $A = 1{,}000(1.08)^2 = 1{,}000(1.1664) = \$1{,}166.40$.

The relation given in Example 1.2 is

$$W(x) = 3x$$

where x is an element of the domain and $W(x)$ an element of the range. For each element x_1 in the domain, the formula $W(x) = 3x$ assigns exactly one element $W=3x_1$ in the range. So, $W(x)=3x$ is also a function. For example, when $x_1=10$, $W=30$.

To illustrate that there are relations that are not functions consider the following example.

Example 1.5 Show that the relation $y^2=x$ is not a function. Assume that x is an element of the domain and y is an element of the range.

Solution Since the domain is not restricted, we can assume that x is any real number as long as the rule $y^2=x$ associates a real number y with the selected value of x. If $x=4$, the corresponding range values are found by solving the equation $y^2=4$; that is,

$$y^2-4=0$$
$$(y+2)(y-2)=0$$
$$y=2, y=-2$$

Hence the ordered pairs (4,2) and (4, −2) belong to the relation, and by Definition 1.4, the relation is not a function.

In Example 1.5 we elected to write the formula for the relation in terms of the variables x and y. The selection of the variables in the formula is arbitrary. To determine if the relation is a function, we must specify which of the variables is an element of the domain and which is a member of the range. It is customary to assume that if x and y are used, x is always the first component of the ordered pair of the relation and hence is an element of the domain, and y is an element of the range. Frequently it is more desirable to denote the second component of the ordered pair as $f(x)$. We call this notation, which we used earlier, *function notation*. As noted, the function notation has the advantage that an equation such as $f(2)=6$ provides us with both the domain value 2 and the range value 6.

Example 1.6 Determine the domain and range of the function

$$f(x)=x^2+4$$

where it is assumed that x is an element of the domain.

Solution Since x is an element of the domain, we are considering the ordered pairs $(x,f(x))$ where $f(x)$ is an element of the range. To determine the domain we can assume that x is any real number as long as its associated value $f(x)$ is also a real number. The expression x^2+4 is a real number whenever x is a real number; therefore, the domain, D, of the function is

$$D=\mathbb{R}, \quad \text{where } \mathbb{R} = \text{the set of real numbers}$$

The range, R, of the function is determined by considering the values of $f(x)$ as x assumes various values in the domain. For any value of x, $x^2 \geqslant 0$;

therefore, $x^2+4 \geqslant 4$ implying that the range, R, is given by:

$$R = \{f(x)\colon f(x) \geqslant 4\}$$

Example 1.7 Determine the domain of the function

$$g(x) = 4 + \sqrt{(x+2)}$$

Solution Since the square root of $(x+2)$ is a real number only if $(x+2)$ is nonnegative we select those values of x that satisfy

$$x + 2 \geqslant 0$$

or

$$x \geqslant -2$$

For these values of x the expression $4 + \sqrt{(x+2)}$ is a real number since the sum of two real numbers is a real number. Hence the domain is given by

$$D = \{x\colon x \geqslant -2\}$$

In real-world functions the domain is often restricted by limitations imposed by the problem. For example, in biology there is a relation called the bioclimatic rule for temperate climates that states that in spring and early summer periodic phenomena, such as the appearance of certain insects, usually occur about 4 days later for each 500 ft of altitude increase from any given base location. This is expressed as a relation as

$$d(h) = 4\left(\frac{h}{500}\right)$$

where $d(h)$ is the delay in appearance of insects measured in days, and h is the increase in altitude from the base location. If we assume that h is an element of the domain, it is obvious that there is some upper limit on h where the relation no longer will hold. If one assumed the base location to be sea level and the other location to be 50,000 ft above sea level, the number of days the insects would appear at the 50,000 ft location after they appeared at sea level would be

$$\begin{aligned} d(50{,}000) &= 4\left(\frac{50{,}000}{500}\right) \\ &= 400 \text{ days!} \end{aligned}$$

This of course is a ridiculous solution. Realistically, a restriction on the domain would be given to indicate the limitations placed on the function. In this case the function might have been given as

$$d(h) = 4\left(\frac{h}{500}\right), \qquad \text{where } 0 \leqslant h \leqslant 3{,}000 \text{ ft}$$

Many real-world problems will be more extensive than those already considered in this section and the appropriate mathematical models for these problems will likewise be more involved. Unfortunately there is no simple procedure that will provide us with a mathematical model of every real-world problem. However, when required to construct a function as a model for a particular problem there are certain steps we can follow that will help us reach our goal.

Before listing these steps it should be noted that building mathematical models requires a great deal of practice and exposure to new problems. It is for this reason that you will find opportunities to practice this art throughout this textbook. As new models are needed we will introduce the mathematical background that is required and show by example how real-world problems can be modeled and solved.

To construct a function as a model for a problem the following steps should be followed:

Step 1. Read the problem through carefully (as many times as necessary) to determine exactly what is asked for in the problem.

Step 2. Determine which quantity will be the independent variable and which one the dependent variable. (Frequently problems are worded "express — — — — as a function of ________." In this case, the ________ is the independent variable and the — — — is the dependent variable.)

Step 3. Denote the independent variable by x or some other suitable variable and the dependent variable as $f(x)$, $W(x)$, or by some other appropriate notation. [For example, when expressing cost as a function of x use $C(x)$; when expressing volume as a function of e use $V(e)$, etc.].

Step 4. Analyze the given problem and write an equation relating the dependent variable to some expression containing the independent variable.

This last step is crucial and the most difficult. It can involve any or all of the following processes:

1. carefully translating verbal sentences into mathematical sentences,
2. looking at data for patterns,
3. drawing a sketch of the problem and relating the dependent and independent variables by a geometric formula implied by the sketch,
4. calculating ordered pairs that belong to the function and generalizing a rule or formula based on specific calculations.

The following two examples and their solutions illustrate the application of this approach.

Example 1.8 Flight Expenses A charter airline advertisement reads "Ski excursions to Austria for the spring vacation. Cost \$95.00 per passenger for 80 passengers with a refund of 50¢ each for each passenger in excess of 80." Assuming that at least 80 passengers make the trip and that the plane holds at most 100 passengers, determine the revenue that the airline receives as a function of the number of passengers taking the trip.

Solution The independent variable is the *number of passengers*: let x = the number of passengers. The dependent variable is the *revenue*: let $R(x)$ = the revenue. Our goal is to express $R(x)$ in terms of x. Revenue = (number of passengers on board)·(cost per passenger) where (the cost per passenger) = \$95 − \$.50·(number of passengers in excess of 80) and (number of passengers in excess of 80) = $(x-80)$.

By substitution

$$\begin{aligned} R(x) &= x[95-.50(x-80)] \\ &= x[95-.5x+40] \\ &= 135x-.5x^2 \end{aligned}$$

Hence we have obtained the desired function. By the limitation set forth in the problem

$$80 \leqslant x \leqslant 100$$

Therefore the domain of this function is

$$D = \{x:\ 80 \leqslant x \leqslant 100, \quad \text{where } x \text{ is an integer}\}$$

Example 1.9 Rectangular Box Configurations A rectangular box is made with a square base whose edge is x cm in length. The height of the box is twice the length of an edge of the base.

(a) Determine the volume of the box as a function of the length of the edge x.
(b) Determine the domain of this function if the length of each edge of the box is at least 1 cm and at most 10 cm.

Solution Drawing a sketch of the box and labeling the given dimensions we have

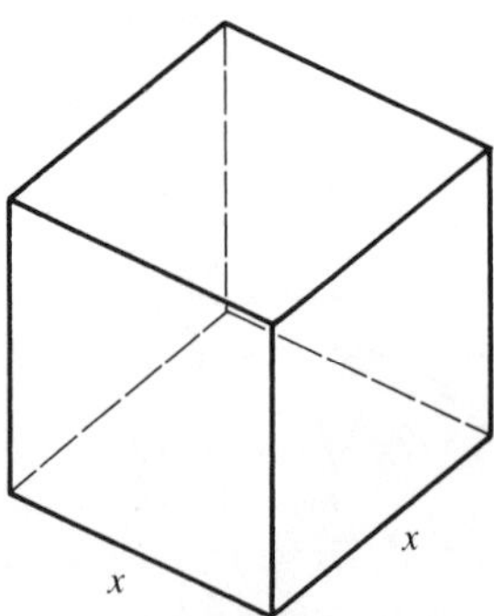

The length of the base edge is the independent variable x. The volume of the box is the dependent variable $V(x)$. Since the volume of any rectangular box is given by

$$V = l \cdot w \cdot h$$

where l = length of one edge of the base

w = length of the other edge of the base

h = length of the height of the box

By substitution we have

$$V(x) = x \cdot x \cdot h$$

Since $h = 2x$, as denoted in the problem, the rule we are seeking is

$$\begin{aligned} V(x) &= x \cdot x \cdot 2x \\ &= 2x^3 \end{aligned}$$

Since x is the smallest edge of the box and $2x$ is the largest we have

$$x \geqslant 1 \quad \text{and} \quad 2x \leqslant 10$$

or

$$x \geqslant 1 \quad \text{and} \quad x \leqslant 5$$

which can be written as $1 \leqslant x \leqslant 5$. Hence the domain for the function described in part (a) is

$$D = \{x\colon\ 1 \leqslant x \leqslant 5\}$$

EXERCISES

1. Given the sets $A = \{-2, -1, 0\}$, $B = \{3, 4, 5\}$, and $C = \{2, 4\}$, determine the following cartesian products.

(a) $A \times B$ (b) $B \times A$ (c) $A \times C$
(d) $C \times A$ (e) $B \times C$ (f) $C \times B$

2. If $f(x) = 3x + 10$ for any real value of x, determine the values $f(1)$, $f(2)$, $f(5)$, $f(-1)$, and $f(-5)$.

3. If $g(x) = x^2 + x + 3$ for any real values of x, determine the values $g(-10)$, $g(-3)$, $g(0)$, $g(3)$, and $g(4)$.

4. If $f(x) = x^3 + 3x^2 + 3x + 2$ for any real values of x, determine the values $f(-5)$, $f(-3)$, $f(0)$, $f(3)$, and $f(5)$.

For each of the relations in Exercises 5 to 15, determine the range for the given domain, the ordered pairs that belong to the relation, and whether the relation is a function.

5. $f(x) = 2x + 6$ for $D = \{1, 2, 3\}$

6. $f(x)=-3x+10$ for $D=\{-3,-2,0\}$
7. $W(x)=\sqrt{x}$ for $D=\{16,25,36\}$
8. $A(x)=x^3$ for $D=\{-1,0,1,2\}$
9. $y=\dfrac{1}{x}$ for $D=\{1,2,3,4\}$
10. $s(x)=-\sqrt{x}$ for $D=\{16,25,36\}$
11. $f(x)=2-x^2$ for $D=\{-2,-1,0,1,2\}$
12. $g(x)=10+\sqrt{x-1}$ for $D=\{2,10,26\}$
13. $y=x^3+2x^2-3x+5$ for $D=\{-2,-1,0,1,2\}$
14. $f(x)=8\pm\sqrt{x}$ for $D=\{0,1,4,9\}$
15. $y^2=4x+5$ for $D=\{0,1,5\}$
16. For each of the relations given in Exercises 6, 11, 14, and 15 draw a sketch to illustrate that the relation can be pictured as a mapping of elements from the domain to the range. Classify each relation as a one-to-one, one-to-many, many-to-one, or many-to-many mapping.
17. Assume that each of the following sets contains all of the ordered pairs that belong to the indicated relation.

$$R_1 = \{(1,2),(2,4),(3,6),(4,8),(5,10)\}$$
$$R_2 = \{(1,1),(2,1),(3,1),(4,1),(5,1)\}$$
$$R_3 = \left\{(1,1),(1,2),(2,5),(3,8),(4,10),(5,-7),\left(6,\tfrac{3}{2}\right)\right\}$$
$$R_4 = \{(1,1),(1,2),(2,1),(3,1),(3,2),(4,4),(5,5)\}$$

(a) Which of these relations are functions?
(b) Classify each relation as one-to-one, many-to-one, one-to-many, or many-to-many.

In Exercises 18 *to* 24 *determine the domain of each function.*

18. $f(x)=3x-6$
19. $h(x)=2-x^2$
20. $g(x)=10+\sqrt{x-1}$
21. $f(x)=\dfrac{1}{x+2}$
22. $f(x)=\dfrac{1}{\sqrt{5-x}}$
23. $f(x)=x^3+2x^2-3x+5$
24. $y=\sqrt[3]{x}$

25. Assume the length of the side of a square is x as illustrated.

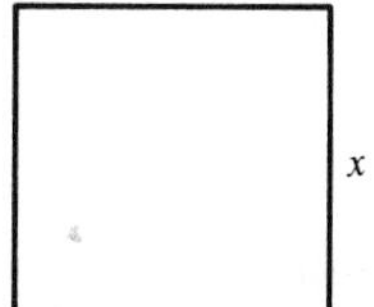

EXERCISE 25

(a) What is the rule or formula that relates the area A of the square to the length of the side x?
(b) Is this formula a relation? A function?

26. Assume the length of an edge of a cube is x as illustrated.

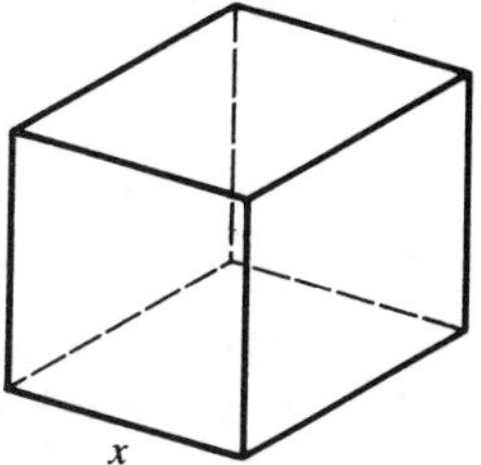

EXERCISE 26

(a) What is the rule or formula that relates the volume of the cube V to the length x?
(b) Is this formula a relation? A function?

27. Assume that length of the edge of a cube is e.

(a) What is the rule or formula that relates the total surface area S of the cube to the length e?
(b) Is this formula a relation? A function?

28. Assume that the length of the edge of a cube is e.

(a) What is the rule or formula that relates the diagonal d, illustrated here, to the length e?

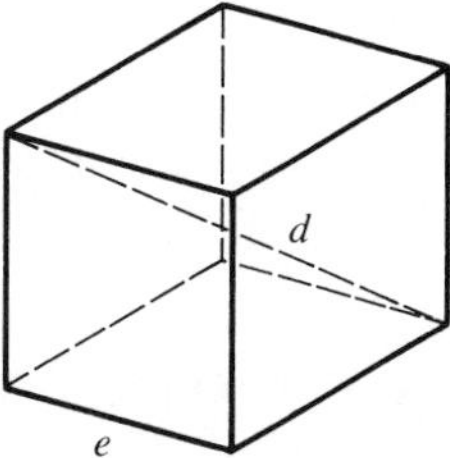

EXERCISE 28

(b) Is this formula a relation? A function?

29. **Banker's Rule** Simple interest is often charged for short-term borrowing. If the amount of money borrowed is P, then the amount A that must be repaid with simple interest is given by the formula

$$A = P\left(1+i\frac{n}{360}\right)$$

where n=the number of days of the loan and i=the annual interest rate. This formula is referred to as the Banker's rule.

(a) If \$3,600 is borrowed at an annual interest rate of 10%, express A in terms of n.
(b) Is this formula a function if n is assumed to be the independent variable?
(c) How much money must be repaid at the end of 180 days?

30. **Cricket Chirps and Temperature** It has been observed that crickets chirp more rapidly on warm evenings than on cool ones. Data show that a close approximation of the temperature in degrees Fahrenheit is given by

$$T = x + 37$$

where x is the number of chirps in any 15-second interval.

(a) Is this formula a relation? A function?
(b) If a cricket chirps 120 times per minute, what is the approximate temperature?

31. Assume you work x hr/week at a rate of \$4/hr.

(a) Determine your gross salary $S(x)$ as a function of x; i.e., determine a formula that relates to $S(x)$ to x.
(b) Assume that 20% of your gross salary is withheld for taxes, etc. What is your take-home pay $p(x)$ as a function of x?
(c) Evaluate $p(40)$.

32. A Florida orange grower now can ship 20 tons of fruit to New England at a profit of \$80/ton. If he waits, he can add 5 tons/week; however, his profit/ton will be reduced by \$10/week. Assume he waits x weeks to ship the fruit.

(a) Determine the number of tons T he will ship as a function of x.
(b) Determine the profit *per ton* as a function of x.
(c) Determine the profit as a function of x.
(d) What is a reasonable domain for this function?

33. A manufacturer has found that the cost of producing x units of a commodity is given by the formula

$$C(x) = .01x^2 + 10x + 5$$

where $C(x)$ is in dollars.

(a) Is this formula a relation? A function?
(b) What is the cost for producing 100 units?
(c) What is the cost for producing the one-hundred-first unit?

34. Jane Lawless, a psychometrician, has lost the formula for converting raw scores on the *Atlantis Test of Mental Maturity* to equivalent IQ scores. However, she does have several scores that were formerly converted (ATMM, IQ): (56, 127), (34, 83), (51, 117), (40, 95), (45, 105), (50, 115).

(a) Determine the function that expresses the relationship between ATMM (domain) and IQ (range). *Hint*: Look for a pattern between the first and second components of the ordered pairs.
(b) Convert an ATMM score of 63 to an IQ score.

35. Auditory Reaction Time Auditory reaction time in an experiment was found to be equal to an irreducible muscle response time of 0.04 sec plus 0.02 sec times the square of the number of letters in the stimulus word, minus a correction factor of 0.01 sec/letter to account for additional stimulus duration.

(a) Express reaction time as a function of word length in letters (x).
(b) What is the predicted reaction time for a 3-letter word? A 6-letter word?

36. Mary Caldwell, the sales manager of a small corporation, is paid a salary of \$600/month plus a commission of 3% of the gross sales for the month. Determine the function that expresses Ms. Caldwell's total monthly earnings as a function of the gross sales, x.

37. Bob Collins sells office copiers to local retailers. Each month the manufacturer pays him \$50 times the *square* of the number of copiers he sells, plus a fixed amount of \$200. However, he incurs a cost to himself of \$100/copier sold. Determine the function that expresses Mr. Collins' monthly net income as a function of x, the number of copiers sold.

38. A manufacturer has decided that the company's operating income is \$25/unit sold minus monthly fixed expenses of \$2,000 and other expenses equal to 10¢ times the square of the number of units sold. Determine the function that expresses operating income in terms of units sold.

1.3 FUNCTIONS AND GRAPHS

Suppose a sales representative rents a car for one day and pays \$15 per day and 12 cents per mile. The daily charge is a fixed cost and will not change regardless of whether the car is driven extensively or not at all; the total mileage charge is directly related to the number of miles driven. To obtain a relation that will yield the total cost for a given day, we must add the total variable cost and the fixed cost. The total variable cost can be obtained by multiplying the mileage cost of 12¢ by the number of miles driven, which varies and is represented by the variable x. Thus, the total cost in dollars, represented by y, for a day in which x miles are driven, is given by

$$y = .12x + 15 \tag{1.2}$$

The sales representative, of course, is interested in how the total cost, y, depends upon x, the number of miles driven per day. This can be determined by inspecting the set of ordered pairs (x,y) that belong to the relation and hence make Eq. (1.2) a true statement. In many situations such as this it is desirable to represent geometrically the set of ordered pairs to provide a picture of this relation. This is done by means of associating the ordered pairs belonging to the relation with points in a plane.

To do this, we construct a graph using two perpendicular lines called the *coordinate axes*. The point of intersection of the two axes is called the *origin*, or zero point. Each line is scaled as shown in Fig. 1.3. It is customary to label the horizontal line as the x axis and the vertical line as the y axis. Note in the figure that the y axis is scaled so that positive values lie above the origin and negative values are below it. On the x axis, the positive values are to the right and negative values to the left of the origin. These represent all possible values of the two variables whose relationship we wish to graph. However, the units of length need not be the same for both axes. The four regions formed by the axes are known as *quadrants*, numbered as indicated by the roman numerals in Fig. 1.3. There we see that if x is *positive*, we go to the *right* on the x axis and if it is *negative*, we go to the *left*. Similarly, if y is *positive*, we go *up* the y axis, and if it is *negative*, we go *down*.

From Fig. 1.3 we see that there are four directions in which we can go: up, down, left, or right. The ordered pairs that were introduced previously are used to indicate these movements. The ordered pair (x,y) tells us how many units to move and in which direction for x and y, and the values assigned to x

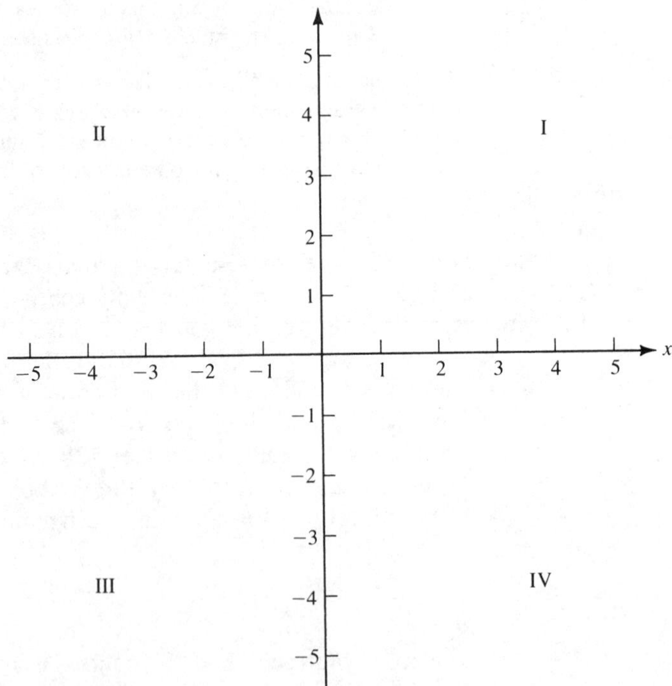

FIGURE 1.3

and y are called the x *coordinate* and y *coordinate*, respectively. For example, the ordered pair (2,3) tells us to move two units to the right of the origin on the x axis and three units above the origin on the y axis. In order that our movements will be coordinated (remember, they are called *coordinate* axes), we first move two units to the right and then move up three units parallel to the y axis. The ordered pair (2,3) represents the *point* at which our moves terminate; for the point (2,3) we say that its *abscissa*, or x coordinate, is 2 and its *ordinate*, or y coordinate, is 3. We note that the ordered pairs that could be represented in Fig. 1.3 are ordered pairs of real numbers. The coordinate system represented in Fig. 1.3 is the cartesian product of the set of real numbers with itself. Thus, the coordinates are called *cartesian coordinates*, or *rectangular coordinates*.

Figure 1.4 presents four different points with one in each quadrant. The point (2,3) is in quadrant I because both coordinates are positive. The point $(-4,2)$ is in quadrant II because the first coordinate is negative, which means go left parallel to the x axis, and the second coordinate is positive, which means go up parallel to the y axis. The point $(-3,-5)$ is in quadrant III because the coordinates indicate that we should move to the left and down. The point $(4,-3)$ is in quadrant IV because the coordinates indicate that we should move to the right and down. The signs associated with the respective coordinates are given in each quadrant. Thus, any point in quadrant I will be

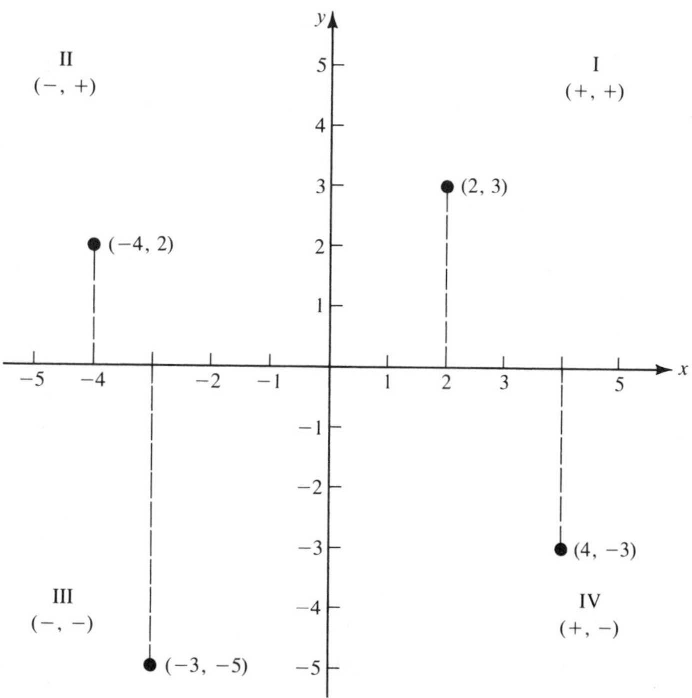

FIGURE 1.4

positive for both coordinates, i.e., $(+,+)$. Similarly, any points with signs $(-,+)$ will appear in quadrant II, those with signs $(-,-)$ will appear in quadrant III, and those with signs $(+,-)$ will appear in quadrant IV.

In the definition of a function, recall that for each element in the domain there exists one and only one element in the range. Thus, for a function where x is an element of the domain, the second component y will receive one and only one value. In order to graph a function, we select a value for x and use the function rule to determine a value for y; next, we determine the point (x,y) on the rectangular coordinates and plot the point. We continue this process for several x values until enough points are obtained to give the specific pattern of the function. In theory we should determine all possible (x,y) pairs and plot them in order to ensure that the graph of the function is correct. In practice, generally we determine enough points to sketch the graph by connecting the points by smooth curves. In sketching the graph of a function we must keep in mind the following definition.

Definition 1.5 The *graph of a function* is the set of points corresponding to the set of ordered pairs belonging to the function.

Consider the function for the car rental example

$$y = .12x + 15$$

In order to graph this function, we need to compute the value of y for each of several different values of x. Since x represents the number of miles driven, it is not feasible for x to be negative, and so we shall consider only nonnegative values of x. We compute the following values of y for the corresponding x values:

x	0	50	100	150	200
y	15	21	27	33	39

Plotting the values in the table and connecting the points with a solid line, we obtain the graph given in Fig. 1.5. The solid line can be considered only as an estimate of where the function should be plotted because generally the domain consists of all real numbers of which there are an infinite number and the function value y has not been determined for each of these values. The graph of the function $y=.12x+15$ as shown in Fig. 1.5 provides the sales representative with a concrete picture of the relationship between the total daily cost of renting the car and the mileage driven per day.

The graph of a function also provides us with an alternate means of determining the domain and range of the function. Consider the four functions whose graphs are shown in Fig. 1.6. In each of the graphs in Fig. 1.6 we have plotted the points whose associated ordered pairs belong to the function.

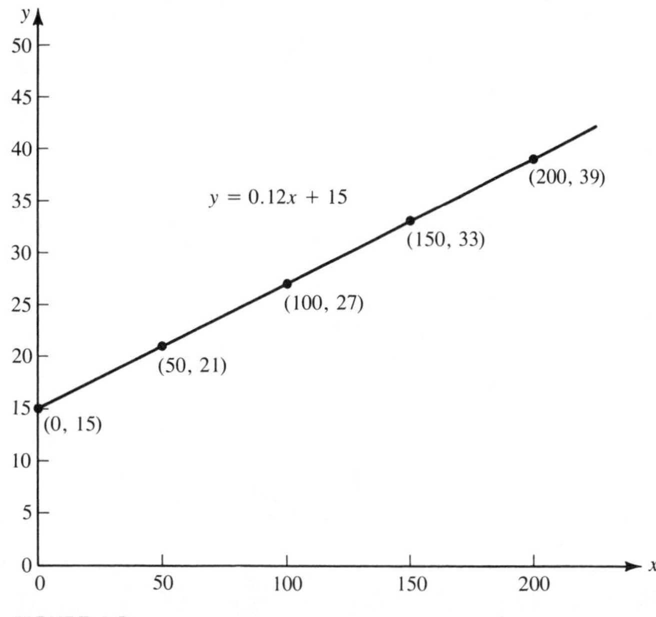

FIGURE 1.5

Since the abscissa of each point is an element of the domain of the function, the domain can be determined by assuming that each point on the curve projects vertically to the x axis as illustrated in Fig. 1.6*a*. The set of all the x values of these projections is the domain of the function. The domains of the functions shown in Fig. 1.6 determined by an inspection of the graphs are

(a) $D=\{x:\ -3 \leqslant x \leqslant 7\}$
(b) $D=\{x:\ 0 \leqslant x \leqslant 10\}$
(c) $D=\mathbb{R}$
(d) $D=\{x:\ x \geqslant 0\}$

Analogously, the ordinate of each point is an element of the range of the function, and the range can be found by assuming that each point on the curve projects horizontally to the vertical axis. The set of all the y values of these projections is the range of the function. The ranges for the functions shown in Fig. 1.6 determined by inspection of the graphs are

(a) $R=\{y:\ 0 \leqslant y \leqslant 5\}$
(b) $R=\{y:\ 1 \leqslant y \leqslant 7\}$
(c) $R=\{y:\ y \geqslant 0\}$
(d) $R=\{y:\ y \leqslant 0\}$

Although we have used set notation to express the domain and range of a function, we find it advantageous to use *interval notation* for this purpose. The set $D=\{x:\ -3 \leqslant x \leqslant 7\}$ is called a *closed interval* and in interval notation is written $[-3,7]$. The use of the square bracket denotes that both the end

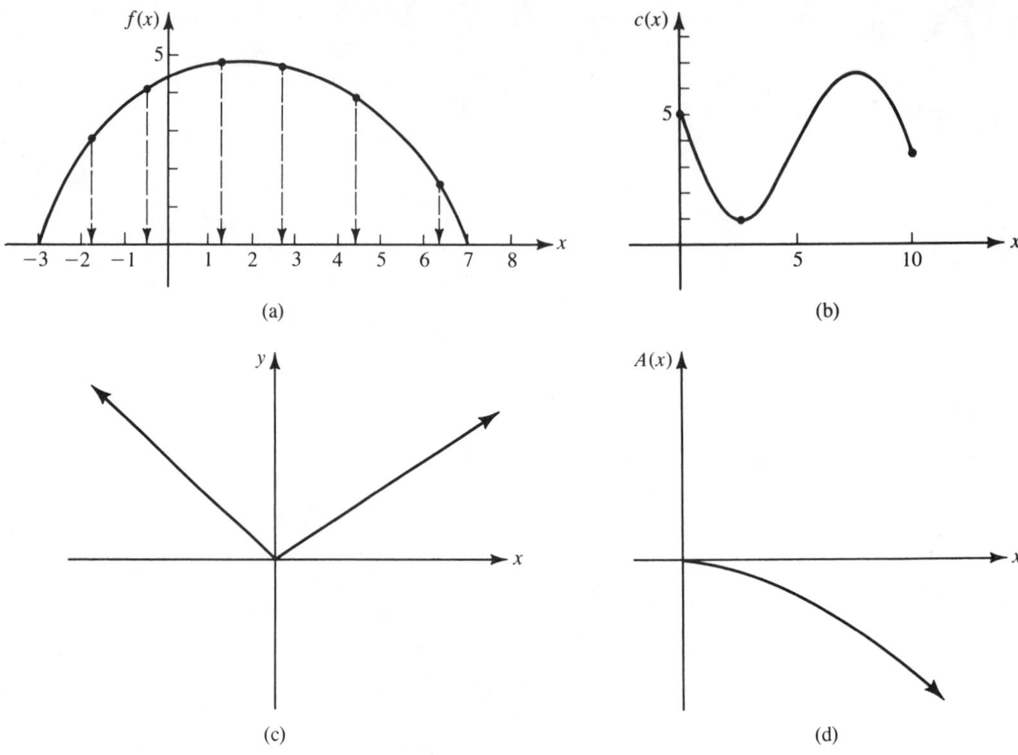

FIGURE 1.6

points -3 and 7 are included in the interval. The set $D=\{x:\ -3<x<7\}$ is called an *open interval* and in interval notation is written $(-3,7)$. The parentheses denote that the end points -3 and 7 are not included in the interval. Although this is the same notation used to indicate an ordered pair, the context in which it is used will make it clear which is intended. An interval with only one end point included, such as $[-3,7)$ or $(-3,7]$ is called a *half-open interval*. A set such as $D=\{x:\ x\geqslant 0\}$ in interval notation is written $[0,\infty)$ and is read, "the set of real numbers equal to or greater than zero." Since infinity, ∞, is not a real number, it is never included as an end point. The domains and ranges of the functions shown in Fig. 1.6 are

Domain	Range
(a) $D=[-3,7]$	$R=[0,5]$
(b) $D=[0,10]$	$R=[1,7]$
(c) $D=(-\infty,\infty)$ or $\mathbb{R}$	$R=[0,\infty)$
(d) $[0,\infty)$	$R=(-\infty,0]$

(*A word of caution*: When using interval notation of any of the forms such as $[a,b]$, (a,b), $[a,b)$, $(a,b]$, etc., note that a must always be less than b).

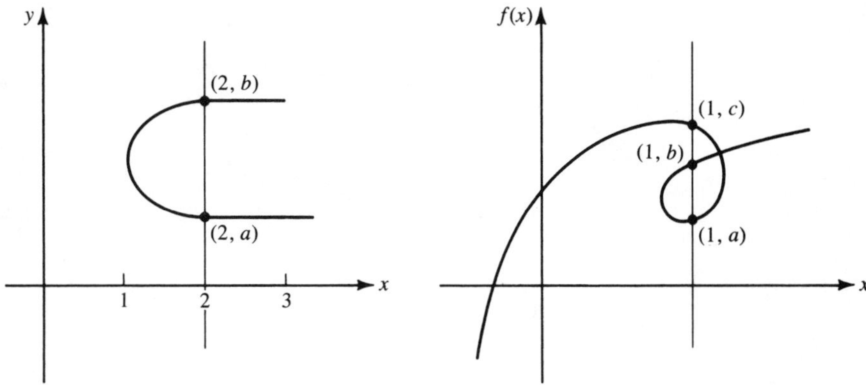

FIGURE 1.7

The graph of a relation can also be used to determine if the relation is a function by applying the following test. If any vertical line intersects the graph at more than one point, the relation is not a function. We can see that this is a valid test by assuming a vertical line intersects the graph in at least two points. This would imply that there exists a domain value that is associated with at least two values in the range and by definition the relation is not a function. Figure 1.7 shows two relations that are not functions and an application of the *vertical-line test*.

Example 1.10 Use the vertical-line test to determine which (if any) of the graphs in Fig. 1.8 represent functions.

Solution Since no vertical line can be drawn to intersect the graphs shown in Fig. 1.8*a* and *b* more than once, these represent functions. In Fig. 1.8*c* any vertical line between $x=-3$ and $x=3$ will intersect the graph at two points; hence, the relation is not a function. In Fig. 1.8*d* any vertical line to the right of $x=1$ or to the left of $x=-1$ will intersect the graph at two points; hence, the relation is not a function.

Many real-world problems can be solved with the aid of a graph. As an illustration, let us consider the following problem that has just been presented to the plant manager of the *XYZ* Box Manufacturing Company. He has been asked to construct open rectangular boxes from flat square sheets of cardboard by cutting out squares from the corners and folding up the edges. The company ordering these boxes provided the sheets of cardboard, which measure 10 in. $\times$ 10 in., and they have requested that the dimensions of each box be selected to maximize its volume. The plant manager, recognizing he has a problem, has just called us in as consultants on the project. Now we have a problem:

First, let us construct a model. A sketch showing how each box is formed is shown in Fig. 1.9. Since each box is made from the same 10 in. $\times$ 10 in. piece of cardboard, we might surmise that the volume of each box will be the

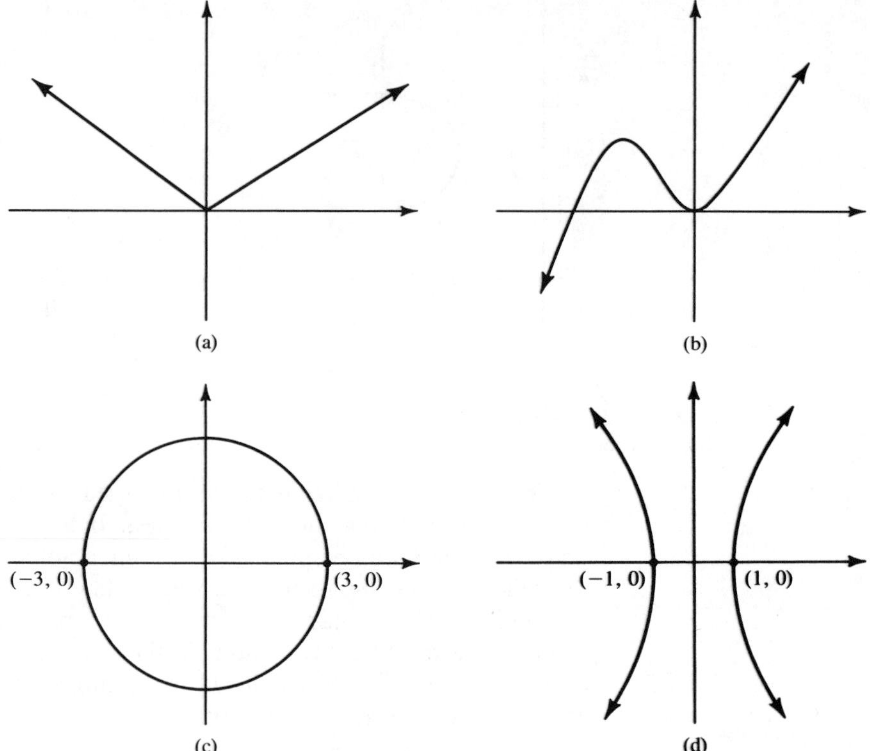

FIGURE 1.8

same, independent of the size of the square cut from each of the four corners of the flat sheet of cardboard. To verify this conjecture, let us calculate the volume of each of the four different configurations shown in Fig. 1.10. The volume of a rectangular box is given by the formula $V = l \cdot w \cdot h$, where l = length of the base, w = width of the base, and h = height of the box as indicated in Fig. 1.9*c*. The volume of each of the boxes constructed from the four configurations shown in Fig. 1.10 is listed in Table 1.2.

These results show that our conjecture is incorrect, as the volume, V, of a box does depend on the size of the square cut from each of the corners. Based

Table 1.2

Configuration	h (in.)	l (in.)	w (in.)	$l \cdot w \cdot h = V$ (cu in.)
(a)	1	8	8	$8 \cdot 8 \cdot 1 = 64$
(b)	2	6	6	$6 \cdot 6 \cdot 2 = 72$
(c)	3	4	4	$4 \cdot 4 \cdot 3 = 48$
(d)	4	2	2	$2 \cdot 2 \cdot 4 = 16$

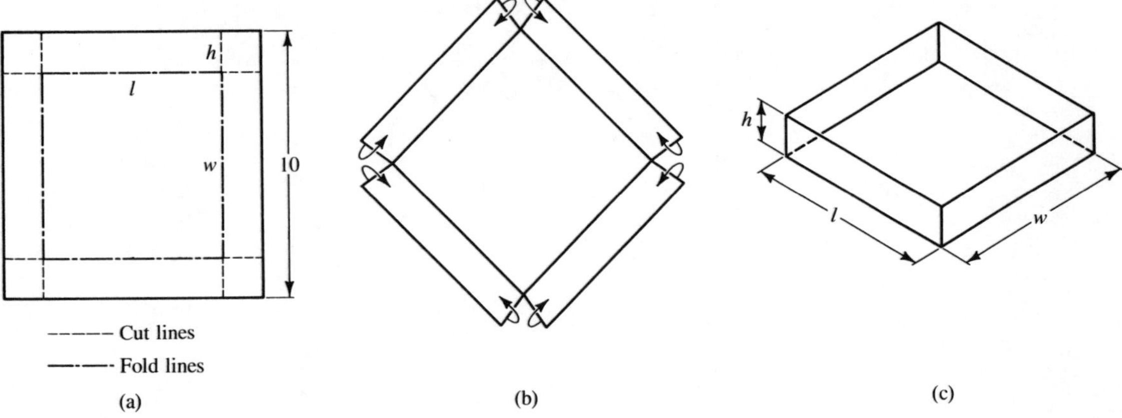

FIGURE 1.9

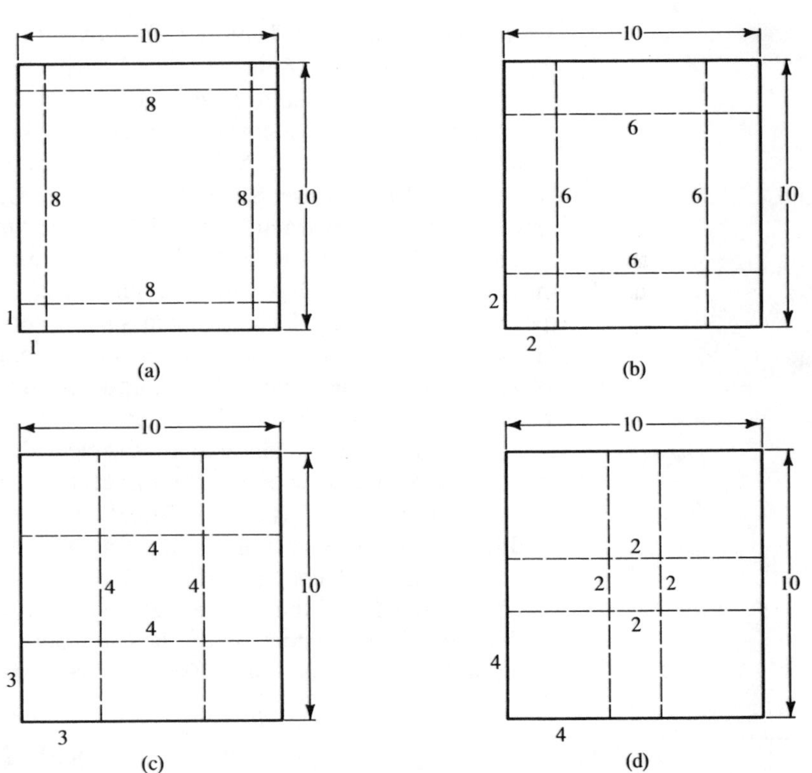

FIGURE 1.10

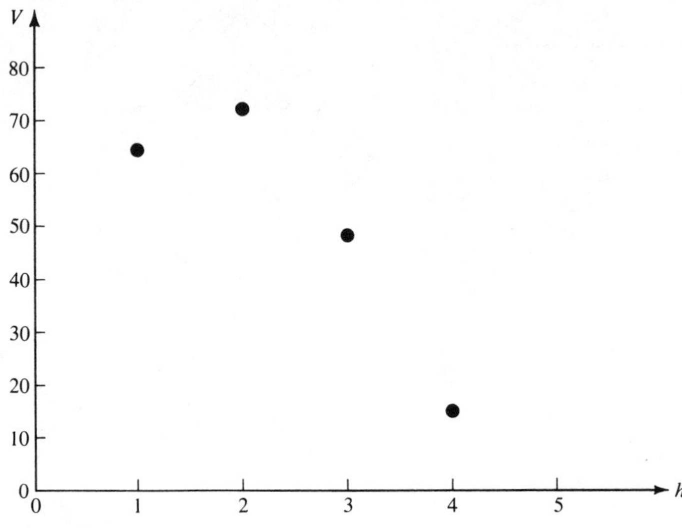

FIGURE 1.11

on the results shown in Table 1.2, the box with maximum volume (72 cu in.) has dimensions 6 in. × 6 in. × 2 in.

To earn our consulting fee, we suggest that a box with even greater volume may exist if we consider nonintegral values for the dimension h. To help select the appropriate value of h, let us graph the data presented in Table 1.2 on a cartesian plane. Since the volume V depends on the choice of h, we graph h on the horizontal axis and V on the vertical axis. The four ordered pairs from Table 1.2 are plotted in Fig. 1.11. Since we have assumed that h is not restricted to integral values, a smooth curve can be drawn through the points shown in Fig. 1.11. Envision such a curve being drawn and estimate the value of h that will produce a box of maximum volume. Using your value of h we can calculate the length, l, and width, w, of the box and collect our consulting fee from the plant manager.

Based on the above discussion, it is clear that for a given value of h there exists exactly one value of V; therefore, there is a function or rule relating the variables V and h. This rule is found by expressing the length, l, and the width, w, of the box in terms of the height, h, and rewriting the right side of the equation $V = l \cdot w \cdot h$ in terms of h only. In Figure 1.12, let h equal the length of the edge of the square that we cut out from each corner. The length and width of the bottom of the box are both equal to $(10-2h)$ as shown. Substituting into the equation $V = l \cdot w \cdot h$, we find

$$V = (10-2h)(10-2h)h$$

or

$$V(h) = (10-2h)^2 h$$

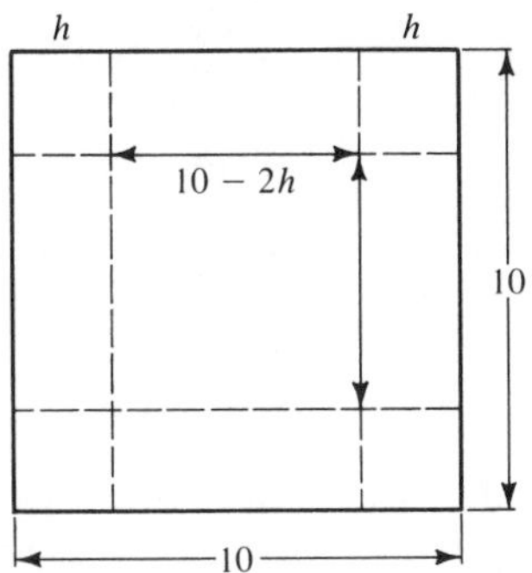

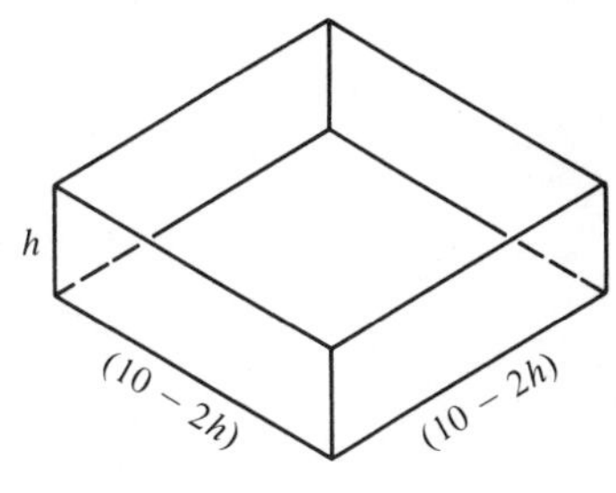

FIGURE 1.12

With this function we can determine the dimensions of the box of maximum volume to the accuracy desired by selecting various values of h. Figure 1.13 indicates that the height of the box of maximum volume is approximately 1.5 in. (In Exercise 13 at the end of this section, you will be asked to find the maximum volume by selecting h values to the nearest hundredth.) In Chapter 5, after developing the tools of calculus, we will show an easier way of solving the plant manager's problem.

To complete our discussion of this problem, we determine the domain, range, and the graph of the function $V(h)=(10-2h)^2h$. The domain of the function is determined by investigating the limitations placed on the variable h in the problem. Since the height, width, and length of the box must be

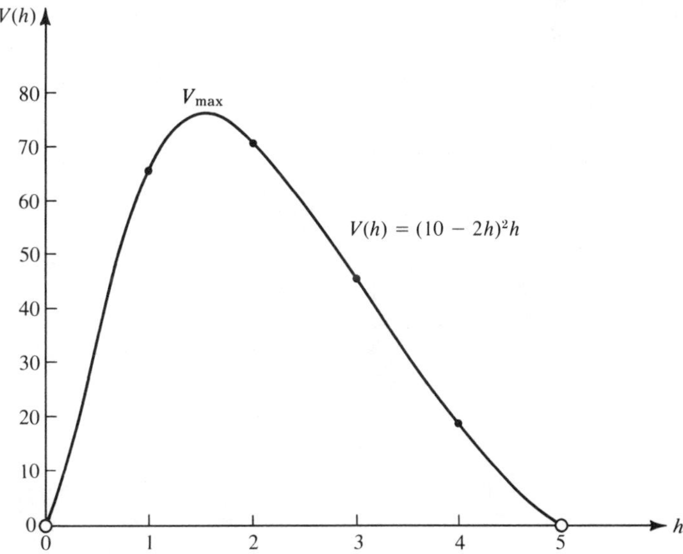

FIGURE 1.13

positive, we can conclude that

$$h > 0$$

and
$$l = w = (10-2h) > 0$$

Solving these inequalities, we find that $h>0$ and $h<5$. Therefore, the domain in interval notation is

$$D = (0,5)$$

The graph of the function $V(h)=(10-2h)^2h$ is shown in Fig. 1.13. The range of the function found by an inspection of Fig. 1.13 in interval notation is

$$R = (0, V_{\max}], \qquad \text{where } V_{\max} \text{ is approximately 74 cu in.}$$

EXERCISES

1. In the accompanying figure, determine the coordinates of

(a) point A, (b) point B, (c) point C, (d) point D,
(e) point E, (f) point F, (g) point G, (h) point H.

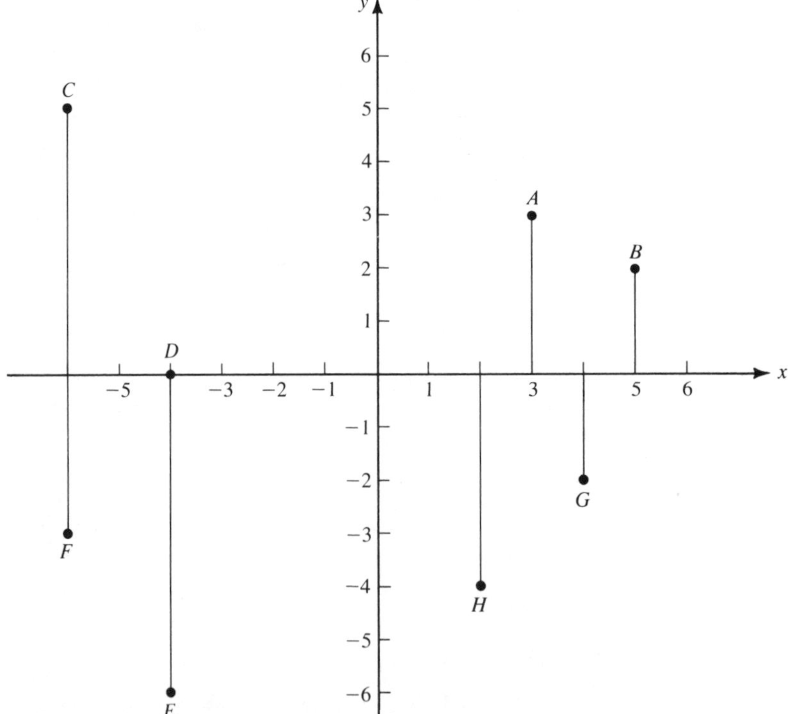

EXERCISE 1

2. If $f(x)=x^2-3$ and the domain of this function is $D=\{-3,-2,-1,0,1,2,3\}$, determine the ordered pairs that belong to this function and plot the points that correspond to these ordered pairs on a cartesian coordinate system.

3. If $f(x)=x^2-3$ and the domain of this function is the closed interval $[-3,3]$, draw a sketch of the graph of this function and determine the range in interval notation.

4. Compare the graphs of Exercises 2 and 3 above.

5. If $g(x)=3x+2$ and x is any real number, draw a sketch of the graph of this function and determine the range.

6. Express the domain and range in interval notation for each of the four functions depicted in the accompanying figure.

7. Use the vertical-line test to determine which (if any) of the graphs in the figure represent functions.

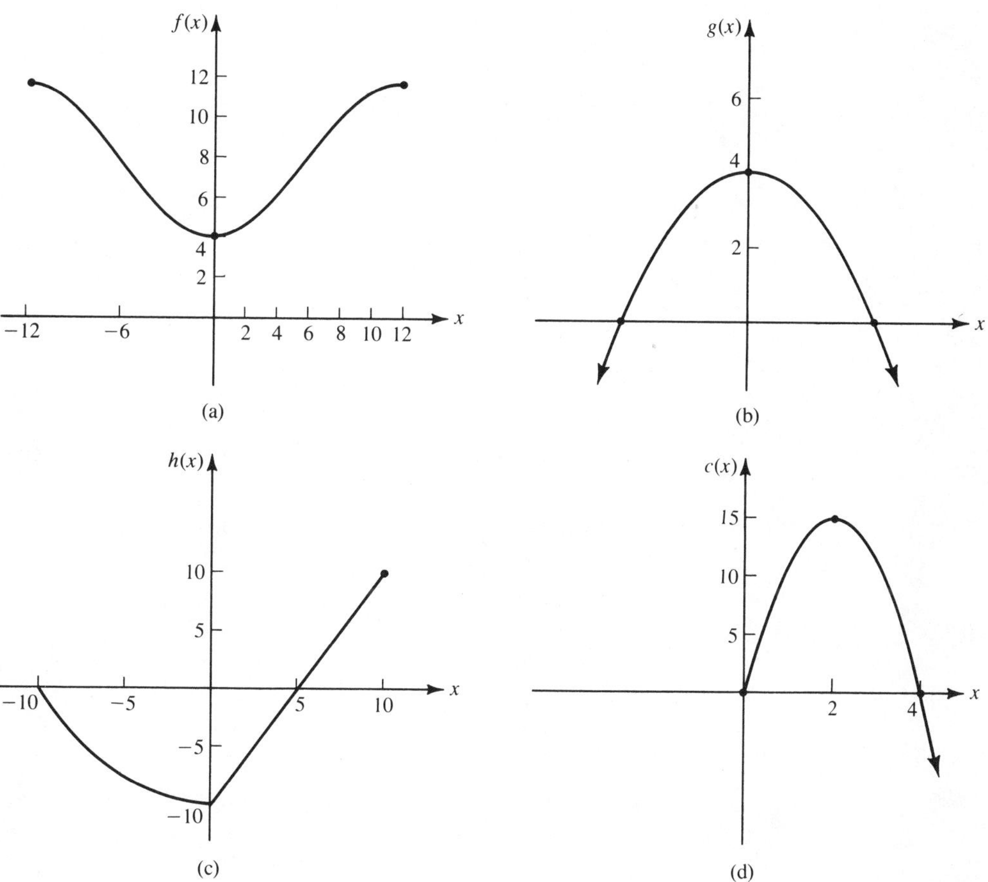

EXERCISE 6

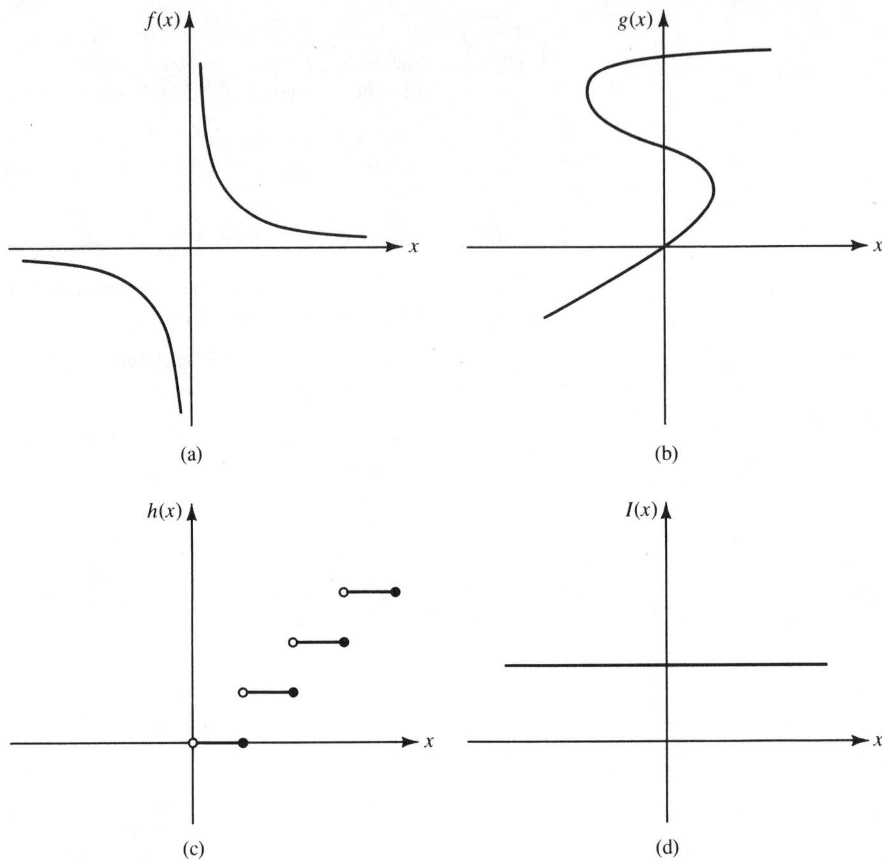

EXERCISE 7

8. Express the domain of each of the following functions in interval notation.

(a) $f(x)=3x^2+5x+2$
(b) $g(t)=\sqrt{t-5}$
(c) $h(u)=\dfrac{1}{u+2}$
(d) $y=\sqrt{x^2+10}$
(e) $y=\sqrt[3]{x-2}$
(f) $f(x)=\dfrac{x-1}{x+2}$
(g) $g(u)=\dfrac{1}{\sqrt{2-u}}$
(h) $u(x)=\dfrac{1}{\sqrt{x^2-4}}$

9. Complete the following table for each of the functions below. Sketch the graph of each function on a separate coordinate system.

x	-5	-4	-3	-2	-1	0	1	2	3	4	5
$f(x)$											

(a) $f(x)=2x-6$
(b) $f(x)=2x^2+x-2$
(c) $f(x)=x^3$
(d) $f(x)=6-x^2$
(e) $f(x)=\dfrac{10}{x}, \quad x\neq 0$
(f) $f(x)=-x+4$
(g) $f(x)=\sqrt{x+5}$
(h) $f(x)=\dfrac{x^2-2x+1}{x-1}, \quad x\neq 1$

10. Draw a sketch of the graph of the profit function described in Exercise 32, Section 1.2. Assume the domain $D=[0,5]$. Based on the graph, when will the profit be a maximum?

11. If \$1,800 is borrowed at 12% simple interest, use the Banker's rule (see Exercise 29, Section 1.2) to express the amount of money A that must be repaid as a function of n, the number of days of the loan, and sketch the graph of this function.

12. A plant has costs of \$50,000 per month plus \$12/unit produced. Determine and graph the function that expresses the monthly costs in terms of units produced.

13. Find the maximum volume of the box discussed in this section (see Fig. 1.13) by selecting h values to two decimal places.

14. An open box is to be made by cutting out square corners from a piece of cardboard that is a 10 in. × 14 in. rectangle.

(a) If the edge of each square is x, express the volume V of the open box as a function of x.
(b) Draw a sketch of the graph of the function determined in part (a).
(c) What are the approximate dimensions (to the nearest tenth of a unit) of the box that has the greatest volume?

15. Express the total surface area S of the open box shown in Fig. 1.9 as a function of h, the length of the edges of the squares cut from the 10 in. × 10 in. square sheet of cardboard.

(a) Determine the total surface area when $h=2$ in.
(b) Sketch the graph of this function.
(c) Is there a box that has a minimum total surface area?

1.4 COMBINING FUNCTIONS

Functions, like algebraic expressions, can be combined under the operations of addition, subtraction, multiplication, and division. Many new functions are obtained by performing these operations on simpler functions. These new functions are useful in building models for more complicated situations.

Perhaps the simplest function to consider as a building block for other functions is the identity function $f(x)=x$. It is called the identity function because it is a formula that pairs any real number with itself. The function $g(x)=x^2$, for example, can be formed from the identity function by writing

$$g(x) = f(x)\cdot f(x)$$

and the function $h(x)=x^2+x$ can now be formed from the function $g(x)$ and $f(x)$ by writing

$$h(x) = g(x) + f(x)$$

In general, given any two functions $f(x)$ and $g(x)$

the sum $f(x)+g(x)$ is denoted by $(f+g)(x)$,
the difference $f(x)-g(x)$ is denoted by $(f-g)(x)$, and
the product $f(x)\cdot g(x)$ is denoted by $(f\cdot g)(x)$.

Each of these new functions has a domain equal to the set of numbers common to both the domains of $f(x)$ and $g(x)$. The *quotient* $f(x)/g(x)$ is denoted by $(f/g)(x)$ and its domain is the set of numbers common to both the domains of $f(x)$ and $g(x)$ excluding any values of x such that $g(x)=0$.

Example 1.11 If $f(x)=2x$ and $g(x)=\sqrt{x}$, find

(a) $(f+g)(x)$
(b) $(f-g)(x)$
(c) $(f\cdot g)(x)$
(d) $\left(\frac{f}{g}\right)(x)$

and determine the domain of each of these new functions.

Solution

(a)
$$\begin{aligned}(f+g)(x) &= f(x) + g(x)\\ &= 2x + \sqrt{x}\end{aligned}$$

The domain of $f(x)$ is the set of real numbers $\mathbb{R}$. The domain of $g(x)$ is $[0, \infty)$. The numbers common to both domains, $[0, \infty)$, is the domain of $(f+g)(x)$.

(b)
$$\begin{aligned}(f-g)(x) &= f(x) - g(x)\\ &= 2x - \sqrt{x}\end{aligned}$$

The domain of $(f-g)(x)$ is the same as in part (a).

(c)
$$\begin{aligned}(f\cdot g)(x) &= f(x)\cdot g(x)\\ &= 2x\sqrt{x}\end{aligned}$$

The domain of $(f\cdot g)(x)$ is the same as in part (a).

(d)
$$\left(\frac{f}{g}\right)(x) = \frac{f(x)}{g(x)}$$
$$= \frac{2x}{\sqrt{x}}$$

The domain of $(f/g)(x)$ is the same as in part (a) with the added restriction that $g(x) \neq 0$. Since $g(x) = \sqrt{x}$, we also require that $x \neq 0$. Therefore, the domain is the open interval $(0, \infty)$.

Example 1.12 Profit-Revenue-Cost If the revenue, $R(x)$, from the sale of x units of a commodity is given by the formula

$$R(x) = 100x$$

and the cost incurred in producing these x units of the commodity is given by the formula

$$C(x) = 400 + 10x,$$

find the difference function: $(R-C)(x)$ and discuss its significance.

Solution

$$(R-C)(x) = R(x) - C(x)$$
$$= 100x - (400 + 10x)$$
$$= 90x - 400$$

If the total revenue from the sale of x items is $R(x)$ and the total cost of producing these x items is $C(x)$, the difference $R(x) - C(x)$ is the profit from the sales of these x items. Therefore, another name for $(R-C)(x)$ is the profit function $P(x)$; hence,

$$P(x) = 90x - 400$$

There is yet another way to form a new function from the two functions $f(x)$ and $g(x)$. Normally we evaluate the function $f(x)$ by replacing the variable x by a real number from its domain. If we replace it instead by the function $g(x)$ we have a new function

$$f[g(x)]$$

which is a function of a function. Such a function is called a *composite function*. To illustrate the characteristics of a composite function, consider the illustration in Fig. 1.14. The function $f[g(x)]$ pairs a number x_1 in the domain of $g(x)$ with a number in the range of $f(x)$ by first mapping the number x_1

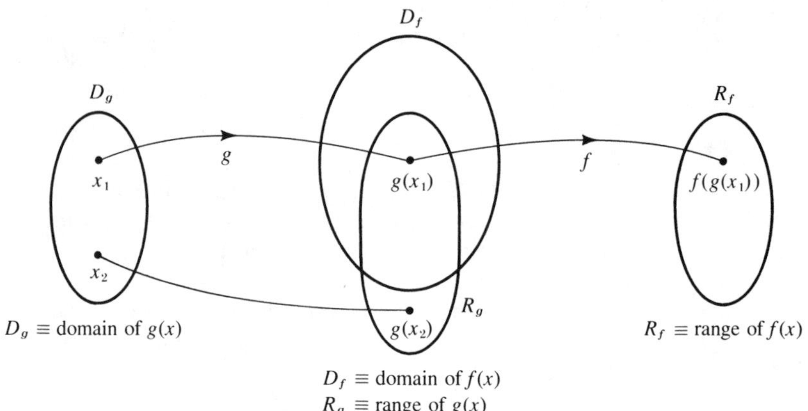

FIGURE 1.14

into the number $g(x_1)$, which must be in the domain of $f(x)$. This number $g(x_1)$ is then mapped into the number $f[g(x_1)]$ in the range of the function $f(x)$.

The domain of the composite function $f[g(x)]$ will be all the x_1 values in the domain of $g(x)$ such that $g(x_1)$ is a value in the domain of $f(x)$. Note that x_2 in the domain of $g(x)$, D_g, in Fig. 1.14 is not in the domain of $f[g(x)]$ since $g(x_2)$ is not in the domain of $f(x)$. Hence, the domain of the composite function $f[g(x)]$ is a subset of the domain of $g(x)$. To illustrate the use of composite functions, let us consider the following physical problem.

Example 1.13 Area of an Oil Spill Suppose an oil spill from an offshore well is circular in shape and is spreading out from its source. If the radius of the circle is increasing at the rate of 10 ft/hr, determine the area of the spill after 24 hr.

Solution The area of a circle is given by the formula $A = \pi r^2$. Since π is constant, approximated by $\frac{22}{7}$, the area A is a function of r, or

$$A = f(r)$$

If the radius of the circle is increasing at the rate of 10 ft/hr, in 1 hr the radius will be 10 ft; in 2 hr, 20 ft; and in t hr, $10t$ ft. Therefore, the radius $r = 10t$. Thus the radius r is a function of t that we write as

$$r = g(t)$$

Combining the function $A = f(r)$ and the function $r = g(t)$, we form the composite function

$$A = f[g(t)]$$

To find the area A when $t=24$ hr, we evaluate

$$f[g(24)]$$

by first evaluating $g(24)$. Replacing $t=24$ in the equation $g(t)=10t$, we find $g(24)=240$. We then evaluate $f(240)$. Since $f(r)=\pi r^2$, $f(240)\doteq \frac{22}{7}(240)^2 \doteq$ 180,955 sq ft. (*Note*: the symbol $\doteq$ means "is approximately equal to.") Therefore, total area covered by the oil slick after 24 hr is

$$A \doteq 180{,}955 \text{ sq ft}$$

Given any two functions $f(x)$ and $g(x)$ we can consider the composite function $f[g(x)]$ and the composite function $g[f(x)]$. These two functions in most cases are not equal as we show below.

Example 1.14 If $f(x)=1/x$ and $g(x)=x^2+2$, find $f[g(x)]$ and $g[f(x)]$ and show they are not equal.

Solution Replacing x by $g(x)$ in the equation $f(x)=1/x$, we have

$$f[g(x)] = \frac{1}{g(x)}$$

Since $g(x)=x^2+2$, by substitution we can rewrite $1/[g(x)]$ as $1/(x^2+2)$; therefore,

$$f[g(x)] = \frac{1}{x^2+2}$$

Replacing x by $f(x)$ in the equation $g(x)=x^2+2$ we have

$$g[f(x)] = [f(x)]^2 + 2$$

Since $f(x)=1/x$, by substitution we can rewrite $[f(x)]^2+2$ as $(1/x^2)+2$; therefore,

$$g[f(x)] = \frac{1}{x^2} + 2 = \frac{1+2x^2}{x^2}$$

Since $1/(x^2+2) \neq (1+2x^2)/x^2$, we conclude that $f[g(x)] \neq g[f(x)]$.

Example 1.15 If y is the function of u, $f(u)=1+u^2$, and u is the function of x, $g(x)=\sqrt{x}$, express y as a function of x and discuss its domain.

Solution Since $y=f(u)=1+u^2$ and $u=g(x)=\sqrt{x}$, we can express y as the composite function of x

$$y = f[g(x)]$$

Substituting $g(x)$ for u in the equation $f(u)=1+u^2$, we find

$$f[g(x)] = 1+[g(x)]^2$$

Since $g(x)=\sqrt{x}$, by substitution we can rewrite $1+[g(x)]^2$ as $1+(\sqrt{x})^2$, which equals $1+x$; therefore,

$$f[g(x)] = 1+x$$

Since $y=f[g(x)]$, by substitution we can express y as a function of x by the formula

$$y = 1+x$$

We might erroneously conclude from this last equation that the domain of the composite function $y=f[g(x)]$ is the set of all real numbers. However, to correctly determine the domain of this composite function we must select all those values in the domain of $g(x)=\sqrt{x}$ such that $g(x)$ is in the domain of $f(x)=1+x^2$. The domain of $g(x)$ is $[0,\infty)$ and $g(x)$ is in the domain of $f(x)$ for these values. Therefore, the domain of the composite function $y=f[g(x)]$ is $[0,\infty)$ and not the set of real numbers.

EXERCISES

Evaluate each of the expressions in Exercises 1 to 12. Assume that $f(x)=x^2-1$, $g(x)=\sqrt{(x+1)}$, and $u(x)=x^3$.

1. $(f+g)(8)$

2. $(f-g)(8)$

3. $\left(\frac{f}{g}\right)(24)$

4. $\left(\frac{f}{u}\right)(1)$

5. $(f\cdot g)(0)$

6. $(f\cdot u)(-1)$

7. $g[f(4)]$

8. $f[g(4)]$

9. $f[u(2)]$

10. $u[f(2)]$

11. $u(f[g(0)])$

12. $g(f[u(0)])$

In Exercises 13 to 24 assume $f(x)=x^2+1$, $g(x)=\sqrt{x-1}$, and $u(x)=1/x$. Express each function as a rule in terms of x and determine the domain of each.

13. $(f+g)(x)$

14. $(f-u)(x)$

15. $(f\cdot g)(x)$

16. $\left(\frac{g}{f}\right)(x)$

17. $\left(\frac{f}{g}\right)(x)$

18. $f[g(x)]$

19. $g[f(x)]$

20. $f[u(x)]$

21. $g[u(x)]$

22. $u[f(x)]$

23. $f(g[u(x)])$

24. $u(g[f(x)])$

In Exercises 25 to 30 each of the expressions is equal to the composite function $f[u(x)]$. Find $f(x)$ and $u(x)$. (There may be more than one acceptable solution.)

25. $\sqrt{x^2+5}$

26. $\dfrac{1}{(x+2)^2}$

27. $\dfrac{1}{\sqrt{x+3}}$

28. $\sqrt[3]{(2x-1)}$

29. $(x^2+2x+5)^3$

30. $6x^2$

31. Profit-Revenue-Cost If the dollar revenue, $R(x)$, from the sale of x units of a certain product is given by $R(x)=200x$ and the cost in dollars, $C(x)$, incurred in manufacturing these x units is given by $C(x)=50x+100$,

(a) determine the profit function $P(x)=(R-C)(x)$.
(b) determine the profit when $x=10$ units.

32. Offshore Oil Leak Oil seeping from an offshore well is spreading on the surface of the water in a circle. The radius of the circle in feet is given by $r(t)=100+20t$ where t is the time in hours measured from $t=0$ when the radius was 100 ft.

(a) If the area of the circle is given by $A(r)=\pi r^2$, where $\pi \doteq \frac{22}{7}$, express the area A as a function of the time by finding $A[r(t)]$.
(b) Determine the area of the circle when $t=10$ hr; that is, find $A[r(10)]$.

33. Air Pollution A study of the air pollution problems of a certain city indicates that the average daily level of carbon monoxide in the air in parts per million will be given by $f(x)=.7x+1$ where x is the number of thousand car miles driven in the city per day. It is estimated that the number of thousands of car miles that will be driven in the city per day in t years will be given by $x(t)=20+2t$.

(a) Determine the carbon monoxide level as a function of the time t; that is, determine $f[x(t)]$.
(b) Determine the level of carbon monoxide when $t=5$ years; that is, find $f[x(5)]$.

1.5 INVERSE FUNCTIONS

In the previous section we created new functions by combining two functions. In this section we will create or generate a new function from a single function. For example, it is easily verified that the ordered pairs $(-3,-2)$, $(-2,0)$, $(-1,2)$, $(0,4)$, $(1,6)$, and $(2,8)$ belong to the function

$$y = 2x + 4 \tag{1.3}$$

To form a new function from the one denoted by Eq. (1.3) we begin by interchanging the first and second components of the above ordered pairs and obtain $(-2,-3)$, $(0,-2)$, $(2,-1)$, $(4,0)$, $(6,1)$, and $(8,2)$. Now we ask, what is the formula or function that is satisfied by these new ordered pairs?

Since these ordered pairs were formed by interchanging the position of x and y, the new rule or formula that will generate these ordered pairs is found by interchanging the x and y in Eq. (1.3). Thus the new function is

$$x = 2y + 4$$

or solving for the dependent variable y we have

$$y = \frac{x-4}{2} \tag{1.4}$$

It is easy to verify that the new ordered pairs satisfy Eq. (1.4).

Writing these two functions using functional notation, we have

$$f(x) = 2x + 4 \qquad \text{and} \qquad g(x) = \frac{x-4}{2}$$

The formulation of $g(x)$ in this manner from the function $f(x)$ necessitates that $f(x)$ be a one-to-one function if $g(x)$ is to be a function. (We will justify this statement shortly.) Since we are interested in relations that are functions we will restrict the formulation of these new relations from one-to-one functions. Therefore, assuming that $f(x)$ is a one-to-one function, the formulation of the function $g(x)$ guarantees that whenever the ordered pair (a,b) belongs to the function $f(x)$, the ordered pair (b,a) belongs to the function $g(x)$, and conversely. Functions exhibiting this characteristic are called *inverse functions*.

Definition 1.6 The function $g(x)$ is called the inverse of the function $f(x)$ and the function $f(x)$ is called the inverse of the function $g(x)$ if and only if whenever the ordered pair (a,b) belongs to $f(x)$, the ordered pair (b,a) belongs to $g(x)$, and conversely.

It is common notation to express the inverse of the function $f(x)$ by $f^{-1}(x)$. For example, for $f(x)=2x+4$, we found $f^{-1}(x)=(x-4)/2$. We read $f^{-1}(x)$ as "the inverse of f of x" or "f inverse of x." The "-1" in this case is not interpreted as an exponent.

Example 1.16 Determine the inverse of the function $f(x)=\frac{1}{2}x+2$. Select five ordered pairs that belong to $f(x)$ and verify that the ordered pairs found by interchanging the 1st and 2nd components belong to $f^{-1}(x)$.

Solution To determine the inverse of $f(x)=\frac{1}{2}x+2$, we rewrite this function as $y=\frac{1}{2}x+2$. This enables us to interchange the variables x and y as follows:

$$x = \tfrac{1}{2}y + 2$$

solving for y we find $$y = 2x - 4$$

or $$f^{-1}(x) = 2x - 4$$

We select five ordered pairs that belong to $f(x)$ by picking at random the x values -2, 0, 2, 4, and 6 and calculating the associated $f(x)$ values.

x	-2	0	2	4	6
$f(x)=\frac{1}{2}x+2$	1	2	3	4	5

Rostering these as ordered pairs, we have

$$\{(-2,1), (0,2), (2,3), (4,4), (6,5)\}$$

We can verify by substitution that the new ordered pairs $\{(1,-2), (2,0), (3,2), (4,4), (5,6)\}$ belong to the function $f^{-1}(x)=2x-4$. For example, $f^{-1}(1)=2(1)-4=-2$; therefore, $(1,-2)$ belongs to $f^{-1}(x)$, etc.

From Definition 1.6 we note that if (a,b) belongs to $f(x)$, then (b,a) belongs to $f^{-1}(x)$, which means a is an element of the domain of $f(x)$ and the range of $f^{-1}(x)$. We conclude that the domain of the function $f(x)$ is equal to the range of $f^{-1}(x)$. In a like manner we can show that the range of $f(x)$ is equal to the domain of $f^{-1}(x)$. This is illustrated schematically in Fig. 1.15.

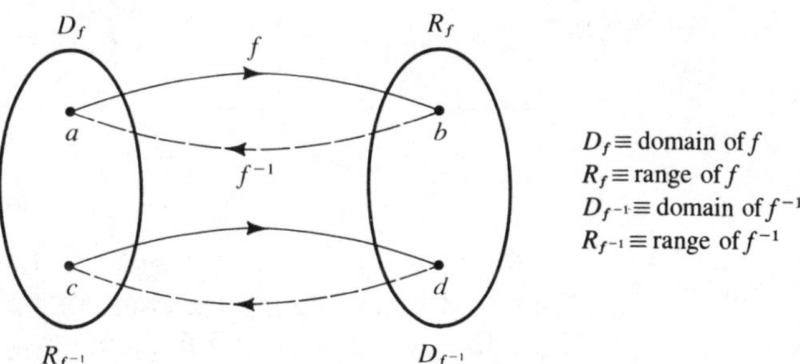

FIGURE 1.15

According to our method of formulating the inverse of a function, Fig. 1.16 shows that the inverse of a many-to-one function, $f(x)$, is not a function. If $f(x)$ is a many-to-one function there exist at least two ordered pairs of the form (a,b) and (c,b) that belong to $f(x)$. Hence the ordered pairs (b,a) and (b,c) belong to the inverse of $f(x)$. Such a relation by Definition 1.4 is not a function. Therefore, only one-to-one functions have inverses that are functions.

Figure 1.17 illustrates that if any horizontal line intersects the graph of a function $f(x)$ more than once, the function $f(x)$ is many-to-one and hence has

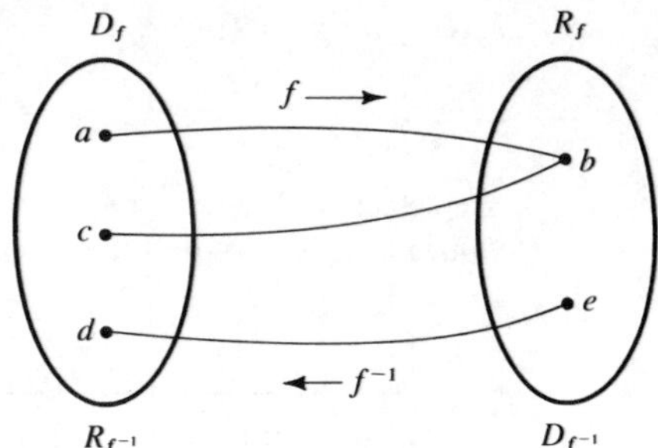

A many-to-one function $f(x)$

FIGURE 1.16

no inverse function. This implies that a function $f(x)$ has an inverse function only if no horizontal line intersects the graph of the function $f(x)$ more than once. [Since (a,b) and (c,b) belong to $f(x)$, the function is many-to-one.]

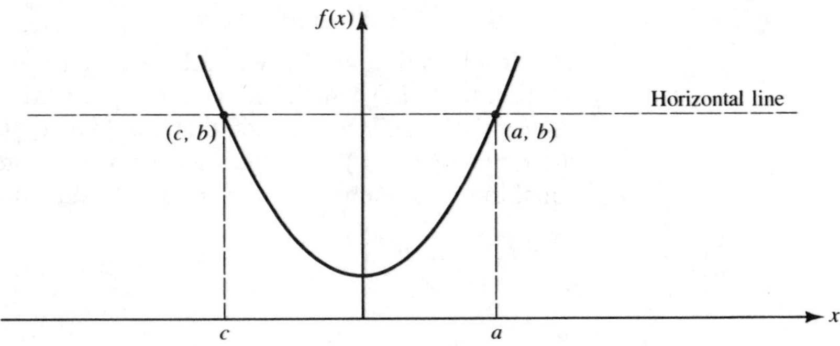

FIGURE 1.17

Another interesting property of functions and their inverses can be discovered by examining the graph of $f(x)$ and its inverse $f^{-1}(x)$. Observe in Fig. 1.18*a* the graphs of $f(x)=2x+4$ and $f^{-1}(x)=(x-4)/2$ and in Fig. 1.18*b* the graphs of $f(x)=\frac{1}{2}x+2$ and $f^{-1}(x)=2x-4$.

We can see in Figs. 1.18*a* and *b* that if we fold the cartesian plane along the dotted line whose equation is $y=x$, the graphs of $f(x)$ and $f^{-1}(x)$ will coincide. The line $y=x$ is called a *line of symmetry* for the graphs of the functions $f(x)$ and $f^{-1}(x)$. Since this can be proved for the graph of any function and its inverse, we formalize the property.

Property 1.1 The line whose equation is $y=x$ is a line of symmetry for the graph of any function $f(x)$, and the graph of its inverse $f^{-1}(x)$ and the graphs of $f(x)$ and $f^{-1}(x)$ are said to be symmetric to each other with respect to this line.

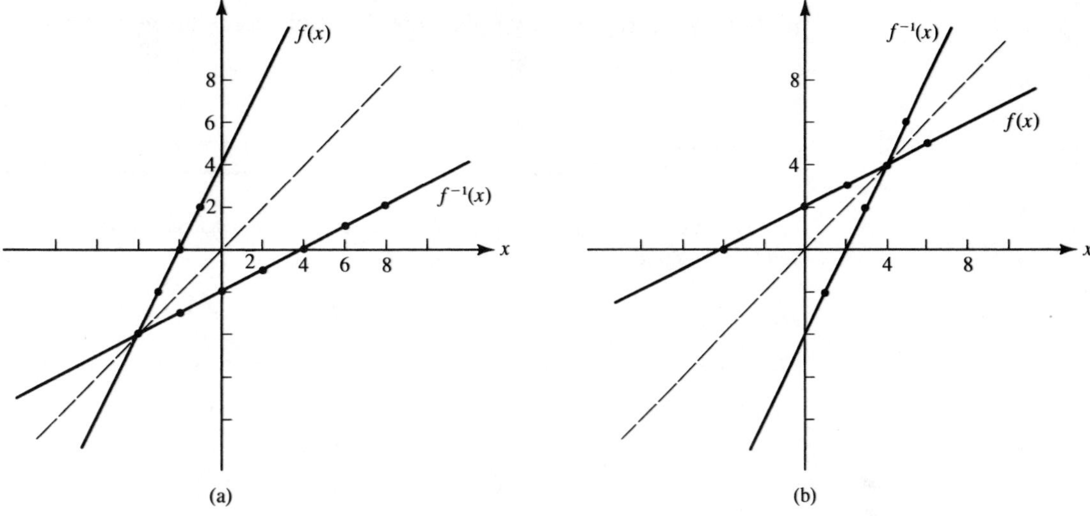

FIGURE 1.18

We can use this property to graph the inverse of a function by "reflecting" the graph of the function across the line $y=x$. For example, to draw the graph of the inverse of the function $g(x)$ shown in Fig. 1.19 we merely draw in the line $y=x$ and reflect the graph of $g(x)$ across this line as shown in the figure. To reflect point P on $g(x)$ to P' on $g^{-1}(x)$, construct the line PO perpendicular to $y=x$ and extend PO to P' such that OP' is equal in length

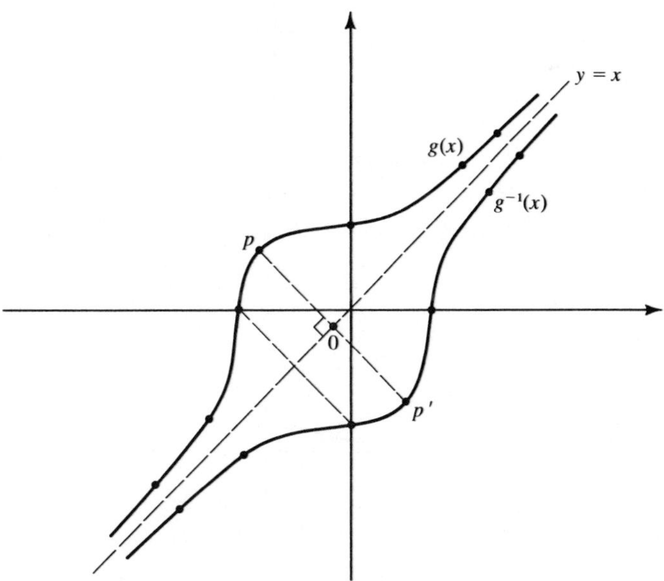

FIGURE 1.19

to PO. If we repeat this procedure for each point on $g(x)$ we will obtain $g^{-1}(x)$.

EXERCISES

In Exercises 1 to 10 determine the inverse function for each given function and graph both on the same cartesian coordinate system.

1. $y=2x-5$

2. $y=-\frac{1}{2}x+3$

3. $f(x)=\frac{1}{2}x+6$

4. $f(x)=-\frac{3}{2}x-4$

5. $g(x)=x-4$

6. $g(x)=-x+3$

7. $f(x)=x^3$

8. $f(x)=x^3-1$

9. $g(x)=\dfrac{1}{x}$

10. $g(x)=\dfrac{2}{x}$

In Exercises 11 to 16 sketch the graph of the inverse function on the same coordinate system as the graph of the given function.

11.

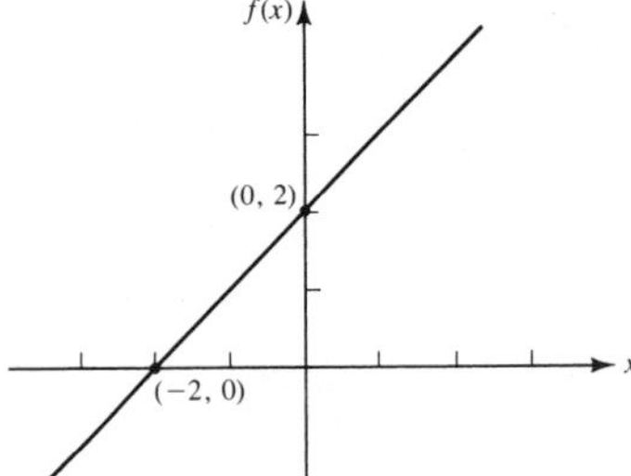

EXERCISE 11

12.

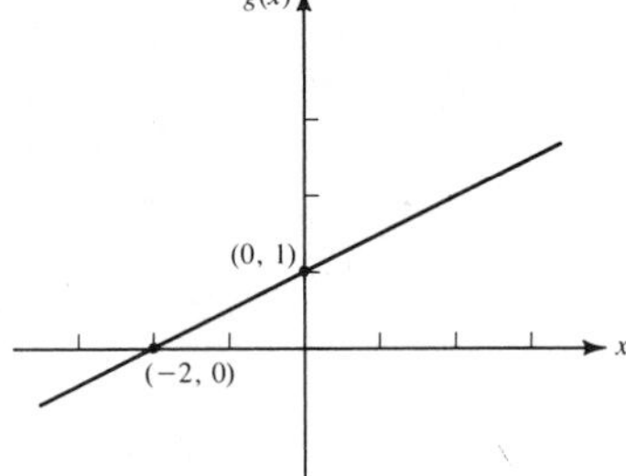

EXERCISE 12

13.

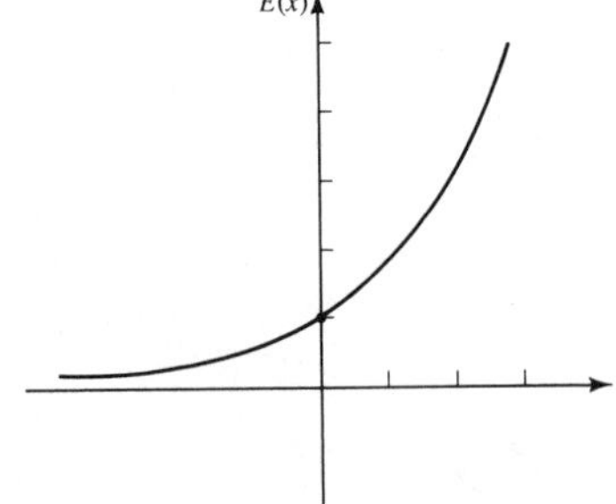

EXERCISE 13

14.

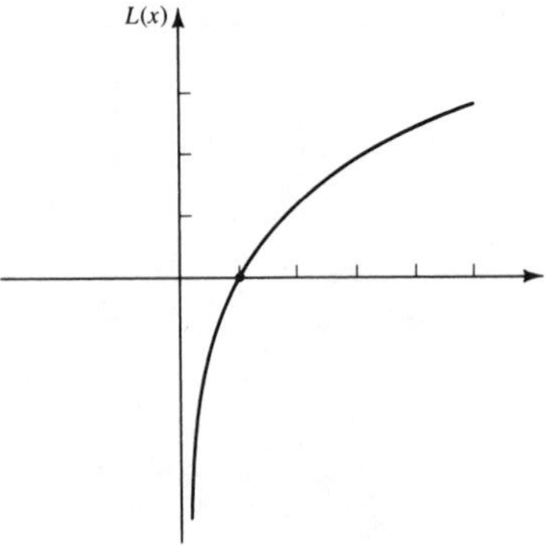

EXERCISE 14

15.

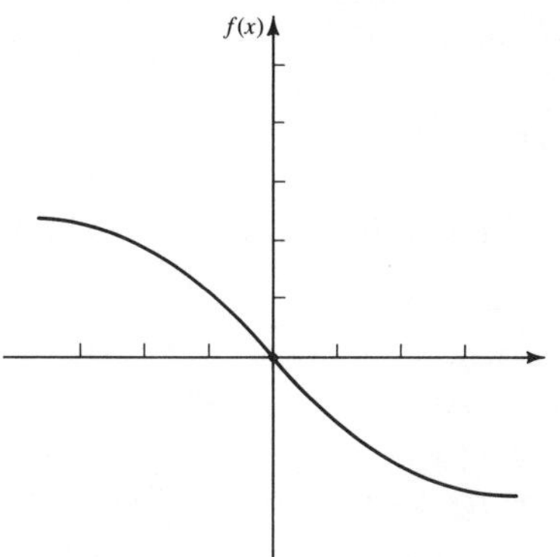

EXERCISE 15

16.

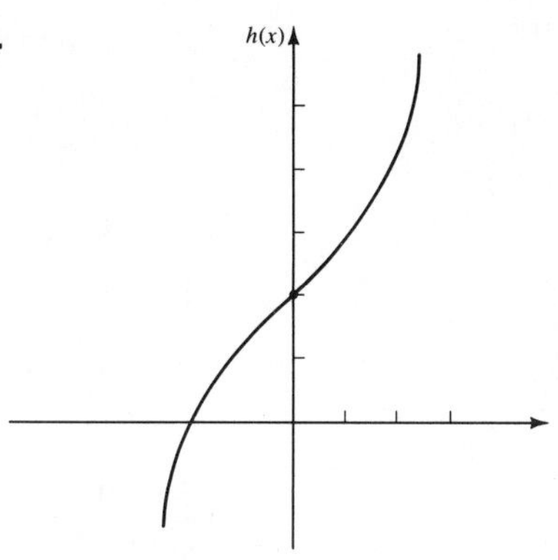

EXERCISE 16

17. The graphs of several functions are shown. Which of these functions have inverse functions?

(a)

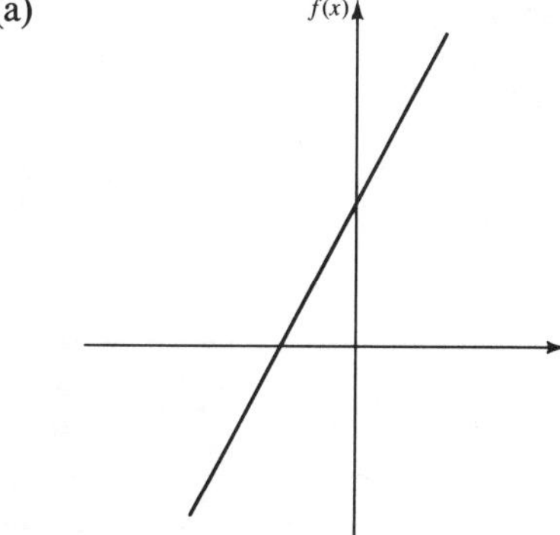

(b)

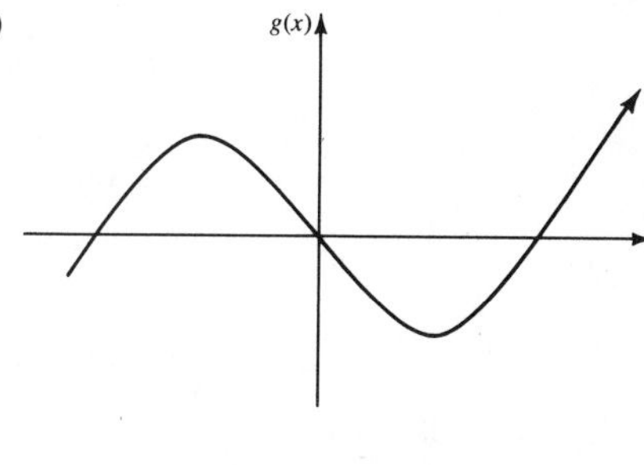

(c)

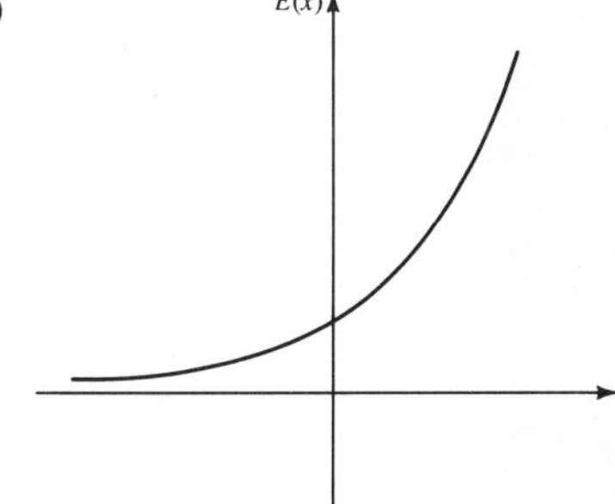

(d)

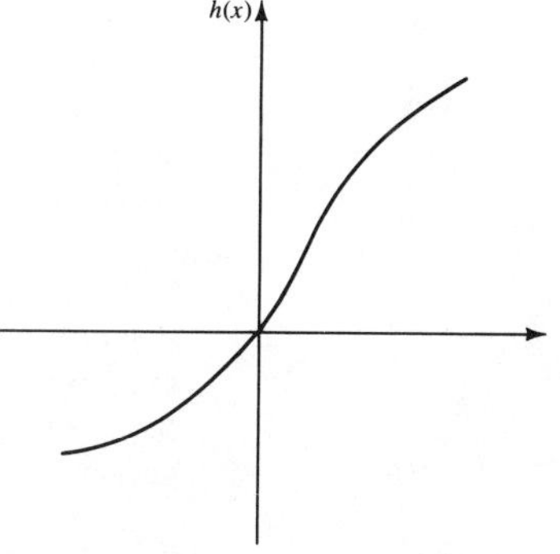

EXERCISE 17

18. Let $f(x)=\frac{1}{2}x-5$.

(a) Determine $f^{-1}(x)$.
(b) Evaluate $f[f^{-1}(4)]$.
(c) Evaluate $f^{-1}[f(4)]$.
(d) Generalize the results indicated by (b) and (c) above.

19. Let $g(x)=-2x+3$.

(a) Determine $g^{-1}(x)$.
(b) Evaluate $g[g^{-1}(6)]$.
(c) Evaluate $g^{-1}[g(6)]$.
(d) Generalize the results indicated by (b) and (c) above.

20. If a machinist makes an hourly wage of \$4 plus \$2 extra for each unit he can produce per hour, the number of dollars that he earns per hour is given by

$$p(x) = 2x + 4$$

where x is the number of units he produces per hour.

(a) Find the inverse function $p^{-1}(x)$.
(b) What is the meaning of x and $p^{-1}(x)$ in the inverse function?

21. Determine the inverse of the function found in Exercise 34, Sec. 1.2 and convert the IQ score of 119 to an ATMM score. When would the inverse function be of more use than the function?

IMPORTANT TERMS AND CONCEPTS

Abscissa
Cartesian coordinate
Cartesian product
Closed interval
Composite function
Coordinate axes
Dependent variable
Domain
Function
Function notation
Graph of a function
Half-open interval
Independent variable
Interval notation
Inverse function
Line of symmetry
Many-to-many relation
Many-to-one relation
Mapping
One-to-many relation
One-to-one relation
Open interval
Ordered pair
Ordinate
Origin
Quadrant
Range
Rectangular coordinates
Relation
Set selector
Vertical-line test
x axis
x coordinate
y axis
y coordinate

---

2 ALGEBRAIC FUNCTIONS

2.1 INTRODUCTION

In Chapter 1 we discussed the importance of functions as mathematical models of real-world problems and some of the general properties of functions. Many of the functions that we use as models are called *algebraic functions*. An algebraic function $f(x)$ is a rule or formula that is formed by combining real numbers and the variable x in addition, subtraction, multiplication, division, or raising to a rational power. The following are examples of algebraic functions.

$$f(x) = 5x - 7, \qquad f(x) = \frac{x^2+3}{x}, \qquad f(x) = \sqrt{(x-5)}$$

In this chapter we will introduce various kinds of algebraic functions, specifically linear, quadratic, polynomial, and rational functions, and investigate the characteristics of each type. We will find that the graphs of functions provide us with many clues to their behavior.

2.2 LINEAR FUNCTIONS

Several of the functions discussed in Chapter 1 were of the general form $f(x) = mx + b$ or $y = mx + b$ where m and b are fixed constants. For example, the function $y = .12x + 15$ relating the total cost y of renting a car to x, the daily mileage driven, is of this form. In this equation, $m = .12$ and $b = 15$. Such a function is called a *linear function* and the equation $y = .12x + 15$ a *linear equation*. We call these functions or equations "linear" because their graphs are always straight lines. Linear functions have many interesting applications because they serve as mathematical models for many real-world problems. We shall study the basic characteristics of these functions in this section and investigate their applications to everyday problems.

Many situations give rise to linear equations that are not in functional form. For example, if the ideal weight of a shipment of two chemicals is 12 tons and each unit of chemical 1 weighs 6 tons while a unit of chemical 2 weighs 4 tons, the number of units shipped of each chemical must satisfy the equation

$$6y + 4x = 12$$

where y and x represent the number of units of chemicals 1 and 2, respectively. This equation is not in functional form but rather in what is commonly known as the *standard form of a linear equation*; i.e.,

$$ax + by = c$$

However, the equation is converted to functional form simply by applying the rules of algebra; namely, by subtracting $4x$ from both sides of the equation and dividing both sides of the resulting equation by 6. The above linear equation is then converted to the *functional form of a linear equation*

$$y = \frac{-2}{3}x + 2$$

In this section, we deal almost exclusively with linear equations that are already in functional form. As seen above, it is a simple matter to convert any linear equation into functional form.

In considering the linear equation $y = mx + b$ let us first consider the relationship of m and b to the graph of the function. The terms m and b are called the *coefficient of* x and the *constant term*, respectively. The constant term b does not create any change in y for different values of x; the coefficient of x indicates how much y changes as x increases by one unit. Thus, we note that y changes by m units for each unit increase in x. Consider, for example, the equation $y = 2x + 5$. Determining the values of y for several specific values of x, we obtain the following table:

x	0	1	2	3	4	5
y	5	7	9	11	13	15

In this table, we note that the y values change by two units for each one-unit change in the value of x. Plotting these points and connecting them with a straight line that extends beyond the points, we obtain the graph of the equation $y = 2x + 5$ (or, more precisely, the graph of the function $y = 2x + 5$), as given in Fig. 2.1. We note without proof that a straight line that passes through any two computed points also passes through all other points that can be plotted for the linear equation $y = 2x + 5$.

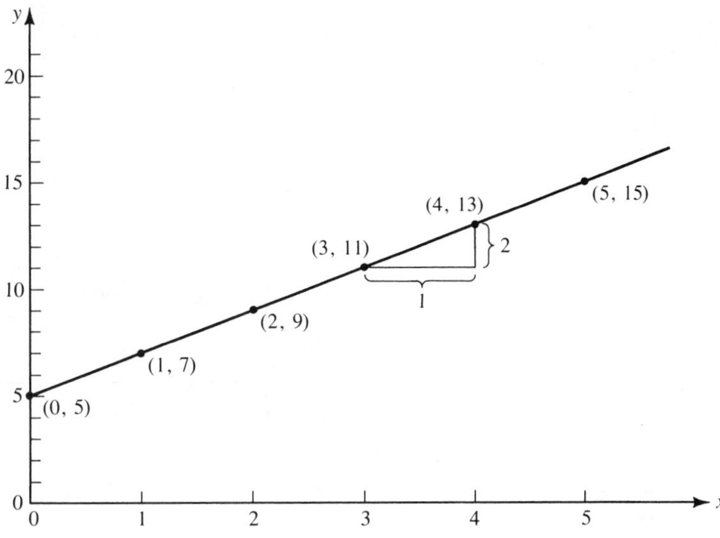

FIGURE 2.1

Slope-Intercept Formula

We see in Fig. 2.1 that changing x by one unit, from 3 to 4, causes y to change by two units; thus, the *rate of change* of y per unit change in x is two units. The rate of change of y per unit change in x for the function $y=2x+5$ is a constant independent of the x and y values selected. For example, changing x by one unit from 1 to 2 causes y to change by two units from 7 to 9, or by changing x by two units from 1 to 3 causes y to change by four units from 7 to 11.

For any two points (x_1,y_1) and (x_2,y_2) belonging to a linear function, the rate of change of the function is given by

$$\frac{y_2-y_1}{x_2-x_1} \tag{2.1}$$

The rate of change is called the *slope* of the straight line, and as shown in Fig. 2.2, it can be interpreted as the ratio of the rise to the run.

In order to see how the slope is related to the constants m and b in the linear function, let us determine the values of y_1 and y_2 as functions of x_1 and x_2, respectively. Letting y_1 and y_2 be the values of y when x equals x_1 and x_2, respectively, by substituting into the equation $y=mx+b$ we have

$$y_1 = mx_1 + b$$

and

$$y_2 = mx_2 + b$$

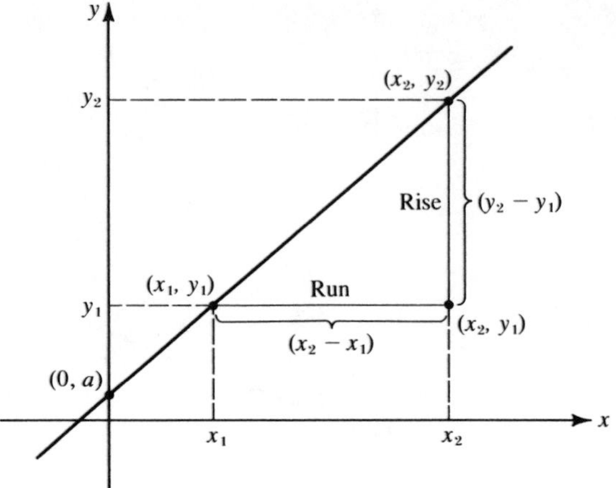

FIGURE 2.2

Determining the appropriate differences for Eq. (2.1), we have

$$\begin{aligned} y_2 - y_1 &= mx_2 + b - (mx_1 + b) \\ &= mx_2 + b - mx_1 - b \\ &= m(x_2 - x_1) \end{aligned}$$

Substituting into Eq. (2.1), we obtain

$$\frac{y_2 - y_1}{x_2 - x_1} = \frac{m(x_2 - x_1)}{x_2 - x_1} = m \tag{2.2}$$

Hence, the slope of the linear function (or, more precisely, the slope of the straight line that is the graph of the linear function) is given by the constant m, which is also the coefficient of x. We now know that the slope of the line segment connecting any two points on the graph of $y = mx + b$ will be constant.

Similarly, the constant b represents the value of the function when x is equal to zero. As we see in Fig. 2.1, for the equation $y = 2x + 5$, the value of b represents the value of y for which the graph of the function *intersects* the y axis; thus, the constant b is called the *y-intercept*. With this knowledge concerning the two constants of the linear equation in functional form, we can determine the slope and y-intercept for the graph of a linear function simply by observing the values of m and b. And, since two points will geometrically determine a straight line, we need only one more point when the linear equation is expressed in functional form because the y-intercept automatically indicates one point, namely, $(0, b)$; the additional point can be computed from the linear equation for an arbitrarily selected x value. It is also true that we can determine the equation of a straight line if we know its slope and y-intercept, and we can determine the slope and y-intercept of a straight line if we have the equation of the line expressed in functional form. For these reasons, a linear equation expressed in functional form is also said to be the *slope-intercept formula* for the equation of a straight line, given by

$$y = mx + b \tag{2.3}$$

Example 2.1 Cost of Metal Fasteners The Stemmons Metal Company produces metal fasteners. From available accounting information, they know that an initial cost of $525 will occur even if they make no fasteners. Furthermore, for all fasteners made, they will incur a material and labor cost of $3/fastener. Determine the linear equation in functional form that will permit us to compute the total cost y of making x fasteners. Determine the cost of making 1,000 fasteners.

Solution The initial cost of $525 occurs even if Stemmons makes no fasteners. Thus, the point $(0, 525)$ must satisfy the linear equation and the y-intercept is 525. The rate of change of the cost for each additional fastener is $3, which is the slope m. Having determined the intercept $b=525$ and the slope $m=3$, we obtain the linear equation through use of the slope-intercept formula:

$$y = 3x + 525$$

Rather than using the variable y for the total cost, sometimes it is preferable to represent the total cost by C. Using this notation the function is written

$$C = 3x + 525$$

or in functional notation, $C(x)=3x+525$. The cost of making 1,000 fasteners, $C(1{,}000)$, is given by

$$\begin{aligned} C(1{,}000) &= 3(1{,}000) + 525 \\ &= \$3{,}525 \end{aligned}$$

Example 2.2 Absorption of Potassium Experimentation has shown that the absorption of potassium by the leaf tissue of corn is a linear function of t, the time. In bright light potassium is absorbed at the rate of 4 units/hr per unit weight of corn. Determine a linear function that will permit us to compute the weight of potassium absorbed, A, per unit weight of corn as a function of time, t, where t is measured in hours.

Solution Since it is given that A is a linear function of time, t, we know that

$$A = mt + b$$

where m and b are constants. When $t=0$, no absorption of potassium has taken place; therefore, $A=0$ and the y-intercept $b=0$. Therefore, we can write $A=mt$. The rate of absorption of potassium is 4 units/hr, which implies that $m=4$. The linear function is

$$A = 4t$$

or in functional notation, $A(t)=4t$. (Actual experiments verify this relationship for the domain $D=[0,4]$.)

Point-Slope Formula

In addition to the slope-intercept formula, two other fairly common methods are available for determining the equation of a straight line. The first, known as the *point-slope formula*, is slightly more general than the slope-intercept formula because it permits the use of any point on the line rather than the point where the line intercepts the y axis. To use the point-slope formula, we must know the coordinates (x_1,y_1) for a specific point P on the line and the slope of the straight line; that is, we must know the coordinates (x_1,y_1) of a point P and the slope m. Given this information, we can select an arbitrary point Q for which the coordinates (x, y) are unspecified and determine the equation of the straight line.

As illustrated graphically in Fig. 2.3, we know that the slope of the line segment connecting the points P and Q is equal to m.

This slope also must be equal to the ratio of the change in y values to the change in x values for the two points because the slope of a straight line is constant for any two points on the line. Thus, from Fig. 2.3 and Eq. (2.2) we have

$$\frac{y-y_1}{x-x_1} = m \qquad \text{or} \qquad y-y_1 = m(x-x_1) \tag{2.4}$$

Using Eq. (2.4) and the given values for (x_1,y_1) and m, we can obtain the

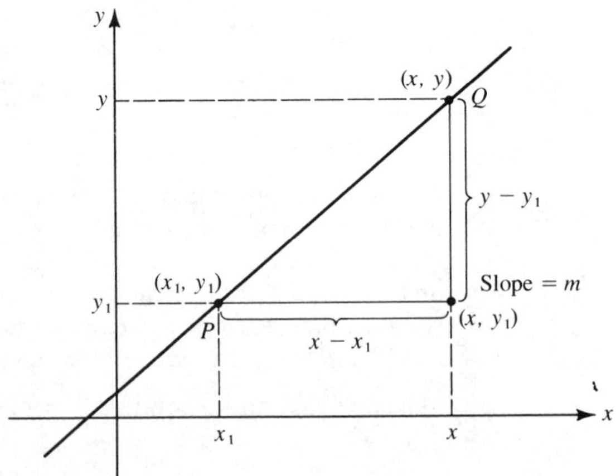

FIGURE 2.3

equation of the straight line having slope m and passing through the point (x_1,y_1). The formula in (2.4) is commonly known as the *point-slope formula.*

Example 2.3 The owner of Spalding Kanine Kennel knows that two dogs in the same pen require 6 lb of dog food per day. Each additional dog put in the pen requires another 2 lb of dog food per day. Obtain the equation in functional form that determines the required number of pounds of dog food for a given number of dogs in the same pen.

Solution Let x represent the number of dogs in the pen and y represent the number of pounds of dog food required. Since the required amount of dog food increases by 2 lb for each additional dog, the slope $m=2$. Therefore, substituting this slope and $x_1=2$ and $y_1=6$ into Eq. (2.4), we have

$$\frac{y-6}{x-2}=2$$

Rewriting the equation in functional form, we have $y=2x+2$.

Two-Point Formula A third method for determining the equation of a straight line is known as the *two-point formula.* The name comes from the fact that it is a formula requiring that we know the coordinates for two points in order to determine the equation of the straight line passing through them. Suppose the points P and Q, as given in Fig. 2.4, are the points with known coordinates and the point R has unknown coordinates. Since a straight line has a constant slope for any two points on the line, we know that the slope of the line segment PQ in Fig. 2.4 must be equal to the slope of the line segment PR. Using the ratio of (2.1) for both line segments, we set the values for the two slopes equal and obtain

$$\frac{y-y_1}{x-x_1}=\frac{y_2-y_1}{x_2-x_1} \tag{2.5}$$

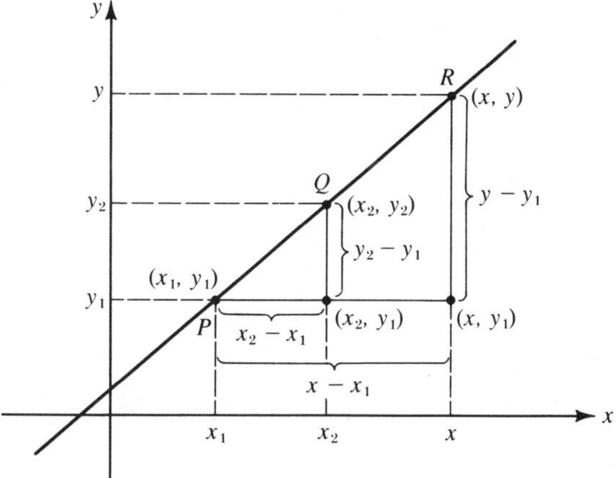

FIGURE 2.4

The formula in (2.5) is the two-point formula for determining the equation of a straight line that passes through (x_1,y_1) and (x_2,y_2). Given the values of the coordinates of any two points, we can determine the equation of the straight line that passes through them.

Example 2.4 Wages from Sale of Pennants Jon Thompson hires college students to sell pennants at the Poindexter University football games. Each student is paid a salary plus a commission. Jim McDaniel and Bruce Woods worked for Jon at one game where Jim sold 1 pennant and received \$6, and Bruce sold 5 pennants and received \$10. Determine the equation Jon uses to find the amount he must pay as the result of the number of pennants sold, under the assumption that it is a linear function.

Solution Letting x and y represent, respectively, the number of pennants sold and the amount paid to the student, we obtain for Jim and Bruce, respectively, the points (x_1,y_1) and (x_2,y_2), which are given by (1,6) and (5,10). Substituting these values into (2.5), we have

$$\frac{y-6}{x-1} = \frac{10-6}{5-1}$$

which reduces to the equation

$$y = x+5$$

Thus, Jon pays each student a salary of \$5 and a commission of \$1/pennant sold. The points and the appropriate differences are given in Fig. 2.5.

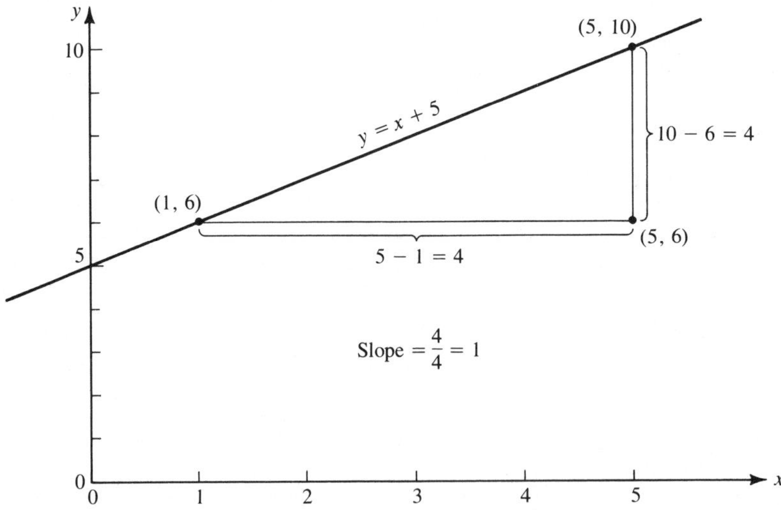

FIGURE 2.5

In summary, we have three methods for determining the equation of a straight line: the slope-intercept formula when the slope and y-intercept are known, the point-slope formula when the slope and the coordinates of a point on the line are known, and the two-point formula when the coordinates for two different points on the line are known.

In each of the linear equations that we have considered, the intercept and slope have been nonnegative numbers. However, either or both of these values may be negative. For an understanding of what it means to have a negative intercept and/or slope, let us consider each briefly. The intercept is merely the value on the y axis where the straight line intercepts the axis. Thus, if the value is positive, the point of intersection is above the x axis (remember from Sec. 1.3 that a positive value of y means it is above the x axis and a negative value of y means it is below the x axis); likewise, if the value is negative, the line intercepts the y axis below the x axis. This is demonstrated by the two equations given in Fig. 2.6, which differ only in the signs of the intercepts.

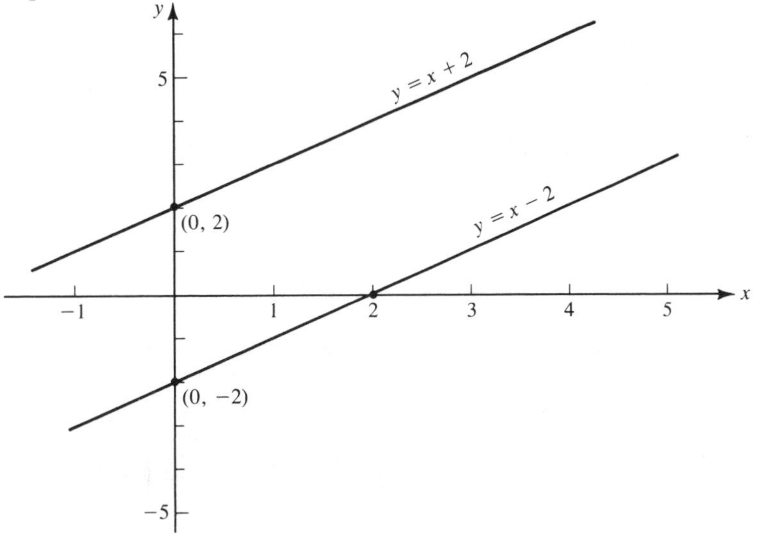

FIGURE 2.6

Since the value of the slope m represents the ratio of the change in y per unit change in x, the sign of the slope indicates simply whether the value of y increases or decreases as the value of x increases. For example, if the slope is positive, the value of y increases as the value of x increases; conversely, if the slope is negative, the value of y decreases as the value of x increases. Two equations that differ only in the sign of m are plotted in Fig. 2.7. The magnitude of rate of change for the two lines is the same, but the direction of change is different because one is increasing and the other is decreasing.

Example 2.5 Fahrenheit and Celsius Temperature If Celsius temperature is a linear function of Fahrenheit temperature, determine the rule that relates these two temperature scales.

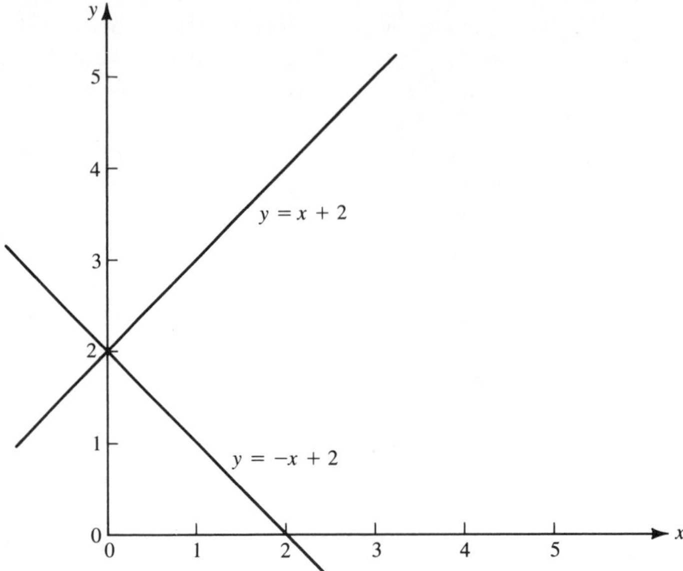

FIGURE 2.7

Solution Let x and y represent, respectively, Fahrenheit and Celsius temperatures. Since water freezes at 32° Fahrenheit and 0° Celsius and boils at 212° Fahrenheit and 100° Celsius, we have two ordered pairs, $(x_1, y_1) = (32, 0)$ and $(x_2, y_2) = (212, 100)$, that belong to our function. When substituted into the two-point formula of Eq. (2.5), we have

$$\frac{y-0}{x-32} = \frac{100-0}{212-32}$$

which reduces to the equation

$$y = \tfrac{5}{9}(x-32)$$

This formula can be written using the more customary variable F for Fahrenheit and C for Celsius as

$$C = \tfrac{5}{9}(F-32)$$

The graph of this function shown in Fig. 2.8 is determined by plotting the point (32, 0) and the y-intercept $-17\tfrac{7}{9}$.

Example 2.6 Depreciation Fred Jones, owner and operator of Mr. Jiffy Duplicating Service, purchased a new typewriter 5 years ago. For income tax purposes he has been assuming straight-line depreciation of the typewriter at the rate of \$50/year. If he currently lists the depreciated value of the typewriter at \$400, determine a linear function that relates the depreciated value to the age of the typewriter and determine the domain of the function.

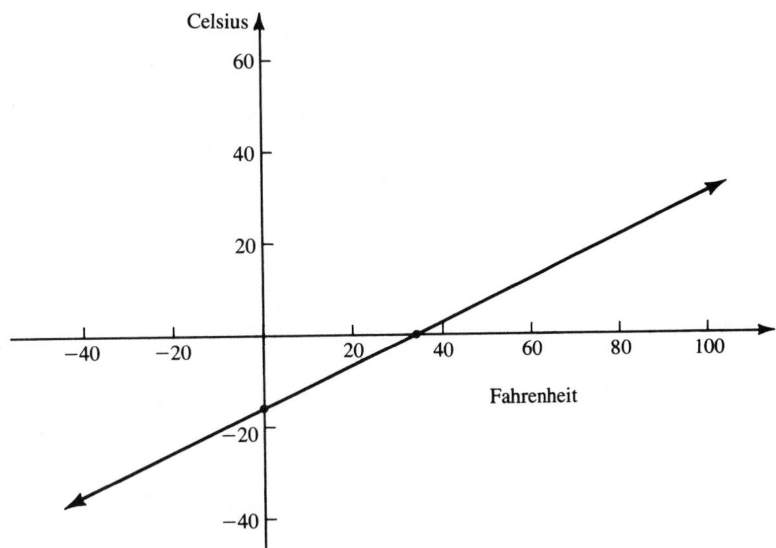

FIGURE 2.8

Solution Let y represent the depreciated value of the typewriter and x its age in years. Since the value of the typewriter is decreasing at the rate of \$50/year, the rate of change of y with respect to x or the slope $m=-50$. The fact that the present depreciated value is \$400 implies that when $x_1=5$, $y_1=400$. Substituting the ordered pair (5,400) and the slope $m=-50$ into Eq. (2.4), we have

$$y-400=-50(x-5)$$

or

$$y=-50x+650 \tag{2.6}$$

If we consider x to be any real number, the domain of the function is

$$D=[0,x_0]$$

where x_0 is the age at which the typewriter is fully depreciated. To determine x_0 we set $y=0$ and solve Eq. (2.6) for x.

$$0=-50x+650$$
$$x=13$$

Therefore, the domain $D=[0,13]$.

The graph of this function, found by plotting the ordered pairs (0,650) and (13,0), is shown in Fig. 2.9.

From the previous example, we see that occasionally we need to determine the value of x that yields a specific value of y. To do this we substitute the specific value of y into the equation and solve for x by applying

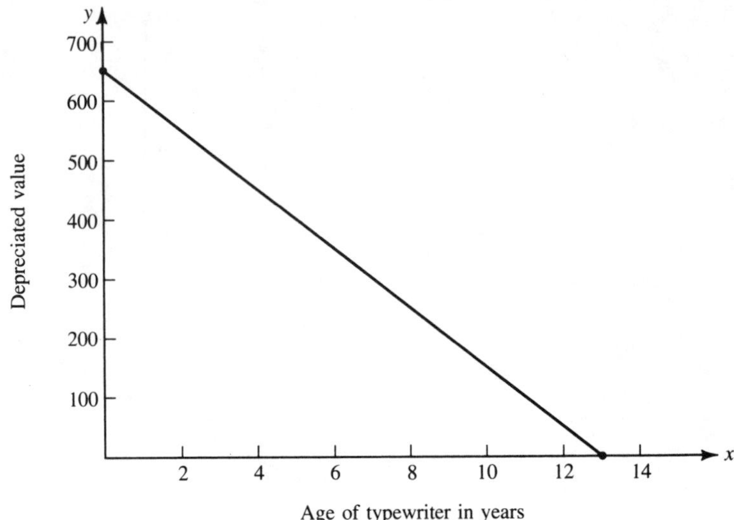

FIGURE 2.9

the rules of algebra. For example, suppose we need the value of x that yields a y value of 10 in the equation $y=2x+6$. We substitute 10 for y, obtaining

$$10 = 2x + 6$$

or

$$x = 2$$

When the y value is set equal to zero, we are seeking the x value for which the straight line cuts the x axis; such a value is called the *x-intercept* of the graph of the function. This value of x is also called a *zero of the function* $f(x)$ since for this x value $f(x)=0$. Linear functions have at most one zero; however, we will see that some functions have many zeros.

In the study of supply and demand of goods, we can often express both the quantity of consumer demand for a commodity and the quantity that can be supplied to the consumer as a function of the price of that commodity. If x is the price of the commodity, generally we denote the *demand function* as $D(x)$ and the *supply function* as $S(x)$. These functions are often linear or can be closely approximated by linear functions. For example, suppose the Tech-Cal Manufacturing Co. has determined that the demand, D, in thousands, for the model T-40 calculator is related to the unit price x in dollars by the function

$$D(x) = 100 - 4x$$

where $10 \leqslant x \leqslant 20$. At a price of $x=\$10$, the demand $D(10)=60$ thousand calculators; whereas at a price of $x=\$20$, the demand $D(20)=20$ thousand

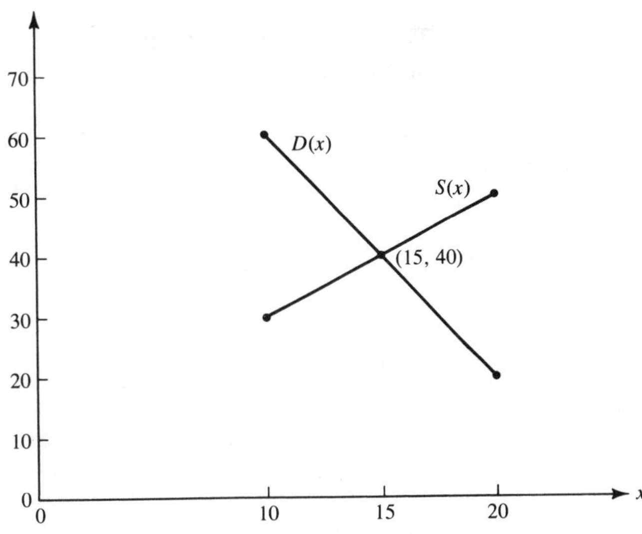

FIGURE 2.10

calculators. Using these two points (10,60) and (20,20) and the fact that $D(x)$ is a linear function of x, we draw the graph shown in Fig. 2.10.

The Tech-Cal Manufacturing Co. has also determined that they are able to manufacture and supply $S(x)$ T-40 calculators at a unit price of x dollars given by the function

$$S(x) = 10 + 2x$$

where $S(x)$ is in thousands and $10 \leqslant x \leqslant 20$. At a price of $x = \$10$, the supply $S(10) = 30$ thousand calculators, and at a price of $x = \$20$, the supply $S(20) = 50$ thousand calculators. Using the two points (10,30) and (20,50), we can draw the graph of the linear supply function as shown in Fig. 2.10.

According to Fig. 2.10 both the supply and demand equal 40 thousand calculators when the unit price $x = \$15$. If the price of the calculators is set at less than \$15, the demand will exceed the supply; whereas the supply will exceed the demand if the unit price is greater than \$15. The price at which the supply and demand are equal is called the *equilibrium price*. The point (15,40) is called the *equilibrium point* and the supply and demand values at this point are called the *equilibrium supply* and *demand*.

The equilibrium price can be determined algebraically by equating the supply and demand functions and solving the resulting equation for the equilibrium price. In our example, $D(x) = S(x)$ when

$$100 - 4x = 10 + 2x$$

or

$$90 = 6x$$

$$x = 15$$

The equilibrium demand or supply can be found by evaluating $D(15)$ or $S(15)$.

$$D(15) = 100 - 4(15)$$
$$= 40 \text{ thousand calculators}$$

This analytical procedure gives us the same result as was determined graphically in Fig. 2.10.

EXERCISES

For each of the linear equations in Exercises 1–10 *express the equation in* functional form, *graph the equation, and specify the values of the slope and y-intercept.*

1. $y-6x=16$

2. $2y+4x=14$

3. $3y-9x=15$

4. $2x+5y=10$

5. $10y-x=50$

6. $.1y-3.7x=.44$

7. $3y-15=0$

8. $\frac{3}{2}y+9x-15=0$

9. $4y-3x+16=0$

10. $2y-7x=-3y+13x$

For each of the linear functions in Exercises 11–20 *determine the x-intercept; that is, the zero of the function, and the y-intercept. Using these two points, sketch the graph of each linear function.*

11. $2y+4x=14$

12. $f(x)=3x-15$

13. $s(x)=\frac{4}{5}x+12$

14. $3y-5x=15$

15. $g(x)=-2x+9$

16. $7x-2y=28$

17. $p(x)=\frac{1}{2}x+5$

18. $f(x)=8-x$

19. $x+\frac{1}{2}y=-6$

20. $C(x)=.1x-.45$

In Exercises 21–26 *find the slope of the line that passes through the given points and find an equation of this line. Express the equation in functional form.*

21. $(0,0)$ and $(3,4)$

22. $(1,2)$ and $(4,5)$

23. $(-1,2)$ and $(3,-6)$

24. $(0,-5)$ and $(5,0)$

25. $(-8,4)$ and $(8,6)$

26. $(-2,-3)$ and $(6,-9)$

In Exercises 27–34 *find an equation of the line satisfying the given conditions. Express the equation in functional form.*

27. $m=4$ and passes through $(2,-3)$

28. y-intercept$=5$ and passes through $(-1,2)$

29. y-intercept$=6$ and x-intercept $=-2$

30. zero of the function$=-5$ and $m=-\frac{3}{2}$

31. x-intercept$=0$ and $m=-4$

32. y-intercept$=-3$ and $m=-\frac{1}{2}$

33. $m=-\frac{3}{2}$ and passes through $(2,6)$

34. $m=0$ and passes through $(1,8)$

35. Draw a horizontal and vertical line through the point $(2,5)$.

(a) What is an equation for the vertical line? The horizontal line?
(b) What is the slope of the horizontal line? The vertical line?
(c) Is the equation of the vertical line a function?
(d) Is the equation of the horizontal line a function?

36. Determine the equation in functional form for the straight line that has slope 4 and passes through the point $(2,6)$. Is the point $(5,18)$ on the line? Is the point $(3,12)$ on the line?

37. Determine the equation in functional form for the straight line that passes through $(-1,1)$ and $(4,-14)$. Is the point $(5,-17)$ on the line? Is the point $(-5,13)$ on the line?

For each of the equations in Exercises 38–43 *determine the value of x that corresponds to the given y or f(x) value.*

38. $3y+4x=15$ when $y=2$

39. $3y-9x=27$ when $y=5$

40. $f(x)=-3x+9$ when $f(x)=30$

41. $f(x)=-2x+5$ when $f(x)=-2$

42. $f(x)=8-2x$ when $f(x)=20$

43. $2y-5x=10$ when $y=-2$

44. The profit y of a Gas-N-Go service station is a function of the number of gallons of gasoline x that it sells. The manager of the service station has determined that for every 5 gallons of gasoline sold he makes 22¢ profit and the relationship between profit and amount sold is linear. Furthermore, he must sell 1,000 gallons per day to break even. Determine the linear equation in functional form that gives the relationship between profit and sales on a daily basis. What is the profit if he sells no gasoline on a particular day? What is the profit if he sells 5,000 gallons on a particular day?

45. A sales representative's reimbursement for expenses is a linear function of the miles traveled. One month she traveled 1,000 miles and was reimbursed \$300. Another month, the reimbursement was \$200 for 800 miles. Determine the linear equation that allows the sales representative to calculate her monthly reimbursement from the miles she travels.

46. For one week, a Fast-Chek store had expenses of \$1,500 and used 10 full-time employees. Each full-time employee averages \$92/week. Determine the linear equation that allows the manager to estimate weekly expenses based on the number of employees. What would be the expenses for the store if 5 employees were employed full-time for one week?

47. The Davis County Coop sells corn for \$4/bushel. They must sell 300 bushels in order to break even. Determine the linear equation in functional form that gives the relationship between profit and bushels sold. How much will they lose if only 200 bushels of corn are sold? What is the profit if 600 bushels are sold?

48. A man borrows \$36,000 at a simple interest rate of 10% and agrees to repay the loan according to the Banker's rule (see Exercise 29 in Section 1.2).

(a) Express A, the amount that must be repaid, as a function of n, the number of days of the loan.
(b) Sketch the graph of the function, plotting n on the horizontal axis.
(c) At what rate is A increasing?

49. Membership in the Woodledge Swim Club costs \$240/family for the 12-week summer season. If a family joins after the start of the season, its fee is reduced proportionately.

(a) Express the membership fee as a function of x, the number of weeks a member is late in joining the club.
(b) What is the cost of a membership that is purchased 2 weeks after the start of the season? 5 weeks?
(b) Sketch a graph of the function determined in part (a) above assuming x is any real number between 0 and 12 inclusive.

50. The demand for a certain product is given by

$$D(x) = \frac{400-3x}{5}$$

where x is the price in dollars/unit and $D(x)$ is the number of units.

(a) Find the demand at a unit price of \$20.
(b) Find the demand at a unit price of \$50.
(c) Find the unit price when the demand is 32 units.

51. The supply function for the product in Exercise 50 is given by

$$S(x) = \tfrac{7}{5}x$$

where $S(x)$ is the number of units that can be supplied at a price of x dollars.

(a) Find the supply at a unit price of \$20.
(b) Find the supply at a unit price of \$50.
(c) Find the unit price when the supply is 84.

52. Assume the domain for the demand and supply functions in Exercises 50 and 51 is $10 \leqslant x \leqslant 100$.

(a) Graph these functions on the same coordinate system.
(b) From the graph estimate the equilibrium price.
(c) Verify your estimate in part (b) algebraically and determine the equilibrium supply and demand.

53. Supply and Demand The supply and demand functions for a certain calculator are given by

$$S(x) = \frac{2}{3}x \qquad \text{and} \qquad D(x) = \frac{300-3x}{3}$$

where x is the unit price in dollars and $S(x)$ and $D(x)$ are in thousands of units. Assuming the domain for these functions is $20 \leqslant x \leqslant 80$

(a) Graph the supply and demand functions on the same coordinate system.
(b) Determine the equilibrium price.
(c) Determine the equilibrium supply and demand.

54. Linear Depreciation A manufacturing company has just purchased a new piece of machinery for $50,000. The accounting department is depreciating this machine at the rate of $2,000/year.

(a) Determine the depreciated value, $D(x)$, of the machine x years from now.
(b) Graph the function determined in part (a) above.
(c) When will the depreciated value be $12,000?

55. Determine the inverse of the function $D(x)$ described in Exercise 50 and

(a) identify the variables x and $D^{-1}(x)$ in the inverse function,
(b) graph the inverse function,
(c) determine the domain and range of the inverse function assuming the domain of $D(x)$ is $10 \leqslant x \leqslant 100$.

2.3 QUADRATIC FUNCTIONS

Although there are many practical applications of linear functions, there are also numerous real-world problems that are described by functions whose graphs are curves rather than straight lines. Perhaps the most elementary example of such a function is the *quadratic function*, which has the general form

$$f(x) = ax^2 + bx + c \tag{2.7}$$

where a, b, and c are real-number constants such that $a \neq 0$. Following are several examples of quadratic functions.

$$f(x) = 2x^2 - 3x + 7$$
$$g(x) = -5x^2 - 9$$
$$y = .1x^2 - 4x + .01$$
$$s(t) = t^2 - 3t + 5$$

Note that each of the above equations is written in *functional form*.

The graphs of quadratic functions are smooth curves called *parabolas*. Figure 2.11a is an illustration of a parabola that opens upward and Fig. 2.11b is an illustration of a parabola that opens downward.

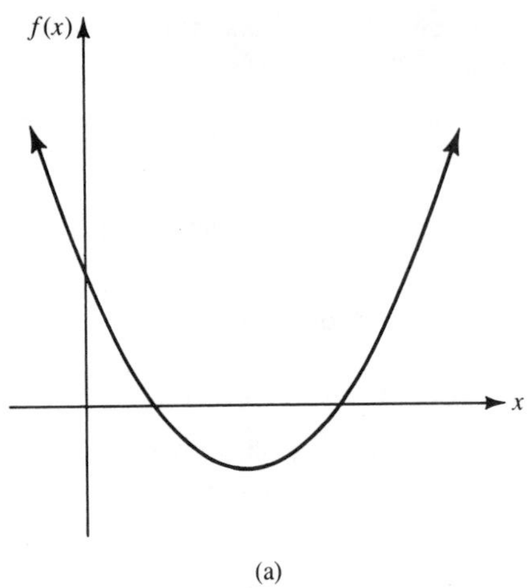

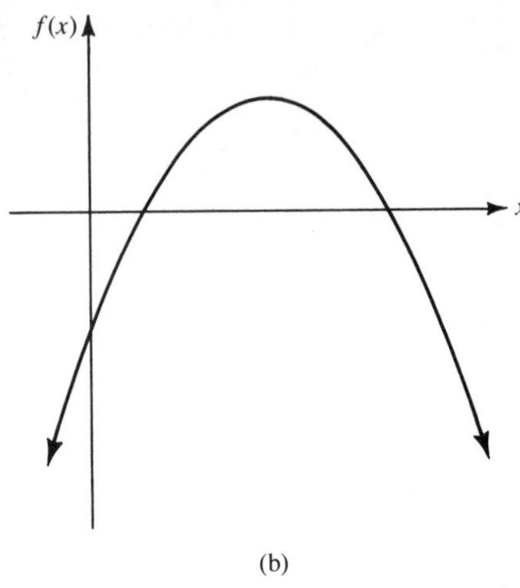

FIGURE 2.11

It is evident from Fig. 2.11 that to draw the graph of a quadratic function we need more than just two points of the graph, as is the case when graphing a linear function.

Let us begin the study of quadratic functions by considering the simplest quadratic function possible; that is, $f(x)=x^2$. [*Note*: We have chosen $a=1$, and $b=c=0$ in Eq. (2.7).] Several ordered pairs that belong to the function $f(x)=x^2$ can be determined by selecting x values of -3, -2, -1, 0, 1, 2, and 3 and computing the corresponding $f(x)$ values as shown in the following table.

x	-3	-2	-1	0	1	2	3
$f(x)=x^2$	9	4	1	0	1	4	9

The graph of the function $f(x)=x^2$ in Fig. 2.12 is found by plotting the points corresponding to the ordered pairs in the table and drawing a smooth curve through them. The parabola in Fig. 2.12 opens upward and the lowest point on the graph is $(0,0)$. Hence $f(0)=0$ and $f(0)<f(x)$ for all $x\neq 0$. To denote this we say the minimum value of $f(x)$ is 0 and occurs when $x=0$.

From a study of this graph, we see that if we fold the page on which the graph is drawn along the vertical coordinate axis, the portion of the graph on the right of the vertical axis will coincide with the portion on the left of this axis. The vertical axis, that is the line whose equation is $x=0$, is called the *axis of symmetry* of this parabola and the parabola is said to be symmetric with respect to this line. The point where the axis of symmetry intersects the

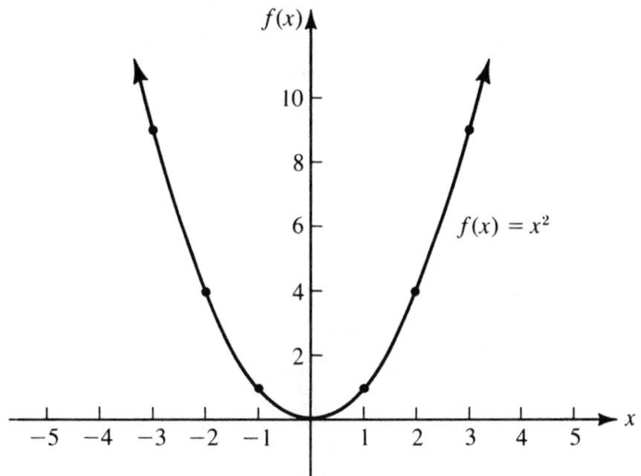

FIGURE 2.12

parabola is called the *vertex of the parabola*. The vertex of the parabola in Fig. 2.12 is (0,0), the lowest point on the curve. Every parabola has an axis of symmetry and the vertex will be either the lowest or the highest point on the curve (see Fig. 2.11). Hence, the value of the ordinate at the vertex will be the minimum value of $f(x)$ or the maximum value of $f(x)$.

Figures 2.13*a* and *b* show the graphs of several quadratic functions of the form

$$f(x) = ax^2$$

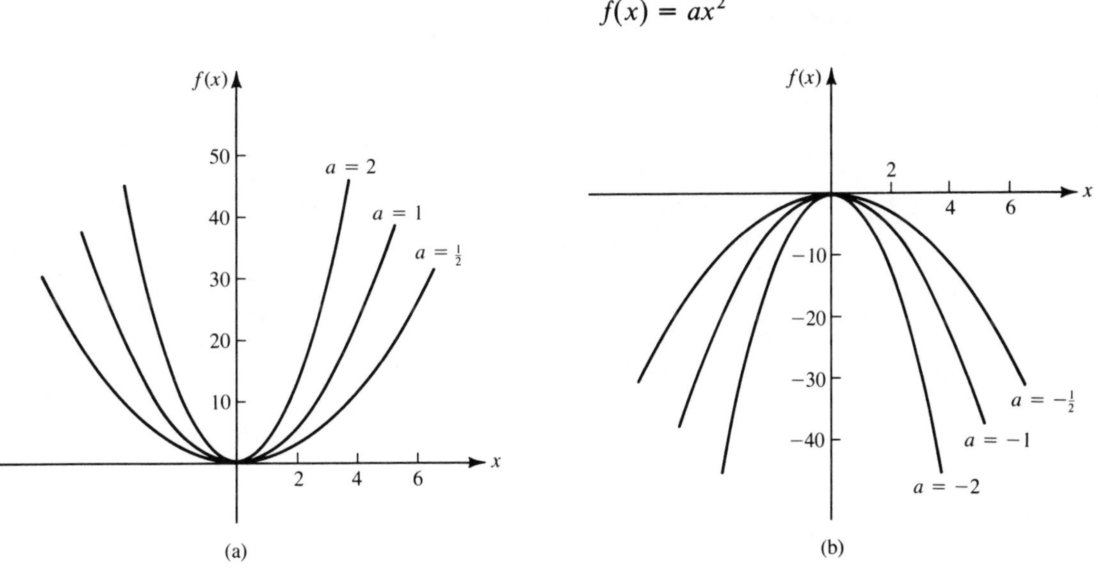

FIGURE 2.13

for various values of a. The graphs are obtained by selecting x values of -4, -2, 0, 2, and 4 and computing the $f(x)$ values as is illustrated for four of these functions in the following tables.

x	-4	-2	0	2	4
$f(x)=2x^2$	32	8	0	8	32

x	-4	-2	0	2	4
$f(x)=\frac{1}{2}x^2$	8	2	0	2	8

x	-4	-2	0	2	4
$f(x)=-2x^2$	-32	-8	0	-8	-32

x	-4	-2	0	2	4
$f(x)=-\frac{1}{2}x^2$	-8	-2	0	-2	-8

Each of the parabolas in Fig. 2.13 is symmetric with respect to the vertical axis and the vertex of each parabola is the point (0,0). The vertex of each parabola in Fig. 2.13*a* corresponds to the lowest point on the graph whereas the vertex of each parabola in Fig. 2.13*b* corresponds to the highest point on the curve. Hence, the minimum value of $f(x)$ is 0 for each function illustrated in Fig. 2.13*a* and the maximum value of $f(x)$ is 0 for each function illustrated in Fig. 2.13*b*.

By inspection of Fig. 2.13 we can see that the greater the absolute value of the coefficient a, the faster the curve rises or falls. We can also conclude from these graphs that if $f(x)=ax^2$, the parabola will open upward whenever $a>0$ and it will open downward whenever $a<0$. Later in our discussion we will show that this statement is also true for the general quadratic function $f(x)=ax^2+bx+c$ where b and c are not necessarily zero.

Another simple form of a quadratic function is $f(x)=ax^2+c$ where $b=0$ and $c\neq 0$. An example of such a function is $f(x)=x^2-4$. Several ordered pairs that belong to this function are shown in the following table.

x	-4	-2	0	2	4
$f(x)=x^2-4$	12	0	-4	0	12

The graph of this function is shown in Fig. 2.14.

The axis of symmetry of the parabola shown in Fig. 2.14 is the vertical axis and the vertex is the point $(0, -4)$. From this graph we note $x=2$ and

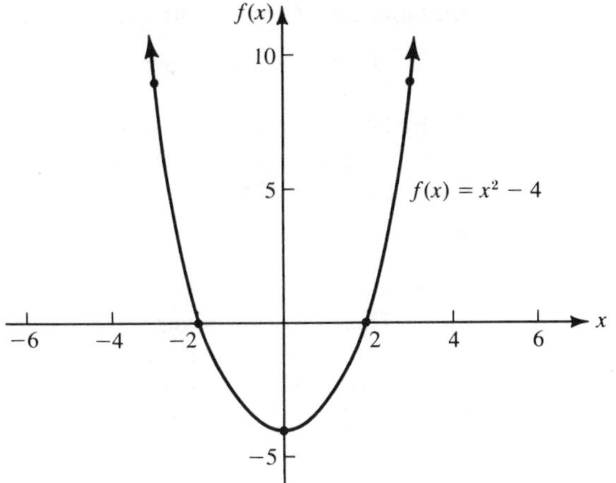

FIGURE 2.14

$x = -2$ are zeros of the function. We can also determine the zeros of this function algebraically by letting $f(x) = 0$ and solving the quadratic equation

$$x^2 - 4 = 0$$

Factoring the left-hand side of this equation, we have

$$(x+2)(x-2) = 0.$$

Setting each factor equal to zero, we obtain

$$x = -2 \qquad \text{or} \qquad x = 2$$

Hence, the zeros of the function are $x = 2$ and $x = -2$ in agreement with the graphical solution.

The vertex of the parabola in Fig. 2.14 can also be determined algebraically. Since x^2 is nonnegative for all real values of x, its smallest value is 0, which occurs when $x = 0$. Hence, the minimum value of $x^2 - 4$ or $f(x)$ is -4, which occurs when $x = 0$. Therefore, the point $(0, -4)$ is the lowest point on the graph of $f(x) = x^2 - 4$ and must be the vertex.

Similarly, the vertex of the graph of $f(x) = 8 - x^2$ must be at the highest point on the curve since we are always subtracting a nonnegative value from 8. Therefore, the maximum value of $8 - x^2$ or $f(x)$ is 8, which occurs when $x = 0$. Hence, the vertex of the parabola is $(0, 8)$ and the parabola opens downward.

Example 2.7 Consider the quadratic function $f(x)=2x^2-32$.

(a) Determine the vertex of the parabola which is the graph of the function.
(b) Determine the zeros of the function.
(c) Sketch the parabola by finding two additional points on the graph on each side of the axis of symmetry.

Solution

(a) Since $x^2 \geqslant 0$ for all real values of x, $2x^2 \geqslant 0$ for each value of x. Therefore, the minimum value of $2x^2-32$ or $f(x)$ is -32, which occurs when $x=0$. Hence, the vertex of the parabola is $(0, -32)$ the lowest point on the curve.
(b) We find the zeros of $f(x)=2x^2-32$ by setting $f(x)=0$ and solving the quadratic equation

$$2x^2-32=0$$

Factoring we obtain

$$2(x^2-16)=0$$
$$2(x-4)(x+4)=0$$
$$x=4 \qquad \text{or} \qquad x=-4$$

Therefore, $x=4$ and $x=-4$ are zeros of the function. [*Note*: The ordered pairs $(4,0)$ and $(-4,0)$ are plotted on the graph in Fig. 2.15.]
(c) Since the axis of symmetry is a vertical line passing through the vertex, its equation is $x=0$. We find two points on either side of the axis of symmetry by selecting x values of -3, -1, 1, and 3 and computing $f(x)$ as shown below.

			(Vertex)		
x	-3	-1	0	1	3
$f(x)=2x^2-32$	-14	-30	-32	-30	-14

Plotting these points and drawing a smooth curve through them, we obtain the graph of $f(x)$ in Fig. 2.15.

Now we consider the general quadratic function $f(x)=ax^2+bx+c$, where $a \neq 0$ and b and c are any real numbers. First we verify that if $a>0$, the graph of this function will open upward and if $a<0$, the graph will open downward. Then we develop a formula for finding the coordinates of the vertex of any quadratic function.

As x assumes large positive values such as 100, 1,000, 10,000, etc., eventually the term ax^2 will dominate the lower degree terms on the right-hand side of the equation $f(x)=ax^2+bx+c$. Hence, $f(x)$ will be approximated by ax^2 as x increases without bound. Similarly, as x assumes values

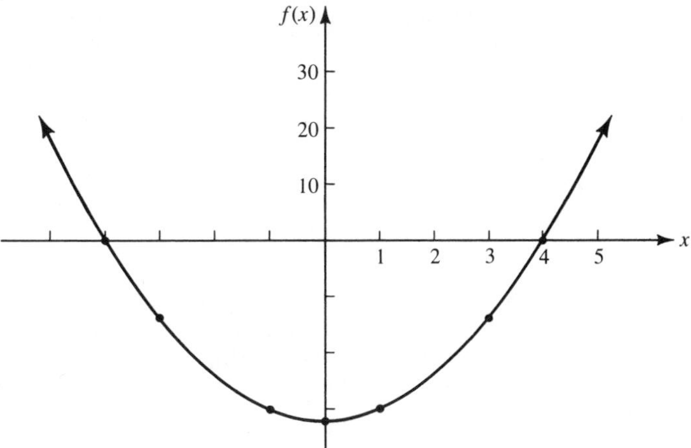

FIGURE 2.15

such as -100, $-1{,}000$, $-10{,}000$, etc., eventually the term ax^2 will dominate bx and c. Hence $f(x)$ will be approximated by ax^2 as x decreases without bound. Therefore, if $a>0$, $f(x)$ approximated by ax^2 will increase without bound as x increases or decreases without bound. Therefore, the graph of $f(x)=ax^2+bx+c$ where a is positive will open upward as shown in Fig. 2.11a. Analogously, if $a<0$, $f(x)$ approximated by ax^2 will decrease without bound as x increases or decreases without bound. Therefore, the graph of $f(x)=ax^2+bx+c$ where a is negative will open downward as shown in Fig. 2.11b.

To develop a formula for the coordinates of the vertex of the graph of any given quadratic function $f(x)=ax^2+bx+c$, where $a\neq 0$, we shall use the method of completing the square. Factoring out the coefficient a, we have

$$f(x) = a\left(x^2+\frac{b}{a}x+\frac{c}{a}\right)$$

To complete the square we add and subtract the square of $\frac{1}{2}$ the coefficient of the x term; that is, $[(1/2)(b/a)]^2$. Hence,

$$f(x) = a\left(x^2+\frac{b}{a}x+\left(\frac{b^2}{4a^2}\right)+\frac{c}{a}-\left(\frac{b^2}{4a^2}\right)\right)$$

Substituting $[x+(b/2a)]^2$ for the trinomial $[x^2+(b/a)x+(b^2/4a^2)]$ and writing $(c/a)-(b^2/4a^2)$ as $(4ac-b^2)/4a^2$, we have

$$f(x) = a\left[\left(x+\frac{b}{2a}\right)^2+\frac{4ac-b^2}{4a^2}\right]$$

Distributing a, we obtain

$$f(x) = a\left(x+\frac{b}{2a}\right)^2+\frac{4ac-b^2}{4a} \tag{2.8}$$

Since the parabola will open upward ($a>0$) or downward ($a<0$), we consider two cases.

Case 1 Assume $a>0$: parabola opens upward $\cup$. Since $[x+(b/2a)]^2 \geqslant 0$ for all real values of x and $a>0$, then $a[x+(b/2a)]^2 \geqslant 0$ for each value of x. Hence, the minimum value of $f(x)$ in Eq. (2.8) is $(4ac-b^2)/4a$, which occurs when $a[x+(b/2a)]^2=0$; that is, when $x=-b/2a$. Therefore, the lowest point on the graph, the vertex, has the coordinates

$$\left(\frac{-b}{2a}, \frac{4ac-b^2}{4a}\right) \tag{2.9}$$

Case 2 Assume $a<0$: parabola opens downward $\cap$. Since $[x+(b/2a)]^2 \geqslant 0$ for all real values of x and $a<0$, then $a[x+(b/2a)]^2 \leqslant 0$ for each x. Hence, the maximum value of $f(x)$ in Eq. (2.8) is $(4ac-b^2)/4a$, which occurs when $a[x+(b/2a)]^2=0$; that is, when $x=-b/2a$. Therefore, the highest point on the graph, the vertex, has the coordinates

$$\left(\frac{-b}{2a}, \frac{4ac-b^2}{4a}\right)$$

Hence, the formulas for the coordinates of the vertex are the same in both cases.

Rather than memorizing both coordinates of the vertex, it is sufficient to know that the first coordinate is $-b/2a$. The second coordinate can be found by evaluating $f(-b/2a)$ by substituting $x=-b/2a$ into the formula for $f(x)$. For example, when $f(x)=2x^2+8x-7$, $a=2$, $b=8$, and $c=-7$, and the x coordinate of the vertex by formula (2.9) is

$$x = \frac{-b}{2a} = \frac{-(8)}{2(2)} = -2$$

the second component of the vertex found by evaluating $f(-2)$ is

$$\begin{aligned} f(-2) &= 2(-2)^2+8(-2)-7 \\ &= -15 \end{aligned}$$

The same value is found by evaluating $(4ac-b^2)/4a$, the second component in formula (2.9). When $a=2$, $b=8$, and $c=-7$,

$$\frac{4ac-b^2}{4a}=\frac{4(2)(-7)-(8)^2}{4(2)}=-15$$

Since the axis of symmetry of a parabola is the vertical line passing through the vertex, the equation of the axis of symmetry of the graph of $f(x)=ax^2+bx+c$ is

$$x=\frac{-b}{2a}$$

To determine the possible zeros of the quadratic function $f(x)=ax^2+bx+c$, we let $f(x)=0$ and solve the resulting *quadratic equation*

$$ax^2+bx+c=0 \qquad (2.10)$$

It is advisable to attempt to solve Eq. (2.10) by factoring the trinomial ax^2+bx+c. If this expression is nonfactorable or difficult to factor, roots of this equation can be determined as follows.

Since $f(x)=ax^2+bx+c$ is equivalent to Eq. (2.8), we can find the zeros of the function by letting $f(x)=0$ in this equation and solving for x as shown below.

$$a\left(x+\frac{b}{2a}\right)^2+\frac{4ac-b^2}{4a}=0$$

$$a\left(x+\frac{b}{2a}\right)^2=-\left(\frac{4ac-b^2}{4a}\right)$$

$$\left(x+\frac{b}{2a}\right)^2=\frac{b^2-4ac}{4a^2}$$

Taking the square root of the left- and right-hand side of this equation, we find

$$\left(x+\frac{b}{2a}\right)=\pm\sqrt{\frac{b^2-4ac}{4a^2}}$$

$$=\pm\frac{\sqrt{(b^2-4ac)}}{2a}$$

Solving for x by adding $-b/2a$ to the left- and right-hand side of this equation and expressing the right-hand side with the common denominator

$2a$, we have

$$x = \frac{-b \pm \sqrt{b^2 - 4ac}}{2a} \tag{2.11}$$

Equation (2.11) is known as the *quadratic formula*, and it yields the two possible roots of the quadratic equation (2.10).

The expression $(b^2 - 4ac)$ in the quadratic formula is called the *discriminant*. Whenever the discriminant is negative there are no real numbers that will satisfy Eq. (2.11) and hence there are no zeros of the associated quadratic function. This implies that the parabola lies either entirely above or below the x-axis as is illustrated in Figs. 2.16a and b, respectively.

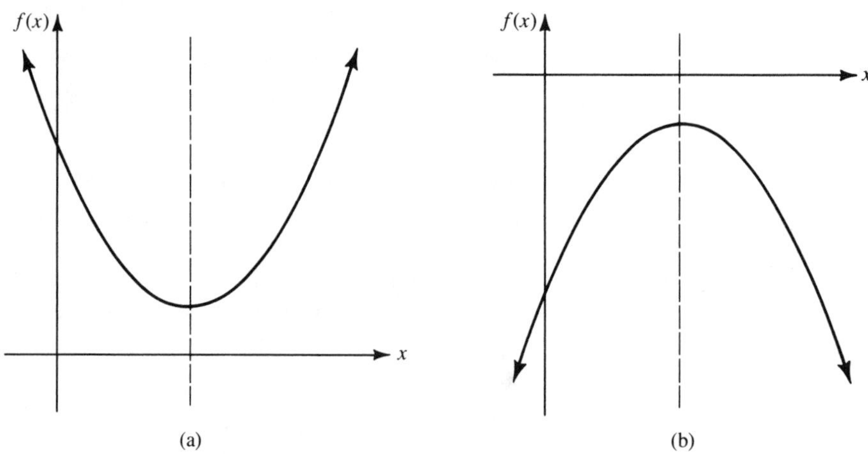

FIGURE 2.16

Example 2.8 Graph the function $f(x) = x^2 - 8x + 7$.

Solution First we determine the coordinates of the vertex. Since $f(x) = x^2 - 8x + 7$ is written in the general form, we have

$$a = 1, \qquad b = -8, \qquad \text{and } c = 7$$

Therefore,

$$\frac{-b}{2a} = \frac{-(-8)}{2(1)} = 4$$

and $f(4) = (4)^2 - 8(4) + 7 = -9$. Hence, the vertex has coordinates $(4, -9)$ and the axis of symmetry is the vertical line whose equation is $x = 4$. Next we note that since $a = 1 > 0$, the parabola opens upward. To find the zeros of the function we let $f(x) = 0$ and solve the quadratic equation

$$x^2 - 8x + 7 = 0$$

Factoring, we have

$$(x-7)(x-1) = 0$$
$$x = 7 \quad \text{or} \quad x = 1$$

Hence, the zeros of the function are $x=7$ and $x=1$. Next we determine a few ordered pairs on either side of the axis of symmetry that belong to the function.

x	-1	0	2	4 (Vertex)	6	8	9
$f(x)=x^2-8x+7$	16	7	-5	-9	-5	7	16

We plot the vertex, the points corresponding to the ordered pairs in the above table, and the zeros of the function. If we draw a smooth curve through these points, we obtain the parabola illustrated in Fig. 2.17. Note that the dotted vertical line in this figure is the axis of symmetry and its equation is $x=4$.

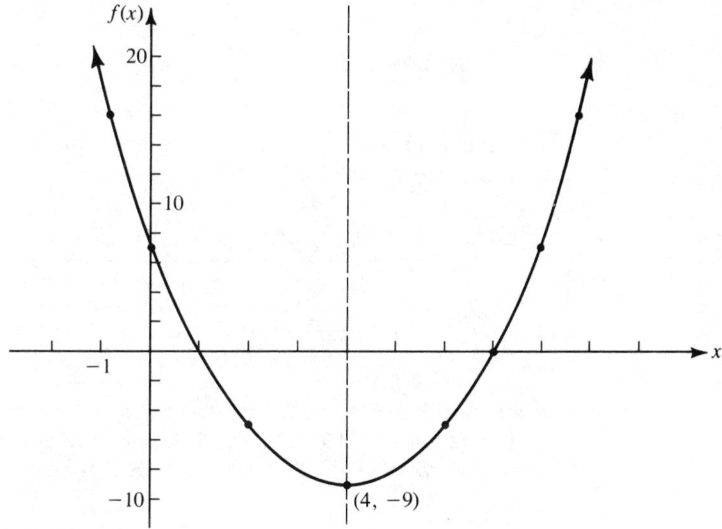

FIGURE 2.17

Example 2.9 Graph the function $f(x)=-2x^2-12x-11$.

Solution We have $a=-2$, $b=-12$, and $c=-11$. Therefore, $-b/2a=-(-12)/2(-2)=-3$ and $f(-3)=-2(-3)^2-12(-3)-11=7$. Hence, the vertex is at $(-3,7)$. Since $a=-2<0$, the parabola opens downward. The zeros of the function are found by solving

$$-2x^2-12x-11=0$$

or

$$2x^2+12x+11=0$$

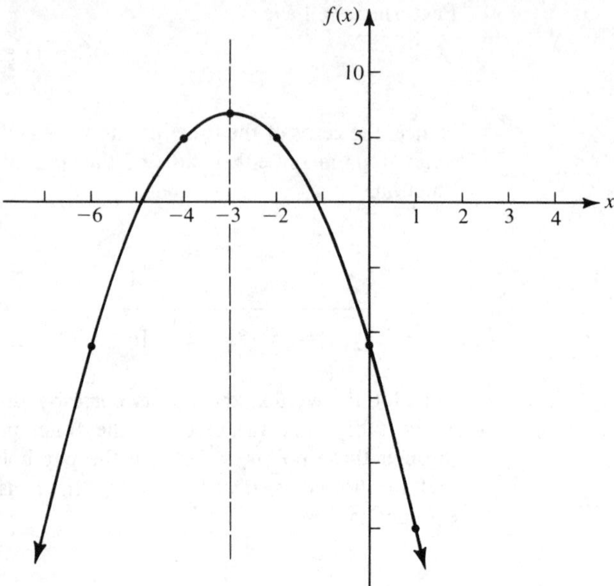

FIGURE 2.18

The trinomial $2x^2+12x+11$ is not factorable so we use the quadratic formula to find solutions.

$$x = \frac{-b \pm \sqrt{b^2-4ac}}{2a} = \frac{-(12) \pm \sqrt{(12)^2-4(2)(11)}}{2(2)}$$

$$= \frac{-12 \pm \sqrt{56}}{4}$$

Approximating $\sqrt{56}$ by 7.5, the two zeros of the function are approximately $(-12 \pm 7.5)/4$ or -4.9 and -1.1. Selecting x values on either side of the line of symmetry, we find a few ordered pairs that belong to the function.

			(Vertex)		
x	-6	-2	-3	0	1
$f(x) = -2x^2 - 12x - 11$	-11	5	7	-11	-25

The graph of the parabola is shown in Fig. 2.18.

We conclude that to graph any quadratic function $f(x) = ax^2 + bx + c$ we should

1. determine the coordinates of the vertex: $(-b/2a, f(-b/2a))$;

2. draw the axis of symmetry as a dotted line: the line whose equation is $x=(-b/2a)$;
3. note if the parabola opens upward: $a>0$, or downward: $a<0$;
4. determine the zeros of the function: solutions of $ax^2+bx+c=0$;
5. find two or three additional points on the graph [The y-intercept $(0, f(0))$ is often a good point to plot.];
6. use the symmetry property of parabolas when drawing the curve.

The range and domain of a quadratic function can easily be determined. For any real value of x, the expression ax^2+bx+c is a real number. Therefore, unless the x values are restricted by other considerations, the domain of every quadratic function is the set of real numbers. In interval notation we denote the domain by

$$D=(-\infty, \infty)$$

or

$$D=\mathbb{R}$$

The range of a quadratic function is determined by finding the $f(x)$ value of the vertex and noting whether the parabola opens upward or downward. For example, since the graph of the quadratic function $f(x)=x^2-8x+7$, as shown in Fig. 2.17, opens upward and the coordinates of the vertex are $(4, -9)$, the range, R, is given by

$$R=[-9, \infty)$$

Example 2.10 Determine the range of the quadratic function $f(x)=-2x^2+3x-5$.

Solution We find the $f(x)$ value at the vertex by evaluating $f(-b/2a)$. Since $a=-2$ and $b=3$, $-b/2a=-3/2(-2)=\frac{3}{4}$ and $f(\frac{3}{4})=-2(\frac{3}{4})^2+3(\frac{3}{4})-5=-\frac{31}{8}$. Since a is negative, the parabola opens downward and the range in interval notation is

$$R=\left(-\infty, \frac{-31}{8}\right]$$

[*A word of caution*: Carefully note the order in which the above interval is written; it is *incorrect* to write $[-\frac{31}{8}, -\infty)$.]

As noted at the beginning of this section, quadratic functions serve as mathematical models for numerous real-world problems. To illustrate the kinds of problems that lead to quadratic functions, let us consider the following examples.

Example 2.11 Calculator Sales and Maximizing Revenue Frank Sama is a vice president of Tech-Cal Manufacturing Company in charge of sales of the T-40 calculators discussed at the end of Section 2.2. Determine the unit price of each calculator if Vice President Sama wants to maximize the revenue from the sale of these calculators. Also determine the maximum revenue.

Solution You will recall from our previous discussion that the demand, $D(x)$, for the T-40 calculators is given by

$$D(x) = 100 - 4x, \qquad 10 \leqslant x \leqslant 20$$

where x is the unit price in dollars and $D(x)$ is the number in thousands that can be sold. If we can sell $D(x)$ calculators at a unit price of x dollars, the revenue, $R(x)$, from these sales is given by

$$R(x) = xD(x)$$

Substituting $(100-4x)$ for $D(x)$, we have

$$\begin{aligned} R(x) &= x(100-4x) \\ &= 100x - 4x^2 \end{aligned}$$

where $10 \leqslant x \leqslant 20$ and $R(x)$ is in thousands of dollars. The revenue, $R(x)$, is a quadratic function of the unit price x. Since the graph of this function is a parabola opening downward, the maximum value of the revenue occurs at the vertex where $x = -b/2a$. From the formula for $R(x)$, we note that $a = -4$ and $b = 100$; therefore,

$$x = \frac{-b}{2a} = \frac{-100}{2(-4)} = \$12.50$$

is the unit price to maximize the revenue. The maximum revenue is

$$\begin{aligned} R(12.50) &= 100(12.50) - 4(12.50)^2 \\ &= 1250 - 625 \\ &= 625 \text{ thousand dollars or } \$625{,}000 \end{aligned}$$

Note in this solution that the maximizing unit price of $x = \$12.50$ is in the domain of the function $R(x)$. If this were not the case, the x value in the domain nearest to $x = -b/2a$ would be chosen.

In conclusion, Vice President Sama should sell the T-40 calculators at \$12.50 each to bring in a maximum revenue of \$625,000.

Example 2.12 Rain Gutter Design A long sheet of aluminum 40 cm wide is made into a house gutter by turning up two equal sides of length x at right angles to the bottom as shown in the figure. Express the cross-sectional area A as a function of x. Draw a graph of the function $A(x)$ and discuss the implications of the graph.

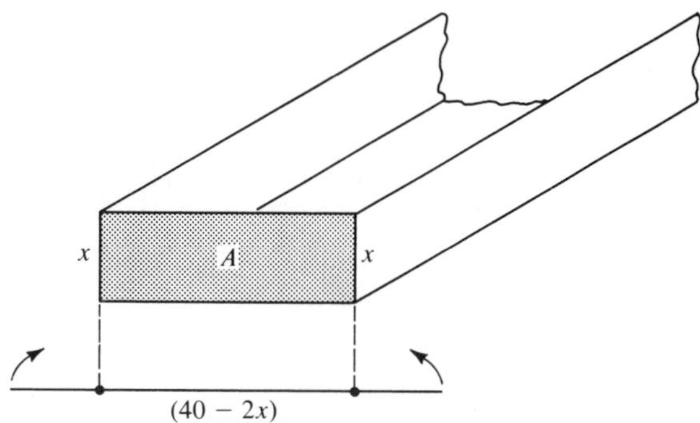

Solution The cross-sectional area, A, equals the length of its base times its height. In terms of the variable x the length is $(40-2x)$ and the height is x; therefore, the area, $A(x)$, is

$$\begin{aligned} A(x) &= x(40-2x) \\ &= 40x - 2x^2 \end{aligned}$$

The domain of this quadratic function is restricted by the physical problem. Since the length $(40-2x)$ and height x are both positive, the inequalities $40-2x>0$ and $x>0$ must be true. Therefore, the domain is

$$D = (0,20)$$

To graph this function we determine the coordinates of the vertex and two points on either side of the axis of symmetry as shown in the following table.

			(Vertex)		
x	4	8	10	12	16
$A(x)$	128	192	200	192	128

[*Note*: The two x values equally distant from the axis of symmetry have exactly the same $f(x)$ value; $f(4)=f(16)$ and $f(8)=f(12)$.]

Plotting the points corresponding to these ordered pairs, we obtain the graph of $A(x)$ as shown in Fig. 2.19. Note that the end point $x=0$ and $x=20$ are not included in the graph as they are not members of the domain. By an inspection of the graph of $A(x)$ we note that the maximum cross-sectional area of 200 sq cm occurs when the height $x=10$ cm. The corresponding cross-sectional dimensions of the gutter are 10 cm×20 cm×10 cm. If the gutter is designed to carry a maximum amount of water, these dimensions would be selected. Fig. 2.19 shows how rapidly the area changes from this maximum if the height of the gutter is either increased or decreased from its optimal value of 10 cm.

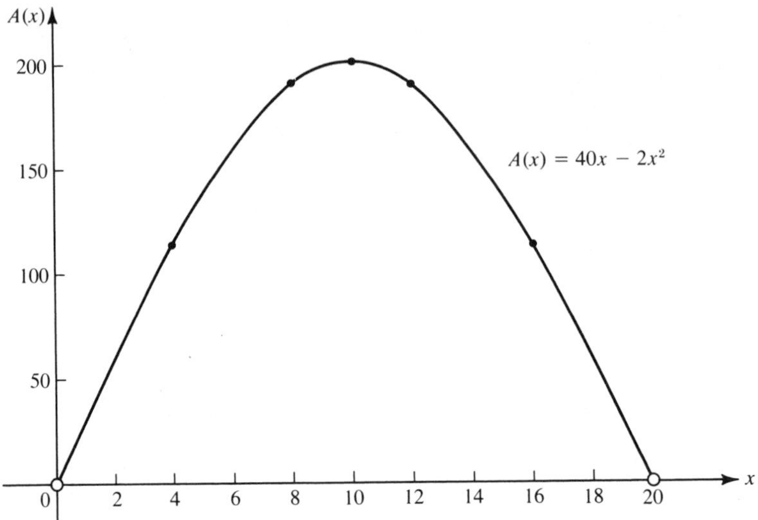

FIGURE 2.19

EXERCISES

Express the quadratic equations in Exercises 1–4 *in functional form.*

1. $y-9x^2+6x-3=0$

2. $2y+5x-3x^2=8+11x^2-9x$

3. $3y+9x^2-4x=5y+2x-7x^2+14$

4. $8y-5x+4x^2+6=6x^2+12+6y+3x$

For each of the quadratic functions in Exercises 5–11, *determine without graphing*

(a) whether the parabola opens downward or upward,
(b) the coordinates of the vertex,
(c) the zeros of the function,
(d) the range of the function,
(e) the equation of the axis of symmetry.

5. $y=x^2+6x-16$

6. $f(x)=2x^2+7x-15$

7. $h(t)=t^2+2t-15$

8. $y=3x^2-15x+12$

9. $C(u)=12u^2-2u-4$

10. $g(x)=-2x^2+11x-12$

11. $f(x)=-2x^2-32$

For each of the quadratic functions in Exercises 12–17 *determine*

(a) which way the parabola opens,
(b) whether the parabola intersects the x axis,
(c) the coordinates of the vertex,

(d) the zeros of the function,
(e) the coordinates of five points belonging to the graph of the function.
(f) Plot the points in (e) and draw a smooth curve.

12. $y=x^2-9x+18$

13. $f(x)=2x^2-12x+18$

14. $g(x)=-x^2+4x-5$

15. $C(x)=2x^2-12x+26$

16. $h(x)=-3x^2-12x-12$

17. $p(x)=-3x^2+36x-105$

18. The demand and supply functions for a certain product are given by

$$D(x) = 400 - x^2 \qquad \text{and} \qquad S(x) = \frac{35}{3}x$$

where x is the unit price in dollars and $D(x)$ and $S(x)$ are the number of units.

(a) Sketch the graph of the demand and supply functions on the same coordinate system. Assume the domain for each function is $[0, 20]$.
(b) Estimate the equilibrium price from the graph and verify your estimate algebraically.
(c) Determine the number of units in demand at the equilibrium price.

19. Maximizing Profits Determine when the Florida orange grower in Exercise 32 in Sec. 1.2 should ship his oranges to maximize his profit. What is the maximum profit?

20. The demand, $D(x)$, for T-51 calculators is given by

$$D(x) = 300 - 2x, \qquad x \geqslant 10$$

where x is the unit price in dollars and $D(x)$ is the number in thousands that can be sold at x dollars per unit.

(a) Determine the unit price that will produce the maximum revenue from the sale of T-51 calculators.
(b) Determine the maximum revenue.

21. Government Budgets Niskanen* argues that government agencies and other nonprofit organizations, in contrast to firms in the private sector, will attempt to expand past the level of profit maximization to the point where their total budgets are as large as possible without running a deficit. For present purposes, this can be assumed to be the largest level at which total budget, B, just covers (i.e., is equal to) total cost, C. Let Q be the level of output, and assume that

$$C = \tfrac{1}{2}Q^2 + 600Q + 50{,}000$$

and

$$B = 1{,}200Q - \tfrac{1}{2}Q^2$$

What level will this agency expand to and what will be its budget at this level of output?

*William A. Niskanen, *Bureaucracy and Representative Government* (Chicago: Aldine Atherton, 1971).

22. A factory has expenses of (x^2+15) hundred dollars where x is the number of units produced. The product can be sold for 16 hundred dollars/unit.

(a) Determine and graph the equation for computing the factory's profit based on the number of units produced and sold.
(b) What is the maximum number of units that the factory should produce?
(c) For what number of units produced will revenue equal expenses?

23. Cakes and Profit The Wee Bakem Bakery bakes birthday cakes that cost \$3 each to make. They know that the number of cakes they sell depends on the price they charge their customers. From past experience, they know that they will be able to sell $600-50x$ cakes per week when they charge x dollars per cake.

(a) Determine and graph the equation for computing the total income from the sale of birthday cakes.
(b) From the equation in part (a), determine the maximum total income and the price per cake that they should charge in order to maximize the total income.
(c) How many cakes will they sell at the price that maximizes the total income?
(d) Determine and graph the equation for computing the total profit from the sale of birthday cakes.
(e) From the equation in part (d), determine the maximum total profit and the price per cake that they should charge in order to maximize the total profit.
(f) How many cakes will they sell at the price that maximizes the total profit?
(g) How much profit will they lose if they charge the price per cake that maximizes total income instead of the one that maximizes total profit?

24. A shipping company had a shipping charge of $3x+20$ dollars where x is the number of pounds shipped. To prevent making shipments too large they have decided to increase the cost by $.06x^2$. Determine and graph the equation for computing the shipping charge from the number of pounds shipped.

25. Neural Impulses A physiological psychologist has found that the number of neural impulses fired after a nerve has been stimulated can be described with a quadratic equation based upon the size of the nerve and intensity of the stimulus. If the equation for a particular nerve is $y=-10x^2+40x$ (where $y=$ the number of responses per millisecond, and $x=$ the time in milliseconds since the stimulus was administered) find the following.

(a) How many milliseconds after the stimulus will the maximum firing rate occur?
(b) What will the maximum firing rate be?
(c) How many milliseconds after administration of the stimulus will it be before the firing rate returns to zero?

26. The Lifeguard's Dilemma The chief lifeguard at Ocean Beach has been given 1,200 ft of rope and instructed to lay out a rectangular restricted swimming area as shown in the following sketch.

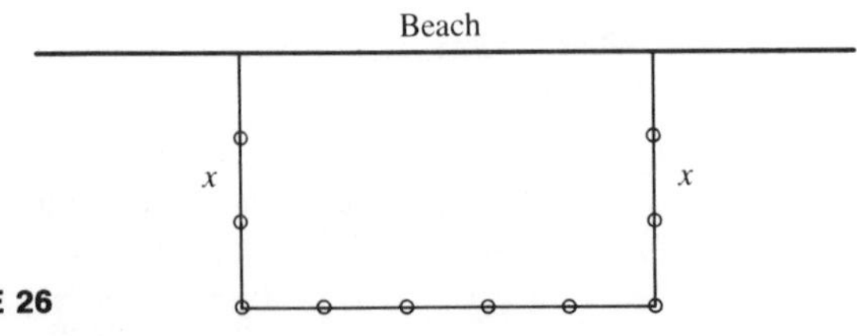

EXERCISE 26

(a) If the length of each side that is perpendicular to the beach is x ft, what is the length of the side parallel to the beach?
(b) Express the area A of the restricted swimming region as a function of the variable x.
(c) Sketch the graph of the function determined in part (b).
(d) Based on the results of parts (b) and (c), determine the dimensions of the rectangular region that will maximize the area for swimming.
(e) What other factors would enter into the selection of the dimensions of the rectangular region?

2.4 POLYNOMIAL FUNCTIONS

The linear and quadratic functions discussed in Sections 2.2 and 2.3 are special cases of *polynomial functions*. Following is a definition of a polynomial function.

Definition 2.1 A polynomial function is a function of the form

$$f(x) = a_n x^n + a_{n-1}x^{n-1} + a_{n-2}x^{n-2} + \cdots + a_1x^1 + a_0 \qquad (2.12)$$

where n is a nonnegative integer, $a_n \neq 0$, and $a_n, a_{n-1}, \ldots, a_1, a_0$ are real constants.

Equation (2.12) is a polynomial function of degree n. Its right-hand side is a polynomial expression in the variable x. You may recall similar expressions in previous courses. The following are examples of polynomial functions.

$f(x) = x^2 - 5x + 6,$ where $n=2, a_2=1, a_1=-5$ and $a_0=6$

$f(x) = 4x^3 - 3x^2 + 4x + 5,$ where $n=3, a_3=4, a_2=-3,$ $a_1=4$, and $a_0=5$

$f(x) = 2x^4 + \frac{5}{2}x + 0.7,$ where $n=4, a_4=2, a_3=a_2=0,$ $a_1=\frac{5}{2}$, and $a_0=0.7$

A linear function is a polynomial function of degree 1, because when $n=1$, $f(x)=a_1x^1+a_0$ [this is equivalent to $f(x)=mx+b$, where $a_1=m$ and $a_0=b$]. Similarly, a quadratic function is a polynomial function of degree 2, because when $n=2$, $f(x)=a_2x^2+a_1x^1+a_0$ [this is equivalent to $f(x)=ax^2+bx+c$, where $a_2=a$, $a_1=b$, and $a_0=c$].

A polynomial function of degree 3 is also called a cubic polynomial function. As in the study of the behavior of linear and quadratic functions, we are interested in the graphs of cubic polynomial functions and higher-degree polynomial functions. The graphs of these functions are nice smooth curves as illustrated in Fig. 2.20; the points A, B, and C on this graph are called *turning points*. A turning point on the graph of a function is a point

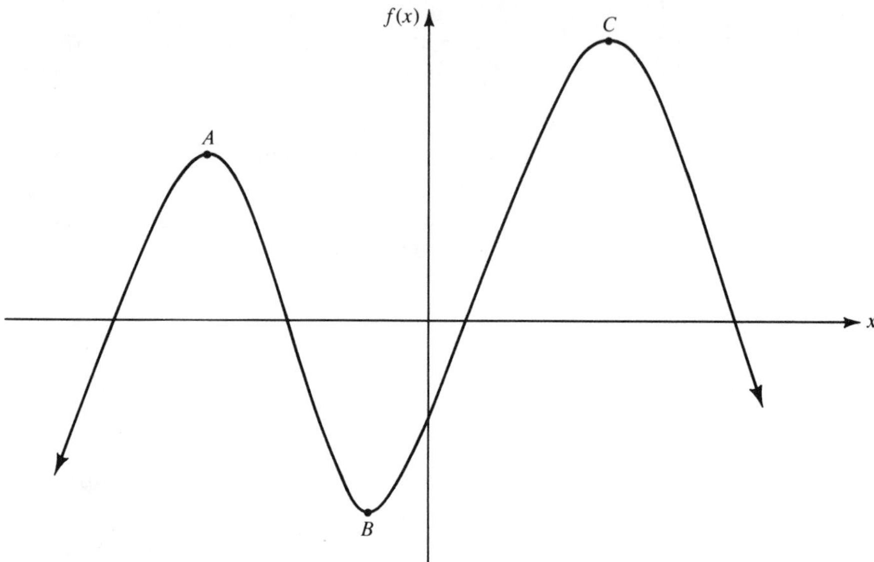

FIGURE 2.20

where the curve stops rising and starts descending or stops descending and starts rising. We shall see shortly that the number of turning points is related to the degree of the polynomial function.

Determining the coordinates of the turning points such as A, B, and C shown in Fig. 2.20 would aid in graphing polynomial functions. Unfortunately, there is not a simple algebraic technique, as in the case of quadratic functions, to determine these coordinates for a third or higher-degree polynomial function. (We will find that once we have developed the tools of calculus in Chapter 4 we will have the power to do this.) We can, with our present capabilities, sketch the graph of any polynomial function by determining a "sufficient number" of ordered pairs that belong to the function. However, how many points are sufficient, and which ones should be determined? To illustrate a procedure that will help answer this question, let us consider graphing the function

$$f(x) = x^3 - 2x^2 - 5x + 6 \tag{2.13}$$

We begin by determining the behavior of $f(x)$ as x increases without bound. To denote this process we write $x \to +\infty$, which is read "x approaches positive infinity." As x assumes large positive values, the x^3 term will dominate the lower-degree terms $-2x^2$, $-5x$, and 6. [Actually, when $x = 10$, $f(10) = 1{,}000 - 200 - 50 + 6$ and the last three terms are small compared to 10^3.] Therefore, as $x \to +\infty$, $f(x)$, approximated by x^3, increases without bound. To denote this we write $f(x) \to +\infty$ as $x \to +\infty$ or

$$\lim_{x \to +\infty} f(x) = +\infty$$

which is read "the *limit* of $f(x)$ as x approaches positive infinity is positive infinity."

Analogously, as x decreases without bound or as $x \to -\infty$, $f(x)$, approximated by x^3, also decreases without bound. We denote this by writing

$$\lim_{x \to -\infty} f(x) = -\infty$$

The behavior of $f(x)$ as $x \to +\infty$ and as $x \to -\infty$ is called the *extent of the function*. The extent of $f(x)$ is illustrated by sketching the graph of $f(x)$ as $x \to +\infty$ and as $x \to -\infty$, as shown in Fig. 2.21.

To complete the graph of $f(x) = x^3 - 2x^2 - 5x + 6$ we must determine the rest of the curve between $x = -10$ and $x = 10$. Based on the partial graph of $f(x)$ shown in Fig. 2.21, there exists at least one value of x in the interval $[-10, 10]$ where the curve crosses the x axis. This implies that there is at least one zero of this function in the interval, and hence at least one solution of the equation

$$x^3 - 2x^2 - 5x + 6 = 0 \tag{2.14}$$

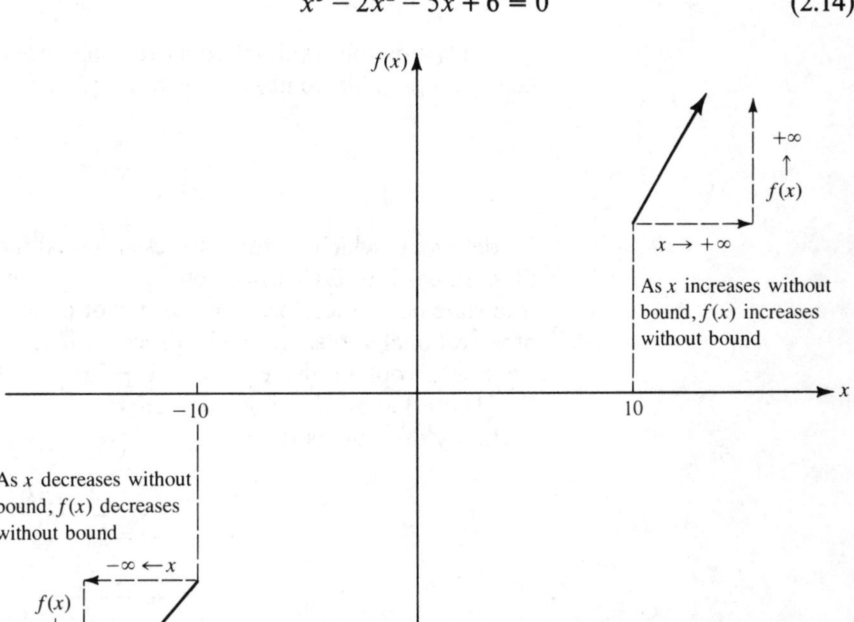

FIGURE 2.21

Unlike a quadratic equation, there is no easy computational formula to solve a third or higher-degree polynomial equation such as Eq. (2.14). There is, however, a rule that can be used to determine all the *rational roots* of such polynomial equations. (An approximation method to determine both rational and irrational roots will be discussed in Chapter 5.) The rule used to determine the rational roots (that is, roots that can be written as the ratio p/q where p and q are integers and $q \neq 0$) is called the *rational roots theorem*, which we state without proof.

Rule 2.1 The rational roots theorem: If the polynomial equation $a_n x^n + a_{n-1} x^{n-1} + \cdots + a_1 x^1 + a_0 = 0$ (where each a_i is an integer) has any rational roots they will be of the form p/q, where p is a divisor of a_0 and q is a divisor of a_n.

(*Note*: a is a "divisor" or "divides" b if and only if a divides b with a remainder of 0.)

Applying the rational roots theorem to Eq. (2.14), where $a_0 = 6$ and $a_n = a_3 = 1$, we note that

$$p \text{ divides } 6 \text{ or } p = \pm 6, \pm 3, \pm 2, \text{ or } \pm 1$$

and

$$q \text{ divides } 1 \text{ or } q = \pm 1$$

The only possible rational roots (or solutions) of Eq. (2.14) are formed by taking all possible combinations of p/q. The possible roots are

$$\left\{ \pm\frac{6}{1}, \pm\frac{3}{1}, \pm\frac{2}{1}, \pm\frac{1}{1} \right\}$$

To determine which, if any, of these rational numbers are roots, we substitute these values into Eq. (2.14). Letting $x = 1$, we find $(1)^3 - 2(1)^2 - 5(1) + 6 = 0$ is a true statement; therefore, $x = 1$ is a root of the equation. To find if there are other rational roots, we could proceed to test the other values. However, if $x = 1$ is a root of the equation $x^3 - 2x^2 - 5x + 6 = 0$, it can be shown that $(x - 1)$ is a factor of the polynomial $x^3 - 2x^2 - 5x + 6$. The other factor can be found by division as follows:

$$\begin{array}{r|l}
 & x^2 - x - 6 \\
\hline
x - 1 & x^3 - 2x^2 - 5x + 6 \\
 & \underline{x^3 - x^2} \\
 & \quad -x^2 - 5x \\
 & \quad \underline{-x^2 + x} \\
 & \qquad -6x + 6 \\
 & \qquad \underline{-6x + 6}
\end{array}$$

Therefore, the equation $x^3-2x^2-5x+6=0$ is equivalent to the equation

$$(x-1)(x^2-x-6) = 0 \tag{2.15}$$

Factoring (x^2-x-6) we can write Eq. (2.15) as

$$(x-1)(x-3)(x+2) = 0$$

The roots of this equation and the zeros of the polynomial function $f(x)= x^3-2x^2-5x+6$ are

$$x = 1, \qquad x = 3, \qquad x = -2$$

(*Note*: The two additional roots, 3 and -2, were listed as possible rational roots.)

If the quadratic (x^2-x-6) were not factorable, irrational roots could have been found by using the quadratic formula.

Since $x=1$, 3, and -2 are zeros of the polynomial function, the ordered pairs $(1,0)$, $(3,0)$, and $(-2,0)$ belong to the function. If these ordered pairs are plotted together with the partial graph of $f(x)$ shown in Fig. 2.21, we obtain the set of points shown in Fig. 2.22.

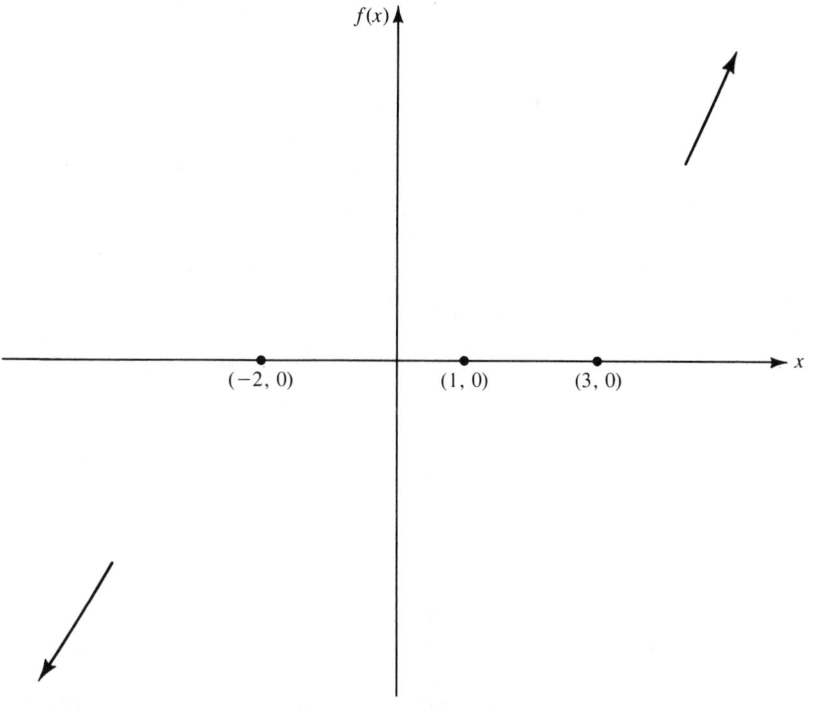

FIGURE 2.22

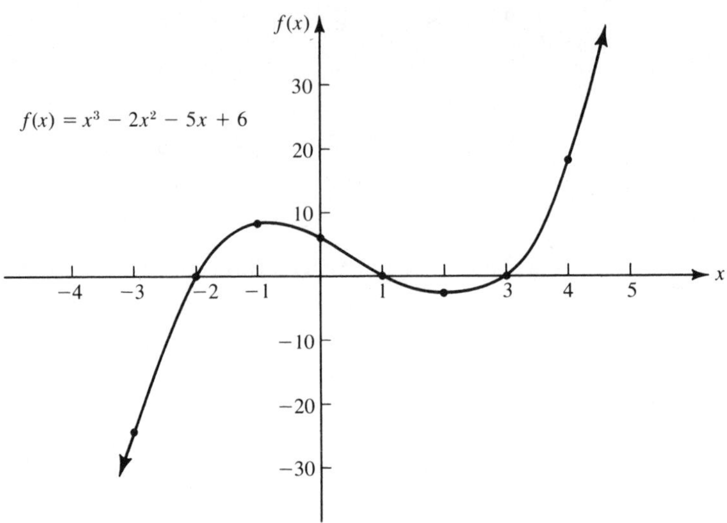

FIGURE 2.23

We choose other x values by picking integral values to the left and right of the points shown in Fig. 2.22; i.e., -3, -1, 0, 2, and 4. The ordered pairs are tabulated below and the completed graph of $f(x)$ is shown in Fig. 2.23.

x	-3	-1	0	2	4
$f(x)$	-24	$+8$	6	-4	18

To calculate the $f(x)$ values, it is far easier to use the factored form of the polynomial $f(x)=(x-1)(x-3)(x+2)$ rather than $f(x)=x^3-2x^2-5x+6$.

In summary, we conclude that as an aid to graphing third or higher degree polynomial functions we should determine

1. The extent of the function $f(x)$.
2. The zeros of the function.
3. The ordered pair $(0, f(0))$; i.e., the y-intercept.
4. A sufficient number of points to establish the curve.

The following properties, which we state without proof, will also aid us to graph any polynomial function.

Property 2.1 An nth degree polynomial function has at most n real zeros.

Property 2.2 If n is odd, there are an odd number of real zeros. (For example, if $n=3$, there are either 3 real zeros or 1 real zero of the function.)

Property 2.3 If n is even, there are an even number of real zeros of the function. (For example, if $n=4$, there are 4, 2, or no real zeros.)

A word of caution: A double zero must be counted as two zeros in Properties 2.2 and 2.3. For example, $f(x)=x^2-4x+4=(x-2)(x-2)$ is an even polynomial having only *one* distinct zero $x=2$; however, since the factor $(x-2)$ appears twice in the factorization, it is a double zero.

Property 2.4 There are at most $(n-1)$ turning points.

Property 2.5 If n is even and $a_n>0$, then $f(x)\to+\infty$ as $x\to+\infty$ or as $x\to-\infty$. Again, if n is even and $a_n<0$, then $f(x)\to-\infty$ as $x\to+\infty$ or as $x\to-\infty$.

Property 2.6 If n is odd and $a_n>0$, then $f(x)\to+\infty$ as $x\to+\infty$ and $f(x)\to-\infty$ as $x\to-\infty$. Again, if n is odd and $a_n<0$, then $f(x)\to-\infty$ as $x\to+\infty$ and $f(x)\to+\infty$ as $x\to-\infty$.

Properties 2.5 and 2.6 are generalizations of our discussion of extent. Rather than memorizing these two properties, it is easier to examine the extent of each function on an individual basis.

Example 2.13 Draw a graph of the polynomial function $f(x)=x^3-16x$.

Solution We first consider extent. For positive large values of x, $f(x)\doteq x^3$; therefore, as $x\to+\infty$, $f(x)\to+\infty$. For negative large values of x, $f(x)\doteq x^3$; therefore, as $x\to-\infty$, $f(x)\to-\infty$.

Second, we determine the zeros of the function by setting $f(x)=0$ and solving the equation

$$x^3-16x=0 \tag{2.16}$$

Since we can factor this polynomial, we need not resort to the rational roots theorem. Equation (2.16) is equivalent to

$$x(x^2-16)=0$$

or

$$x(x-4)(x+4)=0$$

The zeros are $x=0, 4, -4$. To complete the graph, we select x values near the zeros of the function and compute $f(x)$ and tabulate as follows

x	-5	-4	-2	0	2	4	5
$f(x)$	-45	0	24	0	-24	0	45

We plot these ordered pairs and draw a smooth curve through these points as shown in Fig. 2.24.

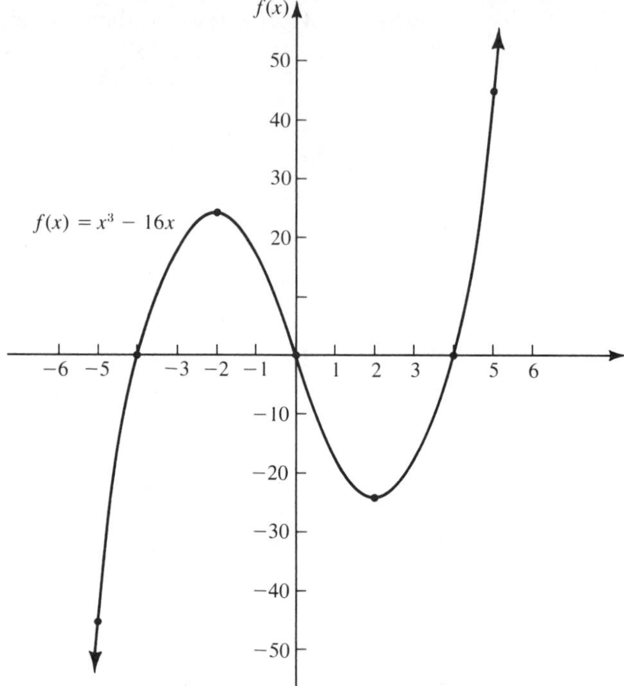

FIGURE 2.24

The domain of any polynomial function is the set of real numbers. We can justify this by noting that whenever x is any real number the polynomial $a_nx^n + a_{n-1}x^{n-1} + \cdots + a_1x^1 + a_0$ is also a real number; hence, as required, the range value $f(x)$ is also real.

If the domain is unrestricted, the range of an odd-degree polynomial function is the set of reals. This is illustrated in Fig. 2.25a and b where the graphs of two simple cubic polynomials are drawn.

For an unrestricted domain, the range of an even-degree polynomial is of the form $R=[b,+\infty)$ or $(-\infty,b]$ where b is a real number. Consider the illustrations in Fig. 2.26a and b, where two simple fourth-degree polynomials are graphed.

When polynomial functions are used as models for real-world problems, the domain and hence the range may be restricted by the physical problem under consideration. For example, the domain of the cubic polynomial function $V(h)=(10-2h)^2(h)$, which we considered in solving our box problem in Section 1.3, was restricted to the interval $(0,5)$.

Example 2.14 Graphing a Profit Curve A manufacturer has determined that if she manufactures x items in one week, her profit, $P(x)$, is given by the formula

$$P(x) = x^2(10-x)^3$$

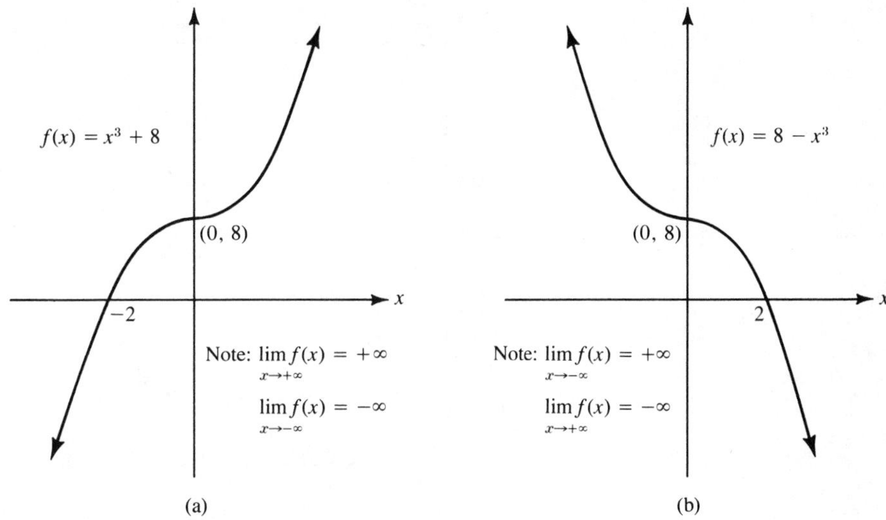

FIGURE 2.25

where P is measured in dollars and x is the number of items in *hundreds*. Plant capacity limits the number of items produced in one week to 1,000 items. Draw a graph of the function $P(x)$ and discuss the profit picture based on the graph.

Solution $P(x)$, the mathematical model for our problem, has a restricted domain. Since we cannot manufacture a negative number of items, we will assume that $x \geqslant 0$. Due to the limited plant capacity, $x \leqslant 10$. Therefore, the domain is the interval $[0, 10]$.

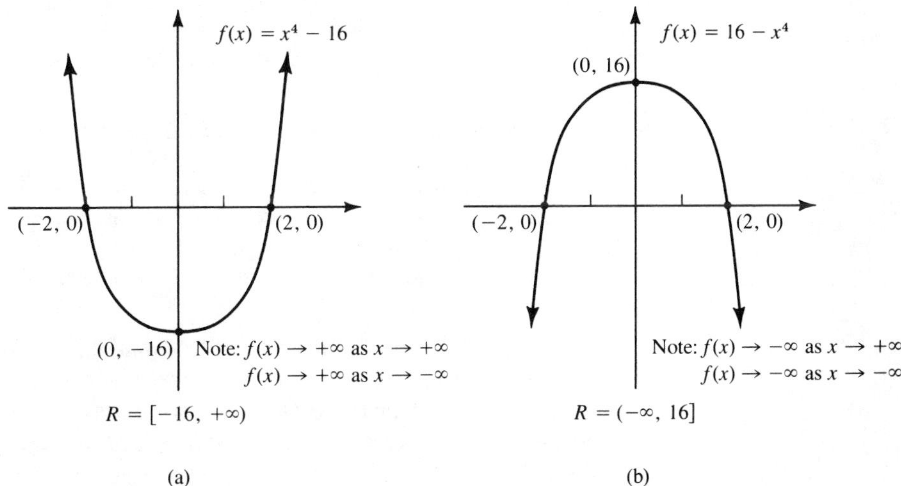

FIGURE 2.26

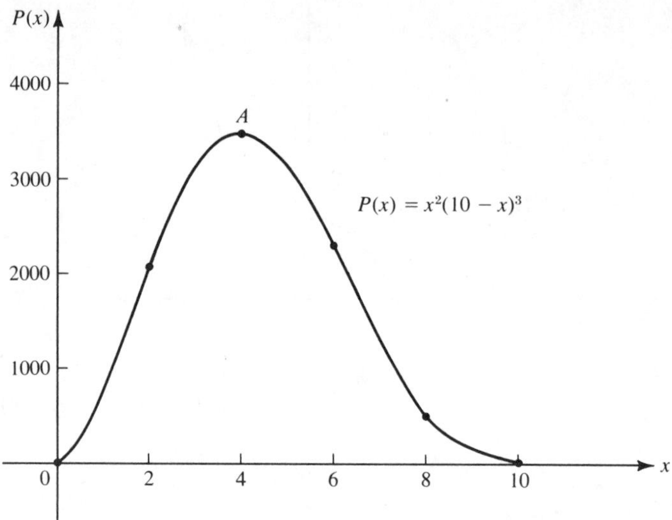

FIGURE 2.27

We find the zeros of the function by setting $P(x)=0$ and solving the equation

$$x^2(10-x)^3 = 0$$

By setting each of the factors, x^2 and $(10-x)^3$, equal to zero we find the roots of this equation are $x=0$ and $x=10$. Selecting x values of 2, 4, 6, and 8 we find the $P(x)$ values as shown in the following table.

x	0	2	4	6	8	10
$P(x)$	0	2,048	3,456	2,304	512	0

Plotting this set of ordered pairs and drawing a smooth curve through the points, we find the graph shown in Fig. 2.27.

Based on Fig. 2.27, we can conclude that the weekly profits are maximized when approximately 400 units are manufactured each week. The maximum profit based on our graph is approximately \$3,500. (In Chapter 5 we will find a procedure to give the exact coordinates of the turning point A.)

Frequently when considering polynomial expressions in the variable x we will be interested in determining those values of x that will make the expression positive or those values that will make the expression negative. For example, suppose we wish to determine those values of x such that $x^3-2x^2-5x+6 \geqslant 0$. You will note that this expression is the right-hand side of Eq. (2.13) and is equal to $f(x)$. Therefore, finding the solution of the cubic inequality

$$x^3 - 2x^2 - 5x + 6 \geqslant 0 \tag{2.17}$$

is equivalent to determining all the intervals in the domain of the function $f(x)$ that correspond to range values that are positive or zero. By inspection of Fig. 2.23 we note that $f(x) \geqslant 0$ whenever x is in the interval $[-2, 1]$ or $[3, \infty)$. For all other x values, $f(x) < 0$. Therefore, the solution of inequality (2.17) in interval notation is

$$[-2, 1] \cup [3, \infty)$$

A method based on this discussion can be used as a short-cut technique to solve polynomial inequalities. The solution of these inequalities is basic to determining where a polynomial function is positive or negative. To illustrate this method, which we call the *cut method*, let us redetermine the solution of the inequality (2.17)

$$x^3 - 2x^2 - 5x + 6 \geqslant 0$$

To apply this technique, the first step is to factor the polynomial. (Fortunately, in most examples that we will encounter this will not be a difficult task.) In this example we already have the factors and inequality (2.17) is equivalent to

$$(x-1)(x-3)(x+2) \geqslant 0 \tag{2.18}$$

The second step is to determine those values of x that are zeros of the polynomial and mark these points on the real line. These points, called *cuts*, are shown in Fig. 2.28.

The three cuts $x = -2$, 1, and 3 divide the real line into the four regions A, B, C, and D as shown in Fig. 2.28. The product $(x-1)(x-3)(x+2)$ is either positive or negative in each of these regions. To determine if the product is positive or negative in region A we select a representative value $x = 4$ from region A and substitute it into $(x-1)(x-3)(x+2)$ and find that

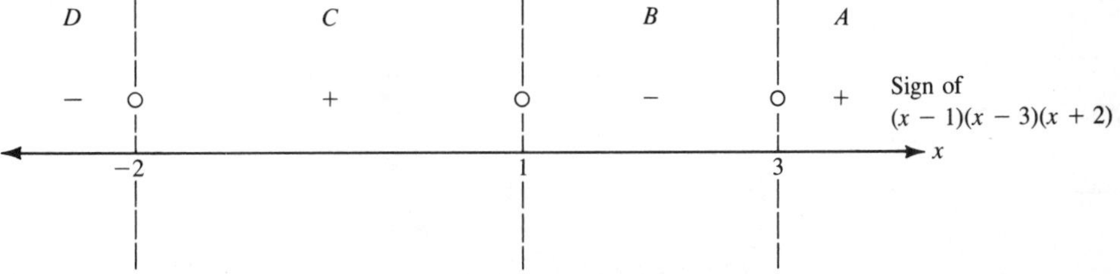

FIGURE 2.28

$(4-1)(4-3)(4+2)$ is positive. We note this by putting a + sign in region A as shown in Fig. 2.28. To determine the sign of $(x-1)(x-3)(x+2)$ in region B we select a representative value $x=2$ from this region and substitute it into this product and find that $(2-1)(2-3)(2+2)$ is negative. We mark region B negative in Fig. 2.28. In a similar manner we pick a value of $x=0$ from region C and find that the product $(x-1)(x-3)(x+2)$ is positive in region C, and a value of $x=-3$ from region D and find that the product is negative in region D. We have marked these regions accordingly.

To determine the solution of $(x-1)(x-3)(x+2) \geqslant 0$ we inspect Fig. 2.28 and determine where the indicated product is + or 0. By inspection, the solution is

$$[-2,1] \cup [3,\infty)$$

(*Note*: The cuts -2, 1, and 3 are included since for these x values the product is 0.)

Example 2.15 Find the solution in interval notation of the quadratic inequality $x^2-x-6<0$ by the cut method.

Solution To employ the cut method, we factor (x^2-x-6) and obtain the equivalent inequality

$$(x-3)(x+2) < 0 \tag{2.19}$$

The cuts, found by setting each factor equal to zero, are $x=3$ and $x=-2$. They are indicated on the real line in Fig. 2.29.

Selecting the value $x=4$ from region A, we find by substituting into $(x-3)(x+2)$ that the product $(4-3)(4+2)$ is positive. Selecting the value $x=0$ from region B, we find by substituting into $(x-3)(x+2)$ that the product $(0-3)(0+2)$ is negative. Selecting the value $x=-4$ from region C, we find by substituting into $(x-3)(x+2)$ that the product $(-4-3)(-4+2)$ is positive.

These results are indicated in Fig. 2.29, where by inspection we determine the solution of inequality (2.19). The product $(x-3)(x+2)$ is less

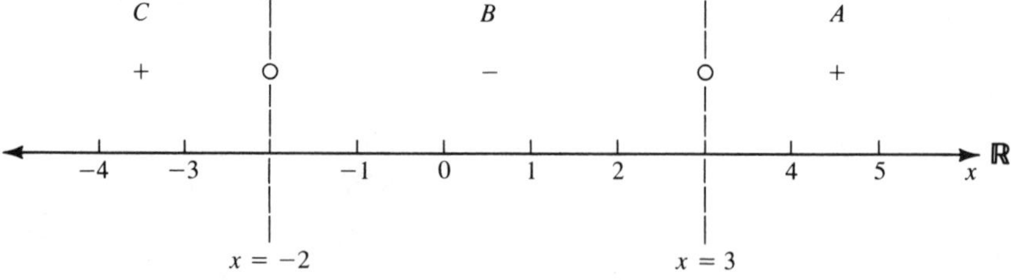

FIGURE 2.29

than zero in region B; therefore, our solution in interval notation is $(-2,3)$. (*Note*: The cuts are not included in this case as the product described in inequality (2.19) is strictly less than zero.)

EXERCISES

List all the possible rational zeros of each function in Exercises 1–5.

1. $f(x)=x^4-3x^3+2x^2-6x+8$

2. $g(x)=2x^3-3x^2+4x-4$

3. $h(x)=4x^3+2x^2-6x+6$

4. $f(x)=\frac{1}{2}x^3+5x^2+7x-3$

5. $g(x)=x^3-\frac{1}{2}x^2+2x-4$

For each of the functions in Exercises 6–15

(a) determine the $\lim_{x\to+\infty} f(x)$,
(b) determine the $\lim_{x\to-\infty} f(x)$,
(c) determine the zeros of the function,
(d) draw a sketch of the graph of the function.

6. $f(x)=3x-6$

7. $f(x)=x^2+2x-35$

8. $f(x)=x^3-8$

9. $f(x)=x^3-x^2-16x+16$

10. $f(x)=x^3-5x^2-12x+36$

11. $f(x)=2x^3+5x^2-28x-15$

12. $f(x)=x^3-6x^2+11x-6$

13. $f(x)=x^4+8$

14. $f(x)=16-x^4$

15. $f(x)=x^4-10x^2+9$

Solve each of the inequalities in Exercises 16–25. *Express the solution in interval notation.*

16. $(x-5)(x+3)\geqslant 0$

17. $(2x+1)(3x-4)\leqslant 0$

18. $(x+1)(x-1)(x-3)<0$

19. $(x+1)^2(x-1)(x+2)>0$

20. $x^3-16x\geqslant 0$

21. $x^2-x\leqslant 6$

22. $7x\geqslant 6x^2-5$

23. $-x^3+25x>0$

24. $15x^2+11x-14\leqslant 0$

25. $x^3-x^2-16x+16\geqslant 0$
(*Hint*: Use rational roots theorem to determine cut values.)

26. A manufacturer has determined that if he produces x items in one month his profit $P(x)$ for the month is given by

$$P(x) = x^2(15-x)^3$$

where $P(x)$ is in dollars and x is the number of items in hundreds. Assuming plant capacity is limited to 1,400 units/month, draw a sketch of $P(x)$ and estimate the maximum profit.

2.5 RATIONAL FUNCTIONS

We can further expand our collection of functions by introducing *rational functions*, which we define below.

Definition 2.2 A rational function is any function that can be expressed as the ratio of two polynomial functions.

For example, the functions

$$f(x) = \frac{x^2 - 1}{x + 2}, \qquad f(x) = \frac{3}{x}, \qquad \text{and} \quad f(x) = \frac{x^2 + 3x + 5}{1}$$

are rational functions. Note that polynomial functions themselves are rational functions having as a denominator the constant polynomial function $p(x) = 1$. The function $f(x) = 2 + (3/x)$ is also rational because it is equivalent to the function $f(x) = (2x + 3)/x$, which is the ratio of two polynomials.

As in the case of polynomial functions, we are interested in the graphs of these functions. Rational functions may at first appear harder to graph than polynomial functions. To determine some of the characteristics of these functions, let us consider the completed graph of the function

$$f(x) = \frac{x - 1}{x - 2}$$

as shown in Fig. 2.30.

First let us consider the extent of the function by an inspection of the graph. We note that as x increases without bound, that is as $x \to +\infty$, the value of $f(x)$ gets closer and closer to 1. We can verify this by observing the data shown in the following table.

x	5	10	15	20	...	100	500	1,000
$f(x)$	1.333	1.125	1.077	1.056	...	1.010	1.002	1.001

To denote the behavior of this function when x increases without bound, we write $f(x) \to 1$ as $x \to +\infty$ or $\lim_{x \to +\infty} f(x) = 1$. In general, if a function $f(x)$ gets closer and closer to the number L as x increases without bound, we denote this by

$$\lim_{x \to +\infty} f(x) = L$$

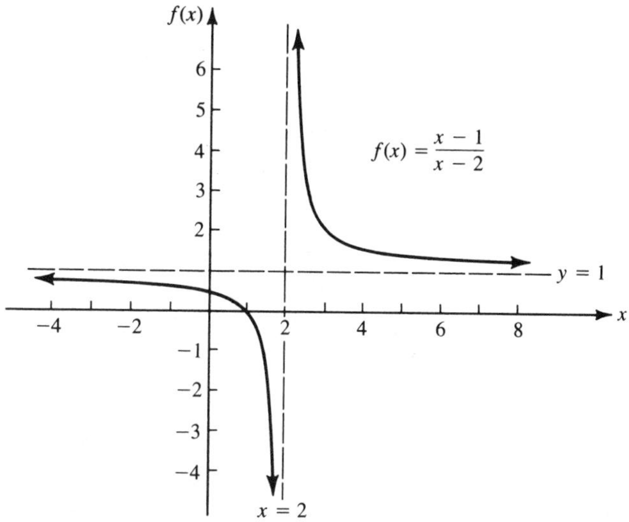

FIGURE 2.30

Analogously, we interpret $\lim_{x \to -\infty} f(x) = L$ to mean that $f(x)$ gets closer and closer to the number L as x decreases without bound. Therefore, by inspecting Fig. 2.30, we can conclude that

$$\lim_{x \to -\infty} f(x) = 1$$

Again we verify this by observing the data presented in the following table.

x	-5	-10	-15	-20	...	-100	-500	$-1{,}000$
$f(x)$	.857	.917	.941	.955	...	.990	.998	.999

The horizontal straight line whose equation is $y = 1$ is called a *horizontal asymptote* of the function. The function value $f(x)$ gets closer and closer to the horizontal asymptote as x increases or decreases without bound.

Similarly, the vertical line whose equation is $x = 2$ is called a *vertical asymptote*. By inspecting Fig. 2.30 we note that as x gets closer and closer to 2 from the right, that is as x assumes values of $2.2, 2.1, 2.01, 2.001$, etc., $f(x)$ increases without bound, or

$$\lim_{x \to 2^+} f(x) = +\infty$$

This is read "the limit of $f(x)$ as x approaches 2 from the right is positive infinity." (*Note*: The raised plus symbol to the right of 2 indicates that x is approaching 2 from the right; this is illustrated in the following table.) We verify the above conclusion by observing the data shown in the table.

x	2.5	2.1	2.01	2.001	2.0001
$f(x)$	3	11	101	1,001	10,001

In general, if a function $f(x)$ increases (or decreases) without bound as x approaches a number *a from the right*, we denote this by

$$\lim_{x\to a^+} f(x) = +\infty \qquad \text{or} \qquad \lim_{x\to a^+} f(x) = -\infty$$

respectively. Analogously, we interpret $\lim_{x\to a^-} f(x) = +\infty$ (or $-\infty$) to mean that $f(x)$ increases without bound (or decreases without bound) as x approaches *a from the left*. Therefore, by inspection of Fig. 2.30, we can conclude that

$$\lim_{x\to 2^-} f(x) = -\infty$$

Again we verify this by observing the data shown in the following table.

x	1.8	1.9	1.99	1.999	1.9999
$f(x)$	-4	-9	-99	-999	$-9,999$

We note from Fig. 2.30 that the graph of $f(x)$ never crosses the vertical asymptote. The reason for this is clear if we look at our function $f(x) = (x-1)/(x-2)$. Since division by zero is undefined, $x=2$ is not in the domain of the function. If the graph of $f(x)$ were to cross the vertical asymptote whose equation is $x=2$, this would imply that there exists a point $(2, f(2))$ belonging to our function, which is a contradiction of the fact that $x=2$ is not in the domain. No function will cross one of its vertical asymptotes; however, it is possible for the graph of a function to intersect a horizontal asymptote [e.g., Fig. 2.31 shows the graph of the rational function $R(x) = (2x-1)/x^2$ crossing its horizontal asymptote at point A].

Now that we have observed some of the characteristics of rational functions, we ask what is the best procedure to determine their graphs? Clearly, picking x values indiscriminately and plotting points is not the best answer. For example, what might we have anticipated as the graph of $f(x) = (x-1)/(x-2)$ had we plotted the points belonging to $f(x)$ shown in Fig. 2.32?

From our discussion and the graph of $f(x) = (x-1)/(x-2)$ shown in Fig. 2.30, it is clear that the determination of the vertical and horizontal asymp-

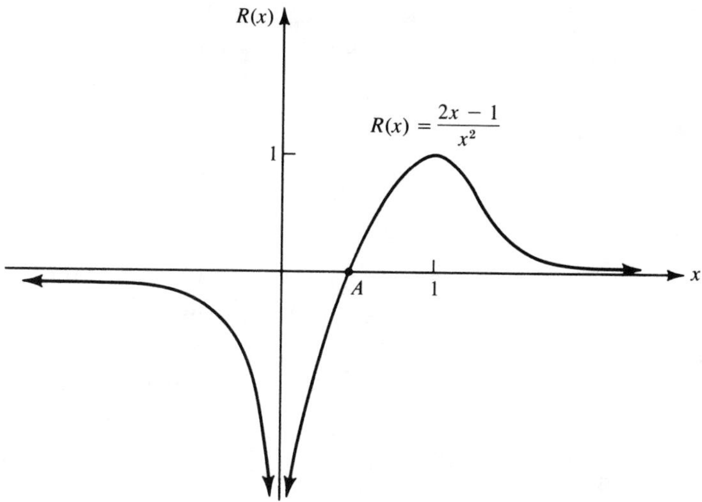

FIGURE 2.31

totes is a key factor in sketching the graph of a rational function. The procedure to determine the equations of these asymptotes is discussed below and summarized in Rules 2.2 and 2.3.

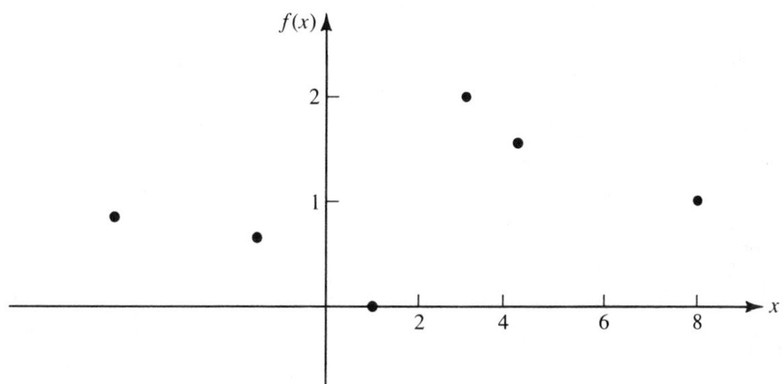

FIGURE 2.32

Vertical Asymptote(s)

Rule 2.2 If a rational function $f(x)=p(x)/q(x)$, where $p(x)$ and $q(x)$ are polynomial functions, then there exists a vertical asymptote for each real x value such that $q(x)=0$ and $p(x)\neq 0$.

For example, if $f(x)=2x/(x-3)$ there is a vertical asymptote of $f(x)$ at $x-3=0$ or $x=3$. We verify this by determining $\lim_{x\to 3^+} f(x)$, $\lim_{x\to 3^-} f(x)$ and graphing the function $f(x)$ in the vicinity of $x=3$. To determine $\lim_{x\to 3^+} f(x)$ we pick x values of $3.1, 3.01, 3.001, \ldots$, and tabulate the ordered pairs that belong to $f(x)$ as shown in the following table.

x	3.1	3.01	3.001
$f(x)=\dfrac{2x}{x-3}$	62	602	6,002

Based on these data, we can conclude that $f(x)$ increases without bound as x approaches 3 from the right; therefore,

$$\lim_{x\to 3^+} f(x) = +\infty$$

The graph of $f(x)$ to the right of $x=3$ would appear as shown in Fig. 2.33.

To determine the $\lim_{x\to 3^-} f(x)$, we pick x values of $2.9, 2.99, 2.999, \ldots$, and calculate $f(x)$ as shown in the next table.

x	2.9	2.99	2.999
$f(x)=\dfrac{2x}{x-3}$	-58	-598	$-5{,}998$

Based on these data, we conclude that $f(x)$ decreases without bound as x

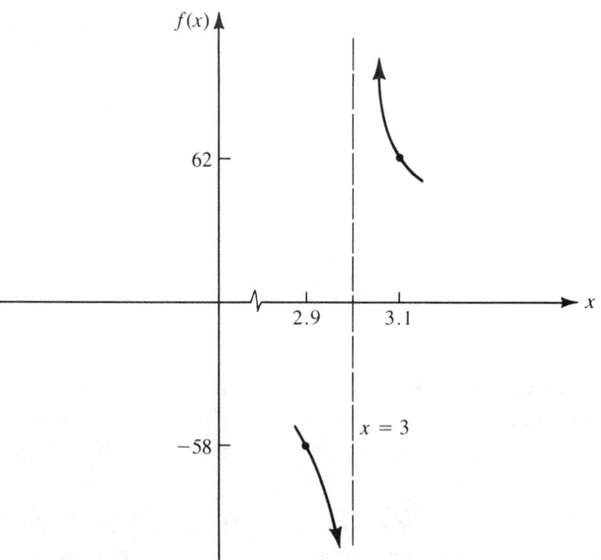

FIGURE 2.33

approaches 3 from the left. The graph of $f(x)$ to the left of $x=3$ is shown in Fig. 2.33.

In general, once we have determined that $x=a$ is a vertical asymptote of a function $f(x)$ by Rule 2.2, it is relatively easy to determine

$$\lim_{x \to a^+} f(x)$$

Rather than evaluating several values of $f(x)$ to the right of $x=a$, we need only evaluate for one x value "very close to" but greater than a. [In many cases a good value to determine is $f(a+.1)$.] If f of this x value is a positive number, then $\lim_{x \to a^+} f(x) = +\infty$; if it is negative, then $\lim_{x \to a^+} f(x) = -\infty$. In a like manner, we evaluate $f(x)$ for an x "very close to" but less than a to determine the $\lim_{x \to a^-} f(x)$. (Again in many cases a good value to consider is $x=a-.1$.)

You perhaps have noted that Rule 2.2 does not state that there is a vertical asymptote at $x=a$ if both $p(a)$ and $q(a)=0$. In this case the polynomials $p(x)$ and $q(x)$ contain a common factor of $(x-a)$. By factoring out the common factor $(x-a)$ in the numerator and denominator, we can write the rational function $p(x)/q(x)$ as $p^*(x)/q^*(x)$, where $p(x) = (x-a)p^*(x)$ and $q(x) = (x-a)q^*(x)$. To graph $p(x)/q(x)$ we graph $p^*(x)/q^*(x)$ for all $x \neq a$. Now either the line whose equation is $x=a$ is a vertical asymptote [$p^*(a) \neq 0$ and $q^*(a)=0$] or the point where this line intersects the graph of $p^*(x)/q^*(x)$ is an "open hole" in the graph of $p(x)/q(x)$ since $x=a$ is not in its domain.

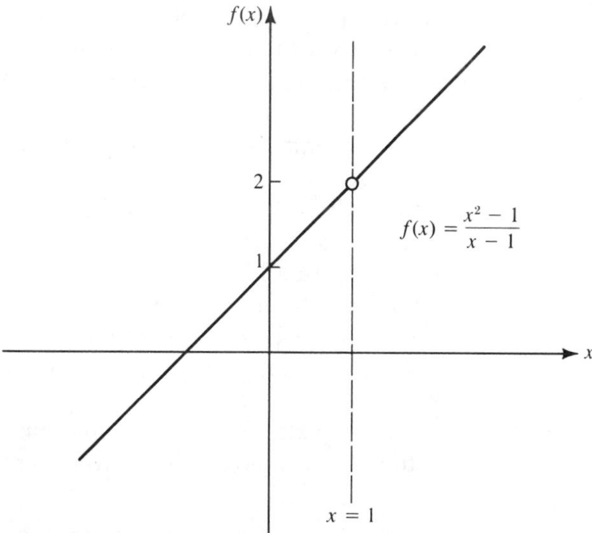

FIGURE 2.34

For example, if $f(x)=(x^2-1)/(x-1)$, we can write $f(x)=[(x+1)(x-1)]/1(x-1)$. Therefore $p^*(x)/q^*(x)=(x+1)/1$. Hence the graph of the function $f(x)=(x^2-1)/(x-1)$ is the same as the graph of the line $y=x+1$ for all $x \neq 1$. Where the vertical line $x=1$ intersects the graph of the line $y=x+1$ there is an open hole to indicate that $x=1$ is not in the domain of $f(x)$. The graph of $f(x)$ is shown in Fig. 2.34.

Horizontal Asymptote(s) To determine if a rational function $f(x)=p(x)/q(x)$ has a horizontal asymptote,* we consider extent; that is, the behavior of $f(x)$ as x increases or decreases without bound. In limit notation we determine both

$$\lim_{x \to +\infty} f(x) \qquad \text{and} \qquad \lim_{x \to -\infty} f(x)$$

To determine the behavior of $f(x)$ we must determine which terms in the function are significant when x increases or decreases without bound. Since the numerator is a polynomial function, it is closely approximated by the highest-degree term, as we found in Sec. 1.6. Similarly the denominator is closely approximated by its term of highest degree. Therefore, the function $f(x)$ can be approximated by the ratio of these two terms whenever $x \to +\infty$ or $-\infty$. For example, the function $f(x)=2x/(x-3)$ is approximated by $2x/x$. Since $2x/x=2$, we can conclude that

$$\lim_{x \to +\infty} f(x) = 2 \qquad \text{and} \qquad \lim_{x \to -\infty} f(x) = 2$$

This implies that the graph of $f(x)$ gets closer and closer to the horizontal line $y=2$ as $x \to +\infty$ or as $x \to -\infty$. Therefore, the function $f(x)=2x/(x-3)$ has a horizontal asymptote whose equation is $y=2$.

Geometrically this implies that the graph of $f(x)$ is represented by either curve A or B as $x \to +\infty$ and by either C or D as $x \to -\infty$, as shown in Fig. 2.35.

To determine whether the graph of $f(x)$ approaches the asymptote from above (curve A), or below (curve B) as $x \to +\infty$, we evaluate $f(x)$ for a large positive value of x. Selecting $x=100$, we find $f(100)=200/(100-3)=\frac{200}{97}$, which is greater than 2; hence, the graph of $f(x)$ approaches the asymptote from above. In a like manner, we find $f(-100)=\frac{200}{103}$ and determine that $f(x)$ approaches 2 from below as $x \to -\infty$.

If we make a composite of the information we have found from Figs. 2.33 and 2.35, a partial graph of $f(x)=2x/(x-3)$ can be drawn as shown in Fig. 2.36.

To complete the graph of this function, we determine the zeros of the function, the y-intercept, and two or three other points. The zeros of $f(x)$ are

*There are functions having more than one horizontal asymptote, but we will not consider such complex functions in this text.

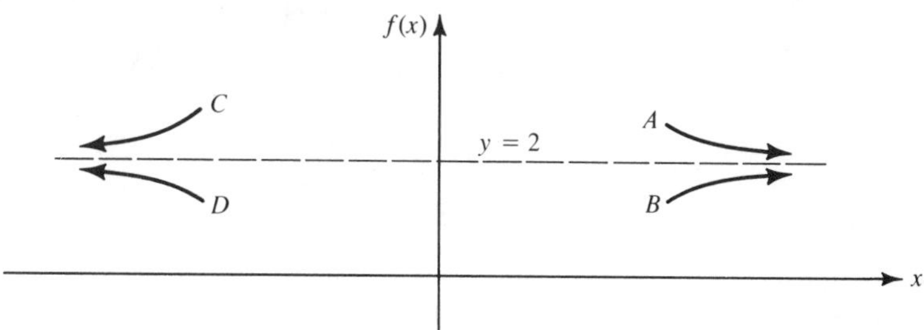

FIGURE 2.35

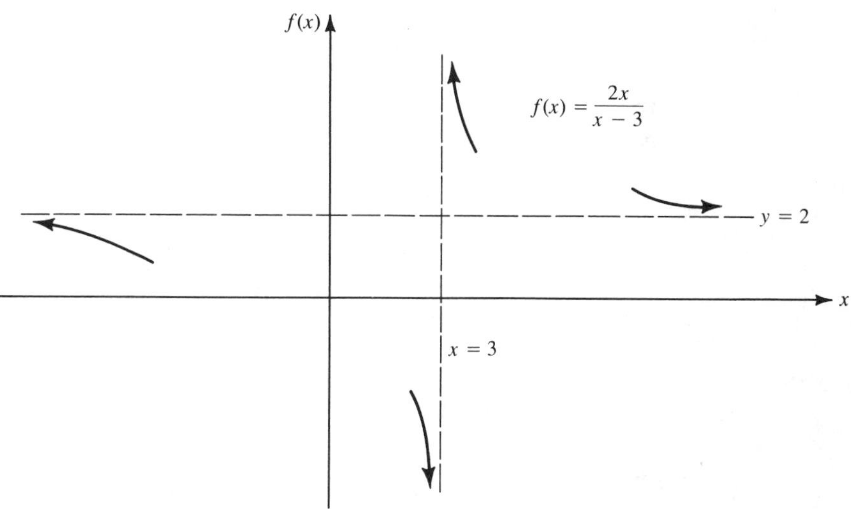

FIGURE 2.36

found by setting $f(x)=0$ and solving the resultant equation $2x/(x-3)=0$ for x. Since $2x/(x-3)=0$ is equivalent to $2x=0$ or $x=0$, there is one zero of the function; i.e., $x=0$, and the ordered pair $(0,0)$ belongs to the function.

Based on the partial sketch shown in Fig. 2.36, we select additional x values of -1, 1, and 4 and compute $f(x)$ as shown in the following table.

x	-1	0	1	4
$\dfrac{2x}{x-3}$	$\dfrac{1}{2}$	0	-1	8

Plotting the points listed in the table together with the information provided by Fig. 2.36 we obtain the graph of $f(x)=2x/(x-3)$ as shown in Fig. 2.37.

To arrive at a general rule for finding horizontal asymptotes let us consider the additional two examples.

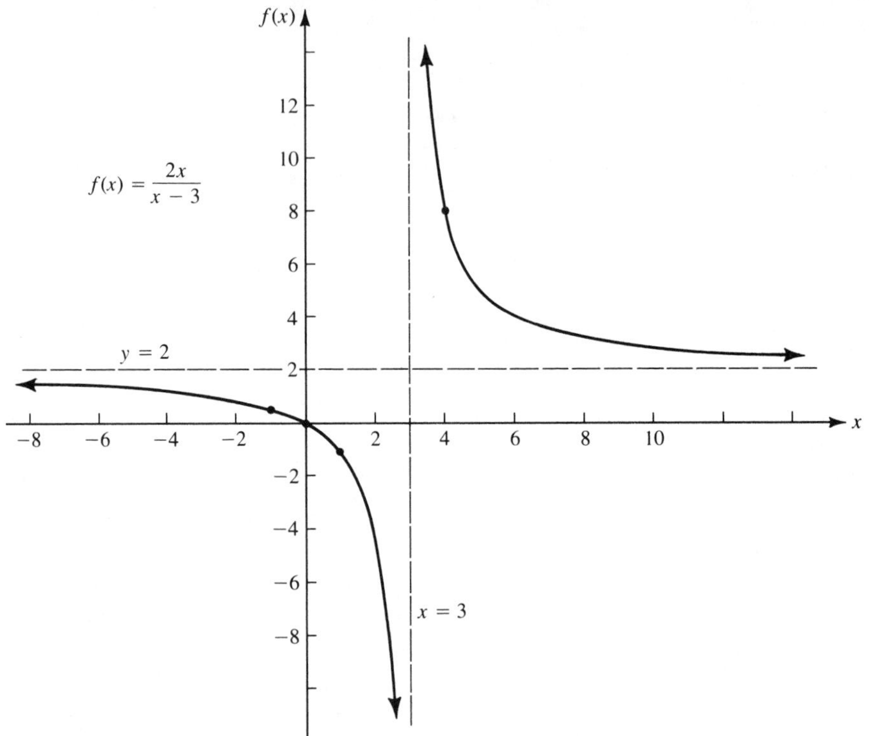

FIGURE 2.37

Example 2.16 Determine the equation of the horizontal asymptote of the rational function

$$f(x) = \frac{3x+2}{x^2+5}$$

Solution As x increases, or decreases without bound, the numerator polynomial $3x+2$ is approximated by $3x$ and the denominator polynomial x^2+5 by x^2; therefore,

$$f(x) \doteq \frac{3x}{x^2} = \frac{3}{x} \quad \text{as} \quad x \to +\infty \quad \text{or} \quad -\infty$$

As $x \to +\infty$ (or $-\infty$), $(3/x) \to 0$; hence we can conclude that $\lim_{x\to+\infty} f(x) = 0$ and $\lim_{x\to-\infty} f(x) = 0$. Therefore, the equation of the horizontal asymptote is $y = 0$.

Example 2.17 Determine the equation of the horizontal asymptote, if one exists, of the rational function

$$f(x) = \frac{x^2+5}{3x+2}$$

Solution As in the above example, we can show that $f(x) \doteq x^2/3x = x/3$ as $x \to +\infty$ or $-\infty$. As $x \to +\infty$, $f(x)$ increases without bound or $f(x) \to +\infty$ and as $x \to -\infty$, $f(x)$ decreases without bound. Therefore,

$$\lim_{x \to +\infty} f(x) = +\infty \qquad \text{and} \qquad \lim_{x \to -\infty} f(x) = -\infty$$

and there is no horizontal asymptote!

Using the above discussion, we can now state and justify the following rule to determine the equation of the horizontal asymptote, if one exists, for any rational function.

Rule 2.3 Consider a rational function of the form

$$f(x) = \frac{a_m x^m + a_{m-1} x^{m-1} + \cdots + a_0}{b_n x^n + b_{n-1} x^{n-1} + \cdots + b_0}$$

(a) If $n > m$, $y = 0$ is the horizontal asymptote of $f(x)$.
(b) If $n = m$, $y = a_m/b_n$ is the horizontal asymptote of $f(x)$.
(c) If $n < m$, no horizontal asymptote exists.

(You will be asked to justify this rule in Exercise 24 at the end of this section.)
To graph a rational function, we suggest the following steps.

Step 1 Use Rule 2.3 to determine if there is a horizontal asymptote. Sketch the horizontal asymptote as a dotted line and determine if $f(x)$ approaches it from above or below as $x \to +\infty$ and as $x \to -\infty$.

Step 2 Factor the numerator and denominator whenever possible.

Step 3 Use Rule 2.2 to determine if there are any vertical asymptotes. Sketch the vertical asymptote(s) as a dotted line and determine $\lim_{x \to a^+} f(x)$ and $\lim_{x \to a^-} f(x)$ where $x = a$ is the equation of a vertical asymptote.

Step 4 Find the zeros of the function. [If $f(x) = p(x)/q(x)$ find those values of x where $p(x) = 0$ and $q(x) \neq 0$]. Plot the corresponding points.

Step 5 Find the y-intercept. [Calculate $f(0)$.] Plot this point.

Step 6 Carefully select a few other points belonging to $f(x)$, and sketch the graph of the function using the information provided by these steps.

The next example illustrates how to apply these steps to graph a rational function.

Example 2.18 Sketch the graph of the rational function $f(x) = (2x-4)/(x+3)$ and discuss its domain and range.

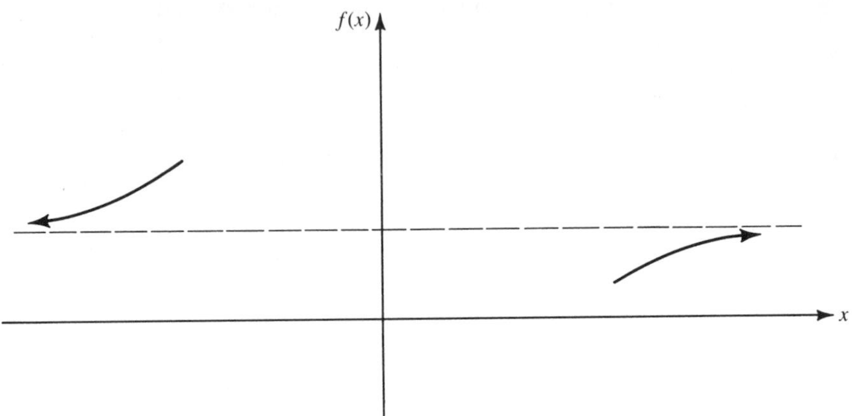

FIGURE 2.38

Solution By Rule 2.3 the equation of the horizontal asymptote is $y=\frac{2}{1}$ or $y=2$. When $x=100$,

$$f(100) = [2(100)-4]/103 \doteq 1.9$$

which implies that when x increases without bound, the graph of $f(x)$ approaches $y=2$ from below. When $x=-100$,

$$f(-100) = \frac{2(-100)-4}{-100+3} \doteq 2.1$$

which implies that when x decreases without bound, the graph of $f(x)$ approaches $y=2$ from above. We begin the graph of $f(x)$ in Fig. 2.38 by drawing the horizontal asymptote as a dotted line and sketching this information.

By Rule 2.2 we note that $x+3=0$ and that $2x-4\neq 0$ when $x=-3$. Therefore, there exists a vertical asymptote at $x=-3$. To evaluate $\lim_{x\to -3^+} f(x)$, we evaluate $f(-2.9)$. We find $f(-2.9)=-9.8/+.1=-98$, which implies $\lim_{x\to -3^+} f(x)=-\infty$. To determine $\lim_{x\to -3^-} f(x)$, we evaluate $f(-3.1)$. We find $f(-3.1)=102$, which implies that $\lim_{x\to -3^-} f(x)=+\infty$.

We continue the graph of $f(x)$ by drawing the vertical asymptote as a dotted line and sketching the function to the left and right of the vertical asymptote (Fig. 2.39). The numerator is zero when $2x-4=0$ or $x=2$. Hence, there is one zero of the function. The y-intercept is found by evaluating $f(0)$, which is $-\frac{4}{3}$. Plotting the x-intercept, $(2,0)$, the y-intercept, $(0,-\frac{4}{3})$, and using the information from Fig. 2.39, we complete a sketch of $f(x)$ as shown in Fig. 2.40.

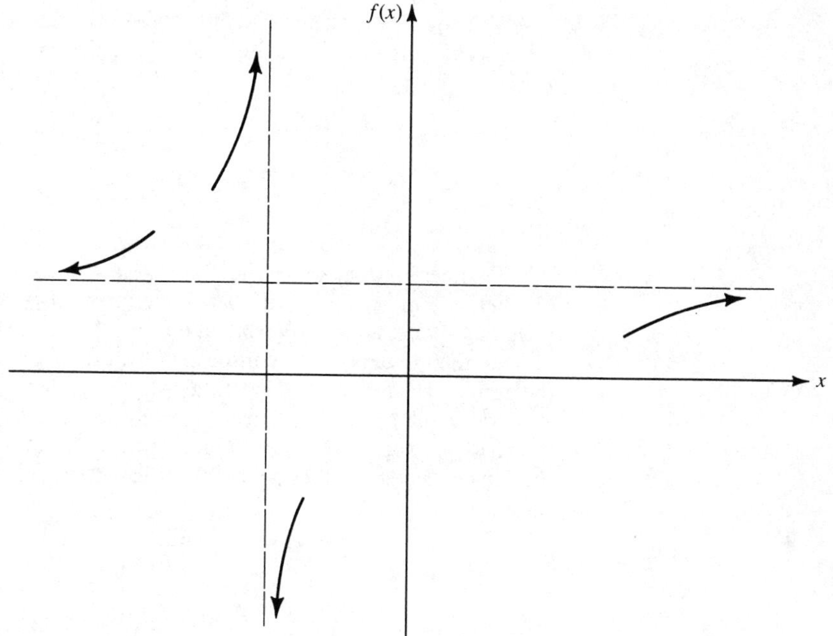

FIGURE 2.39

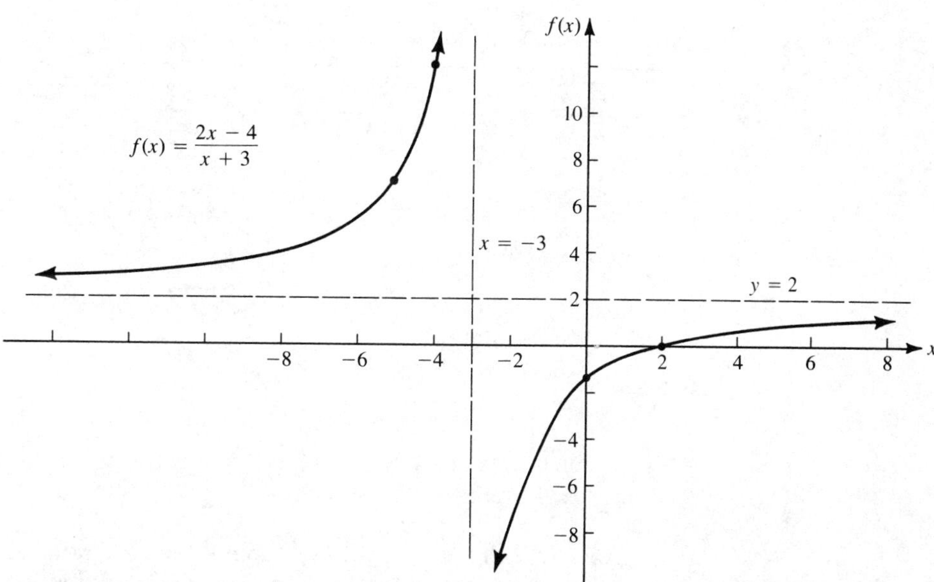

FIGURE 2.40

We should select other x values, e.g., -4, -2, and 4, and check to see if the ordered pairs shown in the following table correspond to points on the graph.

x	-4	-2	4
$f(x)$	12	-8	$\frac{4}{7}$

To determine the domain of $f(x)$, we note that $(2x-4)/(x+3)$ is a real number as long as $x \neq -3$. Hence, the domain of $f(x)$ is the set of reals not including -3, using interval notation: $D=(-\infty, -3)\cup(-3, +\infty)$. We determine the range of $f(x)$ by inspecting the graph. Note that if each point on the graph were translated horizontally to the vertical axis, every ordinate on this axis would be "covered" except $y=2$ (see Fig. 2.41). Since the graph does not cross the horizontal asymptote in this problem, $y=2$ is not an element of the range. Therefore, the range of $f(x)$ in interval notation is $R=(-\infty,2)\cup(2, +\infty)$.

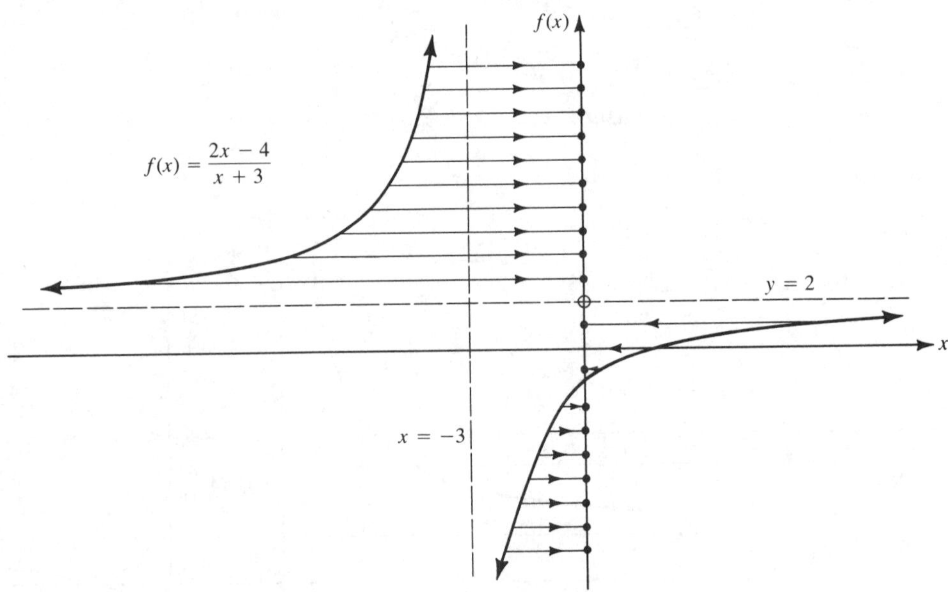

FIGURE 2.41

In our last example, we will develop a rational function as a mathematical model of a real-world problem.

Example 2.19 Minimizing the Surface Area A rectangular closed box with a square base is to be constructed with a fixed volume of 100 cu in. Express the total surface area of the box as a function of x, the edge of the base, and draw the graph of this function.

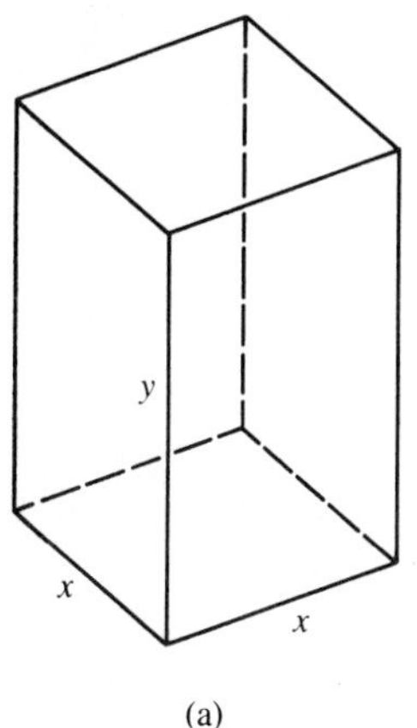

(a)

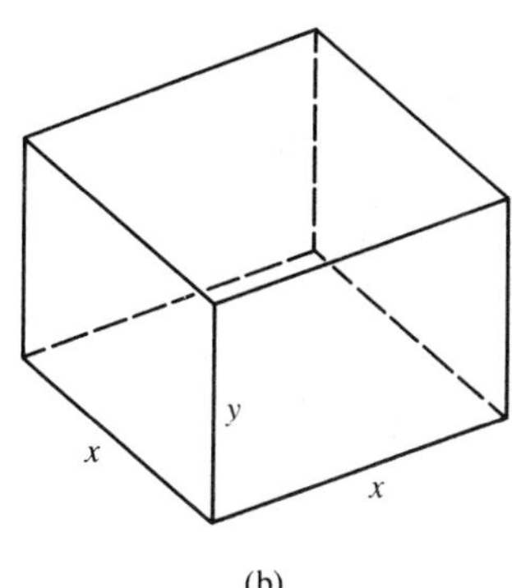

(b)

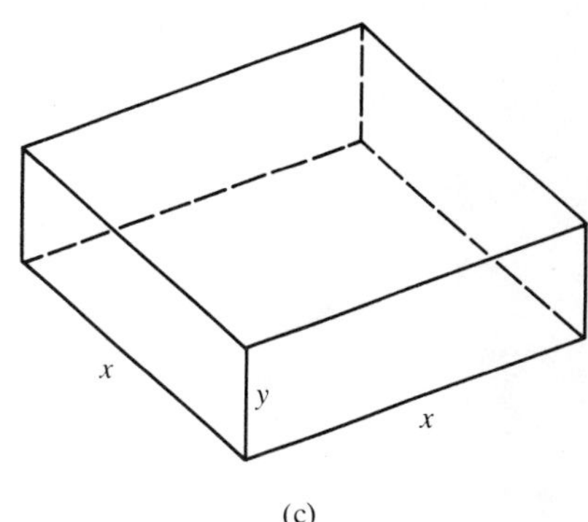

(c)

FIGURE 2.42

Solution Fig. 2.42 shows three different boxes, each with a volume of 100 cu in. and a square base. If we let x = the edge of the base and y = the height of the box, we note in Fig. 2.42 that as the edge x increases in length the height y decreases. To establish the relationship between x and y we recall that the volume of a box is given by $V = l \cdot w \cdot h$. In terms of x and y, the volume of each box in Fig. 2.42 is given by $V = x^2y$. Since each box is 100 cu in. in volume, by substitution: $100 = x^2y$ or $y = 100/x^2$. From this latter equation, we can see that as x increases, y decreases.

The total surface area of any box in Fig. 2.42 equals the sum of the areas of the faces of the box. Using the formula $A = lw$ for the area of a rectangle, we can determine the area of each face of any box in Fig. 2.42 as follows:

Area of the base $= x \cdot x = x^2$

Area of the top $= x \cdot x = x^2$

Area of front face $= x \cdot y$

Area of back face $= x \cdot y$

Area of right-side face $= x \cdot y$

Area of left-side face $= x \cdot y$

Therefore, the total area $A = 2x^2 + 4xy$. Since $y = 100/x^2$, by substituting $100/x^2$ for y in the previous equation we can write the area, A, as a function of x:

$$\begin{aligned} A(x) &= 2x^2 + 4x\left(\frac{100}{x^2}\right) \\ &= 2x^2 + \frac{400}{x} \end{aligned}$$

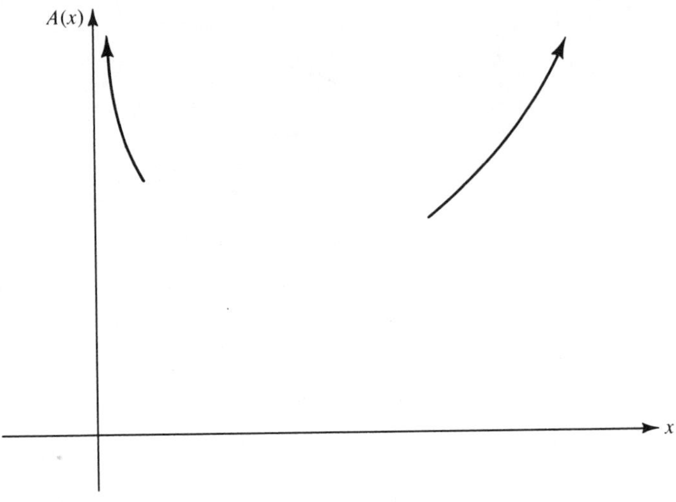

FIGURE 2.43

Rewriting $A(x)$ as the ratio of two polynomials, we have the rational function

$$A(x) = \frac{2x^3+400}{x}$$

Before we graph $A(x)$, we note that the domain is restricted by the problem.

Since x represents an edge of the box, it cannot be zero or negative; therefore, the domain $D=(0, +\infty)$. Since $A(x)$ is a rational function, we follow our step-by-step procedure to graph such a function.

By Rule 2.3, there is no horizontal asymptote. We note that as x increases without bound $A(x)$, approximated by $2x^2$, also increases without bound. By Rule 2.2, $x=0$ (the y axis) is a vertical asymptote. The $\lim_{x\to 0^+} f(x) = +\infty$. Using this information we draw a partial graph of $A(x)$ in Fig. 2.43.

Since $x=0$ is not in the domain of the function, there is no y-intercept and since $2x^3+400\neq 0$ when x is positive, there are no zeros of the function. To complete our graph, we select x values of 1, 5, 10, 15, 20, and compute $f(x)$ as shown in the following table.

x	1	5	10	15	20
$\frac{2x^3+400}{x}$	402	130	240	477	820

Plotting the points from this table and the information shown in Fig. 2.43, we complete the graph of $A(x)$ as shown in Fig. 2.44.

By inspection of the graph of $A(x)$ in Fig. 2.44, we note that the total surface area is very large either when x is small or when x is very large. A minimum value of the surface area occurs somewhere between $x=4$ and

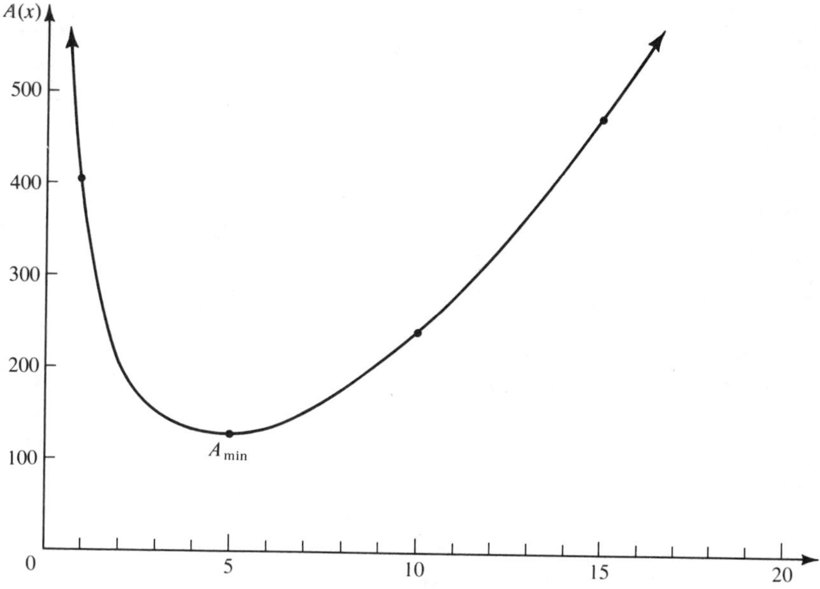

FIGURE 2.44

$x=6$. We could approximately determine this minimum value by plotting many points of the curve in this region. In Chapter 5 we will develop a procedure to find this value with ease. (*Note*: The minimum surface area corresponds to the minimum amount of material to make the box.)

EXERCISES

For each of the rational functions in Exercises 1–10 *determine*

(a) the $\lim_{x\to+\infty} f(x)$,
(b) the $\lim_{x\to-\infty} f(x)$,
(c) the equation of any horizontal asymptote,
(d) the equation(s) of any vertical asymptote(s).

1. $f(x) = \dfrac{x+2}{x-1}$

2. $f(x)=\dfrac{3x+5}{x+1}$

3. $f(x)=\dfrac{5}{2x-5}$

4. $f(x)=\dfrac{x}{x^2-4}$

5. $f(x)=\dfrac{x^2-4}{x}$

6. $f(x)=\dfrac{x^2-4}{x^2+4}$

7. $f(x)=\dfrac{3x-6}{5x+4}$

8. $f(x)=\dfrac{6-x^2}{x^2+1}$

9. $f(x)=\dfrac{3-x}{x+2}$

10. $f(x)=\dfrac{3}{x^2-9}$

For each of the rational functions in Exercises 11–20

(a) sketch any horizontal asymptote,
(b) sketch any vertical asymptote(s),
(c) plot the points corresponding to the zeros of the function,
(d) plot the y-intercept,
(e) sketch the graph of the function,
(f) determine the domain and range of the function.

11. $f(x)=\dfrac{x+2}{x-1}$

12. $g(x)=\dfrac{3x+5}{x+1}$

13. $h(x)=\dfrac{5}{2x-5}$

14. $f(x)=\dfrac{3-x}{x+2}$

15. $S(x)=\dfrac{-3x}{x-4}$

16. $y=\dfrac{2}{x-4}$

17. $C(x)=\dfrac{4x}{3-x}$

18. $f(x)=\dfrac{2}{x^2}$

19. $y=\dfrac{x^2-1}{x}$

20. $p(x)=\dfrac{x^2-9}{2x}$

21. Cost of Removing Auto Exhaust Pollutants An automobile manufacturer has determined that the cost of removing pollutants from the exhaust of one of the new models is given by

$$C(x) = \frac{3{,}000-10x}{100-x} \qquad \text{for} \qquad 50 \leqslant x < 100$$

where $C(x)$ is in dollars and x is the percentage of the pollutants removed.

(a) Determine the cost of removing each of the following percentages of pollutants: 50, 75, 80, 85, 90, and 95.
(b) Sketch the graph of the function $C(x)$ on the indicated domain.
(c) What percentage of the pollutants can be removed for $200?
(d) According to this function, is it possible to remove all of the pollutants?

22. It is estimated that the cost of removing a pollutant from waste water from a factory is given by

$$C(x) = \frac{10x}{100-x}$$

where the cost $C(x)$ is in thousands of dollars and x is the percentage of the pollutant removed.

(a) Determine the cost of removing each of the following percentages of the pollutant: 25, 50, 75, 80, and 90.
(b) Sketch the graph of the function $C(x)$.
(c) What percentage of the pollutant can be removed for $50,000? For $100,000?

23. **Minimizing the Cost of Material** A rectangular box with a square base and an open top is to be constructed with a volume of 500 cu in. If the material of construction costs 10¢/sq in. for the base and 5¢/sq in. for the faces,

(a) Express the cost, C, of the material as a function of x, the length of an edge of the base.
(b) Sketch a graph of this cost function.
(c) From your graph estimate the value of x that will minimize the cost of the material.

24. Justify each of the three parts of Rule 2.3.

2.6 SPECIAL FUNCTIONS

Many real-world problems are so complex that no single formula can be found to serve as a model over the entire domain of the function. In these cases we often use different formulas for various intervals of the domain of the function. Such problems frequently occur in business, where it is common to give discounts for large-quantity sales. For example, consider the following situation.

The Hartford Manifold Forms Company supplies a particular business form to several retail stores according to the following price schedule:

for orders up to and including 1,000: 10¢/form.

for orders over 1,000: \$100 plus 6¢/form over 1,000.

To determine the cost, $C(x)$, to a retail store of a purchase of x forms, we must consider whether x is equal to, less than, or greater than 1,000. If $x \leqslant 1{,}000$, the cost $C(x)$ is given by

$$C(x) = 10x \text{ cents}$$

or

$$= .10x \text{ dollars}$$

However, if the number of forms purchased is greater than 1,000 ($x > 1{,}000$), we have the cost $C(x)$ equal to \$100 plus $.06(x-1{,}000)$. [*Note*: $(x-1{,}000)$ is the number of forms purchased over 1,000.] We can express this as

$$C(x) = 100 + .06(x - 1{,}000) \text{ dollars}$$

or

$$= 40 + .06x \text{ dollars}$$

We can combine these two formulas by introducing the following new notation for the cost function, $C(x)$.

$$C(x) = \begin{cases} .10x & \text{if } \ 0 \leqslant x \leqslant 1{,}000 \\ 40 + .06x & \text{if } \ x > 1{,}000 \end{cases} \tag{2.20}$$

The formula (2.20) for $C(x)$ is described by two different equations each being defined on a different interval of the domain of the function. In the interval $[0, 1{,}000]$ the rule to determine the cost is $C(x) = .10x$ and in the interval $(1{,}000, \infty)$ the rule is $C(x) = 40 + .06x$. This function $C(x)$ clearly is a model of our original problem.

To graph $C(x)$ we first graph the rule $C(x) = .10x$ on the interval $[0, 1{,}000]$, then the rule $C(x) = 40 + .06x$ on the interval $(1{,}000, \infty)$. Since these functions are linear in both intervals, we select two x values in each interval and calculate the corresponding $C(x)$ value as shown in the following table.

x	0	1,000	2,000	4,000
$C(x)$	0	100	160	280

The graph of $C(x)$ is shown in Fig. 2.45.

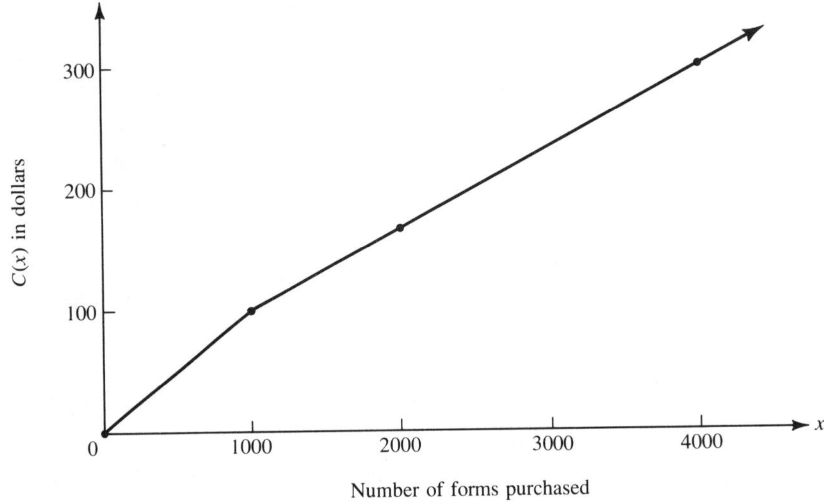

FIGURE 2.45

As another illustration of this type of function consider the following example.

Example 2.20 Sketch the graph of

$$f(x) = \begin{cases} \sqrt{x} & \text{if} \quad 0 \leqslant x \leqslant 16 \\ \frac{1}{2}x - 3 & \text{if} \quad 16 < x \leqslant 32 \end{cases}$$

Solution We first graph $f(x) = \sqrt{x}$ on $[0, 16]$ by selecting x values of 0, 1, 4,

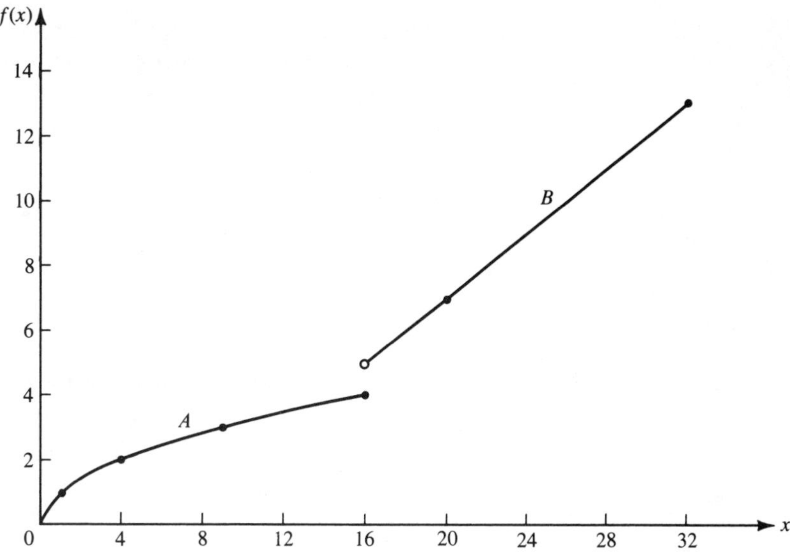

FIGURE 2.46

9, and 16 and calculating the corresponding $f(x)$ values as shown in the following table.

x	0	1	4	9	16
$f(x)=\sqrt{x}$	0	1	2	3	4

Plotting these points and drawing a smooth curve between them, we obtain part A of the graph shown in Fig. 2.46.

To graph $f(x)=\frac{1}{2}x-3$ on $(16,32]$, we note that $f(x)$ is linear on this interval. Selecting x values of 20 and 30 we find $f(20)=7$ and $f(30)=12$, and obtain part B of the graph shown in Fig. 2.46. Note that we have used an open circle at the point $(16,5)$ on part B of the graph to indicate that this point is not included. The point $(16,4)$, however, is included. At $x=16$ we say the function $f(x)$ displays a jump or a break in its graph.

Another function that requires two different algebraic formulas is the *absolute value function* $f(x)=|x|$. Intuitively, the absolute value of a is the distance a is from zero on the number line, or more formally we have the following definition.

Definition 2.3 The absolute value of a is denoted by $|a|$, where

$$|a| = \begin{cases} a & \text{if} \quad a \geqslant 0 \\ -a & \text{if} \quad a < 0 \end{cases}$$

Note in this definition that if a is negative ($a<0$), then $|a|=-a$. For example, if $a=-5$, then $|(-5)|=-(-5)=5$. And of course if a is zero or positive ($a \geqslant 0$), then $|a|=a$.

To graph the absolute value function $f(x)=|x|$ we have, by Definition 2.3,

$$f(x) = |x| = \begin{cases} x & \text{if} \quad x \geqslant 0 \\ -x & \text{if} \quad x<0 \end{cases}$$

Hence, the absolute value function $f(x)=|x|$ is represented by the linear function $f(x)=x$ on $[0,\infty)$ and the linear function $f(x)=-x$ on $(-\infty,0)$. The graph of $f(x)=x$ on $[0,\infty)$ is the ray shown as part A in Fig. 2.47 and the graph of $f(x)=-x$ on $(-\infty,0)$ is the ray shown as part B. The domain of $f(x)=|x|$ is the set of reals and the range is the interval $[0,\infty)$.

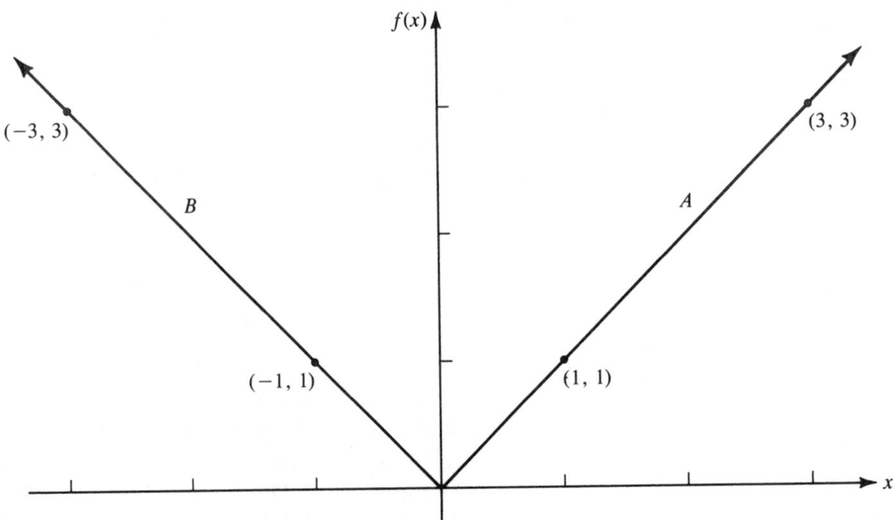

FIGURE 2.47

Example 2.21 Draw the graph of the absolute value function $f(x)=|2x-6|$.

Solution By Definition 2.3,

$$f(x) = |(2x-6)| = \begin{cases} (2x-6) & \text{if} \quad (2x-6) \geqslant 0 \\ -(2x-6) & \text{if} \quad (2x-6)<0 \end{cases}$$

Since $2x-6 \geqslant 0$ is equivalent to $x \geqslant 3$ and $2x-6<0$ is equivalent to $x<3$, we can write $f(x)$ as

$$f(x) = \begin{cases} 2x-6 & \text{if} \quad x \geqslant 3 \\ 6-2x & \text{if} \quad x<3 \end{cases}$$

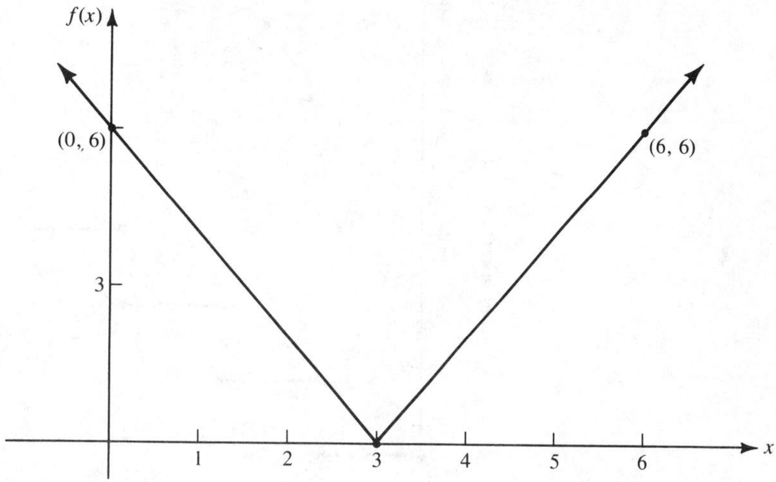

FIGURE 2.48

(*Note*: We have replaced $-(2x-6)$ by $6-2x$.) Graphing the two linear functions $f(x)=2x-6$ and $f(x)=6-2x$ on their respective domains, $[3,\infty)$ and $(-\infty,3)$, we obtain the graph of $f(x)$ as shown in Fig. 2.48.

One other special function of interest is called the *step function*. Consider the cost of mailing any light package, containing written material, weighing up to 6 oz.* The cost of stamps is 15¢ for the first ounce and 13¢ for each additional ounce or fraction thereof. If $C(w)$ is the cost and w is the weight in ounces, we can conclude that

$$C(w) = \begin{cases} 15 & \text{if} \quad 0<w\leqslant 1 \\ 28 & \text{if} \quad 1<w\leqslant 2 \\ 41 & \text{if} \quad 2<w\leqslant 3 \\ 54 & \text{if} \quad 3<w\leqslant 4 \\ 67 & \text{if} \quad 4<w\leqslant 5 \\ 80 & \text{if} \quad 5<w\leqslant 6 \end{cases}$$

Note that the cost is defined by several different rules over the different intervals of the domain. In each case $C(w)$ is a constant. For example, $C(w)=15$ when w is in $(0,1]$, $C(w)=28$ when w is in $(1,2]\ldots$ and finally $C(w)=80$ when w is in $(5,6]$.

The graph of the stamp function $C(w)$ is shown in Fig. 2.49. Note that the open circles are drawn at end points that are not included.

*In 1978 the actual upper limit was 13 oz.

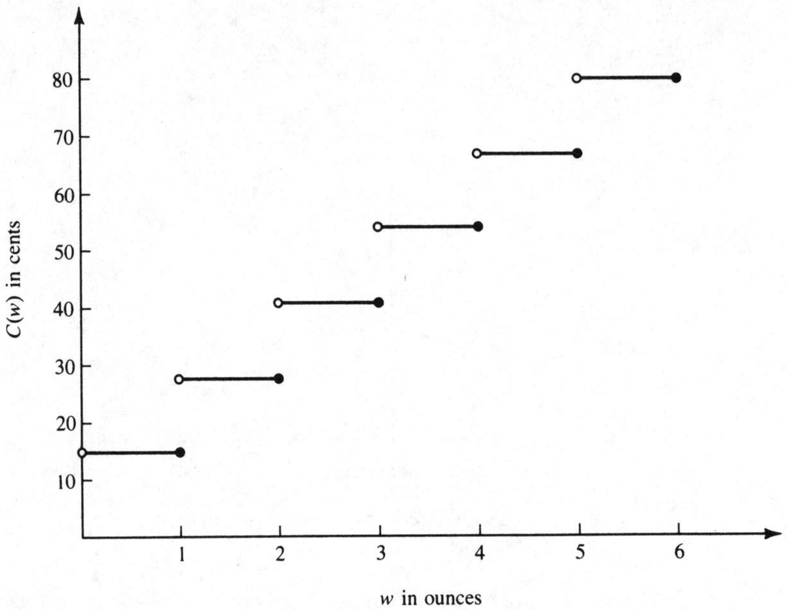

FIGURE 2.49

By inspection of Fig. 2.49 we note that this function "jumps" at each of the values $w=1$, 2, 3, 4, and 5. The domain of $C(w)$ is $(0,6]$ and the range $R=\{15, 28, 41, 54, 67, 80\}$.

EXERCISES

In Exercises 1–12 *sketch the graph of each function and determine its domain and range.*

1. $f(x)=\begin{cases} x+3 & \text{if} \quad x\leq 0 \\ x & \text{if} \quad x>0 \end{cases}$

2. $g(x)=\begin{cases} x & \text{if} \quad x<0 \\ -x & \text{if} \quad x\geq 0 \end{cases}$

3. $h(x)=\begin{cases} \sqrt{x} & \text{if} \quad x\geq 4 \\ -x & \text{if} \quad x<4 \end{cases}$

4. $s(x)=|x-3|$

5. $f(x)=-|x|$

6. $g(x)=|2x+5|$

7. $C(x)=\begin{cases} 0 & \text{if} \quad 0<x\leq 1 \\ 2 & \text{if} \quad 1<x\leq 2 \\ 4 & \text{if} \quad 2<x\leq 3 \\ 6 & \text{if} \quad 3<x\leq 4 \end{cases}$

8. $S(x)=\begin{cases} 10 & \text{if} \quad 0\leq x<5 \\ 20 & \text{if} \quad 5\leq x<10 \\ 30 & \text{if} \quad 10\leq x<20 \end{cases}$

9. $f(x)\begin{cases} x^2-1 & \text{if} \quad x\leq 0 \\ -x^2 & \text{if} \quad x>0 \end{cases}$

10. $g(x)=\dfrac{|x|}{x}$

11. $f(x)=\begin{cases} 4+4x-x^2 & \text{if} \quad x\geq 2 \\ 3x+2 & \text{if} \quad x<2 \end{cases}$

12. $f(x)=|x|+2$

13. **Nurse's Wages** A nurse receives \$6/hr for each hour worked in a regular 8-hr day. If she works more than 8 hr in any one day, she receives time and a half for the next 2 hr and double time for anything over a 10-hr day. Express her daily wage as a function of t, the number of hours she works per day. Assume $0 \leqslant t \leqslant 12$. Graph this function.

14. A company selling notebooks to its retail outlets uses the following rate schedule:
15¢/notebook on orders up to and including 100
5¢ discount on each notebook over 100
2¢ additional discount on each notebook over 500
Determine a function that will give the cost of the purchase of x notebooks and graph this function.

15. Group tennis lessons given by the club pro are available at the following group rates for a 2-hr session.

Size of group	*Rate per person*
2–5	\$5
6–9	\$4
10–14	\$3.50

(a) Express the tennis pro's income, I, from a 2-hr session as a function of n, the number of members in his class.
(b) Graph this function.

16. The charge for a phone call is \$1.00 for the first 3 min and 20¢/min for each additional minute.

(a) Express the charge C as a function of x, the number of minutes of a call.
(b) Graph this function.

17. A machinist earns \$6.00/hr plus incentive pay. His incentive pay is 50¢ per machined part he can turn out per hour over 30.

(a) Express his hourly wage as a function of x, the number of parts he machines in any one hour.
(b) Graph this function.

18. **Computer Rental Fee** The fee for rental of computer time is \$100 for the first hour or any fraction thereof and \$50 for each additional hour or any fraction thereof.

(a) Express the rental fee as a function of x, where x is the number of hours of computer time. Assume that $0 < x \leqslant 10$.
(b) Graph this function.

19. A company's retirement plan provides that each employee at age 65 will receive 2% of his average annual salary for each complete year he has been employed. Determine the retirement benefit for those employees whose average annual earnings are \$20,000 as a function of x, the number of years of employment. Assume that $15 \leqslant x \leqslant 40$.

20. The taxi fare in a certain city is \$2.00 for the first mile or any part thereof and 40¢ per mile thereafter.

(a) Express the fare, F, as a function of x, the number of miles driven.
(b) Graph this function.

IMPORTANT TERMS AND CONCEPTS

Absolute value function
Algebraic function
Asymptote—horizontal
—vertical
Axis of symmetry
Cut method
Demand function
Discriminant
Equilibrium demand
Equilibrium point
Equilibrium price
Equilibrium supply
Extent of a function
Functional form of a linear equation
Functional form of a quadratic equation
Intercept: x-intercept
y-intercept
Limit: $\lim_{x\to\infty} f(x)$, $\lim_{x\to-\infty} f(x)$
$\lim_{x\to a^+} f(x)$, $\lim_{x\to a^-} f(x)$
Linear equation
Linear function
Parabola
Point-slope formula
Polynomial function
Quadratic equation
Quadratic formula
Quadratic function
Rate of change
Rational function
Rational root
Rational roots theorem
Slope
Slope-intercept formula
Standard form of a linear equation
Step function
Supply function
Turning point
Two-point formula
Vertex of a parabola
Zero of a function

SUMMARY OF RULES AND FORMULAS

Standard form of a linear equation

$$ax + by = c$$

Slope of a straight line passing through (x_1, y_1) and (x_2, y_2)

$$m = \frac{y_2 - y_1}{x_2 - x_1}$$

Slope-intercept form of a linear equation

$$y = mx + b$$

Point-slope formula for a linear equation

$$y - y_1 = m(x - x_1)$$

Two-point formula for a linear equation

$$\frac{y-y_1}{x-x_1} = \frac{y_2-y_1}{x_2-x_1}$$

General (or standard) form of a quadratic function

$$f(x) = ax^2 + bx + c$$

Coordinates of the vertex of the parabola, which is the graph of the quadratic function $f(x)=ax^2+bx+c$

$$\left(\frac{-b}{2a}, f\left(\frac{-b}{2a}\right)\right)$$

The quadratic formula

$$x = \frac{-b \pm \sqrt{b^2-4ac}}{2a}$$

Rule 2.1 The rational roots theorem:

If the polynomial equation $a_n x^n + a_{n-1} x^{n-1} + \cdots + a_1 x^1 + a_0 = 0$ (where each a_j is an integer) has any rational roots, they will be of the form p/q, where p is a divisor of a_0 and q is a divisor of a_n.

Rule 2.2

If a rational function $f(x)=p(x)/q(x)$, where $p(x)$ and $q(x)$ are polynomial functions, then there exists a vertical asymptote for each real x value such that $q(x)=0$ and $p(x)\neq 0$.

Rule 2.3

Consider a rational function of the form

$$f(x) = \frac{a_m x^m + a_{m-1} x^{m-1} + \cdots + a_0}{b_n x^n + b_{n-1} x^{n-1} + \cdots + b_0}$$

(a) If $n>m$, $y=0$ is the horizontal asymptote of $f(x)$.
(b) If $n=m$, $y=a_m/b_n$ is the horizontal asymptote of $f(x)$.
(c) If $n<m$, no horizontal asymptote exists.

3 EXPONENTIAL AND LOGARITHMIC FUNCTIONS

3.1 INTRODUCTION

In Chapter 2 we studied the behavior and characteristics of a set of relations we classified as algebraic functions. These functions were considered as mathematical models for several problems from various disciplines, and by studying them we were able to predict the behavior of the variables in our real-life situations. In this chapter we will study two classes of relations that are not algebraic but are of considerable importance as they are used as models for problems in business, biology, social science, and other disciplines. These relations are classified as *transcendental functions*.

Let us consider a classical problem that perhaps you have seen presented as a puzzle or a game.

Example 3.1 The Tower of Hanoi In a temple in India stand three upright posts. At the beginning of time (for our purposes we shall assume this was 10,000 B.C), 64 concentric disks were placed on one of the posts; the largest disk at the bottom, the next largest resting on it, and so on. Priests have been engaged in shifting the disks to other posts, subject to the following laws:

1. Only one disk may be moved at a time.
2. A disk may never be placed over a smaller disk.

When all 64 disks have been moved to one of the other posts the world will come to an end. Assume that the priests have been working in shifts, moving 1 disk per second for 24 hours a day, and have never made a mistake. How much longer will the world exist?

Figure 3.1 shows a physical model of this problem where five disks are considered rather than 64.

Solution Our task is to determine a mathematical model that will predict the number of moves and hence the time required to restack the 64 disks in the same order on another post, while abiding by the rules of the game. This

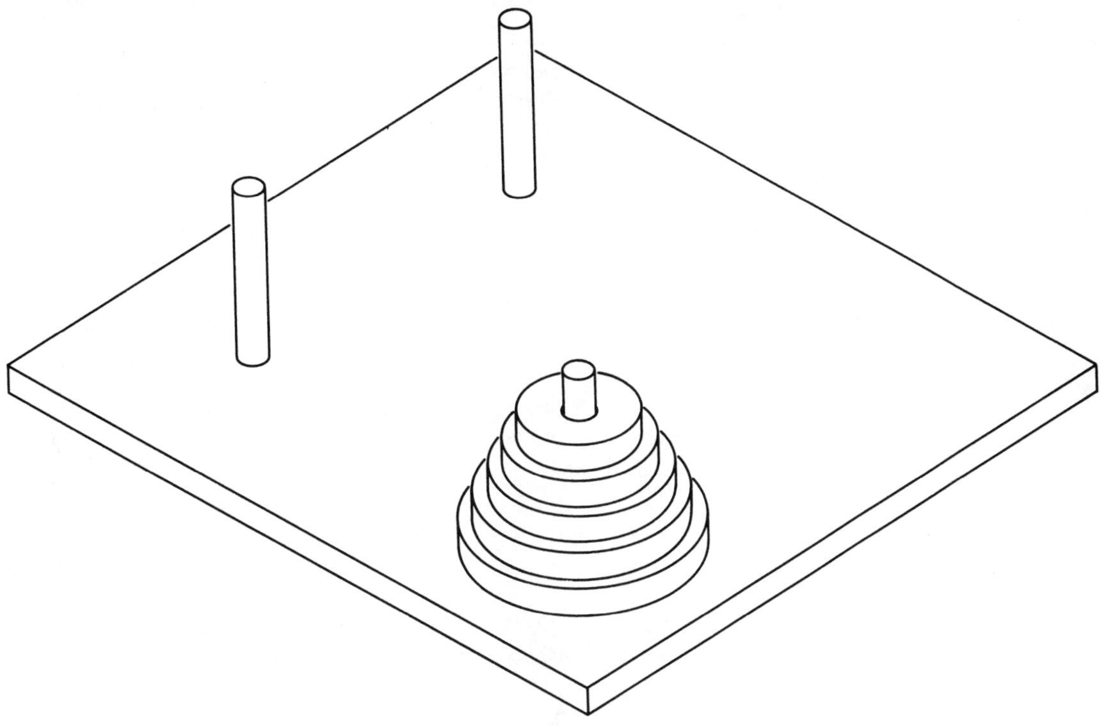

FIGURE 3.1

sounds like a formidable problem and our first reaction might be to give up in despair. However, as is often the case, a solution will become evident by considering a simpler version of the original problem. Instead of 64 disks let us consider how many moves it will take to transfer a fewer number of disks and see if a pattern evolves that will predict the number of moves for 64 disks.

Using a dime, penny, nickel, quarter, and a half dollar for disks you can verify the results shown in Table 3.1. Clearly the number of moves, $M(x)$, is related to the number of disks, x; hence we have a relation. All we need is the right-hand side of the equation, $M(x)=?$, to have a mathematical model for this problem. Can you see a pattern for predicting the formulation for $M(x)$? If you need help, consider the following hint: Rewrite the numbers 1, 3, 7, 15, and 31 as shown in the right-hand column of Table 3.2 and look for a pattern with the left-hand column.

By inductive reasoning we can now conclude that $M(x)=2^x-1$. Therefore, if $x=64$, $M(64)=2^{64}-1$, and the time required to move all 64 disks is $(2^{64}-1)$ sec. Now $2^{64}=(2^8)^8=(256)^8=(2.56\times 10^2)^8=(2.56)^8\times 10^{16}$. Approximating 2.56 by 2 we have

$$\begin{aligned}(2.56)^8\times 10^{16} &\doteq (2)^8\times 10^{16}\\ &\doteq 256\times 10^{16}\text{ sec} = 2.56\times 10^{18}\\ &\doteq 2{,}560{,}000{,}000{,}000{,}000{,}000\text{ sec}\end{aligned}$$

Table 3.1

Number of disks (x)	*Number of moves* $M(x)$
1	1
2	3
3	7
4	15
5	31
⋮	⋮
x	$M(x)=?$

Table 3.2

x	$M(x)$
1	$1=2^1-1$
2	$3=4-1=2^2-1$
3	$7=8-1=2^3-1$
4	$15=16-1=2^4-1$
5	$31=32-1=2^5-1$
⋮	⋮
x	$M(x)=?$

There are $60\times60\times24\times365$ sec in a year, which is approximately 3.2×10^7 sec. Hence the number of years required to move all 64 disks is greater than

$$\frac{2.56\times10^{18}}{3.2\times10^7}\doteq 8\times10^{10}\text{ years!}$$

If you subtract the 12,000 years already used up, the world still has approximately 80,000,000,000 years left to exist.

The function $M(x)$, in the above example can be considered to be the difference of two functions $f(x)-g(x)$, where $f(x)=2^x$ and $g(x)=1$, the constant polynomial function. What kind of a function is $f(x)=2^x$? It does not satisfy our description of an algebraic function as the variable x appears as an exponent and not as a base. Functions that can be written in the form $f(x)=ab^x$, where $b>0$ and $\neq1$, are called *exponential functions*. Before we continue our study of the characteristics of exponential functions, we will pause for a brief review of the laws of exponents, which are fundamental to the understanding of these functions.

3.2 EXPONENTS

In general, mathematical notation is a shorthand notation that saves time and space and simplifies the representation of complex expressions. This is the primary purpose of exponential notation. For example, the term $bbbb$ is denoted by writing b^4, where the number b is called the *base* and 4 is called the *exponent*, or *power*, of b. In general, if n is a positive integer and b is any real number, we write

$$\underbrace{bbb\cdots b}_{n\text{ factors}} = b^n$$

for which the right side is read "b raised to the nth power." By convention, it is understood that the factor b by itself has an exponent of 1; in other words, the absence of an exponent is understood to mean that the exponent is 1. For example,

$$bbb = b^3$$
$$bb = b^2$$
$$b = b^1$$

By extending the above notational concept, we see that

$$b^m b^n = \underbrace{(bb\cdots b)}_{m \text{ factors}}\ \underbrace{(bb\cdots b)}_{n \text{ factors}} = \underbrace{(bb\cdots b)}_{m+n \text{ factors}} = b^{m+n}$$

Thus the product of two factors with the *same* base is simply the "base raised to the sum of the powers." Similarly, if we wish to raise to a power a factor that is already a base raised to a power, we have

$$(b^m)^n = \underbrace{b^m b^m \cdots b^m}_{n \text{ factors}} = b^{mn}$$

Thus when we raise to a power a base that is already raised to a power, we obtain the base raised to the product of the two exponents.

The above two relationships constitute two of the fundamental *laws of exponents*. Simply stated they are

$$b^m b^n = b^{m+n} \tag{3.1}$$

and

$$(b^m)^n = b^{mn} \tag{3.2}$$

In our previous discussion, the exponents m and n were considered as positive integers and the base b was considered as any real number. The laws actually apply for any *real* exponents m and n and for any real number b such that $b \neq 0$. However, when b is negative and/or the exponents are fractional, we run the risk of obtaining roots of negative numbers and/or roots of positive numbers for which there is more than one root. These situations generally require a more extensive understanding of complex numbers than is needed in this text. Therefore, we shall confine our consideration of b to the positive real numbers and our consideration of any factor involving a root to mean the positive real root.

Let us consider some of the special types of exponents, such as *negative* and *fractional* exponents. When a base raised to a negative exponent is encountered, it is merely the reciprocal of the base raised to the absolute value of the exponent; for example,

$$2^{-3} = \frac{1}{2^3}$$

Thus, in general, the meaning of negative exponents is explained by

$$b^{-m} = \frac{1}{b^m} \tag{3.3}$$

From Eq. (3.3), it is apparent that b cannot be zero because division by zero is undefined.

Using (3.1) and (3.3), we note that

$$b^m b^{-m} = b^{m-m} = b^0$$

which means that $b^0 = 1$ since

$$b^m b^{-m} = b^m \frac{1}{b^m} = 1$$

However, this is only true if $b \neq 0$, which is one of the restrictions on b. We can also see that

$$\frac{b^m}{b^n} = b^m b^{-n} = b^{m-n} \tag{3.4}$$

Using the same approach that was employed for obtaining (3.1) and (3.2), we can show that

$$(ab)^m = a^m b^m \tag{3.5}$$

and

$$\left(\frac{a}{b}\right)^m = \frac{a^m}{b^m} \tag{3.6}$$

for any two real numbers a and b, such that $a \neq 0$ and $b \neq 0$.

We have justified the laws of exponents given in Eq. (3.1) through Eq. (3.6) assuming that the exponents m and n are integers. As indicated, these laws also hold if the exponents m and n are any real numbers and the base is a positive real number. For example, applying the law of exponents given in (3.2), we could write

$$(3^{1/2})^2 = 3^1 = 3$$

or

$$(4^{1/3})^3 = 4^1 = 4$$

In both cases, the number given by the base raised to a fractional exponent (that is, $3^{1/2}$ or $4^{1/3}$) represents the number which, when raised to a power equal to the reciprocal of its fractional exponent, is equal to the base. In general, we say that the number represented by $b^{1/m}$ is a number that will yield base b when raised to the mth power. It is also called the mth root of b, and it may be denoted in radical form by $\sqrt[m]{b}$, where m is called the *index* of the radical and b is called the *radicand.*

As noted previously, the reason for stating that the mth root of b represents *a* number rather than *the* number is that b may have more than one real root. For example, for $b=9$ and $m=2$, we see that both 3 and -3 satisfy the above definition of $b^{1/m}$. However, in this text, we shall consider only the positive root of b when dealing with fractional exponents. In some texts this is referred to as the principal root. For example,

$$(4)^{1/2} = \sqrt{4} = 2$$

$$(16)^{1/4} = \sqrt[4]{16} = 2$$

$$(25)^{1/2} = \sqrt{25} = 5$$

If m is a rational number p/q, where p and q are integers and $q \neq 0$, then $b^m = (b)^{p/q}$ may be written in radical form in one of two ways:

$$b^{p/q} = b^{p(1/q)} = (b^p)^{1/q} = \sqrt[q]{b^p} \tag{3.7}$$

$$b^{p/q} = b^{(1/q)p} = (b^{1/q})^p = (\sqrt[q]{b}\,)^p \tag{3.8}$$

In both cases q is the index of the radical; however, in the first case (b^p) is the radicand whereas (b) is the radicand in the second case. For example, by Eq. (3.7)

$$(4)^{3/2} = \sqrt[2]{(4)^3} = \sqrt{64} = 8$$

and by Eq.(3.8)

$$(4)^{3/2} = (\sqrt[2]{4}\,)^3 = (2)^3 = 8$$

Let us consider some examples that utilize the laws of exponents as given in (3.1) to (3.8). Assume the variables represent positive real numbers.

Example 3.2 Write each of the following terms using a single exponent.

(a) $\dfrac{3^3 3^5}{(3^4)^2}$ (b) $\dfrac{5^2 5^3}{(5^4)^{1/2}}$

(c) $\dfrac{6^4 6^{1/2}}{6^{3/2} 6^2}$ (d) $\dfrac{3^2 \cdot 3^{1/2}}{(3^{-5} \cdot 3)}$

Solution

(a) $\dfrac{3^3 3^5}{(3^4)^2} = \dfrac{3^{3+5}}{3^{4 \cdot 2}} = \dfrac{3^8}{3^8} = 3^{8-8} = 3^0 = 1$

(b) $\dfrac{5^2 5^3}{(5^4)^{1/2}} = \dfrac{5^{2+3}}{5^{4\cdot 1/2}} = \dfrac{5^5}{5^2} = 5^{5-2} = 5^3$

(c) $\dfrac{6^4 6^{1/2}}{6^{3/2} 6^2} = \dfrac{6^{4+1/2}}{6^{3/2+2}} = \dfrac{6^{9/2}}{6^{7/2}} = 6^{9/2-7/2} = 6^1 = 6$

(d) $\dfrac{3^2 \cdot 3^{1/2}}{(3^{-5}\cdot 3)} = \dfrac{3^{2+1/2}}{3^{-5+1}} = \dfrac{3^{5/2}}{3^{-4}} = 3^{5/2+4} = 3^{13/2}$

Example 3.3 Using the laws of exponents, simplify the following expressions and express the answers in terms of positive exponents.

(a) $\dfrac{2x^2y(x^{-4})}{3xy^2(y^{-3})}$ (b) $\dfrac{4x^{-2}y^2(x^{1/2}y^{-1})}{2xy(x^{-1/2}y)}$

Solution

(a) $\dfrac{2x^2y(x^{-4})}{3xy^2(y^{-3})} = \dfrac{2x^{-2}y}{3xy^{-1}} = \dfrac{2}{3}x^{-3}y^2 = \dfrac{2y^2}{3x^3}$

(b) $\dfrac{4x^{-2}y^2(x^{1/2}y^{-1})}{2xy(x^{-1/2}y)} = \dfrac{4x^{-3/2}y}{2x^{1/2}y^2} = 2x^{-2}y^{-1} = \dfrac{2}{x^2y}$

Example 3.4 Using the laws of exponents, simplify the following expressions and determine the numerical answers.

(a) $(\frac{2}{3})^4(2\cdot 3)^{-4}\cdot 2^3\cdot 3^6$ (b) $(2\cdot 3)^2(4\cdot 5)^{-2}(\frac{5}{3})^3$

Solution

(a) $(\frac{2}{3})^4(2\cdot 3)^{-4}\cdot 2^3\cdot 3^6 = 2^4 3^{-4} 2^{-4} 3^{-4} 2^3 3^6$
$= 2^3\cdot 3^{-2}$
$= \frac{8}{9}$

(b) $(2\cdot 3)^2(4\cdot 5)^{-2}(\frac{5}{3})^3 = 2^2\cdot 3^2\cdot (2^2)^{-2}\cdot 5^{-2}\cdot 5^3\cdot 3^{-3}$
$= 2^{-2}3^{-1}5$
$= \frac{5}{12}$

Example 3.5 Evaluate the following expressions.

(a) $(128)^{3/7}$ (b) $(243)^{3/5}$ (c) $(125)^{4/3}$ (d) $(64)^{2/3}$

Solution

(a) $(128)^{3/7} = (2^7)^{3/7} = 2^3 = 8$ (b) $(243)^{3/5} = (3^5)^{3/5} = 3^3 = 27$
(c) $(125)^{4/3} = (5^3)^{4/3} = 5^4 = 625$ (d) $(64)^{2/3} = (2^6)^{2/3} = 2^4 = 16$

Example 3.6 Rewrite each of the following expressions in radical form.

(a) $x^{2/3}$ (b) $(x^2+5)^{3/2}$ (c) $(x+3)^{-2/3}$

Solution

(a) $x^{2/3} = \sqrt[3]{x^2}$ or $(\sqrt[3]{x}\,)^2$

(b) $(x^2+5)^{3/2}=\sqrt{(x^2+5)^3}$ or $\left(\sqrt{(x^2+5)}\,\right)^3$

(c) $(x+3)^{-2/3}=\dfrac{1}{(x+3)^{2/3}}=\dfrac{1}{\sqrt[3]{(x+3)^2}}$ or $\dfrac{1}{\left(\sqrt[3]{(x+3)}\,\right)^2}$

Example 3.7 Rewrite each radical expression in exponential notation.

(a) $\sqrt[3]{x^5}$ (b) $\left(\sqrt{(x+2)}\,\right)^3$ (c) $\dfrac{1}{\sqrt[3]{(x^2+3)^2}}$

Solution

(a) $\sqrt[3]{x^5}=x^{5/3}$

(b) $\left(\sqrt{(x+2)}\,\right)^3=(x+2)^{3/2}$

(c) $\dfrac{1}{\sqrt[3]{(x^2+3)^2}}=\dfrac{1}{(x^2+3)^{2/3}}=(x^2+3)^{-2/3}$

Example 3.8 If $f(x)=2^x$ and $g(x)=3^x$, simplify each of the following functions.

(a) $f(x)\cdot f(x)$ (b) $f(x)\cdot g(x)$ (c) $\dfrac{f(x)}{g(x)}$ (d) $f(g(x))$

Evaluate

(e) $f(-1)$ (f) $g(2)$ (g) $f(g(1))$ (h) $g(f(1))$

Solution

(a) $f(x)\cdot f(x)=2^x\cdot 2^x=2^{2x}=(2^2)^x=4^x$
(b) $f(x)\cdot g(x)=2^x\cdot 3^x=(2\cdot 3)^x=6^x$
(c) $\dfrac{f(x)}{g(x)}=\dfrac{2^x}{3^x}=\left(\dfrac{2}{3}\right)^x$
(d) $f(g(x))=f(3^x)=2^{(3^x)}$
(e) $f(-1)=2^{-1}=\frac{1}{2}$
(f) $g(2)=3^2=9$
(g) Since $g(1)=3$, $f(g(1))=f(3)=2^3=8$, or using part (d) directly, $f(g(1))=2^{(3^1)}=2^3=8$
(h) Since $f(1)=2$, $g(f(1))=g(2)=3^2=9$

EXERCISES

In the following exercises assume that all variables represent positive real numbers.

1. Use the laws of exponents and evaluate each of the following expressions.

(a) $\dfrac{2^3 3^2}{(3^6)^{1/2}}$ (b) $\dfrac{3^4 2^{-2}}{\left(\frac{3}{2}\right)^3}$

(c) $\dfrac{\left(\frac{4}{3}\right)^2\left(\frac{3}{2}\right)^3}{\left(\frac{2}{3}\right)^2}$ (d) $\dfrac{\left(\frac{3}{8}\right)^2\left(\frac{2}{9}\right)^3}{(8\cdot 9)^{-2}}$

2. Use the laws of exponents and simplify each of the following expressions.

(a) $\dfrac{x^2y^3}{xy^2}$ (b) $\dfrac{(3xy^2)^2}{(x^2y^2)^2}$

(c) $\dfrac{(x^2)^3(y^2)^4}{(x^3)^3(y^3)^2}$ (d) $\dfrac{x^2y^3}{(y/x)^3(x^3)^2}$

3. Simplify each of the following expressions and express the answers in terms of positive exponents only.

(a) $\dfrac{3x^2y^3z}{x^3yz^2}$ (b) $\dfrac{(x^2y)^3(x/y^2)^4}{(x^3/y^2)^2}$

(c) $\dfrac{(2x+1)^3(y+3)^2}{(2x+1)^2(y+3)^{-4}}$ (d) $\dfrac{(x^3yw^2)^2(xy^2w)^3}{(x^2y/w)^3}$

4. Evaluate each of the following expressions.

(a) $2^3(1+2^{-4})$ (b) $\dfrac{3+5^{-2}}{2^2}$

(c) $\dfrac{2^{-3}+2^{-2}}{2^{-1}}$ (d) $2^{-2}(1+3^{-2})$

5. Simplify each of the following expressions.

(a) $\dfrac{(x^{1/2})^3(y^{2/3})^2}{(x^{3/8})^4(y^{2/3})^3}$ (b) $\dfrac{(x^{3/4})^2(y^{1/2})^3}{(x^{1/2}y^{1/3})^6}$

(c) $\dfrac{(x^2y)^{1/4}(y^3)^{1/2}}{(xy^2)^{1/3}(x^2)^{1/3}}$ (d) $\dfrac{(xy^{1/2})^{1/2}(x^{1/4}y^{1/2})^2}{(xy^2)^{1/4}}$

6. Evaluate each of the following expressions.

(a) $9^{3/2}$ (b) $(256)^{3/4}$

(c) $(16)^{-5/4}$ (d) $\left(\dfrac{9\cdot 16}{25}\right)^{3/2}$

7. Rewrite each radical expression in exponential form.

(a) $\sqrt{x^3}$ (b) $\sqrt[3]{(x^2+5x+3)}$

(c) $\dfrac{1}{\sqrt{(x+2)^3}}$ (d) $\dfrac{1}{\left(\sqrt[3]{x+5}\right)^2}$

8. Rewrite each of the following expressions in radical form.

(a) $x^{3/2}$ (b) $(x^2+3x+9)^{2/3}$
(c) $(x^2+9)^{-2/3}$ (d) $(x+3)^{-3/2}$

9. If $f(x)=3^x$ and $g(x)=2^x$, simplify each of the following functions.

(a) $f(x)\cdot f(x)$ (b) $f(x)\cdot g(x)$ (c) $\dfrac{f(x)}{g(x)}$ (d) $f(g(x))$

Evaluate

(e) $f(-2)$ (f) $g(0)$ (g) $f(g(2))$ (h) $g(f(2))$

10. Resource Allocation Model in Presidential Campaigning Brams and Davis* argue that in allocating resources to a state under the electoral college system, candidates should allocate campaign funds not in proportion to the size of the states, but rather in proportion to the 3/2 power of the size of their electoral votes. Thus if E_1 and E_2 are the electoral votes of two states, the ratio of expenditures in state 1 to state 2 should be

$$\frac{E_1^{3/2}}{E_2^{3/2}}$$

Ohio, Washington, and Maine have 25, 9, and 4 electoral votes, respectively. Assuming Brams and Davis to be correct,

(a) How much more money should be spent in Ohio than in Washington, and in Washington than in Maine?
(b) If a candidate has \$800,000 to be allocated to the above three states, how much should be allocated to each state?

3.3 EXPONENTIAL FUNCTIONS

At the end of Sec. 3.1 we concluded that any function of the form

$$f(x) = ab^x \tag{3.9}$$

where a and b are nonzero real numbers such that b is positive and not equal to 1, is called an exponential function because the independent variable x appears as an exponent in the relation. Exponential functions are important in economics and business research because they can be used to describe growth and decline relationships. In psychology they are sometimes used to represent learning curves, and in biology bacterial growth and decay can be described by these functions.

Let us investigate the behavior and characteristics of these exponential functions by considering as an example the first term of the mathematical model for the Tower of Hanoi puzzle, $f(x)=2^x$. You will recall that the variable x represents the number of disks; hence the domain for this function

*Steven J. Brams and Morton D. Davis, "Resource Allocation Models in Presidential Campaigning: Implications for Democratic Representation," *Annals of the New York Academy of Sciences*, Vol. 219, pp. 105–123, November, 1973.

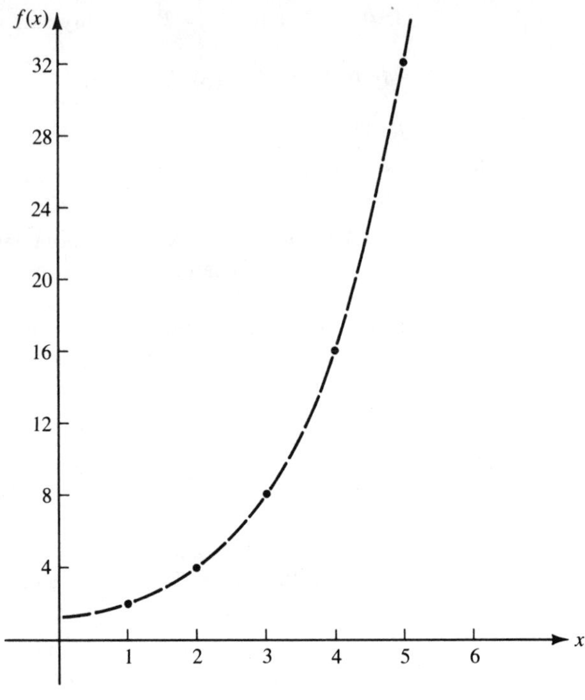

FIGURE 3.2

Table 3.3

x	$f(x)=2^x$
1	2
2	4
3	8
4	16
5	32
6	64
⋮	⋮

is the set of positive integers. A graph of $f(x)$ appears in Fig. 3.2. An abbreviated list of ordered pairs belonging to the function is given in Table 3.3.

Let us remove the restriction on the domain and ask ourselves for what real values of x will the corresponding second component, $f(x)$, be a real number. Clearly x can be replaced by any integer as shown in the following table.

x	...	-4	-3	-2	-1	0	1	2	3	4	...
$f(x)$	...	$\frac{1}{16}$	$\frac{1}{8}$	$\frac{1}{4}$	$\frac{1}{2}$	1	2	4	8	16	...

Now if x is a rational number of the form p/q, $q>0$, is $f(x)$ defined? From our previous discussion of exponents, $2^{p/q}$ means the qth root of (2^p), which is well-defined for all integers p and q, $q>0$. The domain, in fact, can be expanded to include all the real numbers including the irrationals. Since in computation we approximate irrational numbers by rationals, it is possible to

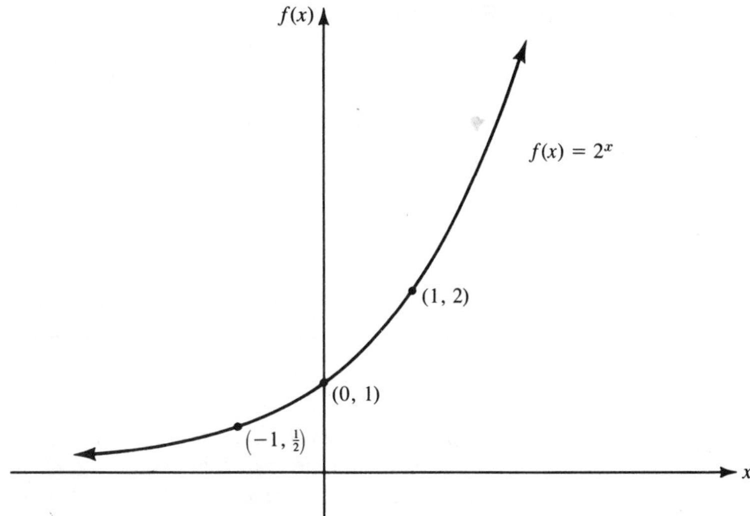

FIGURE 3.3

approximate an irrational exponent with a rational exponent. For example, $2^{\sqrt{2}}$ will be approximated by $2^{1.414}=2^{1414/1000}$ or $\sqrt[1000]{2^{1414}}$.

The graph of $f(x)=2^x$ for x being any real number is shown as a smooth curve in Fig. 3.3. The graph of any exponential function of the form $f(x)=b^x$, where $b>1$, is similar in shape and passes through the point (0, 1) since $b^0=1$. In all cases where $b>1$, $f(x)=b^x$ is such that as a point on its graph moves from left to right, the value of $f(x)$ increases as can be verified by inspecting Fig. 3.3 where $b=2$. Since this is true on the entire interval $(-\infty,\infty)$, we say that $f(x)$ is an increasing function on $(-\infty,\infty)$.

In Fig. 3.3 we note that the graph of the exponential function to the left of $x=0$ approaches the x axis as x decreases without bound, that is, as $x\to-\infty$. However, it never touches the x axis. As we discussed in Chapter 2, the x axis, or the line $y=0$, is a horizontal asymptote of the curve, and the function is said to asymptotically approach this line. For any exponential function of the form $f(x)=ab^x$, the x axis will always be a horizontal asymptote of its graph.

Example 3.9 Sketch the graphs of the functions $f(x)=3\cdot2^x$ and $g(x)=-3\cdot2^x$.

Solution To sketch the graphs of these functions we only need four or five ordered pairs that belong to each function, and knowing the general shape of each curve we can draw it. It is usually a good idea to select from the domain the integers 0, ±1, ±2 for abscissas, and calculate their respective ordinates.

For example,

$$f(x)=3\cdot 2^x$$

x	...	-2	-1	0	1	2
$f(x)$	...	$\frac{3}{4}$	$\frac{3}{2}$	3	6	12

$$g(x)=-3\cdot 2^x$$

x	...	-2	-1	0	1	2
$g(x)$	...	$-\frac{3}{4}$	$-\frac{3}{2}$	-3	-6	-12

Graphing these ordered pairs and connecting these points with a smooth curve, we obtain the graphs of $f(x)$ and $g(x)$ as shown in Fig. 3.4. Note that $f(x)$ and $g(x)$ are symmetrical to each other with respect to the x axis. The function $f(x)$ is increasing on $(-\infty, \infty)$. Since $g(x)$ decreases as a point on the graph of $g(x)=-3\cdot 2^x$ moves from left to right, we say $g(x)$ is a decreasing function on $(-\infty, \infty)$.

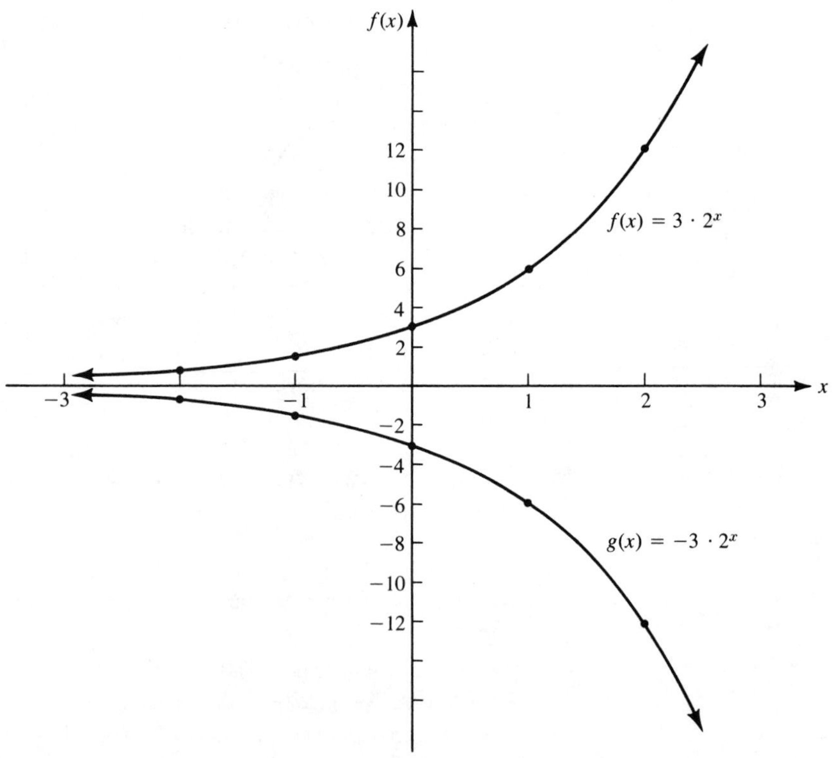

FIGURE 3.4

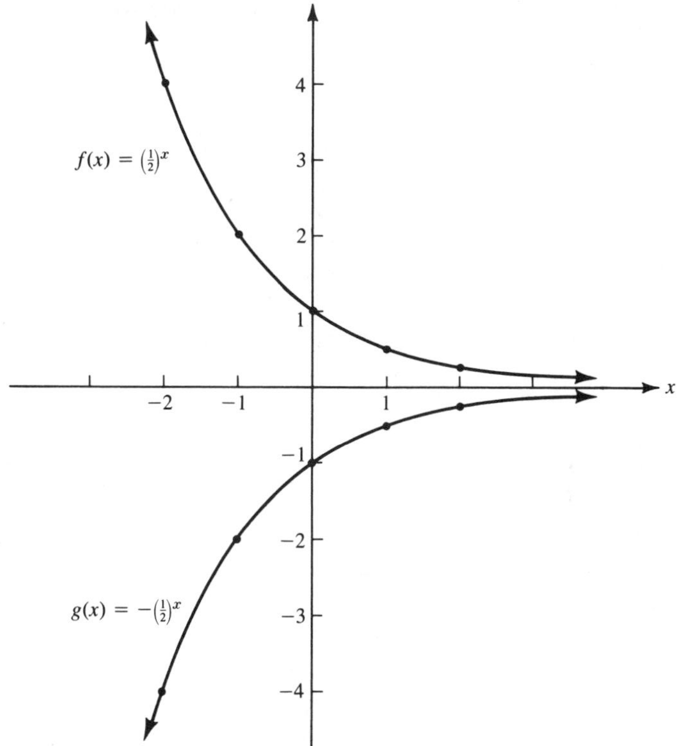

FIGURE 3.5

Example 3.10 Sketch the graph of the functions $f(x)=(\frac{1}{2})^x$ and $g(x)=-(\frac{1}{2})^x$.

Solution As in the last example, we select five ordered pairs to establish the curve, and knowing the general shape of the graph of an exponential function we can draw a sketch (Fig. 3.5).

$f(x)=(\frac{1}{2})^x$

x	-2	-1	0	1	2
$f(x)$	4	2	1	$\frac{1}{2}$	$\frac{1}{4}$

$g(x)=-(\frac{1}{2})^x$

x	-2	-1	0	1	2
$g(x)$	-4	-2	-1	$-\frac{1}{2}$	$-\frac{1}{4}$

Note that $f(x)$ and $g(x)$ are symmetrical with respect to the x axis and that $f(x)$ is decreasing on $(-\infty,\infty)$, whereas $g(x)$ is increasing on $(-\infty,\infty)$.

Based on these examples we have the general situations depicted in Fig. 3.6. Note that we have restricted b to be greater than zero and not equal to 1. If $b<0$, the relation is not defined for all reals; for example, $(-4)^{1/2}$ is not a real number. If $b=1, f(x)=ab^x=a$, which is a linear function.

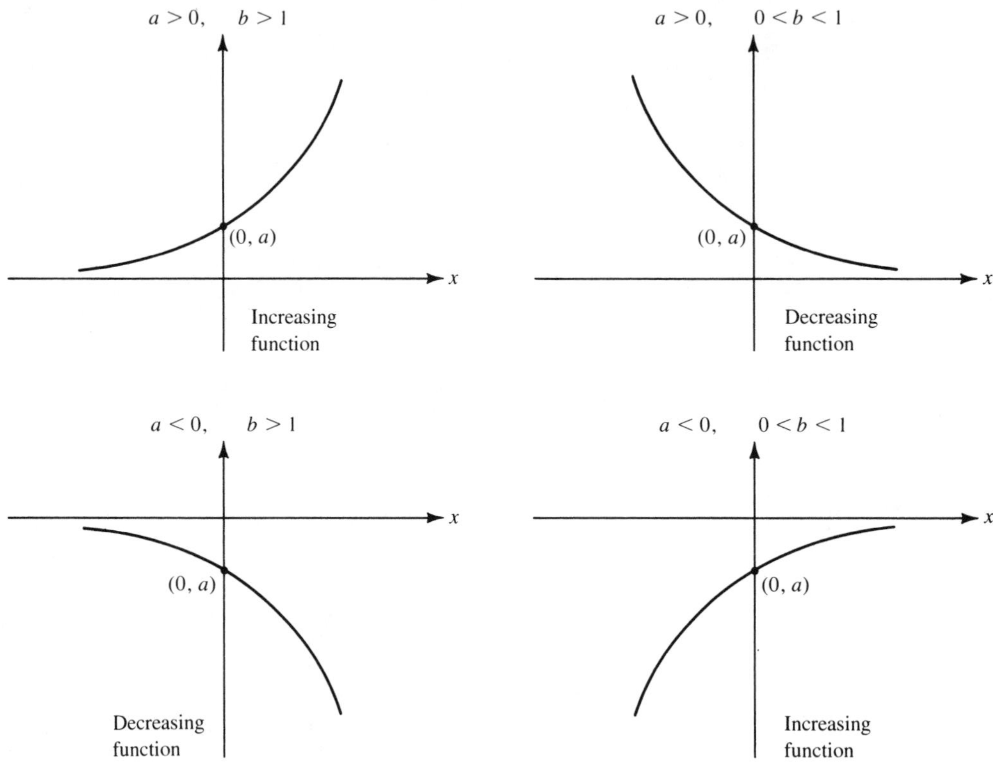

FIGURE 3.6

Another characteristic of the exponential function $f(x)$ is that the values of $f(x)$ change by a constant percentage when the independent variable x changes by a constant amount. For example, consider the function $f(x) = 3 \cdot 2^x$, and the following data.

x	1	2	3	4
$f(x)$	6	12	24	48

When x increases by 1 from 1 to 2, 2 to 3, 3 to 4, etc., the values of $f(x)$ increase from 6 to 12, 12 to 24, 24 to 48, etc. The $f(x)$ values double, or the percent change in $f(x)$ for a unit change in x is a constant 100% regardless of which two successive values of x we choose. In making this statement we have used the definition of *percent change* given by

$$\% \text{ change} = \left| \left(\frac{\text{new value} - \text{old value}}{\text{old value}} \right) (100) \right| \%$$

For example, the percent change in $f(x)$ when x changes from 1 to 2 would be

$$\begin{aligned}\%\text{ change} &= \left|\left(\frac{f(2)-f(1)}{f(1)}\right)(100)\right|\% \\ &= \left|\left(\frac{12-6}{6}\right)100\right|\% = 100\%\end{aligned}$$

Similarly, you can verify that the percent change in $f(x)$ when x changes from 2 to 3, or 3 to 4 will be 100%.

Given a general exponential function $f(x)=ab^x$, the following property can be used to calculate the percent change in $f(x)$ for a unit change in x.

Property 3.1 The percent change in the function $f(x)=ab^x$ will equal $|(b-1)100|\%$ for a unit change in x.

The justification for this statement is based on our previous numerical example. Instead of letting x change from 1 to 2, 2 to 3, etc., we assume a general case and let x change from x_0 to (x_0+1) and calculate

$$\%\text{ change} = \left|\left(\frac{f(x_0+1)-f(x_0)}{f(x_0)}\right)100\right|\%$$

Substituting $f(x_0+1)=ab^{x_0+1}$ and $f(x_0)=ab^{x_0}$ into this equation we have

$$\begin{aligned}\%\text{ change} &= \left|\left(\frac{ab^{x_0+1}-ab^{x_0}}{ab^{x_0}}\right)100\right|\% \\ &= \left|\left(\frac{ab^{x_0}(b^1-1)}{ab^{x_0}}\right)100\right|\% \\ &= |(b-1)100|\%\end{aligned}$$

The percent change will be a percent increase if the function is increasing, and a percent decrease if the function is decreasing. The magnitude of the percent change is given by $|(b-1)100|\%$. We must keep in mind the principles that were used to develop the four sketches of Fig. 3.6 to determine whether the percent change is an increase or a decrease. To better understand this characteristic of exponential functions, let us consider some examples.

Example 3.11 Determine the percent change in the exponential function $f(x)=-2\cdot 3^x$ obtained for a one-unit increase in x, and state whether this change is an increase or decrease.

Solution First determine the percent change by using Property 3.1. We note $b=3$, therefore

$$|(b-1)100|\% = 200\%.$$

Now, since 3^x is an increasing function, $(-2)3^x$ is the opposite, hence the 200% change is a 200% decrease.

Example 3.12 Determine the percent change in the exponential function $g(x)=(\frac{1}{2})^x$ obtained for a unit change in x, and state whether this is a percent increase or decrease.

Solution Since $b=\frac{1}{2}$, the percent change by Property 3.1 is

$$\left|\left(\tfrac{1}{2}-1\right)100\right|\% = 50\%$$

To determine if this is a percent increase or decrease, we consider the function $(\frac{1}{2})^x$ and note that as x increases, $(\frac{1}{2})^x$ decreases. Thus the function is decreasing and our change is a 50% decrease.

Example 3.13 Exponential Growth of Sales Consider the exponential function that arises in determining the size of the market S, n years from now, for a product that is currently doing a sales volume of P dollars per year and growing at a rate of r per year. The formula for determining S is

$$S = P(1+r)^n$$

This formula fits the general exponential function form given in (3.9), where $a=P$, $b=1+r$, and $x=n$. Although n represents the number of years for which the sales are permitted to grow, it can actually assume any positive real value. It could also assume negative real values, but this gives a different meaning to the problem.

For the above exponential function with $P=100$ and $r=0.05$, sketch the graph of the function for nonnegative values of n and determine the constant percent change for a unit change in n.

Solution Substituting the appropriate values into the function, we obtain the exponential function

$$S = 100(1.05)^n$$

Computing and plotting the following values, we obtain the graph given in Fig. 3.7.

n	0	1	2	3
S	100	105	110.25	115.76

The graph does not appear to differ much from a straight line because b is *close* to 1. The constant percent change from one year to the next (i.e., a one-unit increase in n) is $|(b-1)100|\%=(1.05-1)100\%=5\%$.

Example 3.14 Inflation and Loss of Purchasing Power Determine the purchasing power of \$1 in 10 years if it decreases due to inflation at a constant rate of 8%/year.

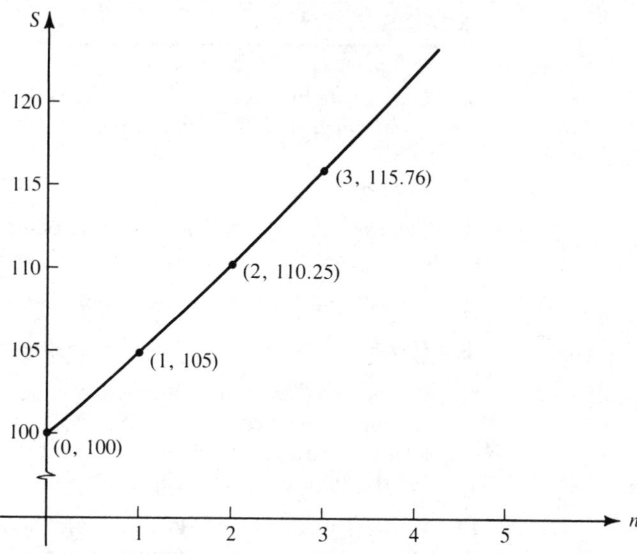

FIGURE 3.7

Solution The function $P(x)$ describing this situation is analogous to the one given in Example 3.13, and is given by

$$P(x) = A(1+r)^x$$

where

$$\begin{aligned} P(x) &= \text{purchasing power in } x \text{ years} \\ x &= \text{number of years} \\ r &= \text{rate of change} \\ A &= \text{initial dollar value} \end{aligned}$$

For our relation $A = \$1.00$, $x = 10$ years, and r, the rate of change, $= -8\%$ since the dollar is decreasing in value each year. Solving for $P(10)$

$$\begin{aligned} P(10) &= 1(1-.08)^{10} \\ &= (.92)^{10} \\ &= 43¢ \text{ (rounded to the nearest cent)} \end{aligned}$$

We note that an 8% decrease in purchasing power is more significant than most people realize.

In the next section we will verify that the functions used in Examples 3.13 and 3.14 are the correct mathematical models for the given situations.

EXERCISES

Sketch the graph and determine the percent change per unit change in x for each of the exponential functions in Exercises 1–10.

1. $f(x)=4\cdot 3^x$ **2.** $f(x)=2\cdot 10^x$ **3.** $f(x)=-2\cdot 5^x$

4. The exponential function $S(x)$ that arises in determining the accumulated value of an investment of \$50 compounded annually at an interest rate of 30%

5. $f(x)=3\cdot(6.82)^x$ **6.** $f(x)=-(11.64)^x$ **7.** $f(x)=4\cdot 12^x$

8. $f(x)=-3\cdot(6.5)^x$ **9.** $f(x)=6\cdot(.64)^x$ **10.** $f(x)=-2\cdot(.8)^x$

11. Awareness Testing and Response Time In a certain awareness test, a subject's response time increases by 40% of his previous score for each additional unit of an experimental stimulant. Determine the exponential function that represents the response time of a subject whose initial response time is 6 sec; 3 sec; 10 sec.

12. Errors and Learning Trials A psychologist has determined that for a complex card-sorting task, the number of errors (E) made on any learning trial (t) is related to the number of categories (C) into which the cards are sorted and the difficulty rating (d) of the arm movements required. If the relationship between these factors is summarized by the function $E=C\cdot d^t$, and for this task there are 100 categories with a difficulty rating of 0.9, determine the following.

(a) The predicted value of the dependent variable when the value of the independent variable is 5.
(b) Is the predicted reduction in the number of errors from trial 4 to trial 5 as great as it is from trial 1 to trial 2?
(c) What is the percentage of reduction in errors from trial 4 to trial 5? From trial 1 to trial 2?
(d) Sketch a partial graph of the predicted effects of practice on errors. Describe the expected effects of continued practice.

13. Bacterial Decay At 30°C, bacteria were subjected to a 5% solution of phenol disinfectant. The number of bacteria decreases at a constant rate of 10%/hr. Assuming you can use a relation similar to the ones given in Examples 3.13 and 3.14, determine how many bacteria are left out of an initial 1,000 after 24 hr.

14. At a fixed temperature the number of bacteria in milk doubles every 3hr. If we initially have N_0 bacteria in a given quantity of milk, after t hr there will be $N=N_0 2^{t/3}$ bacteria in the milk. Draw a graph of this exponential function by graphing N/N_0 on the vertical axis.

15. Penetration of Light through Water For relatively clear bodies of fresh water, light intensity is recorded according to the exponential function

$$I = I_0 3^{-kd}$$

where I is the intensity at d feet below the surface, and I_0 is the intensity at the surface; k is called the ccefficient of extinction. If $k=0.05$, find the depth where the light intensity is one-third that at the surface.

3.4 EXPONENTIAL GROWTH AND DECAY

In the last section we learned that exponential functions have the characteristic that the percent change of the dependent variable is constant for a unit change of the independent variable. In this section, we will justify the statement that if a unit change in one variable produces a constant percent change in another variable, the relation between these variables is an exponential function. Many real-life situations follow this pattern. For example, money placed in a savings institution increases at a fixed percentage for a given unit of time; population growth or decline may be at a fixed percentage for a given unit of time; and even pollutants in a stream are known to decrease at a constant percentage in each mile of water downstream from the source. If the exponential function describing the real-world situation is increasing, this increase is known as *exponential growth*; and if the exponential function is decreasing, this decrease is known as *exponential decay*.

Let us consider the following problem.

Example 3.15 Double Your $ at a Constant Rate How many years will it take to double an initial investment of $10,000 if it grows at a constant rate of 8% per year?

Solution The conditions of this problem fit the pattern discussed above; that is, the rate of change is a constant 8% per unit change of time. A function may exist to make our work easier, but first let us calculate how much money we will have at the end of each year for several years and record the amounts systematically (see Table 3.4). We can continue this process, and once the figure in the fourth column equals or exceeds $20,000 we will have found our solution. This procedure is tedious work. Fortunately there is a pattern here that will give us a relation between A, the amount of money we have at the end of each year, and t, the number of years we have been accumulating interest. This pattern will be evident if we reconstruct Table 3.4 using P dollars as our initial investment (see Table 3.5).

Based on our calculations we see that after t years the amount of our investment will be

$$A(t) = P(1+8\%)^t$$

Table 3.4

Year	*Amount at Beginning of the Year*	*Interest Earned*	*Amount at End of the Year*
1	$10,000	8%(10,000) = $800	$10,800
2	$10,800	8%(10,800) = 864	$11,664
3	$11,664	8%(11,664) = 933	$12,597
4	$12,597	8%(12,597) = 1,008	$13,605
5	$13,605	8%(13,605) = 1,088	$14,693

Table 3.5

Year	*Amount at Beginning of the Year*	*Interest Earned*	*Amount at End of the Year*
1	P	$8\%P$	$P+8\%P=P(1+8\%)$
2	$P(1+8\%)$	$8\%P(1+8\%)$	$P(1+8\%)+8\%P(1+8\%)$ $=P(1+8\%)(1+8\%)$ $=P(1+8\%)^2$
3	$P(1+8\%)^2$	$8\%P(1+8\%)^2$	$P(1+8\%)^2+8\%P(1+8\%)^2$ $=P(1+8\%)^2(1+8\%)$ $=P(1+8\%)^3$
⋮			⋮
t			$A(t)=$?

Therefore, if our initial investment is \$10,000, after t years we will have an amount $A(t)$ given by

$$\begin{aligned} A(t) &= 10{,}000(1+8\%)^t \\ &= 10{,}000(1.08)^t \end{aligned}$$

We see that this relation is an exponential function where the constant percent change per year is $|(b-1)100|\%=8\%$. To solve our problem of doubling our money, we need only find the solution of

$$20{,}000 = 10{,}000(1.08)^t$$

or

$$(1.08)^t = 2$$

With the use of a hand-held calculator, we find that $(1.08)^9=1.999$ and $(1.08)^{10}=2.159$, hence it takes 10 years to double your investment.

Based on the example above, we can conclude that if P dollars is invested at a rate of r percent per year and left to accumulate for t years the dollar value of the investment is given by

$$A(t) = P(1+r)^t \tag{3.10}$$

In business applications formula (3.10) is the *compound interest formula*, and $A(t)$ is the maturity value. Table A.4 in the Appendix presents values of the term $(1+r)^n$ for values of n from 1 to 50 for each value of r from 0.01 to 0.08. The use of Table A.4 is not absolutely necessary with today's inexpensive hand-held calculators, but it is convenient and presents many data for easy comparison.

Frequently interest is compounded semiannually, quarterly, or for other periods requiring a modification of Eq. (3.10). If interest is compounded n times per year, the interest rate per period is given by r/n where r is called the *nominal yearly interest rate* and the number of periods it is compounded in t years is nt, hence Eq. (3.10) is modified for this situation as follows:

$$A(t) = P\left(1+\frac{r}{n}\right)^{nt} \tag{3.11}$$

Example 3.16 Compound Interest Mr. Joe Friendly, a local insurance salesman, deposited \$6,000 with a savings and loan firm. The money is to accrue interest at the rate of 6%/year compounded semiannually. Determine the compound maturity value at the end of $2\frac{1}{2}$ years if no further deposits or withdrawals are made.

Solution Since the interest is compounded semiannually, $n=2$, $r=.06$, and $t=2.5$. Substituting these values into Eq. (3.11) and locating the appropriate value from Table A.4, we obtain the compound maturity value

$$\begin{aligned} A(2.5) &= 6{,}000\left(1+\frac{.06}{2}\right)^{2(2.5)} \\ &= 6{,}000(1.03)^5 \\ &\doteq 6{,}000(1.1592) \\ &\doteq 6{,}955 \end{aligned}$$

Therefore the maturity value is \$6,955.

For many years the greatest number of interest periods per year were offered by institutions that paid daily interest. (Most banks used 360 days per year to facilitate calculation.) Today a savings and loan advertisement might look like this:

GOLDEN NUGGET SAVINGS & LOAN ASSOCIATION Highest Rate of Return $7\frac{3}{4}$% Nominal Interest *Compounded Continuously*

What does the term "compounded continuously" mean? Is interest compounded every hour, every minute, or every second? *Continuous compounding* assumes that interest is computed more often than every second, more often than every tenth of a second, and even more often than every hundredth of a second, etc. This implies that n, the number of interest periods per year, approaches infinity or is not bounded above. To determine the maturity value $A(t)$ we must consider

$$P\left(1+\frac{r}{n}\right)^{nt} \text{ as } n \to \infty$$

or in terms of a limit

$$A(t) = \lim_{n\to\infty} P\left(1+\frac{r}{n}\right)^{nt} \tag{3.12}$$

To evaluate this limit we make the following substitution. Let

$$\frac{1}{h} = \frac{r}{n} \qquad \text{or} \qquad h = \frac{n}{r}$$

This implies that as $n\to\infty$, $h\to\infty$, and substituting $1/h=r/n$ and $n=hr$ into Eq. (3.12), we have

$$A(t) = \lim_{h\to\infty} P\left(1+\frac{1}{h}\right)^{hrt}$$

Since P, r, and t are not dependent on h, we can write

$$A(t) = P\left[\lim_{h\to\infty}\left(1+\frac{1}{h}\right)^{h}\right]^{rt} \tag{3.13}$$

Values of $[1+(1/h)]^h$ for various values of h are shown in the following table.

h	1	10	20	30	40	50	60	100
$(1+\frac{1}{h})^h$	2	2.59	2.65	2.67	2.685	2.691	2.696	2.705

From these data it appears that as h gets larger and larger, $[1+(1/h)]^h$ approaches a constant value approximately equal to 2.71. It can be shown that $\lim_{h\to\infty} [1+(1/h)]^h$ to ten decimal places is equal to 2.7182818285. This limit is an irrational number and is of such significance in mathematics that it is assigned a special letter, e. Summarizing,

$$\lim_{h\to\infty}\left(1+\frac{1}{h}\right)^{h} = e$$

$e = 2.7182818285$ to 10 decimal places

e is an irrational number

By substituting $\lim_{h\to\infty}[1+(1/h)]^h = e$ we can rewrite Eq. (3.13) as

$$A(t) = Pe^{rt}$$
$$\doteq P(2.718)^{rt} \tag{3.14}$$

Eq. (3.14) is the mathematical model we can use to calculate the maturity value $A(t)$ when interest is *compounded continuously* for t years at a nominal annual interest rate of r on an initial investment of P dollars. Table A.3 in the Appendix presents the values of e^x for various values of x ranging from 0 to 5.0.

Example 3.17 The $ Conscious Professor A college professor invests $10,000 in her school's credit union, which is currently paying 8% nominal annual interest compounded continuously. Determine the maturity value of her investment in 10 years and compare this investment to one where the rate of return was 8% compounded annually.

Solution Substituting the values $P=10{,}000$, $r=.08$, and $t=10$ into Eq. (3.14), we can calculate the maturity value $A(10)$ by using Table A.3.

$$\begin{aligned} A(10) &= 10{,}000e^{(.08)(10)} \\ &= 10{,}000e^{.8} \\ &= (10{,}000)(2.2255) \\ &= \$22{,}255 \end{aligned}$$

In comparison, if the investment were made at 8%/year compounded annually we make the appropriate substitutions into Eq. (3.10), and find, using Table A.4,

$$\begin{aligned} A(10) &= 10{,}000(1+.08)^{10} \\ &= 10{,}000(1.08)^{10} \\ &= 10{,}000(2.1589) \\ &= \$21{,}589 \end{aligned}$$

The college professor will realize a return of $666 more if she invests her $10,000 at the continuously compounded interest rate of 8%.

Example 3.18 Effective Annual Interest Rate Being an old established institution, MegaBucks Savings Bank compounds interest on an annual basis only. To meet the competition, their advertisements claim that their annual rate is chosen to effectively yield the same return as their competitors. If their competitors compound continuously at 8%, what will MegaBuck's annual interest rate be to substantiate their advertising?

Solution Since MegaBucks compounds only once a year, the maturity value for any investment is

$$A(t) = P(1+r)^t$$

where r is the annual interest rate. The same P dollars invested at a nominal rate of 8% compounded continuously will yield at maturity

$$A(t) = Pe^{.08t}$$

If MegaBucks is to match this investment, the maturity values must be the same.

$$P(1+r)^t = Pe^{.08t}$$

or

$$(1+r)^t = e^{.08t} = (e^{.08})^t$$

Using the property of exponents that implies if $a^x = b^x$ and a and b are positive, then $a = b$, we have

$$(1+r) = e^{.08}$$

or

$$r = e^{.08} - 1$$

Using Table A.3 and evaluating,

$$r = 1.0833 - 1 = .0833 \quad \text{or} \quad 8.33\%$$

To match an 8% interest rate compounded continuously, MegaBucks must pay 8.33% annual interest rate.

The annual interest rate required to yield the same maturity value of a given investment is called the *effective annual interest rate*. One way to compare various investments is to compute their effective annual rate of return.

Example 3.19 If you had \$10,000 to invest, would you place it in an institution offering 8% compounded quarterly, or one offering $7\frac{3}{4}\%$ compounded continuously?

Solution Calculate the effective annual interest rate of each investment, and select the higher rate of return. First we equate an annual investment at r_{eff} (effective annual rate) to an investment compounded quarterly at a nominal rate of 8%, and find

$$P(1+r_{eff})^t = P\left(1+\frac{.08}{4}\right)^{4t}$$

or

$$(1+r_{eff})^t = \left[(1+.02)^4\right]^t$$

which implies that

$$(1+r_{eff}) = (1.02)^4$$

or

$$\begin{aligned} r_{eff} &= (1.02)^4 - 1 \\ &= .0824 \\ &= 8.24\% \end{aligned}$$

Next we equate an annual investment at r_{eff} to an investment of $7\frac{3}{4}\%$ compounded continuously, and find

$$P(1+r_{eff})^t = Pe^{.0775t}$$

or

$$\begin{aligned} (1+r_{eff}) &= e^{.0775} \\ r_{eff} &= e^{.0775} - 1 \\ &= .0806 \\ &= 8.06\% \end{aligned}$$

Since the effective annual interest rate of the investment at 8% compounded quarterly is greater, we would select it to give us the greater return.

You perhaps have noted in the above calculations that the effective annual interest rate can be determined by letting $P = \$1.00$ and equating the investments for $t = 1$ year.

Example 3.20 Population Growth The population of a certain city has been increasing at a constant rate of 3%/year from 1960 to 1977. Assuming a population of 1 million in 1960, what will the population be in 1980 if the rate of increase remains the same?

Solution This problem is analogous to the problems involving compound interest. The variable, population, is increasing at a constant rate per unit change in the variable time. Our equation relating these variables will be of the same form as Eq. (3.10), or

$$P(t) = P_0(1+r)^t$$

where

$$\begin{aligned} P_0 &= \text{the initial population} \\ P(t) &= \text{the population in } t \text{ years} \\ r &= \text{annual growth rate (or decline)} \end{aligned}$$

Substituting $P_0 = 1$ million, $r = 3\%$, and $t = 20$ yr into this equation and using Table A.4 we find

$$\begin{aligned} P(20) &= 1 \text{ million}(1.03)^{20} \\ &= 10^6(1.806) \\ &= 1{,}806{,}000 \text{ people} \end{aligned}$$

This is nearly double the 1960 population figure.

EXERCISES

1. Determine the compound maturity value of \$1,500 invested for 19 years at a rate of 8%/year compounded semiannually. What is the amount of interest due?

2. **Compound Maturity Values** The Airtight Bank pays its depositors an annual interest rate of 5.5% compounded quarterly. The Goldbrick Bank pays its depositors an annual rate of 5.2% compounded monthly. For a \$1,000 investment for 1 year, determine the compound maturity value from each bank and determine which bank offers the more favorable option.

3. Mrs. Betty Cash borrowed \$1,500 from the Best National Bank on a 3 month note at 7.5% annual interest. Determine the compound maturity value of the note if the interest is compounded monthly.

4. The sales of the Electric Power Company are increasing at the rate of 8%/year. How many years will it take for the sales to double? Triple?

5. The Bluestreak Motor Company has been steadily increasing its business over the past 10 years. If 10 years ago the sales were \$300,000 and today's sales are \$600,000, what is the annual growth rate compounded annually?

6. **Crude Oil Production** The Slick Oil Company has produced 2.36 million barrels of crude oil this year, and it has been increasing production each year by 6% over the previous year. How many million barrels of crude oil did Slick produce 10 years ago?

7. **Effective Annual Interest Rate** What is the effective annual interest rate of an investment made at 10%

   (a) compounded semiannually?
   (b) compounded continuously?

8. Compare the following two investment opportunities by determining the effective annual interest rate of each.

   (a) 12% compounded semiannually
   (b) $11\frac{1}{2}\%$ compounded continuously

9. **Exponential Bacterial Growth** A colony of bacteria has 3×10^6 individuals initially and 12×10^6 individuals 2 hr later. Assuming that the bacteria grow continuously at a constant rate, find the growth rate and determine the number of individuals in the colony at the end of an additional 2 hr.

10. **The Fruit Fly Puzzle** One hundred fruit flies are placed in a large glass jar and are observed to double in number each day.
(a) How many fruit flies are there at the end of the first day?
(b) How many fruit flies are there at the end of 5 days?
(c) If the jar holds 100,000 flies, on what day will it be filled?

11. **Moose Population Growth** The moose population in the state of Maine has been growing at a constant rate of 6%/year over the last 10 years. In how many years will the population double from its present value if this rate continues?

12. On the same coordinate axes draw a graph of $f(x)=e^x$ and $g(x)=e^{-x}$. (*Hint*: Use Table A.3.)

13. **Absorption of Ultraviolet Light** A thin sheet of plastic was found to cut out 5% of harmful ultraviolet light entering through the windows of a solarium. If each additional sheet cuts out 5% of the ultraviolet light reaching it, what percentage of the ultraviolet light will pass through 10 sheets?

14. **Radium Decay** In physical chemistry there are many applications of Eq. (3.14). One is to the disintegration of radioactive substance. In this case we usually write Eq. (3.14) in the form

$$Q = Pe^{-rt}$$

where P is the original amount, Q the terminal amount, t the time in seconds, and r the decay rate per second. The decay rate of radium A is known by experiment to be $r=.00385$ or .385%/sec. In 180 sec find the terminal amount Q in terms of the initial amount P.

15. **Carbon-14 Decay** Carbon-14 decays at a rate of .0124%/year. Using the equation discussed in Exercise 14, determine what percentage of the original amount remains after 10,000 years. (*Hint*: Use Table A.3 to evaluate e^{-x}.)

16. **World Population Growth** The world's population in 1980 is estimated to be 4.5×10^9 persons. The yearly growth rate is approximately 2%. Under the assumption that the current growth rate remains constant, how large would the world's population be in 1990 and 2000?

17. The quantity of usable timber (B board feet) in a young forest grows at a yearly rate of 4%.

(a) If there are B_0 board feet available for cutting now, how much will be available 10 years from now?
(b) What is the percent increase in 10 years?
(c) If we assume an annual growth rate of 4%, in how many years will the quantity of available timber double?

18. **Reaction to a Stimulus—Weber's Law** Ernest Heinrich Weber (1795–1878), German anatomist and physiologist, noted while studying the response of humans to physical stimuli that discrimination is possible if the magnitude of the stimulation is increased by a constant percentage of the original value. For example, if a person holding a weight of 20 g in his hand is tested for the ability to distinguish between this weight and a slightly higher weight, he can make this distinction only when the new weight is at least 21 g; that is, a 5% increase. If

you plan an experiment to test this theory, what is the largest weight you will need if the first weight is 100 g and you want your subjects to be able to make 10 successive discriminations?

19. **The Logistic Function** Another relation that describes growth is called the logistic function. It has been demonstrated that animal populations, under certain conditions, follow the logistic law of growth given by

$$P = \frac{a}{1+be^{-ct}}$$

where P represents the number of animals in the population, t denotes time in years, and a, b, and c represent constants determined by environmental conditions. Draw a sketch of the graph of

$$P = \frac{1000}{1+19e^{-.5t}}$$

where $0 \leqslant t \leqslant 20$.

20. **P.C.B. Contamination** The concentration of the pollutant P.C.B. (polychlorinated biphenyl) in grams per liter in the Housatonic River is approximated by

$$P(x) = .01e^{-.5x}$$

where x is the number of miles downstream from the source. Find

(a) $P(0)$
(b) $P(1)$
(c) $P(2)$

3.5 LOGARITHMIC FUNCTIONS

In Secs. 3.3 and 3.4 we studied the properties and characteristics of exponential functions. In this section we will consider relations that are the inverses of exponential functions.

You will recall from Sec. 1.5 that if $y=2^x$ the inverse of this function can be obtained by interchanging the x and y in the equation $y=2^x$. Hence, the inverse function is

$$x = 2^y$$

The graphs of $y=2^x$ and $x=2^y$ are shown in Fig. 3.8.

The graph of $x=2^y$ can be obtained in several ways. We could use the property that a function and its inverse are symmetric with respect to the line $y=x$, or we could find several ordered pairs that belong to $x=2^y$ and draw the graph in the usual manner. To find ordered pairs that belong to $x=2^y$ we find it more convenient to select y values first and compute the x values as shown in the following table.

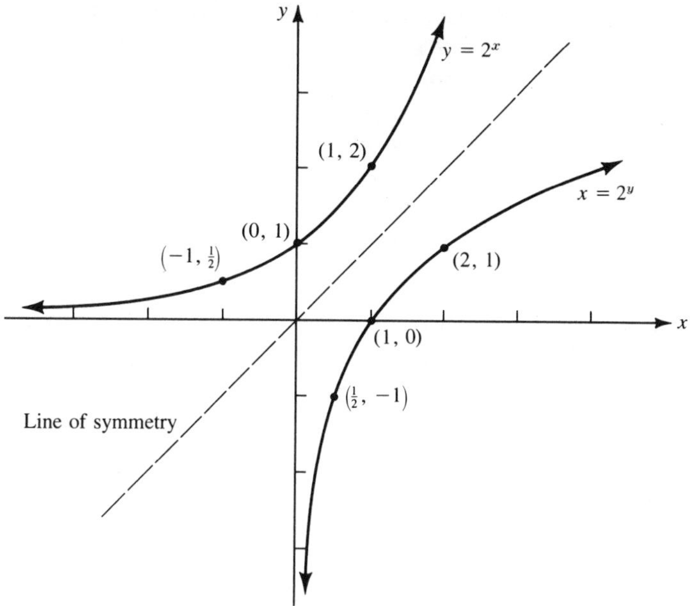

FIGURE 3.8

$x = 2^y$:

y	0	1	-1	2	-2	3	-3	4	-4
x	1	2	$\frac{1}{2}$	4	$\frac{1}{4}$	8	$\frac{1}{8}$	16	$\frac{1}{16}$

Based on the graph of $x = 2^y$ in Fig. 3.8 we conclude that this relation is a function defined on the domain $(0, \infty)$, and its range is the set of reals, $\mathbb{R}$. As we discussed in Sec. 1.5, the domain of $x = 2^y$ is the range of $y = 2^x$ and the range of $x = 2^y$ is the domain of $y = 2^x$.

In previous courses you may have seen graphs of functions similar to $x = 2^y$, but you know them by a different name. Note that in our equation x is expressed in terms of y. Normally, we prefer to think of x as the independent variable and express y in terms of x, or say y is a function of x. Unlike the linear cases discussed in Sec. 1.5, we cannot easily solve the equation $x = 2^y$ for the variable y. This can be accomplished, however, by introducing a new name for this function. When $x = 2^y$, we solve for y by saying y equals the *logarithm* of x base 2, and write $y = \log_2 x$. This agreement leads to the following definition.

Definition 3.1 If $b > 0$ and $\neq 1$, $M = b^N$ is equivalent to $\log_b M = N$.

In terms of this notation, the inverse of the exponential function $y = b^x$ is the *logarithmic function* $x = b^y$ or $y = \log_b x$. Thus, the graph of $x = 2^y$ in Fig.

3.8 is a graph of the logarithmic function $y = \log_2 x$ or $f(x) = \log_2 x$. Let us review Definition 3.1 before we proceed with our discussion of logarithmic functions.

From Definition 3.1 we see that the logarithm of a number M to the base b is the power N to which b is raised to obtain M. Thus, we may write the following. If $b > 0$ and $\neq 1$

$$\log_b b^3 = 3$$
$$\log_b b^2 = 2$$
$$\log_b b^1 = 1$$
$$\log_b b^0 = 0$$
$$\log_b b^{-1} = -1$$
$$\log_b b^{-2} = -2$$
$$\log_b b^{-3} = -3$$

In the above relationships, we used only integers for the exponents, but the same is true for any real-number exponent of b. For example,

$$\log_b b^{1/2} = \tfrac{1}{2}$$
$$\log_b b^x = x$$
$$\log_b b^{x+y} = x + y$$
$$\log_b b^{\sqrt{3}} = \sqrt{3}$$

The important fact that we should keep in mind from Definition 3.1 is that the logarithm of a number is an exponent.

You perhaps may now recall having studied logarithms where the base b was 10. These are called *common logarithms*, which were often used in computational work before the availability of inexpensive hand-held calculators.

Using Definition 3.1 and the decimal representation of 10^n for integer values of n, we can write the following:

$$\log_{10} 1000 = \log_{10} 10^3 = 3$$
$$\log_{10} 100 = \log_{10} 10^2 = 2$$
$$\log_{10} 10 = \log_{10} 10^1 = 1$$
$$\log_{10} 1 = \log_{10} 10^0 = 0$$
$$\log_{10} 0.1 = \log_{10} 10^{-1} = -1$$
$$\log_{10} 0.01 = \log_{10} 10^{-2} = -2$$
$$\log_{10} 0.001 = \log_{10} 10^{-3} = -3$$

As a matter of convention, the logarithm of a number to base 10 is generally written without the base 10; that is, whenever we encounter $\log N$, it is understood to mean $\log_{10} N$.

For all practical applications of logarithms, the base b is greater than 1. The general shape of the log function $f(x) = \log_b x$ where $b > 1$ can be found by reflecting the exponential function $y = b^x$, $b > 1$ across the line of symmetry, $y = x$, as shown in Fig. 3.9.

Since the general shape of the logarithmic function is known, it is relatively easy to draw the graph of a particular function as discussed in the following example.

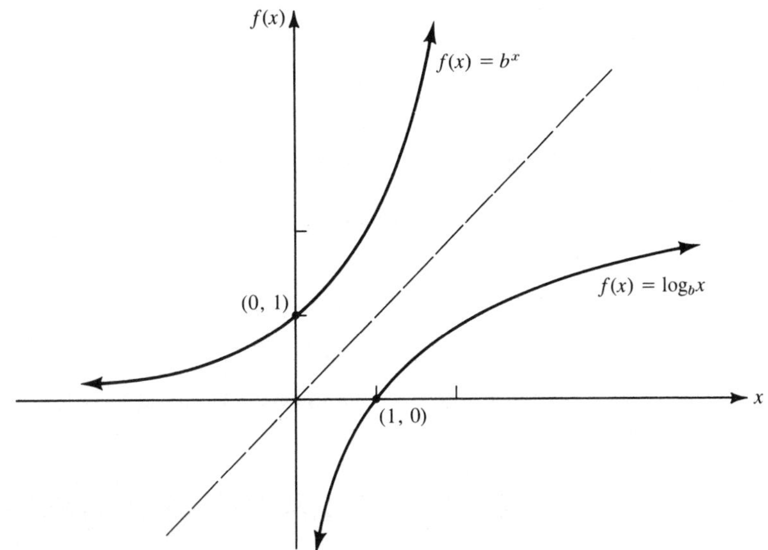

FIGURE 3.9

Example 3.21 Find several ordered pairs that belong to $y = \log_3 x$ and draw a sketch of the function.

Solution To graph the function $y = \log_3 x$, we use Definition 3.1 to obtain the exponential form $x = 3^y$. In this form it is easy to select four or five values of y and calculate the associated x values as shown:

$x = 3^y$:	y	0	1	-1	2	-2	3	-3	4	-4
	x	1	3	$\frac{1}{3}$	9	$\frac{1}{9}$	27	$\frac{1}{27}$	81	$\frac{1}{81}$

The graph of $y=\log_3 x$ is shown in Fig. 3.10.

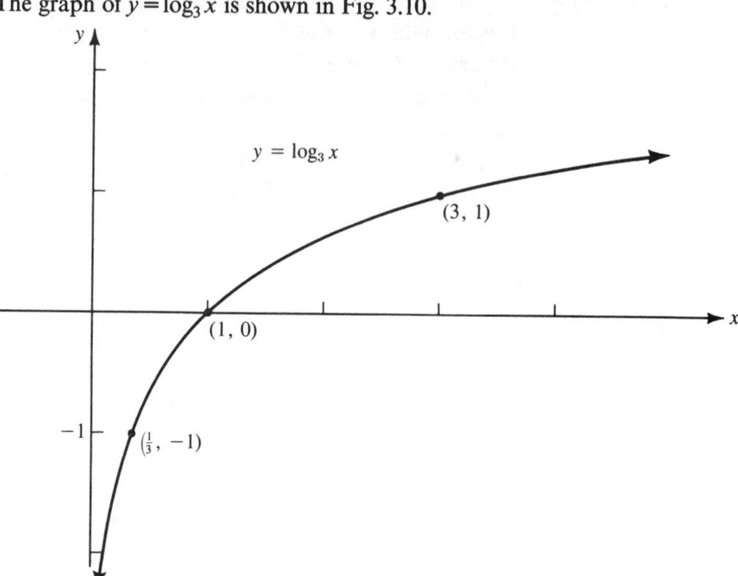

FIGURE 3.10

The following characteristics of logarithmic functions of the form $f(x)=\log_b x$ where $b>1$ can be verified by examining Figs. 3.9 and 3.10.

1. The graph passes through (1,0).
2. The function is increasing on $(0,\infty)$.
3. The vertical axis is an asymptote.
4. The domain is $(0,\infty)$.
5. The range is the set of reals, $\mathbb{R}$.

Two particular bases of logarithmic functions merit special consideration. As noted earlier, base 10 is the most commonly used base for computational work. The graph of $f(x)=\log_{10} x$ is similar to those shown in Figs. 3.9 and 3.10; however, we show it in Fig. 3.11 to draw attention to several key points.

Note in Fig. 3.11 that if $1 \leqslant x \leqslant 10$, then $0 \leqslant \log x \leqslant 1$. Table A.1 in the Appendix presents values of the $\log_{10} N$ for values of $1.0 \leqslant N \leqslant 9.99$. The first table of such logarithms was published in 1624 by the English mathematician Henry Briggs. For simplicity, no decimal points appear in Table A.1. For example, to find log 2, locate $N=20$ in the first column, and to the right of the 20 appears 3010, which means $\log 2=.3010$. In a like manner we find

x	3	4	5	7	8	9
$\log x$	.4771	.6021	.6990	.8451	.9031	.9542

These ordered pairs are graphed in Fig. 3.11.

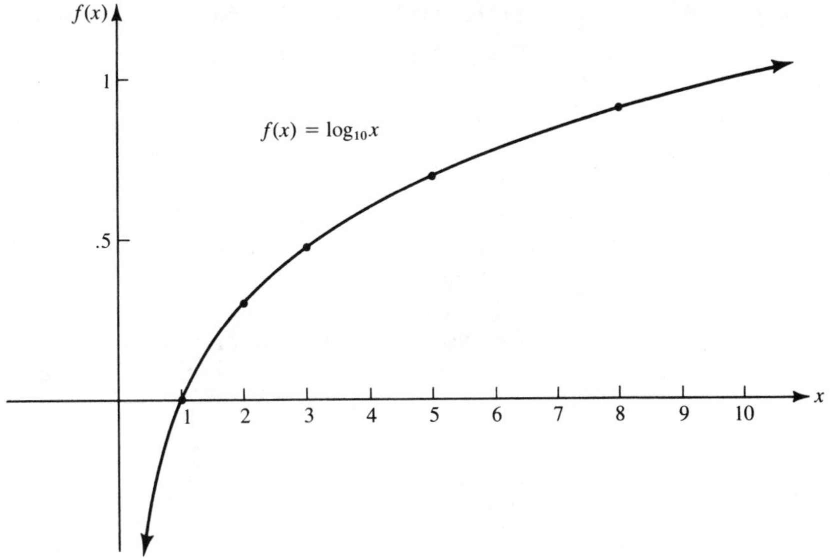

FIGURE 3.11

The other base of particular significance is e. You will recall our introduction of e as a base in the discussion of exponential functions in Sec. 3.4. Logarithms to the base e are called *natural or Napierian logarithms* and frequently we abbreviate $\log_e x$ as $\ln x$. In the study of calculus we will find that the choice of base e for the logarithmic function will simplify certain results. Tables A.2 and A.3 present values for $\ln x$ and e^x respectively for selected values of x. Table A.2 of natural logarithms shows all decimal points, unlike Table A.1 of common logarithms.

We will now consider several *basic laws of logarithms*. If we let

$$M = b^x$$
$$N = b^y$$

we have by Eq. (3.1)

$$(MN) = b^{x+y}$$

From the above three relations and the definition of a logarithm, we have

$$\log_b M = x$$
$$\log_b N = y$$
$$\log_b(MN) = x + y$$

Since the sum of the right side of the top two equations is equal to the right side of the third equation, the same must be true for the terms on the left side of the equations. Thus, we have

$$\log_b(MN) = \log_b M + \log_b N \tag{3.15}$$

This is one of the most important laws of logarithms: It states that *the logarithm of a product is simply the sum of the logarithms of the individual factors*. It can be extended to include the product of any number of factors for which the logarithm of the product will be the sum of the logarithms of the individual factors.

Another important law of logarithms is obtained using the notation $M = b^x$. By raising both terms to the Nth power, we obtain

$$M^N = (b^x)^N = b^{Nx}$$

Thus we have

$$\log_b M^N = Nx$$

However, since $x = \log_b M$, the above equation becomes

$$\log_b M^N = N \log_b M \tag{3.16}$$

Hence we obtain another basic law concerning logarithms that states that *the logarithm of a number raised to a power is equal to the power times the logarithm of the number*.

We now have two basic laws of logarithms from which we can derive other laws. For example, we note that

$$\frac{M}{N} = MN^{-1}$$

Taking the logarithm to the base b of both sides, we obtain

$$\log_b \frac{M}{N} = \log_b MN^{-1}$$

Applying the rules given in (3.15) and (3.16) in that order, we obtain

$$\log_b \frac{M}{N} = \log_b MN^{-1} = \log_b M + \log_b N^{-1} = \log_b M - \log_b N$$

Thus we obtain a law pertaining to a quotient of two numbers that states that *the logarithm of a quotient is the logarithm of the numerator minus the logarithm of the denominator*. In equation form, we have

$$\log_b \frac{M}{N} = \log_b M - \log_b N \tag{3.17}$$

Another property of interest is that every term of the form b^x, where $b>0$, can be expressed as a power of e. To establish this relationship, consider the following: Since $b>0$ there exists a y such that

$$\ln b = y \quad \text{or, equivalently,} \quad b = e^y$$

Substituting $\ln b$ for y in this last equality, we have,

$$b = e^{\ln b}$$

Therefore,

$$b^x = (e^{\ln b})^x$$
$$= e^{x \ln b} \tag{3.18}$$

Example 3.22 Convert the following exponential functions to exponential functions with base e:

(a) 8^x (b) 6^x (c) 3.4^x

Solution

(a) Determining from Table A.2 that $\ln 8 = 2.07944$ and substituting into Eq. 3.18, we obtain

$$8^x = e^{2.07944x}$$

(b) $6^x = e^{x \ln 6} = e^{1.79176x}$
(c) $3.4^x = e^{x \ln 3.4} = e^{1.22378x}$

EXERCISES

1. Express each exponential equation in logarithmic notation.

(a) $3^2 = 9$ (b) $10^3 = 1{,}000$
(c) $10^{-1} = 0.1$ (d) $8^{-1/3} = \frac{1}{2}$

2. Express each logarithmic equation in exponential notation.

(a) $\log_6 36 = 2$ (b) $\log_{10} 10{,}000 = 4$
(c) $\log_5 125 = 3$ (d) $\log_3 81 = 4$

3. Evaluate each of the following logarithms without the use of tables.

(a) $\log_{10} 100$ (b) $\log_5 5^3$ (c) $\log_3 \sqrt{3}$
(d) $\log_2 16$ (e) $\ln e^{10}$

4. Justify by the basic laws of logarithms that

(a) $\log 4{,}000 = \log(4\times 10^3) = \log 4 + \log 10^3 = (\log 4) + 3$
(b) $\log .004 = \log(4\times 10^{-3}) = \log 4 + \log 10^{-3} = (\log 4) - 3$
(c) $\log 80 = \log(8\times 10^1) = \log 8 + \log 10^1 = (\log 8) + 1$

5. Evaluate the following common logarithms by the use of Table A.1.

(a) $\log 4$ (b) $\log 8$ (c) $\log 3$ (d) $\log 9$

6. By considering Exercises 4 and 5 above, evaluate the following common logarithms.

(a) $\log 4{,}000$ (b) $\log .004$
(c) $\log 80$ (d) $\log .09$

7. Determine the natural logarithm for each of the following numbers by using Table A.2.

(a) 10 (b) 5 (c) 7 (d) 2 (e) 1.2 (f) 4.32

8. Convert each of the following exponential functions to an exponential function base *e*. (Use the results of Exercise 7).

(a) 10^x (b) 5^x (c) 7^x (d) 2^x (e) 1.2^x (f) 4.32^x

9. Sketch the graph of the exponential function $y=5^x$ and its inverse $y=\log_5 x$ on the same coordinate system. Determine the domain and range of $y=5^x$ and the domain and range of $y=\log_5 x$.

10. If $f(x)=\ln x$ and $g(x)=e^x$, evaluate

(a) $f(g(1))$ (b) $g(f(1))$ (c) $f(g(2))$ (d) $g(f(2))$

11. Let $f(x)=\ln x$ and $g(x)=e^x$ and define $F(x)=f(g(x))$ and $G(x)=g(f(x))$. Express $F(x)$ and $G(x)$ as algebraic functions and determine their respective domains.

12. If $A=Pb^x$, show by the basic laws of logarithms that

$$\log_{10} A = \log_{10} P + x\log_{10} b$$

13 **Audibility and Loudness** The unit of measure of loudness is the decibel. It can be shown that loudness, L, measured in decibels is given by

$$L = 10\log\left(\frac{I}{I_0}\right)$$

where I is the intensity of sound being measured and I_0 is the threshold of audibility or the lowest intensity that can be heard at a frequency of 1,000 cycles/sec.

(a) Determine the intensity of sound, I, in terms of I_0 to produce a loudness L of 10 decibels.
(b) Determine the intensity of sound, I, in terms of I_0 to produce a loudness of 20 decibels. Is this intensity twice that of the answer in part (a)?

IMPORTANT TERMS AND CONCEPTS

Base
Common logarithm
Compound interest formula
Continuous compounding
Effective annual interest rate
Exponent
Exponential decay
Exponential function
Exponential growth
Index
Laws of exponents
Laws of logarithms
Logarithm
Logarithmic function
Natural logarithm
Nominal yearly interest rate
Percent change
Radicand
Transcendental function

SUMMARY OF RULES AND FORMULAS

Laws of exponents (Assume a and b are positive real numbers.)

$$b^m \cdot b^n = b^{m+n}$$

$$(b^m)^n = b^{mn}$$

$$b^{-m} = \frac{1}{b^m}$$

$$b^0 = 1$$

$$\frac{b^m}{b^n} = b^{m-n}$$

$$(ab)^m = a^m b^m$$

$$\left(\frac{a}{b}\right)^m = \frac{a^m}{b^m}$$

$$b^{1/m} = \sqrt[m]{b}$$

$$b^{p/q} = \sqrt[q]{b^p} = (\sqrt[q]{b}\,)^p$$

Exponential function base b

$$f(x) = ab^x$$

Compound interest formula

$$A(t) = P\left(1 + \frac{r}{n}\right)^{nt}$$

Continuous growth formula

$$A(t) = Pe^{rt}$$

Logarithmic function base b

$$f(x) = \log_b x$$

Logarithmic function base e

$$f(x) = \ln x$$

Laws of logarithms

$$\log_b(MN) = \log_b M + \log_b N$$

$$\log_b M^N = N \log_b M$$

$$\log_b \frac{M}{N} = \log_b M - \log_b N$$

$$b^x = e^{x \ln b}$$

4 DIFFERENTIAL CALCULUS: BASIC METHODOLOGY

4.1 INTRODUCTION

In many situations where we wish to analyze a function it is helpful to first graph the function and then try to obtain certain information from the graph. Such graphs, which appear in our daily newspapers, show at a glance what is happening to the Dow Jones stock averages, weather, air and water pollution levels, and economic indicators. In many cases we are more interested in the relative changes of the function than in the function value. In other words, the shape is at least as important as the actual data supplied by the graph. For example, if a function describes a company's profit in terms of the number of years it has been in operation, we might be as interested in how fast the profit is increasing (or decreasing) in the 10th year of operation as we are in the profit figure for that year. Similarly if we have a function that describes the amount of pollutant in each cc of water in terms of the number of miles downstream from a factory, we may be just as interested in how fast the pollutant is decreasing as we are in how much pollutant there is in each cc at this point. In order to investigate problems of this nature, we shall need to develop the idea of rate of change of a function—a basic concept in the study of differential calculus. You will find calculus interesting because it involves new concepts not previously covered in algebra and it has many powerful applications.

The general definition of *calculus* is *any method of calculating or investigating by algebraic symbols*; however, the more common usage refers to *infinitesimal calculus*, which encompasses the study of two types of limits: *derivatives* and *integrals*. Thus, calculus is divided into two distinct sections called *differential calculus* and *integral calculus*. Since calculus involves the study of two applications of the notion of limits, we shall begin by examining the concept of a *limit*.

4.2 LIMIT OF A FUNCTION

The concept of *limit* is one of the most fundamental concepts of calculus. Previously, in Chapter 2, when we investigated the extent of a function $f(x)$, we considered the behavior of the second component of the ordered pair $(x,f(x))$ as x assumed large positive values. If $f(x)$ approached the number L as x increased without bound, we denoted this by

$$\lim_{x\to\infty} f(x) = L$$

Analogously if $f(x)$ approached the number L as x decreased without bound, we denoted this by

$$\lim_{x\to-\infty} f(x) = L$$

We again used the concept of limit when we investigated behavior of a function near its vertical asymptote. As you recall, if $x=a$ was the equation of a vertical asymptote for the function $f(x)$, we studied the behavior of $f(x)$ as x approached a from the right and as x approached a from the left. In such cases we found that as x approached a from the right, $f(x)$ either increased without bound (i.e., $\lim_{x\to a^+} f(x)=+\infty$) or $f(x)$ decreased without bound (i.e., $\lim_{x\to a^+} f(x)=-\infty$). Analogously, as x approached a from the left, we found that either $\lim_{x\to a^-} f(x)=+\infty$ or $\lim_{x\to a^-} f(x)=-\infty$.

In this chapter we are going to examine the *limit of a function*, $f(x)$, as x approaches a number a from the right and from the left. In general these will be cases where $x=a$ is not necessarily an equation of a vertical asymptote for the function $f(x)$. We want to determine if $f(x)$ approaches some number L as x approaches a from the right and from the left.

Considerable effort has been exerted to make the concept of a limit mathematically precise and rigorous. This effort was necessary in order to establish a sound foundation to build the concepts of calculus. But, for our purposes, we shall use a more informal approach by giving a general intuitive definition of the limit of a function and then stating without proof some properties concerning the limit.

Consider a function $f(x)$ and let the independent variable assume values near but not equal to a given constant a; the second component $f(x)$ of the ordered pair $(x,f(x))$ will then yield a corresponding set of values. For example, let $f(x)=x^2+2$ and let $a=4$. Choosing values of x that are near 4 on the left we obtain, for example, the ordered pairs shown in the following table.

x	3.9	3.99	3.999	3.9999
$f(x)$	17.21	17.9201	17.992001	17.99920001

Choosing values of x that are near 4 on the right we obtain, for example, the ordered pairs shown in the next table.

x	4.1	4.01	4.001	4.0001
$f(x)$	18.81	18.0801	18.008001	18.00080001

In this example, the function value, $f(x)$, corresponding to the x values that are near but not equal to 4 are near the value 18; furthermore, the closer the x values are to 4 from either the left or the right, the closer the corresponding function values $f(x)$ are to the value 18. The values of $f(x)$ can be made to differ from 18 by as little as we please by taking the values of x sufficiently close to 4. Under these conditions we say the *left-hand limit* of $f(x)$ as x approaches 4 from the left is 18 and the *right-hand limit* of $f(x)$ as x approaches 4 from the right is 18. Since the left-hand limit and right-hand limit equal the same number, we say that the limit of $f(x)$ as x approaches 4 is 18. In the concept of limit, we are interested not in the value of $f(x)$ when x is equal to 4, but rather in the number that $f(x)$ approaches as x gets closer and closer to 4 from the left and the right.

Another example may help clarify this. Consider $f(x)=(x^2-1)/(x-1)$. Now $f(1)$ does not exist as division by zero is undefined. However, we can still ask if there is a limit of $f(x)$ as x approaches 1. Consider the following table.

x	.9	.99	.999	$\cdots$	1.0001	1.001	1.01
$f(x)$	1.9	1.99	1.999	$\cdots$	2.0001	2.001	2.01

By inspection of the data it appears that $f(x)$ is getting closer and closer to 2 as x approaches 1 from the left and from the right. We certainly can get $f(x)$ as close to 2 as we want by selecting the values of x close enough, but not equal to 1. If we can get $f(x)$ as close to L as we want by selecting our x's close to but not equal to a, we say that the number L is the limit of $f(x)$ as x approaches a. Hence in this example the limit L is 2.

Let us return to our first example, the limit of $f(x)=x^2+2$ as x approaches 4, and investigate this question by considering what is happening graphically. Fig. 4.1a illustrates the behavior of $f(x)$ as x approaches 4 from the right, and Fig. 4.1b the behavior of $f(x)$ as x approaches 4 from the left. As x approaches 4 from the right, $x=x_1$, $x=x_2$, and $x=x_3$, the corresponding values of $f(x)$, $f(x_1)$, $f(x_2)$, and $f(x_3)$ approach 18 from above. As x approaches 4 from the left as in Fig. 4.1b, $f(x)$ approaches 18 from below. To denote that $f(x)$ approaches 18 as x approaches 4 from the left, we write

$$\lim_{x\to 4^-} f(x) = 18$$

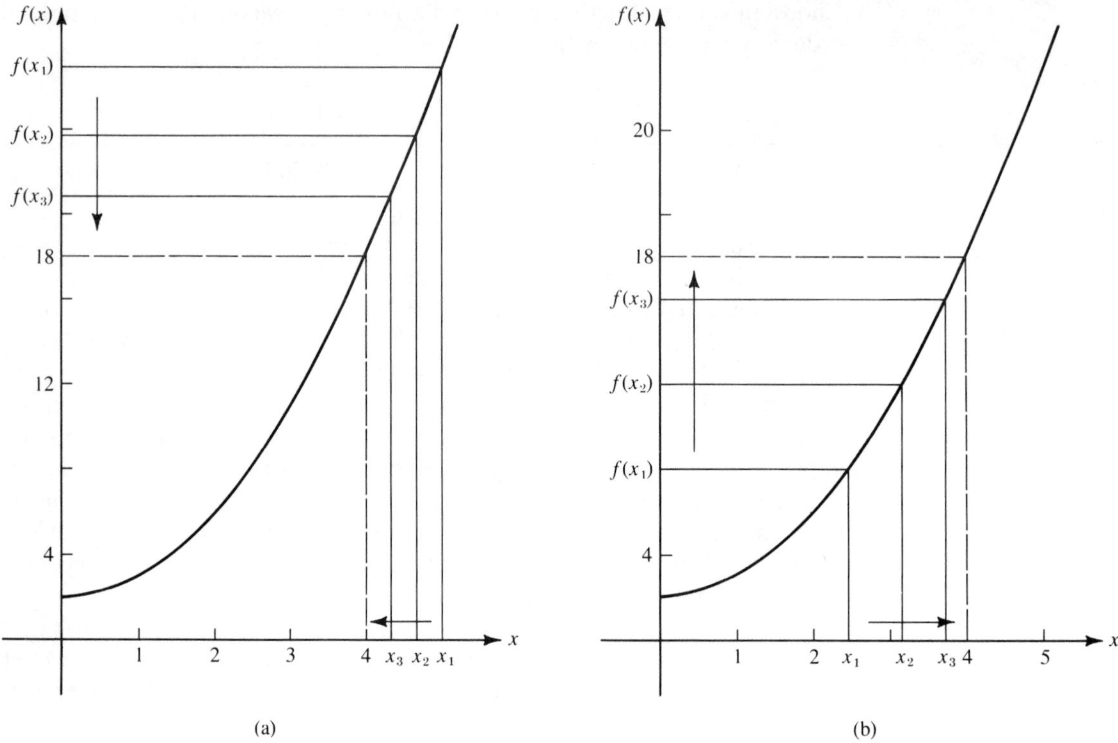

FIGURE 4.1

(the 4^- means that x approaches 4 from the left) and to denote that $f(x)$ approaches 18 as x approaches 4 from the right, we write

$$\lim_{x \to 4^+} f(x) = 18$$

(the 4^+ means that x approaches 4 from the right). Since these limits equal the same number, 18, we say the limit of $f(x)$ as x approaches 4 exists and write:

$$\lim_{x \to 4} f(x) = 18$$

which is read "the limit of $f(x)$ as x approaches 4 is 18."

In general, when a function $f(x)$ approaches a number L as x approaches a from the left and from the right, we write

$$\lim_{x \to a} f(x) = L$$

This is often abbreviated $f(x) \to L$ as $x \to a$.

Based on our discussion, we conclude: $\lim_{x\to a} f(x) = L$ if and only if the $\lim_{x\to a^-} f(x) = L$ and the $\lim_{x\to a^+} f(x) = L$. If the $\lim_{x\to a} f(x)$ is not some number L, then we say the limit of $f(x)$ as x approaches a *does not exist* or is undefined, and write

$$\lim_{x\to a} f(x) \text{ D.N.E.}$$

Example 4.1 Consider the four sketches in Fig. 4.2 and determine if $f(x) \to$ some number L as $x \to a$.

Solution

(a) As $x \to a$ from the right, $f(x)$ gets closer and closer to 4; and as $x \to a$ from the left, $f(x)$ gets closer and closer to 4. Therefore,

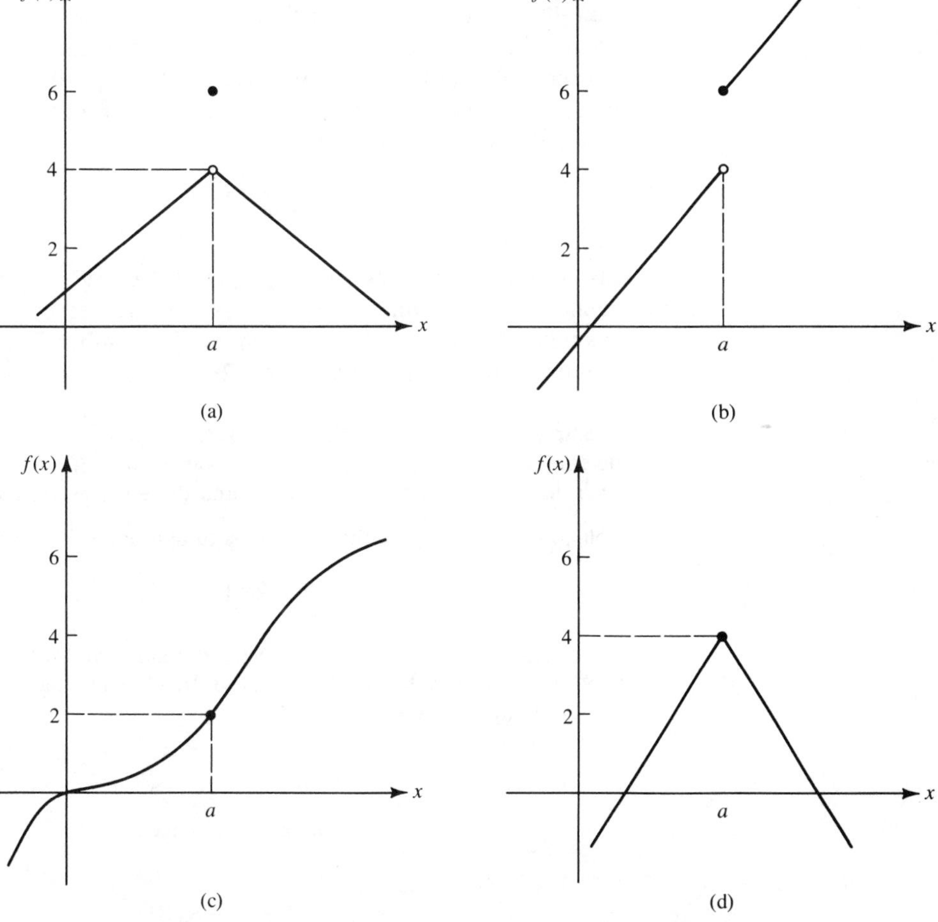

FIGURE 4.2

$\lim_{x \to a} f(x) = 4$. (*Note*: The functional value $f(a) = 6 \neq \lim_{x \to a} f(x)$ in this case.)

(b) As $x \to a$ from the right, $f(x)$ gets closer and closer to 6; and as $x \to a$ from the left, $f(x)$ gets closer and closer to 4. Since $6 \neq 4$, the left- and right-hand limit are not equal, hence the $\lim_{x \to a} f(x)$ D.N.E.

(c) As $x \to a$ from the right, $f(x)$ gets closer and closer to 2; and as $x \to a$ from the left, $f(x)$ gets closer and closer to 2. Therefore, $\lim_{x \to a} f(x) = 2$. (*Note*: In this example $f(a) = 2$ also.)

(d) As $x \to a$ from the right, $f(x)$ gets closer and closer to 4; and as $x \to a$ from the left, $f(x)$ gets closer and closer to 4. Therefore, $\lim_{x \to a} f(x) = 4$. (*Note*: In this example $f(a) = 4$ also.)

Our discussion should make the following definition of a limit seem reasonable.

Definition 4.1 A function $f(x)$ is said to approach a limit L as x approaches a if and only if the absolute value of the difference between $f(x)$ and L is less than an arbitrary positive number for all values of x that are sufficiently close to a and for which $x \neq a$; this is denoted mathematically as

$$\lim_{x \to a} f(x) = L$$

This definition would be very significant if our goal were to prove all our statements regarding limits. However, as indicated earlier, our major concern in the study of limits will be to determine the number L, if it exists, as $x \to a$. As an illustration consider Example 4.2.

Example 4.2 Determine the limit of $f(x) = 2x + 4$ as x approaches 1 by determining the function values for six values of x that are within 0.01 of 1 such that three values are less than 1 and three values are greater than 1.

Solution Symbolically, the problem is to determine L, where

$$\lim_{x \to 1} (2x + 4) = L$$

Choosing values of x that are symmetrical about 1 and within 0.01, we select 0.99, 0.999, 0.9999, 1.0001, 1.001, and 1.01. Determining the corresponding values of $f(x)$ we have

$$f(0.99) = 5.98$$
$$f(0.999) = 5.998$$
$$f(0.9999) = 5.9998$$
$$f(1.0001) = 6.0002$$
$$f(1.001) = 6.002$$
$$f(1.01) = 6.02$$

Thus, we see that $f(x)$ approaches 6 as x approaches 1; hence, $L=6$.

The conclusion reached in Example 4.2 can also be arrived at in a slightly different manner. As x gets closer and closer to 1, from the left or the right, clearly $2x$ gets closer to 2 and $2x+4$ gets closer to 6; hence

$$\lim_{x\to 1} (2x+4) = 6$$

This kind of analysis can be used in many problems to determine the limit if it exists and is obviously much shorter than choosing many values of x near a as was done previously. Consider the following example.

Example 4.3 Find the limit, if it exists, for each of the following:

(a) $\lim_{x\to 3} (x^2+1)$
(b) $\lim_{x\to 4} \left(\frac{x}{x-2}\right)$
(c) $\lim_{h\to 0} (h^2+2h+5)$
(d) $\lim_{h\to 0} (h^3-5h^2-h-3)$
[*Note:* h is the variable in parts (c) and (d).]

Solution

(a) as $x\to 3$ from the left or right, $x^2\to 9$ and $(x^2+1)\to 10$; therefore, $\lim_{x\to 3} (x^2+1)=10$

(b) as $x\to 4$ from the left or right, $(x-2)\to 2$ and $\left(\frac{x}{x-2}\right)\to 2$; therefore, $\lim_{x\to 4} \left(\frac{x}{x-2}\right)=2$

(c) as $h\to 0$ from the left or right, $h^2\to 0$, $2h\to 0$, and $(h^2+2h+5)\to 5$; therefore, $\lim_{h\to 0} (h^2+2h+5)=5$

(d) as $h\to 0$ from the left or right, $h^3\to 0$, $5h^2\to 0$, and $(h^3-5h^2-h-3)\to -3$; therefore, $\lim_{h\to 0} (h^3-5h^2-h-3)=-3$

We have shown by Example 4.3 that sometimes it is not necessary to actually compute several values of $f(x)$ for values of x near a to evaluate the limit of $f(x)$ as x approaches a.

To facilitate finding the limits for a variety of problems we state several useful rules. These rules can be proved to be true by recourse to Definition 4.1. We are more interested in the application of these rules than in a rigorous proof of each one.

Rule 4.1

(a) $\lim_{x\to a} kx = k\cdot a$ for all real numbers a where k is a constant (4.1)

(b) $\lim_{x\to a} k = k$ where k is a constant (4.2)

In each of the following rules it is assumed that $f(x)$ and $g(x)$ are any functions such that $\lim_{x\to a} f(x)$ and $\lim_{x\to a} g(x)$ exist.

Rule 4.2 The limit of the sum of two functions is equal to the sum of the limits of the functions:

$$\lim_{x\to a}\left[f(x)+g(x)\right] = \lim_{x\to a} f(x) + \lim_{x\to a} g(x) \tag{4.3}$$

Example 4.4 Find the limit of $(6x+5)$ as $x\to 2$ by using Rule 4.1 and Rule 4.2.

Solution If we let $f(x)=6x$ and $g(x)=5$, by Eq. (4.1) and (4.2) we have

$$\lim_{x\to 2} 6x = 6\cdot 2 = 12$$

and

$$\lim_{x\to 2} 5 = 5$$

Therefore, using Rule 4.2,

$$\begin{aligned}\lim_{x\to 2}(6x+5) &= \lim_{x\to 2} 6x + \lim_{x\to 2} 5\\ &= 12+5\\ &= 17\end{aligned}$$

Using Rule 4.2 and Rule 4.1 it is easy to show in an analagous manner that

$$\lim_{x\to a}\left[(mx+b)\right] = ma+b$$

where m and b are constants. Based on this we conclude that if $f(x)$ is a linear function

$$\lim_{x\to a} f(x) = f(a)$$

Rule 4.3 The limit of the product of two functions is equal to the product of the limits of the functions

$$\lim_{x\to a}\left[f(x)\cdot g(x)\right] = \left[\lim_{x\to a} f(x)\right]\left[\lim_{x\to a} g(x)\right] \tag{4.4}$$

Example 4.5 Find the limit of $(6x^2)$ as $x\to 2$ by using Rule 4.3 and Rule 4.1.

Solution If we let $f(x)=6x$ and $g(x)=x$, by Eq. (4.1) we have

$$\lim_{x\to 2} (6x) = 12$$

$$\lim_{x\to 2} x = 2$$

Therefore, using Rule 4.3

$$\lim_{x\to 2} (6x^2) = \lim_{x\to 2} (6x\cdot x) = \Big(\lim_{x\to 2} 6x\Big)\Big(\lim_{x\to 2} x\Big)$$
$$= 12\cdot 2$$
$$= 24$$

Rules 4.2 and 4.3 can be extended to any finite number of functions. This would allow us to show, for example:

$$\lim_{x\to 2} (3x^3-4x^2+5x-3) = 15$$

Rule 4.4 The limit of the quotient of two functions is equal to the quotient of the limits of the functions, provided the limit of the divisor is not zero.

$$\lim_{x\to a}\left[\frac{f(x)}{g(x)}\right] = \frac{\lim_{x\to a} f(x)}{\lim_{x\to a} g(x)} \quad \text{provided } \lim_{x\to a} g(x) \neq 0 \tag{4.5}$$

Example 4.6 Find the limit of $6x^2/(6x+5)$ as $x\to 2$ using Rule 4.4.

Solution If we let $f(x)=6x^2$ and $g(x)=6x+5$, we have shown $\lim_{x\to 2} 6x^2=24$ and $\lim_{x\to 2} (6x+5)=17$. Therefore, by applying Rule 4.4, we have

$$\lim_{x\to 2}\left(\frac{6x^2}{6x+5}\right) = \frac{\lim_{x\to 2} 6x^2}{\lim_{x\to 2} (6x+5)} = \frac{24}{17}$$

By applying Rule 4.3 to any number of equal factors, we obtain the following.

Rule 4.5 The limit of the nth power of a function is equal to the nth power of the limit of that function:

$$\lim_{x\to a} [f(x)]^n = \Big[\lim_{x\to a} f(x)\Big]^n \tag{4.6}$$

where n is a positive integral exponent.

Example 4.7 Find $\lim_{x\to 2}(6x+5)^3$ by using Rule 4.5.

Solution Let $f(x)=6x+5$. We have shown $\lim_{x\to 2}(6x+5)=17$. Therefore, using Rule 4.5, we have

$$\lim_{x\to 2}(6x+5)^3=\left[\lim_{x\to 2}(6x+5)\right]^3=(17)^3=4913$$

Rule 4.6 The limit of the *n*th root of a positive function is equal to the principal *n*th root of the limit of that function:

$$\lim_{x\to a}\left[f(x)\right]^{1/n}=\left[\lim_{x\to a}f(x)\right]^{1/n} \tag{4.7}$$

Example 4.8 Find the $\lim_{x\to 2}(6x+5)^{1/2}$ using Rule 4.6.

Solution Let $f(x)=6x+5$. We have shown that $\lim_{x\to 2}(6x+5)=17$. Therefore, using Rule 4.6, we have

$$\begin{aligned}\lim_{x\to 2}(6x+5)^{1/2}&=\left[\lim_{x\to 2}(6x+5)\right]^{1/2}\\&=(17)^{1/2}\\&=\sqrt{17}\end{aligned}$$

Example 4.9 Apply Rules 4.1 through 4.6 to find

$$\lim_{x\to 2}\left[\frac{(2x^2-4)^{1/2}(x+3)^3}{x^2}\right]$$

Solution

$$\begin{aligned}\lim_{x\to 2}(2x^2-4)&=\left(\lim_{x\to 2}2x\right)\left(\lim_{x\to 2}x\right)+\lim_{x\to 2}(-4)\text{ by Rules 4.2 and 4.3}\\&=(4)(2)+-4\text{ by Rule 4.1}\\&=4\\\lim_{x\to 2}x^2&=\left(\lim_{x\to 2}x\right)^2\text{ by Rule 4.5}\\&=2^2\text{ by Rule 4.1}\\&=4\\\lim_{x\to 2}(x+3)&=\lim_{x\to 2}x+\lim_{x\to 2}3\text{ by Rule 4.2}\\&=2+3\text{ by Rule 4.1}\\&=5\end{aligned}$$

Therefore,

$$\lim_{x\to2}\left[\frac{(2x^2-4)^{1/2}(x+3)^3}{x^2}\right]=\frac{\lim\limits_{x\to2}(2x^2-4)^{1/2}\cdot\lim\limits_{x\to2}(x+3)^3}{\lim\limits_{x\to2}x^2}$$

by Rules 4.3 and 4.4

$$=\frac{\left[\lim\limits_{x\to2}(2x^2-4)\right]^{1/2}\left[\lim\limits_{x\to2}(x+3)\right]^3}{\left(\lim\limits_{x\to2}x\right)^2}$$

by Rules 4.5 and 4.6

$$=\frac{4^{1/2}(5)^3}{4}\quad\text{by substitution}$$

$$=\frac{125}{2}$$

Occasionally we encounter a problem involving the quotient of two functions for which Rule 4.4 does not appear to be applicable. For example, let $f(x)=x^2-9$ and $g(x)=x-3$. If we wish to determine $\lim\limits_{x\to3}f(x)/g(x)$, we obtain

$$\lim_{x\to3}f(x)=\lim_{x\to3}(x^2-9)=0$$

and

$$\lim_{x\to3}g(x)=\lim_{x\to3}(x-3)=0$$

Thus, the divisor has a limit of zero, which nullifies the application of Rule 4.4; however, if we factor the respective functions and divide by the common factors, we have

$$\frac{f(x)}{g(x)}=\frac{(x-3)(x+3)}{(x-3)}=x+3,\text{ provided }x\neq3$$

Since we are only interested in values close to 3, but not equal to 3, we have

$$\lim_{x\to3}\frac{f(x)}{g(x)}=\lim_{x\to3}(x+3)=6$$

Example 4.10 Find the limit, if it exists, of $(h^3+2h^2+7h)/h$ as $h\to0$.

Solution Since $(h^3+2h^2+7h)/h=h^2+2h+7$ for all $h\neq 0$, we have by substitution

$$\lim_{h\to 0}\left(\frac{h^3+2h^2+7h}{h}\right) = \lim_{h\to 0}(h^2+2h+7)$$

and applying the limit rules, we find

$$\lim_{h\to 0}(h^2+2h+7) = 7$$

Limit questions similar to this will play a very important role in the development of differential calculus, which will be discussed in the next section.

In the definition of $\lim_{x\to a} f(x)$, the reader will recall that it is not required to know the value of $f(x)$ at $x=a$. The limit depends only on the values of $f(x)$ in the neighborhood of $x=a$, not on the value of $f(x)$ at $x=a$; thus, the limit may or may not equal $f(a)$. However, the case where $\lim_{x\to a} f(x)=f(a)$ is of special interest.

Definition 4.2 A function $f(x)$ is *continuous* at $x=a$ if the following three conditions are satisfied: (a) $f(a)$ exists, (b) $\lim_{x\to a} f(x)$ exists, and (c) $\lim_{x\to a} f(x)=f(a)$.

It is important to note in our definition that we are talking about *continuity at a point* in the domain of the function. Returning to Fig. 4.2 we note that the function represented in Fig. 4.2*a* is not continuous at $x=a$ since $\lim_{x\to a} f(x)\neq f(a)$. Note $\lim_{x\to a} f(x)=4$ and $f(a)=6$. In Fig. 4.2*b* the function is not continuous at $x=a$ since $\lim_{x\to a} f(x)$ does not exist. In Figs. 4.2*c* and 4.2*d* the functions are continuous at $x=a$. Check the three parts of Definition 4.2 to verify this conclusion.

A function satisfying Definition 4.2 is said to be *continuous at the point* $x=a$, and a function that is continuous at every point in the domain (or an interval) is said to be *continuous in the domain* (or interval). A function that is not continuous at a point $x=a$ is said to be *discontinuous* there. If we reconsider Figs. 4.2*c* and *d*, we can conclude that the functions illustrated are continuous in the domain defined by their graphs. Comparing Figs. 4.2*a* and *b* with 4.2*c* and *d* we intuitively arrive at the conclusion that a continuous function is one whose graph constitutes a smooth curve without "holes" or jumps in it; i.e., it will be continuous if we can trace the curve without having to lift our pens or pencils from the paper in order to reach all segments of the curve.

Example 4.11 Determine whether the function $f(x)=3x^2+2x+1$ is continuous at $x=1$. Is $f(x)$ continuous at every point in its domain?

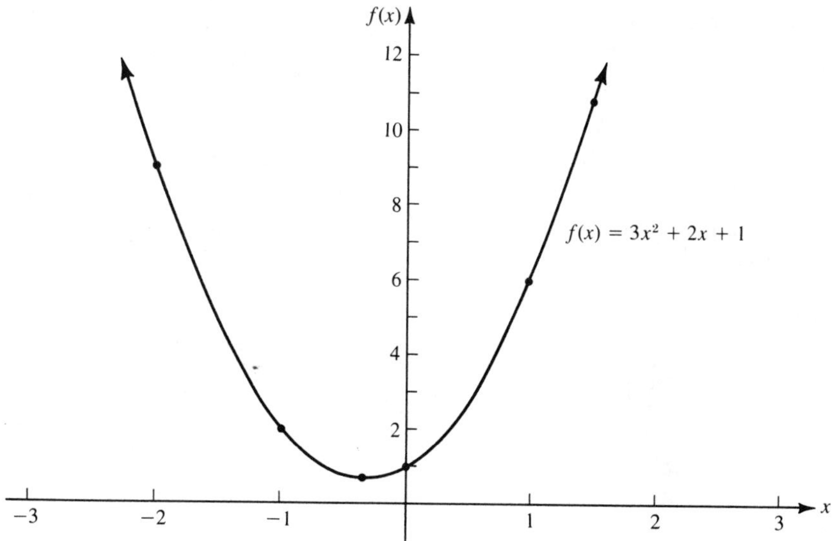

FIGURE 4.3

Solution Since $f(1)=3(1)^2+2(1)+1=6$, the function exists at $x=1$. Also $\lim_{x \to 1} (3x^2+2x+1)=6$, and hence $\lim_{x \to 1} (3x^2+2x+1)=f(1)$. Therefore, $f(x)$ is continuous at $x=1$. By applying our limit rules we conclude that for any $x=a$, where a is a real number, the $\lim_{x \to a} f(x)=f(a)$. Therefore, $f(x)$ is continuous on its entire domain. A check of Fig. 4.3, which shows a graph of this function, will verify that the curve can be traced without lifting the pen, in agreement with the intuitive definition of a continuous function.

Example 4.12 Determine whether the function $f(x)=1/(x-3)$ is continuous at $x=3$.

Solution The function value at $x=3$, $f(3)$, is not defined because division by zero is not permitted. Therefore, the function is not continuous at $x=3$. However, the function is continuous for all values of x for which $x \neq 3$.

Fig. 4.4*a* shows a sketch of this function. Based on our intuitive description of continuity we can easily verify that the function is continuous for all $x \neq 3$ and discontinuous at $x=3$.

Example 4.13 Determine whether the function

$$f(x) = \begin{cases} -x, & x<0 \\ 2, & x=0 \\ x, & x>0 \end{cases}$$

is continuous at $x=0$.

Solution Since $f(0)=2$, we know that $f(0)$ exists. Moreover, as x approaches zero from the left, $f(x)$ approaches zero; likewise, as x approaches zero from

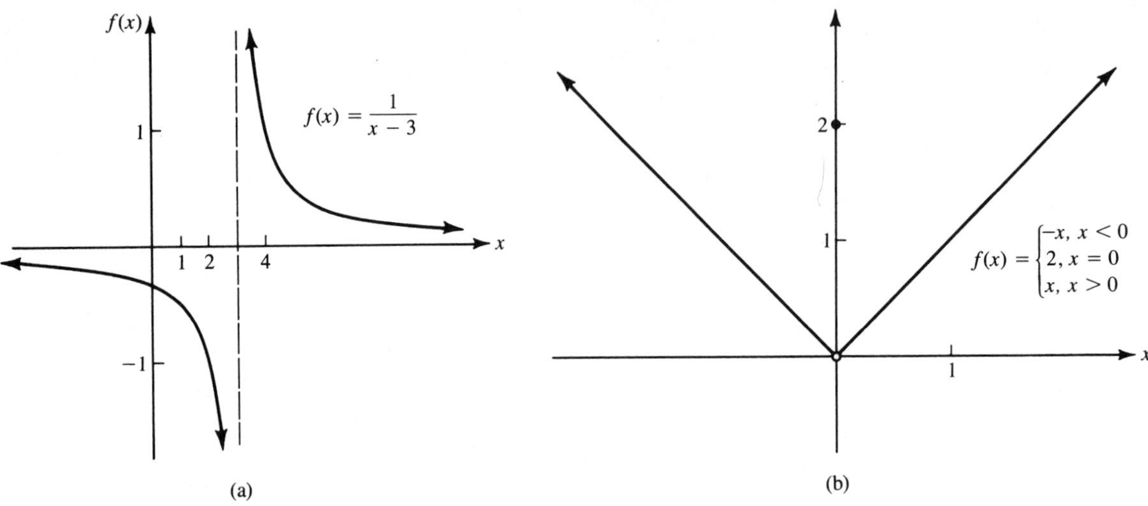

FIGURE 4.4

the right, $f(x)$ approaches zero. Hence, $\lim_{x\to 0} f(x)=0$. However, since $\lim_{x\to 0} f(x)\neq f(0)$, the function is not continuous at $x=0$. Fig. 4.4*b* shows a sketch of this function. We can verify that it is continuous where it is represented by a smooth curve without holes or jumps.

Since continuous functions are of utmost importance in the study of calculus, we list the following three properties, which require proofs that are beyond the scope of this book. For each property the domain D is the closed interval $[a,b]$. You should study them until you *comprehend* their meanings and can sketch appropriate curves to *illustrate* their meanings.

Property 4.1 If a function $f(x)$ is continuous in D, then $f(x)$ has a greatest value and a least value in the range.

Property 4.2 A function $f(x)$ that is continuous in D cannot pass from one value in the range to another value in the range without assuming every intermediate value at least once.

Property 4.3 If a function $f(x)$ is continuous in D and if $f(a)$ and $f(b)$ are opposite in sign, then there exists at least one value $x=c$ in the domain for which $f(c)=0$.

From the definitions of a *limit* and of a *continuous function*, it can be shown that the sums, differences, and products of two continuous functions are themselves continuous functions. Furthermore, the quotient of two continuous functions is continuous at every point for which the denominator does not equal zero. Nearly all of the functions with which we shall be concerned in the remainder of this textbook are continuous. In particular, all *polynomials* and *rational functions* (quotients of polynomials) are continuous, except at points where the denominator equals zero. This statement is of tremendous importance. If $P(x)$ is a polynomial function and we wish to determine $\lim_{x \to a} P(x)$, we can conclude immediately without using our limit rules that $\lim_{x \to a} P(x) = P(a)$ since every polynomial function is continuous for all real numbers a. Furthermore, if $f(x) = P(x)/Q(x)$ is a rational function and we wish to obtain $\lim_{x \to a} f(x)$, we know, since all rational functions are continuous on their domains, that $\lim_{x \to a} f(x) = P(a)/Q(a)$, provided $Q(a) \neq 0$. You can see this is a very powerful result that provides us with the means of finding the limits of continuous functions with ease.

The limit of the quotient $[f(x+h) - f(x)]/h$ as h approaches zero is very important in the development of differential calculus. The quotient $[f(x+h) - f(x)]/h$ is called the *difference quotient* for the function $f(x)$. The limit of the difference quotient, if it exists, as the variable h approaches zero will depend on the function $f(x)$ that is being considered. We will illustrate this in the next three examples.

Example 4.14 If $f(x) = x^2 + 2x + 3$, find

$$\lim_{h \to 0} \left[\frac{f(x+h) - f(x)}{h} \right]$$

Solution $f(x+h)$ is found by replacing x by $(x+h)$ in the equation $f(x) = x^2 + 2x + 3$:

$$\begin{aligned} f(x+h) &= (x+h)^2 + 2(x+h) + 3 \\ &= x^2 + 2xh + h^2 + 2x + 2h + 3 \end{aligned}$$

Since $f(x) = x^2 + 2x + 3$, the difference $f(x+h) - f(x) = 2xh + h^2 + 2h$ and the difference quotient is

$$\left[\frac{f(x+h) - f(x)}{h} \right] = \left(\frac{2xh + h^2 + 2h}{h} \right)$$

Therefore,

$$\lim_{h \to 0} \left[\frac{f(x+h) - f(x)}{h} \right] = \lim_{h \to 0} \left(\frac{2xh + h^2 + 2h}{h} \right)$$

where h is the variable and x is assumed constant independent of h. For all values of $h \neq 0$:

$$\frac{2xh + h^2 + 2h}{h} = 2x + h + 2$$

Therefore, by substitution,

$$\lim_{h\to 0}\left(\frac{2xh+h^2+2h}{h}\right) = \lim_{h\to 0}(2x+h+2)$$

Since $2x$ and 2 are constants independent of h, $(2x+h+2)$ is a polynomial function in terms of h, and

$$\lim_{h\to 0}(2x+h+2) = 2x+2$$

and we conclude: if $f(x)=x^2+2x+3$, then

$$\lim_{h\to 0}\left[\frac{f(x+h)-f(x)}{h}\right] = 2x+2$$

Example 4.15 If $f(x)=1/x$, find

$$\lim_{h\to 0}\left[\frac{f(x+h)-f(x)}{h}\right]$$

Solution $f(x+h)$ found by replacing x by $(x+h)$ in the equation $f(x)=1/x$ is

$$f(x+h) = \frac{1}{x+h}$$

and the difference $f(x+h)-f(x)=\frac{1}{x+h}-\frac{1}{x}$; hence the difference quotient

$$\frac{f(x+h)-f(x)}{h} = \frac{\left(\frac{1}{x+h}-\frac{1}{x}\right)}{h}$$

To simplify this quotient we express the numerator $\frac{1}{x+h}-\frac{1}{x}$ with a common denominator; that is,

$$\begin{aligned}\frac{1}{x+h}-\frac{1}{x} &= \frac{x\cdot 1}{x(x+h)}-\frac{1\cdot(x+h)}{x(x+h)}\\ &= \frac{x-(x+h)}{x(x+h)}\\ &= \frac{-h}{x(x+h)}\end{aligned}$$

Hence,

$$\left(\frac{\frac{1}{x+h}-\frac{1}{x}}{h}\right) = \frac{\frac{-h}{x(x+h)}}{h} = \frac{-h}{x(x+h)}\cdot\frac{1}{h}$$

Since $[-h/x(x+h)]\cdot 1/h=[-1/x(x+h)]$ for all $h\neq 0$, we can write

$$\lim_{h\to 0}\left[\frac{f(x+h)-f(x)}{h}\right]=\lim_{h\to 0}\left[\frac{-1}{x(x+h)}\right]=\lim_{h\to 0}\left(\frac{-1}{x^2+hx}\right)$$

By Rule 4.4

$$\lim_{h\to 0}\left(\frac{-1}{x^2+hx}\right)=\frac{\lim_{h\to 0}(-1)}{\lim_{h\to 0}(x^2+hx)}$$

By Rule 4.1, $\lim_{h\to 0}(-1)=-1$, and since (x^2+hx) is a polynomial in h, $\lim_{h\to 0}(x^2+hx)=x^2$. Hence, for all $x\neq 0$,

$$\lim_{h\to 0}\left(\frac{-1}{x^2+hx}\right)=\frac{-1}{x^2}$$

and we conclude: if $f(x)=1/x$, then

$$\lim_{h\to 0}\left[\frac{f(x+h)-f(x)}{h}\right]=\frac{-1}{x^2}$$

for all $x\neq 0$.

Example 4.16 If $f(x)=\sqrt{x}$, find

$$\lim_{h\to 0}\left[\frac{f(x+h)-f(x)}{h}\right]$$

Solution $f(x+h)=\sqrt{x+h}$; hence, $f(x+h)-f(x)=\sqrt{x+h}-\sqrt{x}$ and the difference quotient

$$\frac{f(x+h)-f(x)}{h}=\frac{\sqrt{x+h}-\sqrt{x}}{h}$$

Therefore,

$$\lim_{h\to 0}\left[\frac{f(x+h)-f(x)}{h}\right]=\lim_{h\to 0}\left[\frac{\sqrt{x+h}-\sqrt{x}}{h}\right]$$

We cannot apply Rule 4.4 at this point to evaluate $\lim_{h\to 0}[(\sqrt{x+h}-\sqrt{x})/h]$ since $\lim_{h\to 0}(h)=0$. To evaluate $\lim_{h\to 0}[(\sqrt{x+h}-\sqrt{x})/h]$, we rationalize the numerator as follows:

$$\frac{\sqrt{x+h}-\sqrt{x}}{h}=\left(\frac{\sqrt{x+h}-\sqrt{x}}{h}\right)\left(\frac{\sqrt{x+h}+\sqrt{x}}{\sqrt{x+h}+\sqrt{x}}\right)$$

$$=\frac{(x+h)-x}{h(\sqrt{x+h}+\sqrt{x})}$$

$$=\frac{h}{h(\sqrt{x+h}+\sqrt{x})}$$

Hence, for all $h \neq 0$,

$$\frac{\sqrt{x+h}-\sqrt{x}}{h}=\frac{1}{\sqrt{x+h}+\sqrt{x}}$$

Therefore,

$$\lim_{h\to 0}\left[\frac{f(x+h)-f(x)}{h}\right]=\lim_{h\to 0}\left(\frac{1}{\sqrt{x+h}+\sqrt{x}}\right)$$

If $x \neq 0$, we can apply Rule 4.4:

$$\lim_{h\to 0}\left(\frac{1}{\sqrt{x+h}+\sqrt{x}}\right)=\frac{\lim_{h\to 0}(1)}{\lim_{h\to 0}(\sqrt{x+h}+\sqrt{x})}$$

By applying Rules 4.2 and 4.6 we find $\lim_{h\to 0}(\sqrt{x+h}+\sqrt{x})=2\sqrt{x}$; hence,

$$\lim_{h\to 0}\frac{1}{\sqrt{x+h}+\sqrt{x}}=\frac{1}{2\sqrt{x}}$$

when $x \neq 0$. Therefore, we can conclude if $f(x)=\sqrt{x}$, then

$$\lim_{h\to 0}\left[\frac{f(x+h)-f(x)}{h}\right]=\frac{1}{2\sqrt{x}}$$

for all $x \neq 0$.

EXERCISES

1. For each of the functions in the accompanying illustration, determine the limit, if it exists, of $f(x)$ as x approaches a.

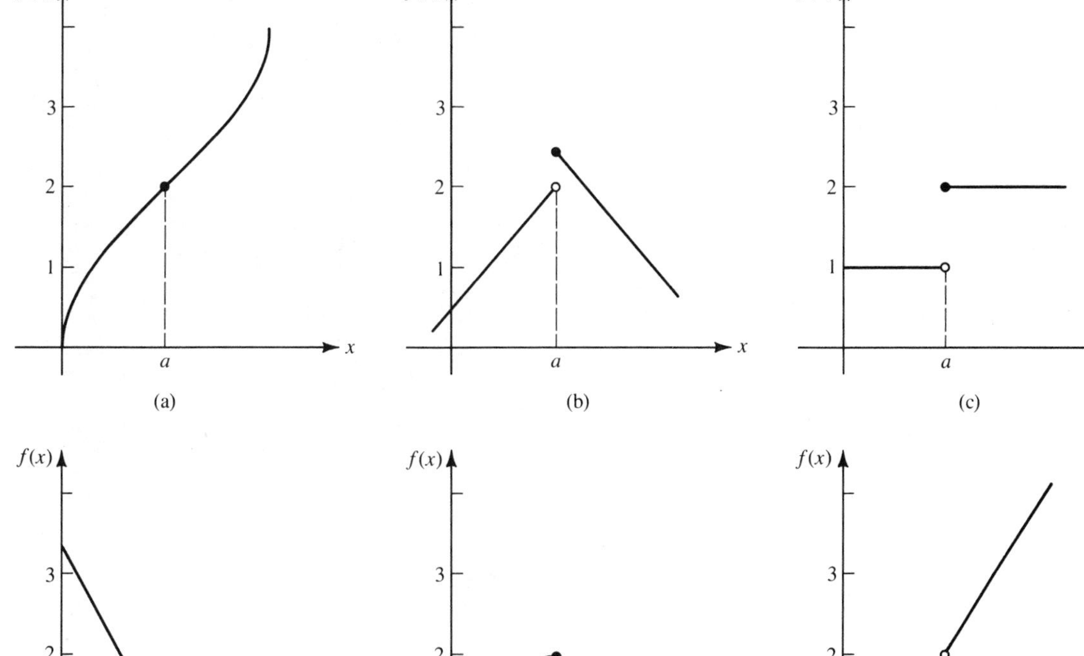

EXERCISE 1

2. Determine each of the following limits by determining the values of the function for 3 values to the left and 3 values to the right of the value a of interest.

(a) $\lim_{x\to 3} (x^2+2)$ (b) $\lim_{x\to -4} (x^2+3x)$

(c) $\lim_{h\to 0} (3-h^2)$ (d) $\lim_{h\to 5} \left(1+\frac{5}{h}\right)$

3. Use the limit rules to determine each limit in Exercise 2.

4. Consider each of the functions in Exercise 2 and determine which are rational functions. (*Note*: Every polynomial function is a rational function.) Using the fact that rational functions are continuous for all real numbers where the denominator is not zero, determine the limit of each of the functions in Exercise 2.

5. Compare your results in Exercises 2, 3, and 4.

6. Using any appropriate technique, determine each of the following limits.

(a) $\lim\limits_{x\to 3}\left(1+\dfrac{1}{x^2}\right)$ (b) $\lim\limits_{h\to 0}(5+2h-3h^2)$

(c) $\lim\limits_{x\to 4}\left(x^2+1-\dfrac{x}{x-2}\right)$ (d) $\lim\limits_{h\to 0}\left(\dfrac{h^2+3h}{h}\right)$

(e) $\lim\limits_{x\to -2}\left(3-\dfrac{x+2}{x-2}\right)$ (f) $\lim\limits_{h\to 0}(h^2+3x+7)$ (assume x is a constant independent of h)

(g) $\lim\limits_{x\to 4}\left(\dfrac{x-4}{\sqrt{x}-2}\right)$ (h) $\lim\limits_{x\to 1}\left(\dfrac{x^3-x^2+2x-2}{x-1}\right)$

(i) $\lim\limits_{x\to 3} f(x)$ where $f(x)=\begin{cases}4, & x>3\\ 2, & x=3\\ 0, & x<3\end{cases}$ (j) $\lim\limits_{x\to 12}\sqrt{x+4}$

(k) $\lim\limits_{x\to 5}\sqrt{2x-10}$ (l) $\lim\limits_{x\to 5} g(x)$ where

$$g(x)=\begin{cases}11-x, & x>5\\ 6, & x=5\\ x+1, & x<5\end{cases}$$

(m) $\lim\limits_{x\to 2} f(x)$ where $f(x)=\begin{cases}5-2x, & x>2\\ 3, & x=2\\ x-1, & x<2\end{cases}$ (n) $\lim\limits_{h\to 0}\dfrac{x^2}{(1+hx)}$ (assume x is a constant independent of h)

7. Determine each of the following limits.

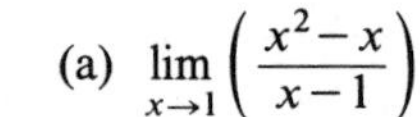

(a) $\lim\limits_{x\to 1}\left(\dfrac{x^2-x}{x-1}\right)$ (b) $\lim\limits_{h\to 2}\left(\dfrac{h^2-4}{h-2}\right)$

(c) $\lim\limits_{x\to 0}\left(\dfrac{x^3+x^2}{x^2+2x}\right)$ (d) $\lim\limits_{h\to 0}\left(\dfrac{h^4+h^2}{h^3+3h^2}\right)$

(e) $\lim\limits_{x\to -3}\left(\dfrac{x^3-9x}{x+3}\right)$ (f) $\lim\limits_{y\to -2}\left(\dfrac{2y^3+4y^2}{y+2}\right)$

(g) $\lim\limits_{x\to 2}\left(\dfrac{x^2-5x+6}{x^2-x-2}\right)$ (h) $\lim\limits_{y\to -1}\left(\dfrac{y^2-3y-4}{y^2-2y-3}\right)$

(*Note*: In parts (b), (d), (f), and (h) we have changed the variable. This, of course, makes no difference in our approach to the problem.)

8. Which of the functions in Exercise 1 are continuous at $x=a$?

9. Determine the values of x for which each of the following functions is discontinuous.

(a) $f(x)=\dfrac{x-1}{x+1}$ (b) $f(x)=\dfrac{2x^2+1}{x^2}$

(c) $f(x)=\dfrac{3x}{4-x^2}$ (d) $f(x)=\dfrac{x}{x^2+3x-4}$

(e) $f(x)=\dfrac{x-1}{x^2-1}$ (f) $f(x)=\dfrac{x^3+x}{x^2}$

(g) $f(x)=\dfrac{3x+1}{x(x^2+4)}$ (h) $f(x)=\dfrac{x+2}{x^2+4x+4}$

10. Determine whether the following functions are continuous at the given point.

(a) $f(x)=\dfrac{x}{x+3}$ (at $x=3$) (b) $f(x)=\dfrac{x^2+2x}{x+2}$ (at $x=-2$)

(c) $f(x)=\dfrac{x^2-x}{x-1}$ (at $x=1$) (d) $f(x)=\dfrac{x^3-9x}{x+3}$ (at $x=-3$)

(e) $f(x)=\dfrac{x^2-5x}{x^3-5x^2}$ (at $x=0$) (f) $f(x)=\dfrac{x^2-2x-3}{x^2-3x-4}$ (at $x=4$)

11. If $f(x)=x^2$

(a) Find the quotient $\left[\dfrac{f(5+h)-f(5)}{h}\right]$

(b) Evaluate $\lim_{h\to 0}\left[\dfrac{f(5+h)-f(5)}{h}\right]$

12. If $g(x)=x^2+3$

(a) Find the quotient $\left[\dfrac{g(a+h)-g(a)}{h}\right]$ where a is a constant.

(b) Evaluate $\lim_{h\to 0}\left[\dfrac{g(a+h)-g(a)}{h}\right]$

13. If $f(x)=3x^2+2x+1$

(a) Find the quotient $\left[\dfrac{f(x+h)-f(x)}{h}\right]$

(b) Evaluate $\lim_{h\to 0}\left[\dfrac{f(x+h)-f(x)}{h}\right]$ (*Hint*: Assume x is a constant independent of h.)

14. If $f(x)=\dfrac{1}{x}+2$

(a) Find the quotient $\left[\dfrac{f(x+h)-f(x)}{h}\right]$

(b) Evaluate $\lim_{h\to 0}\left[\dfrac{f(x+h)-f(x)}{h}\right]$

15. If $f(x)=\dfrac{1}{x^2}$

(a) Find the quotient $\left[\dfrac{f(x+h)-f(x)}{h}\right]$

(b) Evaluate $\lim_{h\to 0}\left[\dfrac{f(x+h)-f(x)}{h}\right]$

16. If $f(x) = \dfrac{1}{\sqrt{x}}$

(a) Find the quotient $\left[\dfrac{f(x+h)-f(x)}{h}\right]$

(b) Evaluate $\lim\limits_{h\to 0}\left[\dfrac{f(x+h)-f(x)}{h}\right]$

17. Sketch a figure to illustrate (a) Property 4.1, (b) Property 4.2, and (c) Property 4.3.

4.3 DERIVATIVE OF A FUNCTION

The concepts of differential calculus are based on the idea of the *rate of change* of a function. Figure 4.5 shows a company's yearly profit in millions of dollars as related to the number of years it has been in operation. Management is not only interested in the fact that the yearly profit was \$20 million at the end of the fifth year, but they are also pleased to note that the profit was increasing rapidly at that time. A measure of this increase is called the rate of change of the function. At the end of 10 years, management may be very satisfied with a profit of \$30 million but disturbed to note that the "rate of increase" is not large.

Intuitively we know what we mean by saying a function is increasing rapidly at some point in its domain or that it shows little or no increase at this point. However, the question arises, how can we define this concept so that we can measure it? One way of doing this would be to define an average rate

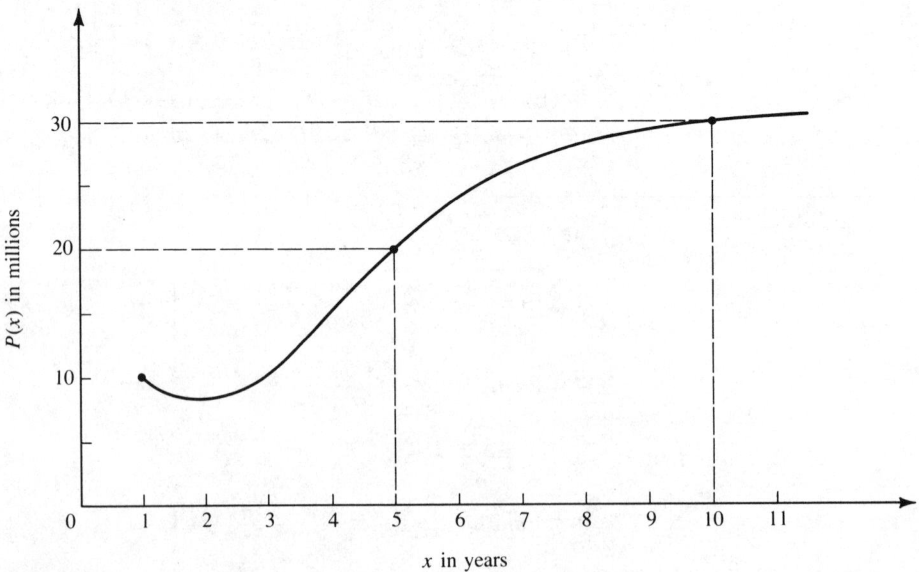

FIGURE 4.5

of change; for example, Fig. 4.5 shows us that our profit increases from 20 million to 30 million from the fifth year to the tenth year, or an average increase of

$$\frac{30-20}{10-5} = \frac{10}{5} = 2; \text{ i.e., } \$2{,}000{,}000/\text{year}$$

If we are interested in the rate of change of the profit at the end of the fifth year, perhaps a better average to consider would be over the four-year period from the fifth to the ninth year:

$$\frac{P(9)-P(5)}{9-5} = \frac{29-20}{4}$$

$$= \frac{9}{4} = 2.25; \text{ i.e., } \$2{,}250{,}000/\text{year}$$

or perhaps over the one-year period from the fifth to the sixth year:

$$\frac{P(6)-P(5)}{6-5} = \frac{23-20}{1}$$

$$= \frac{3}{1} = 3; \text{ ie., } \$3{,}000{,}000/\text{year}$$

Now these are all helpful average figures, but we are interested in how fast the profit is increasing right at the end of the fifth year. This is called the *instantaneous rate of change* or the rate of change of the function. Further inspection of our graph perhaps will help us to define this concept.

Geometrically you may have noted that the average rate of change we have defined is related to the calculation of a slope. In particular if you consider Fig. 4.6, you will note that the average rate of change of the profit is nothing more than the slope of the lines S_1, S_2, and S_3.

Note:

$$\text{Slope } S_1\text{: } m_{s_1} = \frac{P(10)-P(5)}{10-5}$$

$$\text{Slope } S_2\text{: } m_{s_2} = \frac{P(9)-P(5)}{9-5}$$

$$\text{Slope } S_3\text{: } m_{s_3} = \frac{P(6)-P(5)}{6-5}$$

The lines S_1, S_2, and S_3 are called secant lines, similar to the secant lines shown in Fig. 4.7. You are probably familiar with such lines from your study of geometry.

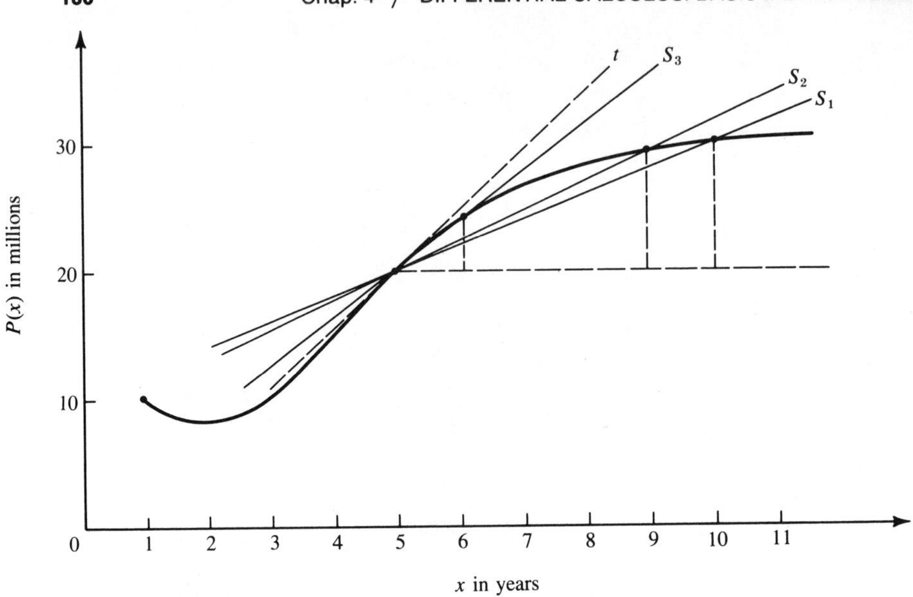

FIGURE 4.6

If the average rate of change is given by the slope of the secant lines, intuitively the instantaneous rate of change of the profit function at $x=5$ should be the slope of the dotted line drawn to the curve shown in Fig. 4.6. This dotted line is called the tangent line to the curve drawn at the point $(5, P(5))$.

From an inspection of the slopes of the secant lines, we will arrive at a method of calculating the *slope of the tangent* line at our point of interest and hence the *rate of change* of the profit at this point. The slope of each of the

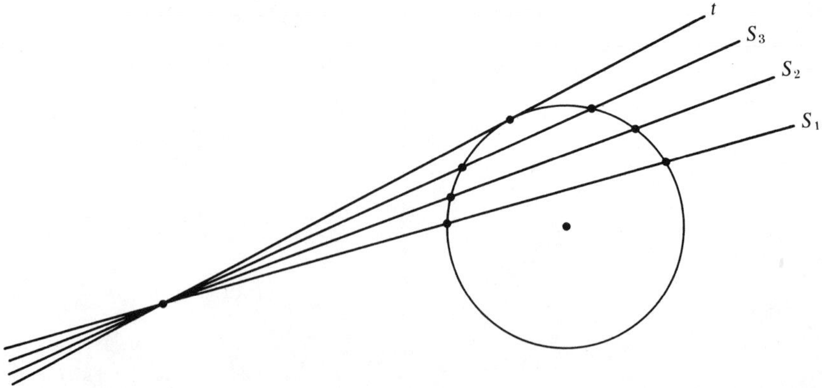

FIGURE 4.7 S_1, S_2, and S_3 are secant lines; t is called a *tangent line.*

secant lines can be expressed as

$$m_s = \frac{P(5+h)-P(5)}{(5+h)-5}$$

where h assumes values of 5, 2, and 1, respectively. Figure 4.8 is a graphical representation of this situation.

The slope, m_s, of a secant line through $(5, P(5))$ can be rewritten as the quotient:

$$\frac{P(5+h)-P(5)}{h}$$

We have referred to this quotient before. It is called the difference quotient for the function $P(x)$. From Fig. 4.8 it can be seen that as Q moves toward R on the curve, the slopes of successive secant lines get closer and closer to the slope of the tangent line t. Note in Fig. 4.8 that as h gets smaller and smaller, $Q \to R$; hence we can conclude that $m_s \to m_t$ as $h \to 0$. We therefore define the slope, m_t, of the tangent line at the point $(5, P(5))$ to be

$$m_t = \lim_{h\to 0}(m_s)$$

$$m_t = \lim_{h\to 0}\left[\frac{P(5+h)-P(5)}{h}\right]$$

This limit gives us the instantaneous rate of change of the profit at the end of the fifth year.

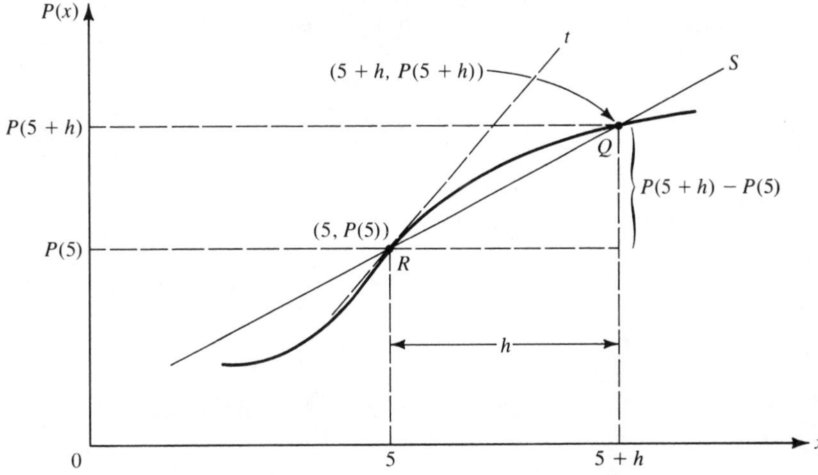

FIGURE 4.8

If we are interested in the instantaneous rate of change of the profit at the end of the third year, we would consider

$$\lim_{h\to 0}\left[\frac{P(3+h)-P(3)}{h}\right]$$

or at the end of the xth year:

$$\lim_{h\to 0}\left[\frac{P(x+h)-P(x)}{h}\right]$$

To evaluate this limit we need an explicit expression for our profit function. For simplicity let us assume our profit function is given by the algebraic expression $P(x)=x^2+5$. The graph of $P(x)$ is shown in Fig. 4.9.

From Fig. 4.9 we can see graphically that the profit function is increasing. What is the rate of change or rate of increase of the profit at the end of the second year; that is, when $x=2$? Based on our discussion, we wish to evaluate

$$\lim_{h\to 0}\left[\frac{P(2+h)-P(2)}{h}\right]$$

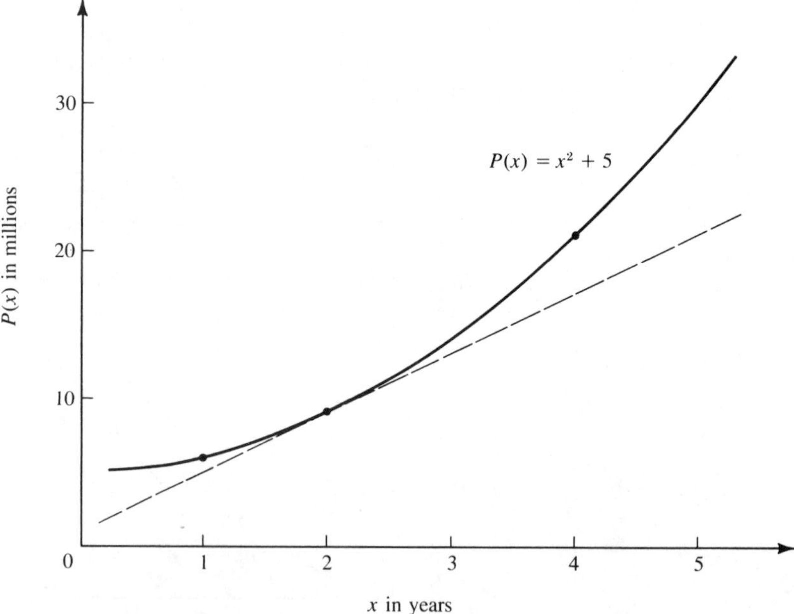

FIGURE 4.9

Since $P(x)=x^2+5$, we evaluate $P(2+h)$ and $P(2)$ as follows:

$$\begin{aligned} P(2+h) &= (2+h)^2+5 \\ &= 4+4h+h^2+5 \\ &= 9+4h+h^2 \\ P(2) &= (2)^2+5 \\ &= 9 \end{aligned}$$

Therefore, $P(2+h)-P(2)=h^2+4h$, and

$$\lim_{h\to 0}\left[\frac{P(2+h)-P(2)}{h}\right] = \lim_{h\to 0}\left(\frac{h^2+4h}{h}\right)$$

From our previous work on limits, we know we cannot apply Rule 4.4 since the limit of the denominator is zero. However, when $h\neq 0$,

$$\frac{h^2+4h}{h} = h+4$$

and by substitution we have

$$\begin{aligned} \lim_{h\to 0}\left(\frac{h^2+4h}{h}\right) &= \lim_{h\to 0}(h+4) \\ &= 4 \end{aligned}$$

Since the profit was measured in millions of dollars, we conclude that at the end of 2 years the profit is *increasing at the rate of* \$4 *million/year* or the rate of change of the profit is \$4 million/year.

Example 4.17 Rate of Change of Profit For the profit function $P(x)=x^2+5$ determine the rate of change of the profit at the end of 4 years.

Solution The rate of change of the profit in millions of dollars/year at the end of 4 years is

$$\lim_{h\to 0}\left[\frac{P(4+h)-P(4)}{h}\right]$$

Since $P(x)=x^2+5$, we evaluate $P(4+h)$ and $P(4)$ as follows:

$$\begin{aligned} P(4+h) &= (4+h)^2+5 \\ &= 16+8h+h^2+5 \\ &= 21+8h+h^2 \\ P(4) &= 21 \end{aligned}$$

Therefore,

$$\frac{P(4+h)-P(4)}{h} = \left[\frac{(21+8h+h^2)-21}{h}\right]$$

$$= \left(\frac{8h+h^2}{h}\right) = 8+h$$

for $h \neq 0$, and by substitution:

$$\lim_{h\to 0}\left[\frac{P(4+h)-P(4)}{h}\right] = \lim_{h\to 0}(8+h) = 8$$

Therefore, the profit is increasing at the rate of 8 million dollars/year. We can see that this is a reasonable solution by considering the sketch drawn in Fig. 4.9. Clearly the profit is increasing faster at the end of 4 years than at the end of 2 years.

Example 4.18 For the same profit function, determine the rate of change of the profit in general by finding it at the end of x years.

Solution The rate of change of the profit at the end of x years is given by:

$$\lim_{h\to 0}\left[\frac{P(x+h)-P(x)}{h}\right]$$

Since $P(x)=x^2+5$, we evaluate $P(x+h)$ and $P(x)$ as follows:

$$P(x+h) = (x+h)^2+5$$

$$= x^2+2\,xh+h^2+5$$

$$P(x) = x^2+5$$

and

$$\frac{P(x+h)-P(x)}{h} = \frac{(x^2+2xh+h^2+5)-(x^2+5)}{h}$$

$$= \frac{h^2+2xh}{h}$$

Therefore,

$$\lim_{h\to 0}\left[\frac{P(x+h)-P(x)}{h}\right] = \lim_{h\to 0}\left[\frac{h^2+2xh}{h}\right]$$

$$= \lim_{h\to 0}[h+2x]$$

$$= 2x$$

Hence, the rate of change of $P(x)=2x$.

Table 4.1 shows the profit and rate of change of the profit for several values of x. Note that when $x=2$ the rate of change of the profit is 4, and when $x=4$ the rate of change of the profit is 8. Thus our general formula verifies the two special cases considered.

Table 4.1

x	$P(x)=x^2+5$ *(millions of \$)*	*Rate of change of* $P(x)=2x$ *(millions of \$/yr)*
1	6	2
2	9	4
3	14	6
4	21	8
5	30	10

From our discussion we have shown that there are two very important interpretations of:

$$\lim_{h\to 0}\left[\frac{f(x+h)-f(x)}{h}\right]$$

The first is that this limit represents the *rate of change of the function* $f(x)$ at the point x, and secondly that this limit represents the *slope of the tangent line* drawn to the graph of $f(x)$ at the point $(x,f(x))$. There are other interpretations that we shall introduce in Section 4.4. This limit is so fundamental and significant that we give it a special name, the *derivative of the function* $f(x)$.

Definition 4.3 The derivative of a function $f(x)$ with respect to x is the limit of the quotient $[f(x+h)-f(x)]/h$ as h approaches zero and is given by the formula

$$f'(x) = \lim_{h\to 0}\left[\frac{f(x+h)-f(x)}{h}\right]$$

whenever the limit exists.

We use the notation $f'(x)$, read "*f prime of x*," to denote the derivative of $f(x)$. Other notations that are frequently used for the derivative of the function $f(x)$ with respect to x are $D_x f(x)$, $(d/dx)f(x)$, and $\dot{f}(x)$. In this text we will primarily use $f'(x)$, but occasionally we will need the $D_x f(x)$ notation.

Example 4.19 Given the function $f(x)=x^3$,

(a) determine the derivative of the function at the point $(2,f(2))$;
(b) find the slope of the tangent line drawn to the curve at this point;
(c) find the rate of change of the function at $x=2$.

Solution

(a) To determine the derivative of $f(x)$ at the point $(2,f(2))$, we have a choice of two different procedures. We could evaluate

$$\lim_{h\to 0}\left[\frac{f(2+h)-f(2)}{h}\right]$$

directly, or we could find

$$f'(x) = \lim_{h\to 0}\left[\frac{f(x+h)-f(x)}{h}\right]$$

and then replace x by 2. Let us show that both give us the same result, which we will denote by $f'(2)$. First we solve for $f'(2)$ by evaluating the following limit.

$$\lim_{h\to 0}\left[\frac{f(2+h)-f(2)}{h}\right]$$

Since $f(x)=x^3$, we have

$$\begin{aligned} f(2+h) &= (2+h)^3 \\ &= (2+h)(2+h)^2 \\ &= (2+h)(4+4h+h^2) \\ &= 8+12h+6h^2+h^3 \end{aligned}$$

and $$f(2) = 8$$

Therefore,

$$f(2+h)-f(2) = h^3+6h^2+12h$$

and $$f'(2) = \lim_{h\to 0}\left(\frac{h^3+6h^2+12h}{h}\right)$$

$$f'(2) = \lim_{h\to 0}(h^2+6h+12)$$

Hence $$f'(2) = 12$$

Secondly, we find $f'(x)$ by evaluating the following limit.

$$\lim_{h\to 0}\left[\frac{f(x+h)-f(x)}{h}\right]$$

Since $f(x)=x^3$, we have

$$f(x+h) = (x+h)^3$$
$$= (x+h)(x+h)^2$$
$$= x^3 + 3x^2h + 3xh^2 + h^3$$

and $$f(x) = x^3$$

Therefore, $$f(x+h) - f(x) = 3x^2h + 3xh^2 + h^3$$

and $$f'(x) = \lim_{h\to 0}\left(\frac{3x^2h+3xh^2+h^3}{h}\right)$$
$$= \lim_{h\to 0}(3x^2+3xh+h^2)$$

Hence $$f'(x) = 3x^2$$

Since $f'(x)=3x^2$ is a function of x, we evaluate $f'(2)$ by substituting $x=2$:

$$f'(2) = 12$$

As expected, both computations lead to the same conclusion, $f'(2)=12$. It is clear that if we want the derivative of the function at several values of x it is easier to find $f'(x)$ first and then substitute the desired value for x.

(b) Since we can interpret $f'(2)$ as the slope of the tangent drawn to the graph of $f(x)$ at $(2,f(2))$, we have

$$m_t = f'(2) = 12$$

(c) Another interpretation of $f'(2)$ is the rate of change of the function $f(x)$ at $x=2$. Therefore, $f'(2)=12$ can be interpreted to mean the function is increasing at the rate of 12 units per unit change in x when $x=2$.

Example 4.20 Slope of a Tangent Line Find the slopes of the tangent lines to $f(x)=x^3$ at each point in the following set

$$\{(0,f(0)), (1,f(1)), (-1,f(-1)), (2,f(2)), (-2,f(-2))\}$$

and compare these slopes to those of the tangent lines drawn in Fig. 4.10.

Solution We have shown that if $f(x)=x^3$, then $f'(x)=3x^2$. The slopes of the tangent lines can be found by evaluating $f'(x)$ at each value of x.

x	0	1	-1	2	-2
$f'(x)=3x^2$	0	3	3	12	12

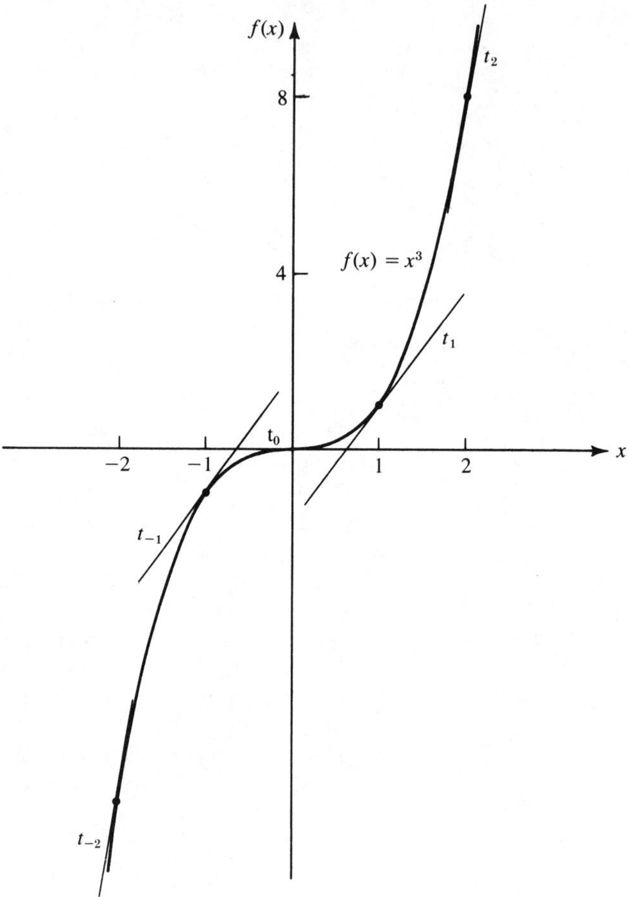

FIGURE 4.10

An inspection of Fig. 4.10 shows that the slopes of the tangent lines drawn agree with our calculated results.

It is common practice to refer to the slope of the tangent line drawn to the graph of a function as the slope of the curve at that point. Let us verify that Definition 4.3 corresponds with the concept of slope for a linear function. For example, suppose $f(x)=3x+2$. From our discussion of the slope of a line in Chapter 2, we know the slope of this line is 3. Using Definition 4.3 to determine the derivative and hence the slope of the graph of the function, we have

$$
\begin{aligned}
f'(x) &= \lim_{h\to 0}\left[\frac{f(x+h)-f(x)}{h}\right]\\
&= \lim_{h\to 0}\left\{\frac{[3(x+h)+2]-(3x+2)}{h}\right\}\\
&= \lim_{h\to 0}\left(\frac{3h}{h}\right)\\
&= \lim_{h\to 0}(3)\\
&= 3
\end{aligned}
$$

This verifies that for the linear function $f(x)=3x+2$ the derivative, given by Definition 4.3, can be interpreted as the slope of the graph of the function.

Now we can generalize this result by considering the general linear function $f(x)=mx+b$ and its derivative. Using Definition 4.3, we have

$$
\begin{aligned}
f'(x) &= \lim_{h\to 0}\left[\frac{f(x+h)-f(x)}{h}\right]\\
&= \lim_{h\to 0}\left\{\frac{[m(x+h)+b]-(mx+b)}{h}\right\}\\
&= \lim_{h\to 0}\left(\frac{mh}{h}\right)\\
&= \lim_{h\to 0}(m) \qquad (\textit{Note}\text{: } m \text{ is a constant.})\\
&= m
\end{aligned}
$$

Therefore $f'(x)=m$ for all x, which verifies that the derivative of any linear function is the same as the slope of the line itself.

Frequently when discussing linear functions we use the form $y=mx+b$. In this case and others where we choose to represent the second component of the ordered pair $(x,f(x))$ by y rather than $f(x)$, we denote the derivative by dy/dx, $D_x y$, or y'. The dy/dx notation, read "dy, dx," is the most common. Note that the function notation, $f'(x)$, has an advantage as it represents the derivative of $f(x)$ at any point x. If we wish to determine the derivative at $x=c$, we evaluate $f'(x)$ at $x=c$ and denote it by $f'(c)$.

We shall find that there are mathematical problems in the fields of business, biological, and social science that can be solved by applying the derivative. Since there are so many applications of the derivative, we will introduce shortcut methods to determine the derivative rather than applying the definition each time. The following problems will help us establish some simple rules that will be generalized in Sec. 4.4.

To determine the derivative for a general function of the form $f(x)=k$, where k is a constant, we apply Definition 4.3.

$$f'(x) = \lim_{h \to 0} \left[\frac{f(x+h)-f(x)}{h} \right]$$

If $f(x)=k$, then $f(x+h)=k$. Therefore,

$$\begin{aligned} f'(x) &= \lim_{h \to 0} \left(\frac{k-k}{h} \right) \\ &= \lim_{h \to 0} \left(\frac{0}{h} \right) = \lim_{h \to 0} (0) = 0 \end{aligned}$$

Therefore, $f'(x)=0$ for all real numbers x. You may have anticipated this result by recalling that the graph of $f(x)=\text{constant}$ is a horizontal straight line whose slope is zero. Formalizing this result, we have the following rule.

Rule 4.7 If k is a constant and $f(x)=k$, then $f'(x)=0$, or

$$D_x f(x) = D_x(k) = 0 \tag{4.8}$$

Determining the derivative of a function by application of rules such as Rule 4.7 is called differentiation by *inspection*, which is the purpose for developing rules for general types of functions. Although we will state our rules in terms of the function $f(x)$, it should be clear that we could use $g(x)$, $h(x)$, $P(x)$, etc. to represent the function, and $g'(x)$, $h'(x)$, $P'(x)$, etc. to represent the derivative.

Example 4.21 Differentiate the following functions by inspection.

(a) $f(x)=8$ (b) $g(x)=-4$
(c) $f(x)=\pi$ (d) $h(x)=\frac{1}{3}$

Solution We obtain the results for each part by applying Rule 4.7; that is,

(a) $f'(x)=0$ or $D_x(8)=0$
(b) $g'(x)=0$ or $D_x(-4)=0$
(c) $f'(x)=0$ or $D_x(\pi)=0$
(d) $h'(x)=0$ or $D_x(\frac{1}{3})=0$

Example 4.22 Find the derivative of each of the following polynomial functions.

(a) $f(x)=x$ (b) $f(x)=x^2$
(c) $f(x)=x^3$ (d) $f(x)=x^4$

Solution

(a) Since $f(x)=x$ is a special case of a linear function where $m=1$, we know that $f'(x)=1$.

(b) To find $f'(x)$ where $f(x)=x^2$, we use Definition 4.3.

$$\begin{aligned} f'(x) &= \lim_{h\to 0}\left[\frac{f(x+h)-f(x)}{h}\right] \\ &= \lim_{h\to 0}\left[\frac{(x+h)^2-x^2}{h}\right] \\ &= \lim_{h\to 0}\left(\frac{x^2+2xh+h^2-x^2}{h}\right) \\ &= \lim_{h\to 0}(2x+h) \\ &= 2x \end{aligned}$$

(c) We have found in Example 4.19 that the derivative of $f(x)=x^3$ is

$$f'(x) = 3x^2$$

(d) The derivative of $f(x)=x^4$ by Definition 4.3 is

$$\begin{aligned} f'(x) &= \lim_{h\to 0}\left[\frac{f(x+h)-f(x)}{h}\right] \\ &= \lim_{h\to 0}\left[\frac{(x+h)^4-x^4}{h}\right] \\ &= \lim_{h\to 0}\left(\frac{x^4+4x^3h+6x^2h^2+4xh^3+h^4-x^4}{h}\right) \\ &= \lim_{h\to 0}(4x^3+6x^2h+4xh^2+h^3) \\ f'(x) &= 4x^3 \end{aligned}$$

From Example 4.22 you may have concluded that if $f(x)=x^n$ where n is any positive integer, then $f'(x)=nx^{n-1}$. This conclusion is true, and furthermore the result can be shown to hold for any real value of n as summarized in the following rule.

Rule 4.8 For any real number n, if $f(x)=x^n$, then $f'(x)=nx^{n-1}$ or

$$D_x f(x) = D_x(x^n) = nx^{n-1} \tag{4.9}$$

The proof of this rule is beyond the scope of this text.

Example 4.23 Differentiate each of the following functions.

(a) $f(x)=x^5$ (b) $g(x)=x^{12}$
(c) $h(x)=x^{-1}$ (d) $y=x^{2/3}$
(e) $R(x)=x^{-3/5}$ (f) $P(x)=\dfrac{1}{\sqrt[3]{x}}$ (*Hint*: $\dfrac{1}{\sqrt[3]{x}}=x^{-1/3}$.)

Solution Applying Rule 4.8 we obtain

(a) $f'(x)=5x^{5-1}=5x^4$ or $D_x(x^5)=5x^4$

(b) $g'(x)=12x^{12-1}=12x^{11}$ or $D_x(x^{12})=12x^{11}$

(c) $h'(x)=-1x^{-1-1}=-1x^{-2}=\dfrac{-1}{x^2}$ or $D_x(x^{-1})=\dfrac{-1}{x^2}$

(d) $\dfrac{dy}{dx}=\dfrac{2}{3}x^{2/3-1}=\dfrac{2}{3}x^{-1/3}=\dfrac{2}{3\sqrt[3]{x}}$ or $D_x(x^{2/3})=\dfrac{2}{3\sqrt[3]{x}}$

(e) $R'(x)=\dfrac{-3}{5}x^{-3/5-1}=\dfrac{-3}{5}x^{-8/5}=\dfrac{-3}{5\sqrt[5]{x^8}}$ or $D_x(x^{-3/5})=\dfrac{-3}{5\sqrt[5]{x^8}}$

(f) $P'(x)=-\dfrac{1}{3}x^{-1/3-1}=\dfrac{-1}{3}x^{-4/3}=\dfrac{-1}{3\sqrt[3]{x^4}}$ or $D_x\left(\dfrac{1}{\sqrt[3]{x}}\right)=\dfrac{-1}{3\sqrt[3]{x^4}}$

Since the derivative of a function at a particular point can be interpreted as the slope of the tangent line drawn to the curve at that point, it is necessary to be able to draw a tangent line to the curve to have a derivative at that point. It can be shown that if a function is discontinuous at a point its derivative at that point does not exist. Figs. 4.11*a* and *b* illustrate functions that are discontinuous at $x=a$ and have no tangent lines at that point.

However, if a function is continuous at $x=a$, this does not guarantee that the derivative exists at $x=a$. For example, the functions illustrated in Figs. 4.11*c* and *d* are continuous at $x=a$; however, no tangent lines can be drawn to the curves at those points and $f'(a)$ does not exist for either function.

Most of the functions we will consider in this text will have derivatives at the points of interest. Such functions are said to be differentiable.

As we have noted, the functional notation $f(x)$ is occasionally replaced by the use of the single variable y. We may also find cases where we are interested in $y=f(t)$ or $z=g(t)$. The use of symbols such as f, x, y, etc. are flexible and we must be sure to use proper notation for the derivative in each case.

Example 4.24 Determine the derivative of the term on the left side with respect to the variable on the right side for each of the following equations (also give the different notations for the derivative).

(a) $y=x^4$ (b) $z=t^{-3}$ (c) $g(x)=x^{1/3}$
(d) $u=v^{10}$ (e) $x=y^2$ (f) $y(t)=t^{-5}$

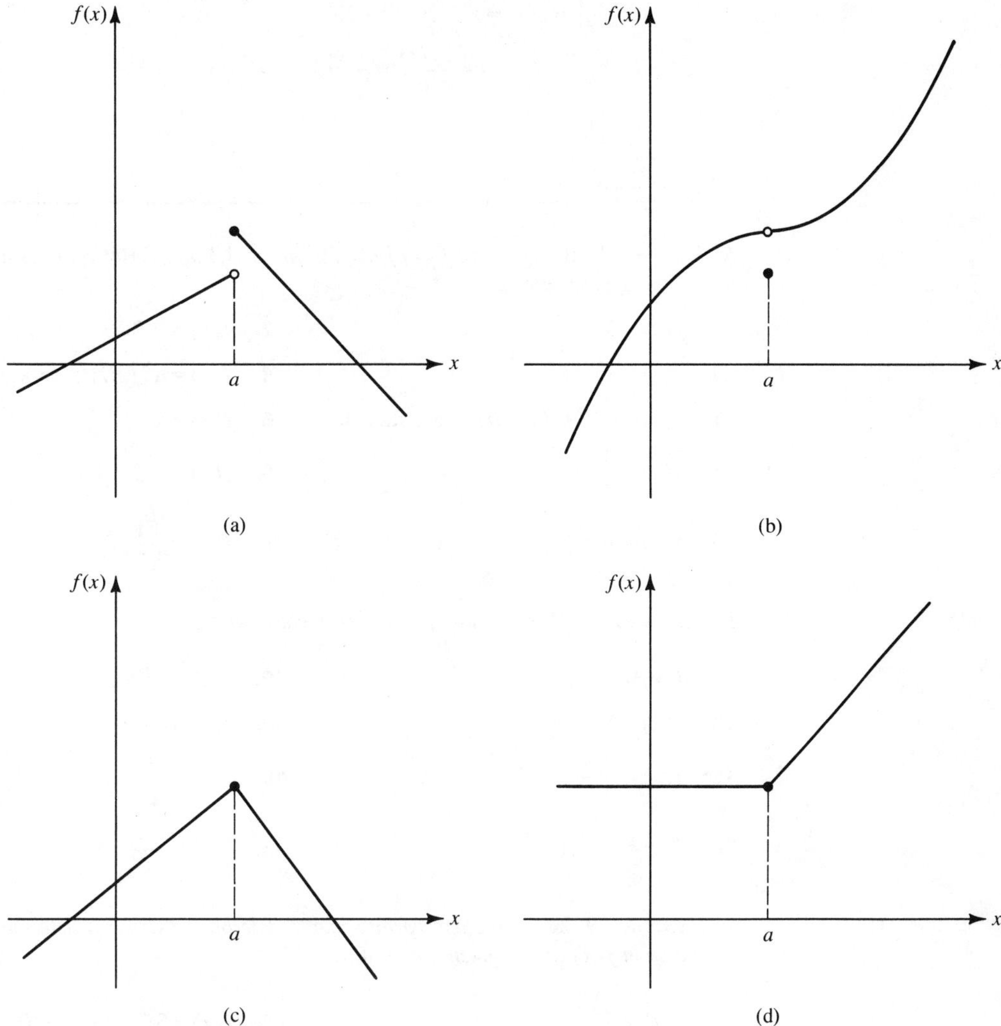

FIGURE 4.11

Solution In each case, we use Rule 4.8 to determine the derivative.

(a) $\dfrac{dy}{dx} = D_x y = y' = 4x^3$

(b) $\dfrac{dz}{dt} = D_t z = z' = -3t^{-3-1} = -3t^{-4}$

(c) $g'(x) = D_x g(x) = \dfrac{dg(x)}{dx} = \dfrac{1}{3}x^{1/3-1} = \dfrac{1}{3}x^{-2/3}$

(d) $\dfrac{du}{dv} = D_v u = u' = 10v^{10-1} = 10v^9$

(e) $\dfrac{dx}{dy} = D_y x = x' = 2y^{2-1} = 2y$

(f) $y'(t) = D_t y(t) = \dfrac{dy(t)}{dt} = -5t^{-5-1} = -5t^{-6}$

EXERCISES

In Exercises 1–10 determine $f'(x)$ from Definition 4.3 and verify your results through the use of the appropriate rule.

1. $f(x)=3$

2. $f(x)=-2$

3. $f(x)=x^2$

4. $f(x)=a^2$ (*Hint*: $a=$ a constant.)

5. $f(x)=b^{-3}$ (*Hint*: $b=$ a constant.)

6. $f(x)=x^4$

7. $f(x)=\dfrac{1}{x}$

8. $f(x)=\dfrac{1}{x^2}$

9. $f(x)=\sqrt{x}$

10. $f(x)=\dfrac{1}{\sqrt{x}}$

In Exercises 11–18 determine $f'(x)$ from Definition 4.3.

11. $f(x)=3x^2+5$

12. $f(x)=x^2+2x+1$

13. $f(x)=9-3x^2$

14. $f(x)=4-2x$

15. $f(x)=\dfrac{1}{x}+5$

16. $f(x)=\dfrac{1}{x^2}-3$

17. $f(x)=\sqrt{x}+7$

18. $f(x)=\dfrac{1}{\sqrt{x}}-5$

In Exercises 19–26 use the appropriate rule to determine $f'(x)$, and determine the slope of the function $f(x)$ at the given value of x.

19. $f(x)=\dfrac{31}{15}$ $(x=4)$

20. $f(x)=5$ $(x=-10)$

21. $f(x)=x^2$ $(x=1)$

22. $f(x)=x^{-1}$ $(x=4)$

23. $f(x)=\dfrac{1}{x^2}$ $(x=-1)$

24. $f(x)=x^5$ $(x=2)$

25. $f(x)=x^{1/2}$ $(x=4)$

26. $f(x)=\dfrac{1}{x^6}$ $(x=-1)$

In Exercises 27–32 determine the derivative of the term on the left side of the equation with respect to the variable on the right side.

27. $y=\dfrac{1}{x^4}$

28. $z=t^{2/3}$

29. $r=s^{0.2}$

30. $h(w)=w^{0.05}$

31. $g(z)=z^3$

32. $f(t)=t^{7/2}$

33. Draw a sketch of the graph of the function $f(x)=x^4$ and

(a) draw tangent lines to the graph of $f(x)$ at the following x values: $\{-2,0,2\}$;
(b) determine the slope of the tangent lines to the curve of $f(x)$ at the x values: $\{-2,0,2\}$;
(c) find an equation of each of the tangent lines in part (a) by using the results of part (b).

34. Population Growth Rate The population of Sanford, Washington has been increasing rapidly since 1970 due to the interest in nuclear power. The equation $P(x)=x^2+40$ gives a good approximation to the population where $P(x)$ is the number of people in thousands and x is the number of years since 1970. For example, $P(0)=40$ implies that the population in 1970 was 40,000.

(a) What was the population in 1977? (*Hint*: $x=7$.)
(b) What was the rate of change of the population in 1977? (Another name for rate of change in this problem would be growth rate.)

35. Mining Stock Your Glitter Gem mining stock has been increasing in value since your purchase in 1960. The market value can be closely approximated by the relation $V(x)=10{,}000+600x$, where $V(x)$ is the dollar value and x is the number of years you have owned the stock.

(a) How much was your stock worth in 1970?
(b) What is the current value?
(c) What was the rate of increase of the value of the stock in 1970? What is the current rate of increase?
(d) Would this be considered a good investment?

36. Growth Rate of an Oil Spill Oil is spreading on the surface of the ocean from an oil tanker leak. The pattern formed by this leak is circular. The radius of the circle, r, measured in feet, is equal to $7t$, where t is the length of the time of the leak in minutes. Find:

(a) The area of the spill after 1 hr.
(b) The rate of increase of the area after 1 hr.
(c) The area and rate of increase after 2 hr.

(*Hint*: The area of a circle of radius r is given by $A=\pi r^2$, where $\pi \doteq \frac{22}{7}$.)

37. Rate of Growth of Sales Suppose your company's sales, $S(x)$, are related to your advertising budget, x, by the following:

$$S(x) = 10 + 50x - x^2$$

where $S(x)$ and x are in hundreds of thousands of dollars. (For example, if $x=2$, this implies an advertising budget of \$200,000.)

(a) What is your company's sales when your advertising budget is \$200,000?
(b) What is the rate of change of the sales when the advertising budget is \$200,000?
(c) Should you continue to increase your advertising budget when you have budgeted \$200,000? \$1 million? \$2.5 million? \$3 million?

4.4 RULES OF DIFFERENTIATION

In Sec. 4.3, we developed the rules for determining the derivatives of two very special functions: a constant and a variable raised to a constant power; that is

$$\text{if } f(x) = k, \text{ then } f'(x) = 0$$

and

$$\text{if } f(x) = x^n, \text{ then } f'(x) = nx^{n-1}$$

These rules were derived on the basis of the definition of derivative. Once a rule has been established, it is obviously easier to find the derivative of a function based on the rule than to start from the definition. In this section, we shall consider rules that permit us to determine derivatives of functions that are obtained by performing algebraic operations (i.e., addition, subtraction, multiplication, and division) on one or more functions. You will recall from Sec. 1.4 that we were able to generate an entire class of functions called the rational functions by performing these operations on the identity function $f(x)=x$. With these differentiation rules at our disposal, we shall easily be able to determine derivatives for a very large group of functions.

In the following development of differentiation rules we will assume that all the functions under consideration are differentiable; that is, their derivatives exist.

Suppose we have a function $h(x)$ that can be represented by a constant k multiplied by another function $f(x)$; that is, $h(x)=kf(x)$. In order to determine $h'(x)$ we apply Definition 4.3 as follows:

$$\begin{aligned} h'(x) &= D_x[kf(x)] \\ &= \lim_{h\to 0}\left[\frac{kf(x+h)-kf(x)}{h}\right] \\ &= \lim_{h\to 0}\left\{k\left[\frac{f(x+h)-f(x)}{h}\right]\right\} \\ &= \left(\lim_{h\to 0} k\right)\lim_{h\to 0}\left[\frac{f(x+h)-f(x)}{h}\right] \\ &= k\lim_{h\to 0}\left[\frac{f(x+h)-f(x)}{h}\right] \end{aligned}$$

By Definition 4.3

$$\lim_{h\to 0}\left[\frac{f(x+h)-f(x)}{h}\right] = f'(x)$$

and therefore,

$$h'(x) = kf'(x) \qquad \text{or} \qquad h'(x) = kD_x f(x)$$

Rule 4.9 The derivative of a constant k multiplied by a function $f(x)$ equals the constant multiplied by the derivative of the function; that is,

$$D_x[kf(x)] = kD_xf(x) = kf'(x) \tag{4.10}$$

Example 4.25 Determine the derivative of each of the following functions.

(a) $f(x)=6x$ (b) $h(x)=2x^5$ (c) $g(x)=\frac{1}{2}x^{-4}$

(d) $f(x)=\dfrac{3}{\sqrt{x}}$ (e) $g(x)=\dfrac{10}{x^5}$ (f) $f(x)=ax^n$

Solution In each of the cases, first we apply Rule 4.9 and then Rule 4.8 to obtain the derivative.

(a) $f'(x)=D_x(6x)=6D_x(x^1)=6(1)x^{1-1}=6(1)=6$
(b) $h'(x)=D_x(2x^5)=2D_x(x^5)=2(5)x^{5-1}=10x^4$
(c) $g'(x)=D_x(\frac{1}{2}x^{-4})=\frac{1}{2}(-4)x^{-4-1}=-2x^{-5}=-2/x^5$
(d) We must rewrite $f(x)=3/\sqrt{x}$ in a form where we can apply Rule 4.8.

$$f(x) = \frac{3}{\sqrt{x}} = \frac{3}{x^{1/2}} = 3x^{-1/2}$$

therefore,

$$f'(x) = D_x(3x^{-1/2}) = 3D_x(x^{-1/2})$$
$$= 3\left(-\frac{1}{2}\right)x^{-1/2-1} = \frac{-3}{2}x^{-3/2} = \frac{-3}{2\sqrt{x^3}}$$

(e) $g(x)=\dfrac{10}{x^5}=10x^{-5}$; therefore

$$g'(x) = D_x(10x^{-5}) = 10(-5)x^{-5-1}$$
$$= -50x^{-6} = \frac{-50}{x^6}$$

(f) $f'(x)=D_x(ax^n)=aD_xx^n=a(n)x^{n-1}=anx^{n-1}$

From Example 4.25 you perhaps have noted that examples (a) through (e) are special cases of (f) and we can state in general:

Rule 4.10 The derivative of a constant k multiplied by a power of x equals the product of the exponent of x and k times x raised to the exponent decreased by 1; that is,

$$D_x(kx^n) = nkx^{n-1}$$

or

$$\text{if } f(x) = kx^n, \text{ then } f'(x) = nkx^{n-1} \tag{4.11}$$

This is called the *power rule* because we refer to kx^n as the power function.

If the function $h(x)$ can be expressed as the sum of two functions $f(x)$ and $g(x)$, then we can write $h(x)=f(x)+g(x)$. In order to determine the derivative of $h(x)$, we write by Definition 4.3,

$$\begin{aligned} h'(x) &= D_x[f(x)+g(x)] \\ &= \lim_{h\to 0}\left\{\frac{[f(x+h)+g(x+h)]-[f(x)+g(x)]}{h}\right\} \\ &= \lim_{h\to 0}\left\{\frac{[f(x+h)-f(x)]+[g(x+h)-g(x)]}{h}\right\} \\ &= \lim_{h\to 0}\left\{\left[\frac{f(x+h)-f(x)}{h}\right]+\left[\frac{g(x+h)-g(x)}{h}\right]\right\} \\ &= \lim_{h\to 0}\left[\frac{f(x+h)-f(x)}{h}\right]+\lim_{h\to 0}\left[\frac{g(x+h)-g(x)}{h}\right] \end{aligned}$$

By Definition 4.3,

$$\lim_{h\to 0}\left[\frac{f(x+h)-f(x)}{h}\right]=f'(x)$$

and

$$\lim_{h\to 0}\left[\frac{g(x+h)-g(x)}{h}\right]=g'(x)$$

therefore,

$$h'(x)=f'(x)+g'(x)$$

Hence, we have a rule for determining the *derivative of the sum of two functions*. If we were to consider the difference of two functions, we would obtain simply the corresponding difference of their respective derivatives.

Rule 4.11 The derivative of the sum (or difference) of two functions equals the sum (or difference) of their respective derivatives; that is,

$$\begin{aligned} D_x[f(x)+g(x)] &= D_xf(x)+D_xg(x) \\ &= f'(x)+g'(x) \qquad (4.12) \\ D_x[f(x)-g(x)] &= D_xf(x)-D_xg(x) \\ &= f'(x)-g'(x) \qquad (4.13) \end{aligned}$$

Rule 4.11 can be extended to include any finite number of functions that are combined by addition and/or subtraction.

Example 4.26 Determine the derivative of each of the following functions.

(a) $f(x)=3x^2+2x$ (b) $g(x)=2x^3-4x^2$

(c) $y=x^3+\dfrac{2}{\sqrt{x}}$ (d) $y=2x-x^{-2}$

(e) $h(x)=\sqrt[3]{x}+\frac{1}{3}x^3$ (f) $g(x)=1+2x+3x^2+x^4-3x^5$

(g) $f(x)=3x^2(3-x)$ (h) $f(x)=x^3(x^2+1)$

Solution In parts (a) to (e) first we apply Rule 4.11, then Rule 4.10.

(a) $f'(x)=D_x(3x^2+2x)=D_x(3x^2)+D_x(2x)=3\cdot 2x^{2-1}+2x^{1-1}=6x+2$

(b) $g'(x)=D_x(2x^3-4x^2)=D_x(2x^3)-D_x(4x^2)=2\cdot 3x^{3-1}-4\cdot 2x^{2-1}=6x^2-8x$

(c) Rewriting $y=x^3+\dfrac{2}{\sqrt{x}}$ as $y=x^3+2x^{-1/2}$

$$\frac{dy}{dx}=D_x(x^3+2x^{-1/2})=D_x(x^3)+D_x(2x^{-1/2})=3x^{3-1}+2\left(-\tfrac{1}{2}\right)x^{-1/2-1}$$

$$=3x^2-x^{-3/2}=3x^2-\frac{1}{\sqrt{x^3}}$$

(d) $\dfrac{dy}{dx}=D_x(2x-x^{-2})=D_x(2x)-D_x(x^{-2})=2-(-2)x^{-2-1}=2+2x^{-3}$
$=2+\dfrac{2}{x^3}$

(e)
$$h'(x)=D_x\left(\sqrt[3]{x}+\frac{1}{3}x^3\right)=D_x\left(x^{1/3}+\frac{1}{3}x^3\right)$$
$$=D_x(x^{1/3})+D_x\left(\frac{1}{3}x^3\right)$$
$$=\frac{1}{3}x^{1/3-1}+\frac{1}{3}\cdot 3x^{3-1}=\frac{1}{3}x^{-2/3}+x^2=\frac{1}{3\sqrt[3]{x^2}}+x^2$$

In part (f), we use the extension of Rule 4.11 to include several terms:

(f)
$$g'(x)=D_x(1+2x+3x^2+x^4-3x^5)$$
$$=D_x(1)+D_x(2x)+D_x(3x^2)+D_x(x^4)-D_x(3x^5)$$
$$=0+2+3\cdot 2x+4x^3-3\cdot 5x^4=2+6x+4x^3-15x^4$$

To solve parts (g) and (h), we multiply the terms and express the result as a sum or difference of power functions.

(g) $f'(x)=D_x[3x^2(3-x)]=D_x[9x^2-3x^3]=D_x(9x^2)-D_x(3x^3)=18x-9x^2$

(h) $f'(x)=D_x[x^3(x^2+1)]=D_x[x^5+x^3]=D_x(x^5)+D_x(x^3)=5x^4+3x^2$

(As the reader develops proficiency in applying the rules concerning derivatives, it will not be necessary to write out each step in as much detail as in Example 4.26; however, when omitting intermediate steps, extreme caution must be exercised in the computations.)

In Sec. 4.3 we noted that the first derivative could be interpreted as the *instantaneous rate of change* of a function. We will refer to it as simply the *rate of change* of a function. The rate of change tells us how much the dependent variable increases or decreases for a unit change in the independent variable.

Example 4.27 Demand for T-Slot Cutters The Mal Machine Tool Co. has found that the number, $D(x)$, of T-slot cutters that it can sell per month is related to the price per cutter, x, by the following equation

$$D(x) = 100 - 4x$$

if the price per cutter is between \$5 and \$20. $D(x)$ is called the demand function. Determine the rate of change of the demand when the price per cutter is \$10 and interpret your results.

Solution The rate of change of the demand function is found by taking the first derivative of $D(x)$ by applying the Rules 4.11 and 4.10.

$$D'(x) = -4$$

The rate of change of $D(x)$ when $x = 10$ is found by evaluating $D'(x)$ when $x = 10$.

$$\begin{aligned} D'(x) &= -4 \\ D'(10) &= -4 \end{aligned}$$

We interpret this result to mean that at a selling price of \$10, the demand for the number of cutters we produce per month is decreasing at the rate of 4 cutters for each dollar increase in the unit price.

It is important to note that regardless of the unit price per cutter, the demand decreases by 4 cutters for each dollar increase in the unit price. A graph of the demand function shown in Fig. 4.12 shows that the demand is decreasing at a constant rate.

Example 4.28 Rate of Change of Revenue Suppose we extend Example 4.27 by asking what is the rate of change of the revenue, $R(x)$, when the selling price, x, per cutter is \$10.

Solution If we are interested in the rate of change of the revenue function $R(x)$ we must express $R(x)$ in terms of x and then find the first derivative. The revenue, $R(x)$, is found by multiplying the unit price, x, by the number sold, $D(x)$.

$$\begin{aligned} R(x) &= xD(x) \qquad \text{where } \$5 \leqslant x \leqslant \$20 \\ &= x(100-4x) \\ R(x) &= 100x - 4x^2 \end{aligned}$$

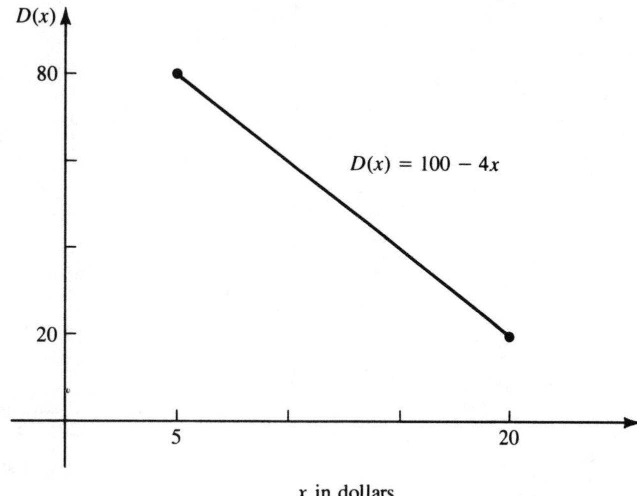

FIGURE 4.12

Applying Rules 4.11 and 4.10, the rate of change of the revenue, $R'(x)$ is given by

$$R'(x) = 100 - 8x$$

The rate of change of the revenue when $x = 10$ is found by evaluating $R'(10)$.

$$\begin{aligned} R'(10) &= 100 - 8(10) \\ &= 20 \end{aligned}$$

We interpret this to mean that when the unit selling price is \$10 the revenue is increasing at the rate of \$20 per dollar increase in the unit price. Based on this result, the management of Mal Machine Tool Co. would increase the unit price to increase the revenue.

Note that the revenue relation in Example 4.28 is a quadratic function and we can easily sketch the graph of $R(x)$ as shown in Fig. 4.13.

From the graph of $R(x)$ it can be seen that the revenue is increasing on the domain $(5, 12.5)$ and decreasing on the domain $(12.5, 20)$. If the revenue is increasing on the interval $(5, 12.5)$, then the rate of change, $R'(x)$, of the revenue should be positive for these values of x. We verify this by determining those values of x that will make $R'(x) > 0$, that is we solve:

$$\begin{aligned} 100 - 8x &> 0 \\ -8x &> -100 \\ x &< 12.5 \end{aligned}$$

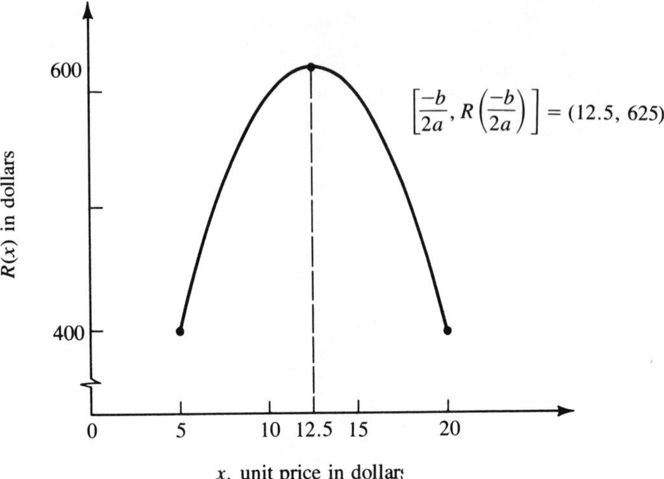

FIGURE 4.13

Since the domain of x was restricted in this problem to be $[5,20]$, we conclude

1. $R'(x)>0$ for $5<x<12.5$.
2. $R(x)$ is increasing on the interval $(5, 12.5)$.

In a like manner we can conclude

1. $R'(x)<0$ for $12.5<x<20$.
2. $R(x)$ is decreasing on the interval $(12.5, 20)$.

Based on our knowledge of quadratic functions, the vertex or turning point of the graph is located at $(-b/2a, f(-b/2a))$. Since our parabola opens downward, the value $R(-b/2a)$ will be a maximum. Since $R(x)=-4x^2+100x$, $-b/2a=-(100)/2(-4)=12.5$, and the maximum value of $R(x)=R(12.5)=$ \$625. The tangent line drawn to the graph of $R(x)$ at the vertex is horizontal and its slope is zero, which implies that $R'(12.5)=0$. We verify this by substituting the value $x=12.5$ into $R'(x)$.

$$R'(x) = 100 - 8x$$
$$R'(12.5) = 0$$

Example 4.29 Given any quadratic function of the form $f(x)=ax^2+bx+c$, verify that the vertex or turning point of the graph of $f(x)$ has coordinates $(-b/2a, f(-b/2a))$.

Solution From our knowledge of the graph of a parabola, we know the vertex or turning point occurs where the tangent line to the curve is horizontal, having a slope $=0$. Since the derivative can be interpreted as the slope of the tangent line, this implies that $f'(x)=0$ at the turning point. From the general form of the quadratic function $f(x)=ax^2+bx+c$, we find $f'(x)=2ax+b$. There is only one solution to the equation $f'(x)=0$ or $2ax+b=0$, and that is

$$x = \frac{-b}{2a}$$

Therefore, we conclude the turning point or vertex of a quadratic function has coordinates $(-b/2a, f(-b/2a))$, which agrees with the results obtained in Chapter 2.

In Example 4.26 parts (g) and (h) we were able to determine the derivative of a product by multiplying terms and expressing the result as the appropriate sum or difference. There are many situations in which this is not possible. For example, if $f(x)=(x^2+2x)^{1/2}(x^3-4)^{1/3}$, we cannot simply multiply the two terms to obtain a sum or difference of two functions; and thus, we need a rule for determining the *derivative of a product of two functions*. Based on Rule 4.11, one might conjecture that the derivative of a product of two functions is the product of their derivatives. For example, if $p(x)= f(x)s(x)$, then $p'(x)=f'(x)s'(x)$. Unfortunately, this is not the case and the rule we are seeking is more complicated. We could employ Definition 4.3 to obtain the derivative of the product $p(x)$; however, since we are more interested in the result, we shall state the rule without deriving it, and illustrate it in several examples.

Rule 4.12 The derivative of the product of two functions, $f(x)$ and $s(x)$, equals the first times the derivative of the second, plus the second times the derivative of the first that is,

$$\begin{aligned} D_x[f(x)s(x)] &= f(x)D_x s(x) + s(x)D_x f(x) \\ &= f(x)s'(x) + s(x)f'(x) \qquad (4.14) \end{aligned}$$

As in the previous cases, it is assumed that the functions $f(x)$ and $s(x)$ have derivatives at the values of x of interest.

Example 4.30 Use Rule 4.12 to determine the derivative of each of the following functions.

(a) $p(x)=3x^2(3-x)$ (b) $p(x)=x^3(x^2+1)$
(c) $f(x)=(x^{3/5}+x^2)(x^{1/2}+x^{-1/2})$ (d) $g(x)=(x^3+2x^2+1)(x^2-x^{-1})$

Solution Note that parts (a) and (b) are identical to parts (g) and (h) of Example 4.26, and would generally be expanded prior to differentiation; hence, we may verify our answers from this example with those from

Example 4.26. First we shall apply Rule 4.12 to each problem and then apply the appropriate combination of the preceding rules to obtain the derivative.

$$\begin{aligned}\text{(a)}\quad p'(x) &= D_x[3x^2(3-x)] = 3x^2D_x(3-x) + (3-x)D_x(3x^2)\\ &= 3x^2(-1) + (3-x)(6x)\\ &= -3x^2 + 18x - 6x^2\\ &= 18x - 9x^2\end{aligned}$$

$$\begin{aligned}\text{(b)}\quad p'(x) &= D_x[x^3(x^2+1)] = x^3D_x(x^2+1) + (x^2+1)D_x(x^3)\\ &= x^3(2x) + (x^2+1)(3x^2)\\ &= 2x^4 + 3x^4 + 3x^2\\ &= 5x^4 + 3x^2\end{aligned}$$

$$\begin{aligned}\text{(c)}\quad f'(x) &= D_x[(x^{3/5}+x^2)(x^{1/2}+x^{-1/2})]\\ &= (x^{3/5}+x^2)D_x(x^{1/2}+x^{-1/2}) + (x^{1/2}+x^{-1/2})D_x(x^{3/5}+x^2)\\ &= (x^{3/5}+x^2)\left(\tfrac{1}{2}x^{-1/2}-\tfrac{1}{2}x^{-3/2}\right) + (x^{1/2}+x^{-1/2})\left(\tfrac{3}{5}x^{-2/5}+2x\right)\end{aligned}$$

We could multiply the indicated factors on the right-hand side of the above equation; however, at this point we choose not to simplify the algebra.

$$\begin{aligned}\text{(d)}\quad g'(x) &= D_x[(x^3+2x^2+1)(x^2-x^{-1})]\\ &= (x^3+2x^2+1)D_x(x^2-x^{-1}) + (x^2-x^{-1})D_x(x^3+2x^2+1)\\ &= (x^3+2x^2+1)(2x+x^{-2}) + (x^2-x^{-1})(3x^2+4x)\end{aligned}$$

The rule for the derivative of a product can also be extended to include the product of a finite number of functions. Suppose we consider the product $p(x)=f(x)\cdot s(x)\cdot t(x)$. Based on Rule 4.12, we have

$$\begin{aligned}p'(x) &= D_x[f(x)\cdot s(x)\cdot t(x)]\\ &= f(x)\cdot D_x[s(x)\cdot t(x)] + [s(x)\cdot t(x)]D_xf(x)\end{aligned}$$

where $[s(x)\cdot t(x)]$ is considered the second function. Applying Rule 4.12 to $D_x[s(x)\cdot t(x)]$ we have

$$\begin{aligned}p'(x) &= f(x)[s(x)\cdot t'(x)+t(x)\cdot s'(x)] + s(x)\cdot t(x)\cdot f'(x)\\ &= f(x)\cdot s(x)\cdot t'(x) + f(x)\cdot t(x)\cdot s'(x) + s(x)\cdot t(x)\cdot f'(x)\end{aligned}$$

and rearranging,

$$= f'(x)\cdot s(x)\cdot t(x) + f(x)\cdot s'(x)\cdot t(x) + f(x)\cdot s(x)\cdot t'(x)$$

Generalizing, we conclude that the derivative of the product of n functions becomes the sum of products, where the first term is the derivative of the first function multiplied by the product of the other $(n-1)$ functions, the second term is the derivative of the second function multiplied by the product of the other $(n-1)$ functions, etc.

Example 4.31 Determine the derivative of each of the following functions.
(a) $f(x)=x^2(x^3+1)(x^2-x)$
(b) $f(x)=x^{1/2}(2-3x)(x^2+1)$
(c) $f(x)=x^{-2}(x^2+3)(x+5)$

Solution

(a)

$$\begin{aligned} f'(x) &= D_x[x^2(x^3+1)(x^2-x)] = \left[D_x(x^2)\right](x^3+1)(x^2-x) \\ &\quad + x^2\left[D_x(x^3+1)\right](x^2-x) + x^2(x^3+1)\left[D_x(x^2-x)\right] \\ &= (2x)(x^3+1)(x^2-x) + x^2(3x^2)(x^2-x) + x^2(x^3+1)(2x-1) \end{aligned}$$

(b)

$$\begin{aligned} f'(x) &= D_x[x^{1/2}(2-3x)(x^2+1)] = \left[D_x(x^{1/2})\right](2-3x)(x^2+1) \\ &\quad + x^{1/2}[D_x(2-3x)](x^2+1) + x^{1/2}(2-3x)\left[D_x(x^2+1)\right] \\ &= \left(\tfrac{1}{2}x^{-1/2}\right)(2-3x)(x^2+1) + x^{1/2}(-3)(x^2+1) + x^{1/2}(2-3x)(2x) \end{aligned}$$

(c)

$$\begin{aligned} f'(x) &= D_x[x^{-2}(x^2+3)(x+5)] = \left[D_x(x^{-2})\right](x^2+3)(x+5) \\ &\quad + (x^{-2})\left[D_x(x^2+3)\right](x+5) + (x^{-2})(x^2+3)[D_x(x+5)] \\ &= (-2x^{-3})(x^2+3)(x+5) + (x^{-2})(2x)(x+5) + (x^{-2})(x^2+3)(1) \end{aligned}$$

Again, to emphasize our differentiation rule, we have elected not to simplify the algebraic expression on the right-hand sides of parts (a), (b), and (c).

Let us verify the results of part (a) by first expanding the product and then finding the derivative.

$$\begin{aligned} f(x) &= x^2(x^3+1)(x^2-x) \\ &= x^7 - x^6 + x^4 - x^3 \end{aligned}$$

therefore,

$$f'(x) = 7x^6 - 6x^5 + 4x^3 - 3x^2$$

The solution to part (a) is $f'(x)=2x(x^3+1)(x^2-x)+x^2(3x^2)(x^2-x)+x^2(x^3+1)(2x-1)$, which upon simplification becomes $7x^6-6x^5+4x^3-3x^2$, and our solution is verified.

Suppose we have a function $q(x)$ that can be expressed as the quotient of two functions $f(x)$ and $g(x)$; that is, $q(x)=f(x)/g(x)$. Again, we could apply Definition 4.3 to the quotient to develop a general formula for determining the *derivative of a quotient of two functions*; however, since this is rather complicated, we shall state the rule without derivation.

Rule 4.13 The derivative of the quotient of two functions equals the denominator times the derivative of the numerator, minus the numerator times the derivative of the denominator, all divided by the square of the denominator; that is,

$$D_x\left[\frac{f(x)}{g(x)}\right]=\frac{g(x)D_xf(x)-f(x)D_xg(x)}{[g(x)]^2} \qquad (4.15)$$

provided $g(x)\neq 0$.

Since subtraction is not commutative, we must take care to memorize this rule correctly or we will end up with a sign error in our answer.

Example 4.32 Determine the derivative of each of the following functions.

(a) $q(x)=\dfrac{x+1}{x-1}$ (b) $q(x)=\dfrac{x+1}{x^2+1}$

(c) $f(x)=\dfrac{x^3-x}{x+1}$ (d) $f(x)=\dfrac{1-x+x^2}{1+x+x^2}$

Solution Applying Rule 4.13, we obtain

(a)

$$\begin{aligned} q'(x) = D_x\left(\frac{x+1}{x-1}\right) &= \frac{(x-1)D_x(x+1)-(x+1)D_x(x-1)}{(x-1)^2} \\ &= \frac{(x-1)(1)-(x+1)(1)}{(x-1)^2} \\ &= -\frac{2}{(x-1)^2} \end{aligned}$$

(b)

$$\begin{aligned} q'(x) = D_x\left(\frac{x+1}{x^2+1}\right) &= \frac{(x^2+1)D_x(x+1)-(x+1)D_x(x^2+1)}{(x^2+1)^2} \\ &= \frac{(x^2+1)(1)-(x+1)(2x)}{(x^2+1)^2} \\ &= \frac{x^2+1-2x^2-2x}{(x^2+1)^2} \\ &= \frac{-x^2-2x+1}{(x^2+1)^2} \end{aligned}$$

(c) $$f'(x) = D_x\left(\frac{x^3-x}{x+1}\right) = \frac{(x+1)D_x(x^3-x)-(x^3-x)D_x(x+1)}{(x+1)^2}$$

$$= \frac{(x+1)(3x^2-1)-(x^3-x)(1)}{(x+1)^2}$$

$$= \frac{3x^3-x+3x^2-1-x^3+x}{(x+1)^2}$$

$$= \frac{2x^3+3x^2-1}{(x+1)^2}$$

(d) $$f'(x) = D_x\left(\frac{1-x+x^2}{1+x+x^2}\right)$$

$$= \frac{(1+x+x^2)D_x(1-x+x^2)-(1-x+x^2)D_x(1+x+x^2)}{(1+x+x^2)^2}$$

$$= \frac{(1+x+x^2)(-1+2x)-(1-x+x^2)(1+2x)}{(1+x+x^2)^2}$$

$$= \frac{-1+2x-x+2x^2-x^2+2x^3-1-2x+x+2x^2-x^2-2x^3}{(1+x+x^2)^2}$$

$$= \frac{-2+2x^2}{(1+x+x^2)^2}$$

The rules for differentiating sums, differences, products, and quotients can be applied in various ways. For example, when differentiating a quotient of two functions, the numerator could be the product of two functions to which we would apply the rule for differentiating products. In the previous examples we have already seen that the rules for a constant multiplied by a function and for the sum and difference of functions may be applied when differentiating products and quotients. Any of these rules may be applied in combination with the other rules in order to obtain the derivative of a function.

Example 4.33 Determine the derivative of each of the following functions.

(a) $f(x)=\dfrac{(x+6)(2x^2+4)}{x+1}$ (b) $f(x)=\dfrac{x^2+3x+2}{(x+1)(x+2)}$

Solution

(a)

$$f'(x) = D_x\left[\frac{(x+6)(2x^2+4)}{x+1}\right]$$

$$= \frac{(x+1)D_x[(x+6)(2x^2+4)] - (x+6)(2x^2+4)D_x(x+1)}{(x+1)^2}$$

$$= \frac{(x+1)\big[(x+6)D_x(2x^2+4) + (2x^2+4)D_x(x+6)\big] - (x+6)(2x^2+4)(1)}{(x+1)^2}$$

$$= \frac{(x+1)(6x^2+24x+4) - (2x^3+12x^2+4x+24)}{(x+1)^2}$$

$$= \frac{(6x^3+30x^2+28x+4) - (2x^3+12x^2+4x+24)}{(x+1)^2}$$

$$= \frac{4x^3+18x^2+24x-20}{(x+1)^2}$$

(b)

$$f'(x) = D_x\left[\frac{x^2+3x+2}{(x+1)(x+2)}\right]$$

$$= \frac{(x+1)(x+2)D_x(x^2+3x+2) - (x^2+3x+2)D_x[(x+1)(x+2)]}{[(x+1)(x+2)]^2}$$

$$= \frac{(x+1)(x+2)(2x+3) - (x^2+3x+2)[(x+1)(1)+(x+2)(1)]}{(x+1)^2(x+2)^2}$$

$$= \frac{(x+1)(x+2)(2x+3) - (x^2+3x+2)(2x+3)}{(x+1)^2(x+2)^2}$$

$$= \frac{(2x+3)[(x+1)(x+2) - (x^2+3x+2)]}{(x+1)^2(x+2)^2}$$

$$= \frac{(2x+3)[(x^2+3x+2) - (x^2+3x+2)]}{(x+1)^2(x+2)^2}$$

$$= 0 \text{ for } x \neq -1 \text{ or } -2$$

This result is not surprising if we look carefully at the original function. If the denominator is expanded, we obtain the numerator; therefore, $(x^2+3x+2)/[(x+1)(x+2)]=1$ for all $x \neq -1$ or -2, and we know that the derivative of the constant 1 is zero. Generally, it is a good idea to first reduce quotients as much as possible, obtaining an equivalent function before taking the derivative.

In applying the differentiation Rules 4.7 through 4.13, it is advisable to study the function carefully to determine which rule or rules apply. Once we have applied our differentiation rules, the subsequent simplification of the algebra becomes easier with practice.

Example 4.34 Rate of Change of Cost Suppose the manufacturer of Excello Widgets has determined that the cost in dollars of manufacturing x widgets is given by $C(x)=2x^2-10x+100$. Determine the rate of change of the cost when 100 widgets are being manufactured and interpret the results.

Solution The rate of change of the cost, $C'(x)$, is found by taking the derivative of $C(x)$,

$$C'(x) = 4x - 10$$

and the rate of change of the cost when $x=100$ is $C'(100)$.

$$\begin{aligned} C'(100) &= 4(100) - 10 \\ &= 390 \end{aligned}$$

We interpret this to mean that when 100 widgets are being manufactured the cost is increasing at the rate of \$390/widget manufactured. We will see that this is a very close approximation to the cost of manufacturing the 101st widget. We can compute the cost of manufacturing the 101st widget by determining the manufacturing cost of 101 widgets, $C(101)$, and subtracting the manufacturing cost of 100 widgets, $C(100)$; that is,

$$\begin{aligned} C(101) &= 2(101)^2 - 10(101) + 100 = \$19{,}492 \\ C(100) &= 2(100)^2 - 10(100) + 100 = \$19{,}100 \end{aligned}$$

Therefore, the cost of manufacturing the 101st widget is

$$C(101) - C(100) = \$392$$

What accounts for the small difference of \$2 in the results? If we examine a portion of the graph of $C(x)$ shown in Fig. 4.14 we will see the reason for this difference. The actual cost of the 101st widget, $C(101)-C(100)$, is shown geometrically in Fig. 4.14 as the measure, $m(\overline{AC})$, of line segment $\overline{AC}$. The measure of line segment $\overline{AB}$, $m(\overline{AB})$, is a close approximation to $m(\overline{AC})$. Now the slope of the tangent line drawn at point D is $m(\overline{AB})/(101-100)=m(\overline{AB})$. Since we also know the slope of the tangent line at D is $C'(100)$, we conclude that $m(\overline{AB})$ or $C'(100)$ is a very close approximation to $C(101)-C(100)$, the cost of manufacturing the 101st widget.

In most practical problems the difference between $C'(x_0)$ and $C(x_0+1)-C(x_0)$ is usually very small, and since it is often easier to calculate $C'(x_0)$ than $C(x_0+1)-C(x_0)$, we use the former to determine the cost of manufacturing the (x_0+1)st item. The function $C'(x)$ is important in the study of economics and is called *marginal cost*. The marginal cost, or the approximate cost of manufacturing the 200th item is found by evaluating $C'(199)$.

If revenue and profit functions are defined in terms of the number of items manufactured, we will consider their derivatives as *marginal revenue* and *marginal profit* functions, respectively. The analysis of problems that consider the behavior of the derivatives of cost, revenue, and profit functions is called *marginal analysis*. We will pursue this topic further in Chapter 5.

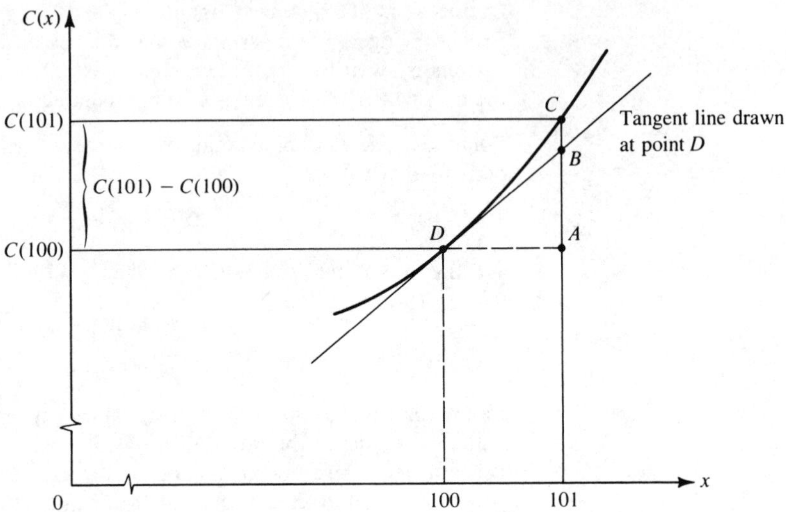

FIGURE 4.14

Example 4.35 Marginal Cost The total cost in dollars for a particular operation is given by $C(x)=3200\sqrt{x}+10{,}000$, where x represents the number of items manufactured. Determine the marginal cost of the sixty-fifth unit.

Solution Marginal cost is the derivative of the cost function; that is,

$$\begin{aligned} MC = C'(x) &= D_x(3200\sqrt{x}+10{,}000) \\ &= D_x(3200x^{1/2}+10{,}000) \\ &= 1600x^{-1/2} \\ C'(x) &= \frac{1600}{\sqrt{x}} \end{aligned}$$

The marginal cost of the sixty-fifth item is found by evaluating $C'(64)$.

$$C'(64) = \frac{1600}{\sqrt{64}} = 200$$

Thus \$200 represents the approximate cost of manufacturing the sixty-fifth unit.

EXERCISES

In Exercises 1–12 *determine the derivative of each of the given functions.*

1. $f(x)=3x^4$

2. $f(x)=\frac{1}{2}x^6$

3. $f(x)=3x+2x^3$

4. $f(u)=3u^2+5u$

5. $g(x)=4\sqrt{x}+5$

6. $g(x)=6\sqrt[3]{x^2}-x^2$

7. $h(x)=10x^2+\dfrac{1}{x^2}-\dfrac{5}{2}$

8. $h(x)=\dfrac{1}{5}x^5-x^3-\dfrac{1}{x}$

9. $g(t)=4t^2+5t+\dfrac{2}{\sqrt[3]{t}}$

10. $g(t)=7+4t^2-3t^3$

11. $u=x^4+3x^3-x^2+\frac{1}{2}x+2$

12. $s=3z+6z^2-\frac{1}{3}z^3$

In Exercises 13–20 determine the value of the derivative of each of the given functions at the given value of the independent variable.

13. $f(x)=x^2+2x$ (at $x=1$)

14. $y=2x^3+4x^2+3x+5$ (at $x=-1$)

15. $u=2z^4+3z^3+z^2+4z+5$ (at $z=0$)

16. $g(t)=t^3-3t^2+3t+5$ (at $t=1$)

17. $g(u)=\dfrac{-5}{u}+4$ (at $u=-2$)

18. $h(x)=\sqrt{x}+\frac{3}{2}x^2$ (at $x=4$)

19. $s(x)=\sqrt[3]{x}-\dfrac{1}{x}$ (at $x=8$)

20. $s(u)=\dfrac{-4}{\sqrt[3]{u}}+2u^{-1}$ (at $u=1$)

In Exercises 21–26 determine an equation of the tangent line to the graph of each of the following functions at the given x value.

21. $f(x)=3x^2-x$ (at $x=1$)

22. $f(x)=\sqrt{x}-4$ (at $x=16$)

23. $g(x)=-2x^2+3$ (at $x=-1$)

24. $g(x)=\dfrac{1}{\sqrt[3]{x}}-2$ (at $x=-1$)

25. $h(x)=x^3-2x^2+3x+1$ (at $x=2$)

26. $h(x)=2x-\sqrt{x}$ (at $x=4$)

Determine the derivative of each of the functions in Exercises 27–36 by expanding each product and simplifying before differentiating. Check your results by applying the rule for the product of functions.

27. $f(x)=(x-4)(x+2)$

28. $y=4x^3(x+3)$

29. $g(t)=\sqrt{t}\,(t^2-3t)$

30. $s=(2w+1)(w-4)$

31. $u=\sqrt[3]{z}\,(4z-2)$

32. $h(t)=3t^4(5t^2+2t+1)$

33. $f(u)=(u^{1/2}-4u)(3u^{-1/2})$

34. $f(x)=(\sqrt{x}+x)\dfrac{1}{\sqrt{x}}$

35. $g(x)=2x^2(x-4)(x^2+x)$

36. $y=4t(2t^2+6)(t^2-3)$

Determine the derivative of each of the functions in Exercises 37–46 by applying the rule for the quotient of functions (Rule 4.13).

37. $f(x)=\dfrac{x+1}{x-1}$

38. $f(u)=\dfrac{u-1}{u+1}$

39. $g(x)=\dfrac{4}{x^2}$

40. $g(z)=\dfrac{3z}{z+3}$

41. $s(t)=\dfrac{3t^2-1}{t+1}$

42. $s(x)=\dfrac{2x^2+3}{x^2+1}$

43. $y=\dfrac{x^2-3x+2}{2x+1}$

44. $y=\dfrac{x+1}{\sqrt{x}}$

45. $h(t)=\dfrac{\sqrt{t}}{t-3}$

46. $f(x)=\dfrac{\sqrt{x}\,(3x-2)}{x+1}$

Determine the slope of each of the curves in Exercises 47–52 at the given value of the independent variable

47. $f(x)=\dfrac{x^2+1}{x-1}$ (at $x=2$)

48. $f(x)=\dfrac{x^3-x^2}{x-1}$ (at $x=0$)

49. $y=\dfrac{3x^2+6x+2}{x^2+2x+1}$ (at $x=2$)

50. $u=\dfrac{w^3}{w+3}$ (at $w=-2$)

51. $s=\dfrac{z^3-4z^2}{z-4}$ (at $z=3$)

52. $g(t)=\dfrac{t^3-t}{t+1}$ (at $t=4$)

Determine the values of the independent variable for which the slope equals zero in each of the functions in Exercises 53–58.

53. $f(x)=x^2+2x+5$

54. $f(x)=x^3+3x^2-9x+7$

55. $g(u)=(u^2+1)(u+2)$

56. $y=(3w^2+4)(2w+5)$

57. $y=\dfrac{x^2+5}{x+2}$

58. $h(t)=\dfrac{t+1}{t^2+3}$

59. Temperatures Downstream from a Steam Plant The increase in the average temperature of a river downstream from the outlet of a steam-generating plant is $T(x)=12/(x+1)$ degrees Celsius, where $x=$ the number of hundreds of meters downstream (i.e., $x=2$ is equivalent to 200 meters downstream).

(a) What is the increase in temperature 200 meters downstream?
(b) What is the rate of change of the temperature 200 meters downstream? Interpret your answer.

60. Votes and Campaign Expenditures Dewey Bayfield is a candidate for the United States Senate. His campaign manager has told Mr. Bayfield that the total number of votes received (in thousands) is related to campaign expenditures x (in hundreds of dollars) by

$$V(x) = 100+4x+2\sqrt{x}$$

Determine the rate of change of $V(x)$ when campaign expenditures are

(a) $x=100$ (i.e., \$10,000)
(b) $x=900$ (i.e., \$90,000)

61. **Marginal Cost** The total cost in dollars to manufacture x items is $C(x) = (0.02)x^2 + 16x + 800$.

(a) What is the total cost of manufacturing 50 items?
(b) What is the marginal cost of the 51st item?
(c) What is the marginal cost when 60 items are being manufactured?

62. **Marginal Revenue** The revenue in dollars from selling wholesale x deluxe tackle boxes is given by

$$R(x) = 10x + 14\sqrt{x}$$

(a) What is the total revenue from the sale of 49 tackle boxes?
(b) What is the marginal revenue from the sale of the 50th tackle box?
(c) What is the marginal revenue when $x = 4$?

63. **Marginal Profit** The Pure-Rena Pet Food Company has determined empirically the relationship between the production volume of dog food and the associated profit to be given by the expression

$$P = 10v - v^2$$

where P = profits (in thousands of dollars)
v = production volume (in thousands of units of production)

(a) Determine the function that represents the rate of change of profit with respect to production (i.e., determine $D_v P$).
(b) Determine the rate of change of the profit function when production volume is 4 and 5. What is happening to the rate of change of P as v increases from 4 to 5?
(c) Determine the rate of change of the profit function when production volume is 6 and 7. What is happening to the rate of change of P as v increases from 6 to 7?
(d) Graph the profit function.
(e) Graph the function that represents the rate of change of P as v increases from 4 to 7. What happens to the sign of $D_v P$ as v goes from 4 to 7?
(f) What do the above results indicate about the optimum production volume for maximizing profit?

64. The Imperial Widget Manufacturing Company produces widgets that sell for \$20 each; thus, the total sales revenue R for widgets is $20x$, where x is the number of units sold; that is, $R = 20x$. Furthermore, it costs \$100 to set up the equipment for a production run, and the material and labor costs are proportional to the square of the number of units manufactured with the constant of proportionality being .05; that is, the total production cost C is given by $C = 100 + .05x^2$. If we denote profit by P, then $P = R - C$.

(a) Determine the equation for P as a function of x and graph it for values of x from 0 to 400.
(b) Determine the marginal profit function with respect to the number of units sold.

(c) What is the marginal profit when $x=150$ units? 200 units? 250 units?
(d) What is the total profit when 150 units are sold? 200 units? 250 units?

65. **Concentration of Pollutant from a Sewer Plant** A pollutant in a river below the overflow outlet from a sewer treatment plant is given by

$$P(x) = \frac{10}{x^2+1}$$

where $P(x)=$ mg of pollutant/liter

$x=$ the distance downstream in hundreds of meters from the plant's outlet

(a) Determine the amount of pollutant/liter in the river at the plant's outlet.
(b) Determine the amount of pollutant/liter in the river 200 meters downstream. (*Hint*: $x=2$.)
(c) Determine the rate of change of the pollutant 200 meters downstream.

4.5 DERIVATIVES OF COMPOSITE FUNCTIONS

We found in the previous section that differentiation Rules 4.7 through 4.13 provide us with the means of finding the derivatives of many of the algebraic functions we studied in Chapters 1 and 2. In this section we will add the last rule necessary to deal with the algebraic functions we will encounter.

Let us consider the derivatives of each of the following rather simple algebraic functions.

(a) $g(x)=(x+5)^2$
(b) $s(x)=(x+5)^3$
(c) $h(x)=(x+5)^{10}$
(d) $t(x)=(x+5)^{1/2}$

We can determine $g'(x)$ by multiplying $(x+5)(x+5)$ and taking the derivative of the product term-by-term by using the power rule. The derivatives $s'(x)$ and $h'(x)$ can be found in a similar manner, although it will be a considerable task to expand $(x+5)^{10}$. The derivative of $t(x)$ cannot be found this way as we cannot expand $(x+5)^{1/2}$ by multiplying. [You will recall that $(x+5)^{1/2}=\sqrt{(x+5)} \neq \sqrt{x} + \sqrt{5}$.] None of our previous rules applies. We could approach this problem by applying the basic definition of a derivative, Definition 4.3; however we are looking for a rule that will apply not only to the function $t(x)$, but to a set of functions similar to it.

You will note these examples above have something in common. Each of the four is of the form u raised to a power, where $u=(x+5)$. From Sec. 1.4 we recall that each of these can be thought of as *composite functions*. Rewriting them in composite form, we have:

(a) $g(x)=f(u)=u^2$ where $u=(x+5)$
(b) $s(x)=f(u)=u^3$
(c) $h(x)=f(u)=u^{10}$
(d) $t(x)=f(u)=u^{1/2}$

Since $u=(x+5)$, u is a function of x, and we write $u=u(x)=(x+5)$. We can then write each of our functions in the standard composite form. For example, $g(x)=(x+5)^2$ can be written $g(x)=f(u(x))=u^2$ where $u=u(x)=(x+5)$. Thus, it is apparent that we need a rule for determining the *derivative of a composite function*. We shall state only the rule because its derivation is beyond the scope of this book (and can be found in almost any textbook on advanced calculus).

Rule 4.14 If f is a function of u for which the derivative exists, and u is a function of x for which the derivative exists, then the derivative of $f[u(x)]$ with respect to the variable x exists and is given by

$$\begin{aligned} D_x[f(u(x))] &= D_u f(u) D_x u(x) \\ &= f'(u)\cdot u'(x) \\ &= \frac{df}{du}\frac{du}{dx} \end{aligned} \tag{4.16}$$

(*Note*: Eq. (4.16) shows three different notational forms for Rule 4.14.)

Rule 4.14 is commonly known as the *chain rule* and can be extended to a chain of composite functions. First, we shall consider a few examples that involve only two functions.

Example 4.36 Find the derivatives of the four functions in the preceding discussion using the chain rule.

Solution

(a) Rewriting $g(x)=(x+5)^2$ in composite form, we have $g(x)=f(u)=u^2$, where $u=u(x)=(x+5)$. Applying Rule 4.14,

$$\begin{aligned} g'(x) &= D_x[f(u(x))] = D_u f(u)\cdot D_x u(x) \\ &= D_u(u^2)\cdot D_x(x+5) \\ &= 2u\cdot 1 = 2u \end{aligned}$$

Since we are looking for $g'(x)$, it is customary to write $g'(x)$ as a function of x; therefore, substituting $(x+5)$ for u we have,

$$\begin{aligned} g'(x) &= 2(x+5) \\ &= 2x+10 \end{aligned}$$

(b) Rewriting $s(x)=(x+5)^3$ in composite form, we have $s(x)=f(u)=u^3$, where $u=u(x)=(x+5)$. Applying Rule 4.14,

$$\begin{aligned} s'(x) = D_x[f(u(x))] &= D_u f(u) \cdot D_x u(x) \\ &= D_u(u^3) \cdot D_x(x+5) \\ &= 3u^2 \cdot 1 = 3u^2 \\ &= 3(x+5)^2 \\ &= 3(x^2+10x+25) \\ &= 3x^2+30x+75 \end{aligned}$$

(c) Rewriting $h(x)=(x+5)^{10}$ in composite form, we have $h(x)=f(u)=u^{10}$, where $u=u(x)=(x+5)$. Applying Rule 4.14,

$$\begin{aligned} h'(x) = D_x[f(u(x))] &= D_u f(u) \cdot D_x u(x) \\ &= D_u(u^{10}) \cdot D_x(x+5) \\ &= 10u^9 \cdot 1 = 10(x+5)^9 \end{aligned}$$

(d) Rewriting $t(x)=(x+5)^{1/2}$ in composite form, we have $t(x)=f(u)=u^{1/2}$, where $u=u(x)=(x+5)$. Applying Rule 4.14,

$$\begin{aligned} t'(x) &= D_x[f(u(x))] = D_u f(u) \cdot D_x u(x) \\ &= D_u(u^{1/2}) \cdot D_x(x+5) \\ &= \tfrac{1}{2}u^{-1/2} \cdot 1 \\ &= \frac{1}{2}(x+5)^{-1/2} \cdot 1 = \frac{1}{2}\frac{1}{(x+5)^{1/2}} \\ &= \frac{1}{2}\frac{1}{\sqrt{x+5}} \end{aligned}$$

As noted earlier, the functions $g(x)$ and $s(x)$ in Example 4.36 can be expanded easily and we can find $g'(x)$ and $s'(x)$ from our previous rules, thus verifying the chain rule in these cases.

(a)
$$\begin{aligned} g(x) &= (x+5)^2 \\ &= x^2+10x+5 \\ g'(x) &= 2x+10 \end{aligned}$$

(b)
$$\begin{aligned} s(x) &= (x+5)^3 \\ &= (x+5)(x+5)^2 \\ &= x^3+15x^2+75x+125 \\ s'(x) &= 3x^2+30x+75 \end{aligned}$$

When composite functions are first encountered, more difficulty is experienced in identifying $u(x)$ than in actually carrying out Rule 4.14. A helpful hint to keep in mind is that in rewriting $g(x)=f(u)$, where $u=u(x)$, the selection of $u(x)$ must be such that both $D_u f(u)$ and $D_x u(x)$ can be found using our previous rules.

Example 4.37 Use Rule 4.14 to determine the derivative of $g(x)=(2x^2+5x+1)^{3/2}$.

Solution We can write $g(x)$ as a composite function of x by letting $u(x)=(2x^2+5x+1)$ and $f(u)=u^{3/2}$. Note that we select $u(x)$ first such that $u'(x)$ and $f'(u)$ can be determined. Applying Rule 4.14.

$$\begin{aligned} g'(x) &= D_x[f(u(x))] = D_u f(u)\cdot D_x u(x) \\ &= D_u(u^{3/2})\cdot D_x(2x^2+5x+1) \\ &= \tfrac{3}{2}(u^{1/2})(4x+5) \end{aligned}$$

Substituting $(2x^2+5x+1)$ for u, we can rewrite $g'(x)$ in terms of the variable x:

$$\begin{aligned} g'(x) &= \tfrac{3}{2}(2x^2+5x+1)^{1/2}(4x+5) \\ &= \tfrac{3}{2}\sqrt{2x^2+5x+1}\,(4x+5) \end{aligned}$$

Example 4.38 Determine $f'(x)$ if $f(x)=(3x^{1/2}+2)^3$.

Solution We can write $f(x)$ as a composite function by letting $u(x)=(3x^{1/2}+2)$ and $g(u)=u^3$. Applying Rule 4.14,

$$\begin{aligned} f'(x) &= D_x[g(u(x))] = D_u g(u)\cdot D_x u(x) \\ &= D_u u^3\cdot D_x(3x^{1/2}+2) \\ &= 3u^2\cdot\left(\tfrac{3}{2}x^{-1/2}\right) \\ &= 3(3x^{1/2}+2)^2\left(\tfrac{3}{2}x^{-1/2}\right) \\ &= \frac{9}{2}\cdot\frac{(3x^{1/2}+2)^2}{\sqrt{x}} \end{aligned}$$

After becoming familiar with the process involved in taking the derivative of a composite function, we see a pattern that leads to a shortcut. However, it is necessary to remember the fundamental Rule 4.14 as we will apply it to different problems later on in Chapter 6.

Each of the problems we have considered are of the form $f(x)=[u(x)]^n$ where n is a real number. From our solutions you have perhaps noted that in each case

$$f'(x) = n[u(x)]^{n-1}\cdot D_x u(x) \tag{4.17}$$

This is often referred to as the *shortcut chain rule* or the *extended power rule*. The shortcut chain rule is much easier to apply than Rule 4.14. To illustrate, consider Example 4.38 where $f(x)=(3x^{1/2}+2)^3$. This function is in the form $[u(x)]^n$ where $u(x)=(3x^{1/2}+2)$ and $n=3$. By the shortcut chain rule,

$$\begin{aligned} f'(x) &= 3(3x^{1/2}+2)^{3-1}\cdot D_x(3x^{1/2}+2) \\ &= 3(3x^{1/2}+2)^2\cdot \tfrac{3}{2}x^{-1/2} \\ &= \frac{9}{2}\cdot\frac{(3x^{1/2}+2)^2}{\sqrt{x}} \end{aligned}$$

This is the exact result found in Example 4.38 by applying Rule 4.14. We will use the shortcut chain rule in lieu of Rule 4.14 whenever our composite function is of the form $[u(x)]^n$.

Example 4.39 Determine $f'(x)$ if $f(x)=(x^2+5x)^3$ by using the shortcut chain rule.

Solution If $f(x)=(x^2+5x)^3$, we have the form $f(x)=[u(x)]^n$ where $u(x)=(x^2+5x)$ and $n=3$. Therefore,

$$\begin{aligned} f'(x) &= 3[x^2+5x]^{3-1}\cdot D_x(x^2+5x) \\ &= 3[x^2+5x]^2\cdot(2x+5) \end{aligned}$$

Example 4.40 Determine dy/dx if $y=(x^2+4)^{-5}$ by the shortcut chain rule.

Solution Since $y=(x^2+4)^{-5}$, we have the form $y=[u(x)]^n$ where $u(x)=(x^2+4)$ and $n=-5$. Therefore,

$$\begin{aligned} \frac{dy}{dx} &= -5(x^2+4)^{-5-1}\cdot D_x(x^2+4) \\ &= -5(x^2+4)^{-6}(2x) \\ &= \frac{-10x}{(x^2+4)^6} \end{aligned}$$

The algebraic functions that we will be differentiating will consist of sums, products, or quotients of functions, some of which will be composite functions.

Example 4.41 Determine $f'(x)$ if $f(x)=(3x^2+2x)^{1/2}+x^3$.

Solution Our function $f(x)$ is the sum of two functions. The first is a composite function, and the second we can handle directly by the power rule.

Therefore,

$$
\begin{aligned}
f'(x) &= \frac{1}{2}(3x^2+2x)^{-1/2}(6x+2)+3x^2 \\
&= (3x^2+2x)^{-1/2}(3x+1)+3x^2 \\
&= \frac{(3x+1)}{\sqrt{3x^2+2x}}+3x^2
\end{aligned}
$$

The final algebraic form of the derivative can be a point of confusion. It is generally agreed that the usual procedures in simplifying algebraic expressions are followed.

Example 4.42 Determine $g'(x)$ if $g(x)=x^3(3x^2+2)^{1/2}$.

Solution The function $g(x)$ is a product of the function x^3 and the composite function $(3x^2+2)^{1/2}$; hence, we will begin by using the product rule.

$$
\begin{aligned}
g'(x) &= x^3D_x(3x^2+2)^{1/2}+(3x^2+2)^{1/2}D_x(x^3) \\
&= x^3\cdot\tfrac{1}{2}(3x^2+2)^{-1/2}(6x)+(3x^2+2)^{1/2}(3x^2) \\
&= 3x^4(3x^2+2)^{-1/2}+3x^2(3x^2+2)^{1/2}
\end{aligned}
$$

We have determined $g'(x)$. All that remains is to simplify the algebra:

$$
\begin{aligned}
g'(x) &= \frac{3x^4}{(3x^2+2)^{1/2}}+3x^2(3x^2+2)^{1/2} \\
&= \frac{3x^4+3x^2(3x^2+2)}{(3x^2+2)^{1/2}} \\
&= \frac{12x^4+6x^2}{(3x^2+2)^{1/2}}
\end{aligned}
$$

It is advisable to write down each step so that if an error is made, a check will show if we made a mistake in applying the differentiation rules or in doing the algebra.

Example 4.43 If $h(x)=(x)/(x^2+5)^3$, determine $h'(x)$.

Solution The function $h(x)$ is a quotient of the function x and the composite function $(x^2+5)^3$; hence, we will begin by using the quotient rule.

$$
\begin{aligned}
h'(x) &= \frac{(x^2+5)^3D_x(x)-xD_x(x^2+5)^3}{\left[(x^2+5)^3\right]^2} \\
&= \frac{(x^2+5)^3(1)-x\cdot 3(x^2+5)^2(2x)}{(x^2+5)^6}
\end{aligned}
$$

Algebraic simplification of this derivative follows:

$$h'(x) = \frac{(x^2+5)^3 - 6x^2(x^2+5)^2}{(x^2+5)^6}$$

$$= \frac{[(x^2+5) - 6x^2](x^2+5)^2}{\left[(x^2+5)^4\right](x^2+5)^2}$$

$$= \frac{5-5x^2}{(x^2+5)^4}$$

Example 4.44 Find the derivative $s'(x)$ if $s(x) = [x/(x^2+1)]^{1/2}$.

Solution The function $s(x)$ is a composite function where $u(x) = x/(x^2+1)$; hence, we will begin with the shortcut chain rule, and find the derivative of $u(x)$ by the quotient rule.

$$s'(x) = \frac{1}{2}\left(\frac{x}{x^2+1}\right)^{-1/2} D_x\left(\frac{x}{x^2+1}\right)$$

$$= \frac{1}{2}\left(\frac{x}{x^2+1}\right)^{-1/2} \cdot \left[\frac{(x^2+1)(1) - x(2x)}{(x^2+1)^2}\right]$$

$$= \frac{1}{2}\left(\frac{x}{x^2+1}\right)^{-1/2} \cdot \left[\frac{1-x^2}{(x^2+1)^2}\right]$$

$$= \frac{1}{2}\left(\frac{x^2+1}{x}\right)^{1/2} \cdot \left[\frac{1-x^2}{(x^2+1)^2}\right]$$

$$= \frac{1-x^2}{2x^{1/2}(x^2+1)^{3/2}}$$

Rule 4.14 can be extended to a chain of composite functions. For a chain of three functions such as $f(g[u(x)])$, Eq. (4.16) becomes

$$D_x\left[f\left(g\left[u(x)\right]\right)\right] = D_g f(g) \cdot D_u g(u) \cdot D_x u(x) \tag{4.18}$$

Eq. 4.18 indicates that when applying the chain rule, we differentiate from the outside in, as we learned in the shortcut chain rule.

Example 4.45 If $h(x) = [(x^2+3x)^3 - 2]^{1/2}$, find $h'(x)$.

Solution We note that $h(x)$ can be expressed as the composite function $f(g) = g^{1/2}$, where $g = (x^2+3x)^3 - 2$ and that g can be expressed as the composite function $g(u) = u^3 - 2$, where $u = u(x) = x^2 + 3x$. Now applying Eq.

(4.18) to the function $h(x)=f(g[u(x)])$, we have

$$\begin{aligned} h'(x) &= D_g f(g)\cdot D_u g(u)\cdot D_x u(x) \\ &= D_g(g^{1/2})\cdot D_u(u^3-2)\cdot D_x(x^2+3x) \\ &= \tfrac{1}{2}g^{-1/2}\cdot 3u^2\cdot (2x+3) \end{aligned}$$

and substituting we have

$$h'(x) = \tfrac{1}{2}\left[(x^2+3x)^3-2\right]^{-1/2}\cdot 3(x^2+3x)^2(2x+3)$$

or, by applying the technique of the shortcut chain rule,

$$\begin{aligned} h'(x) &= \tfrac{1}{2}\left[(x^2+3x)^3-2\right]^{-1/2}\cdot D_x\left[(x^2+3x)^3-2\right] \\ &= \tfrac{1}{2}\left[(x^2+3x)^3-2\right]^{-1/2}\cdot 3(x^2+3x)^2\cdot D_x(x^2+3x) \\ &= \tfrac{1}{2}\left[(x^2+3x)^3-2\right]^{-1/2}\cdot 3(x^2+3x)^2\cdot (2x+3) \end{aligned}$$

To apply differentiation Rules 4.7 through 4.14, we must first determine the form of the function and then apply the rules shown in Table 4.2.

Table 4.2

	Form of the Function	*Example*	*Rule(s) (Applied in Order)*
1.	Simple power, ax^n	$3x^2$	4.10
2.	Sum of powers	$3x^3+2x^2+7x+5$	4.11, 4.10
3.	Composites	$(x^2+3x+5)^3$	4.14, 4.11, 4.10
4.	Products of powers and/or composites	$x^2(x^2+5)^{1/2}$	4.12, 4.14, 4.11, 4.10
5.	Quotient of powers and/or composites	$\dfrac{(x^2+3)^{1/3}}{x}$	4.13, 4.14, 4.11, 4.10
6.	Composite of a quotient	$\left(\dfrac{x}{x^2+1}\right)^{1/2}$	4.14, 4.13, 4.11, 4.10
7.	Composite of a composite	$\left[(x^2+1)^{1/2}+5\right]^{1/3}$	Eq. (4.18), 4.10, 4.11

Now we will apply our differentiation rules in the following geometric example.

Example 4.46 Draw the graph of the function $f(x)=\sqrt{25-x^2}$ by selecting as values of x: $-5, -4, -3, 1, 0, 1, 3, 5$. Draw tangent lines to the graph at $(-3,f(-3))$, $(0,f(0))$, and $(3,f(3))$, and find the slopes of these lines by using the derivative of $f(x)$.

Solution To find the ordered pairs belonging to the function, we fill in the following table.

x	-5	-4	-3	-1	0	1	3	5
$f(x)=\sqrt{25-x^2}$	0	3	4	$\sqrt{24}$	5	$\sqrt{24}$	4	0

We graph the points on the cartesian plane and draw a smooth curve through the points (see Figure 4.15). We draw the tangent lines t_1, t_2, and t_3 at $(-3,4)$, $(0,5)$, and $(3,4)$, respectively, and note $m_{t_1}>0$, $m_{t_2}=0$, and $m_{t_3}<0$. The slopes of the tangent lines are found by first finding the derivative of $f(x)$ by Rule 4.14 or the shortcut chain rule:

$$\begin{aligned} f'(x) &= D_x\sqrt{(25-x^2)} \\ &= D_x(25-x^2)^{1/2} \\ &= \tfrac{1}{2}(25-x^2)^{-1/2}(-2x) \\ &= \frac{-x}{\sqrt{25-x^2}} \end{aligned}$$

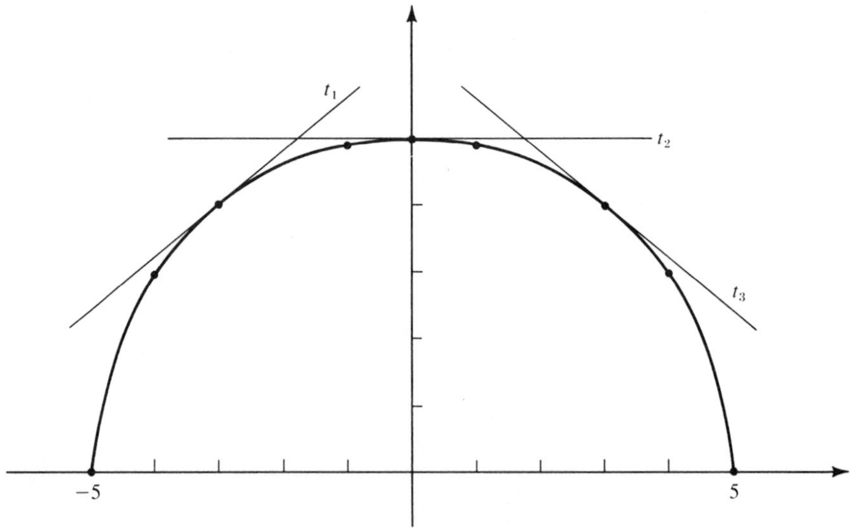

FIGURE 4.15

Then we substitute -3, 0, and 3 for x in the derivative function to find the slopes:

$$m_{t_1} = f'(-3) = \frac{-(-3)}{\sqrt{25-9}} = \frac{3}{4}$$

$$m_{t_2} = f'(0) = \frac{-(0)}{\sqrt{25}} = 0$$

$$m_{t_3} = f'(3) = \frac{-3}{\sqrt{25-9}} = -\frac{3}{4}$$

EXERCISES

For each of the composite functions $f(u(x))$ in Exercises 1–8 determine the component parts $f(u)$ and $u(x)$, and find $D_x[f(u(x))]$ by using the chain rule (Rule 4.14).

1. $(3x+1)^3$

2. $(2x^3-7)^4$

3. $(2x+6)^8$

4. $(12x+3x^2)^{12}$

5. $(10x^2-3x+2)^{3/2}$

6. $\dfrac{1}{(2x+5)^3}$

7. $\dfrac{2}{\sqrt{7x^2+3}}$

8. $(3x^2+2x)^{-4}$

9–16. Verify your results in Exercises 1–8 using the shortcut chain rule.

For each of the composite functions $f(g(u(x)))$ in Exercises 17–20 determine the component parts $f(g)$, $g(u)$, and $u(x)$, and find $D_x[f(g(u(x)))]$ by using Eq. 4.18.

17. $[(x+2)^3+1]^{-2}$

18. $\sqrt[5]{(3x+7)^2+1}$

19. $[(x^2-4)^{2/3}+5]^4$

20. $[(5x+1)^2+2]^{-1/2}$

21–24. Verify your results in Exercises 17–20 using the shortcut chain rule.

Determine the derivative with respect to x for each of the functions in Exercises 25–44.

25. $f(x)=(4x+3)^2$

26. $f(x)=(6x-7)^4$

27. $g(x)=(5x+7)^{-3}$

28. $g(x)=(3x^2+4)^{-1/2}$

29. $h(x)=x\sqrt{3x+5}$

30. $h(x)=x^2\sqrt[3]{9x^2-3}$

31. $g(x)=\dfrac{x}{\sqrt{x^2+1}}$

32. $g(x)=\dfrac{\sqrt{x^2+1}}{x}$

33. $y=\left(\dfrac{x}{x+1}\right)^{1/2}$

34. $y=\left(\dfrac{6x}{x^2+1}\right)^{1/3}$

35. $f(x)=\sqrt[3]{(3x^2+4x-2)^2}$

36. $f(x)=3(x^2+4x)^3+(x-5)^4$

37. $s(x)=x^2(x-100)^3$

38. $s(x)=\left(\sqrt{x}-\frac{.001}{x^2}\right)^2$

39 $f(x)=(3x-5)^{1/3}\cdot x^2$

40. $f(x)=(x^2+1)\cdot\sqrt{4x+2}$

41. $g(x)=\sqrt[4]{x^2+4}$

42. $g(x)=\sqrt{\frac{2x}{x+5}}$

43. $f(x)=\frac{(x^2-4)^3}{(2x+3)^2}$

44. $f(x)=[2(x^2+4)^3-3(x^2+4)^2]^4$

45. **Concentration of a Drug in the Blood** Suppose that the concentration, $C(t)$, of a drug in the blood t hr after injection ($t \geqslant 1$) is given by $C(t)=(t^2+20)^{-2}$. If the concentration, $C(t)$, is in milligrams of drug/gram of blood, determine

(a) the concentration 2 hr after injection,
(b) the rate of change of the concentration 2 hr after injection.

46. If $f(x)=\sqrt{16-4x^2}$,

(a) Sketch a graph of $f(x)$ by completing the following table and plotting the associated points.

x	0	1	2	-1	-2
$f(x)$					

(b) Draw a tangent line to the graph of $f(x)$ at $x=1$.
(c) Determine an equation of the tangent to the graph of $f(x)$ at $x=1$.

47. **Birthrate of a Female Beetle** Biologists have shown that the birthrate, measured in the number of eggs laid a day, of the female beetle *Rhizopertha dominica* decreases as the density of the beetles increase. The number of eggs laid/day/female, $E(x)$, was found to be related to the number of insects, x, found in each gram of wheat by the equation

$$E(x) = 12(2x+1)^{-1/2}$$

(a) Determine the birthrate/female when there are 4 insects/gram of wheat.
(b) Determine the rate of change of the birthrate when there are 4 insects/gram of wheat.

48. **Consumption of Sawfly Cocoons by Deer Mice** Ecologists have shown that the number of prey eaten by a predator is related to the density of the prey. In experiments designed to determine the consumption of sawfly cocoons by deer mice, it was determined that the number of cocoons eaten, $N(x)$, was related to x, the number of cocoons/square meter, by the equation

$$N(x) = 8(\sqrt[3]{9x+100})$$

(a) Determine the number of cocoons eaten when there are 100 sawfly cocoons/square meter.
(b) Determine the rate of change of $N(x)$ when there are 100 cocoons/square meter.

49. Monthly Demand for a Stereo Unit The Stellar Stereo Manufacturing Company has determined from sales experience that the monthly demand, $D(x)$, for their economy set is related to its selling price, x, by the equation

$$D(x) = \frac{1{,}000}{(x+10)^{1/2}}$$

where $D(x)$ is in hundreds of units and x is in dollars.

(a) What is the monthly demand for the economy model when the selling price is \$90?
(b) What is the rate of change of the demand when the selling price is \$90?
(c) Approximately how many will they sell if the unit price is \$91?

50. The cost in dollars of manufacturing a batch of parts is given by the relation $C(x)=(3x^{1/2}+2)^2$ where x is the number of parts produced per batch.

(a) Determine the total cost of producing a batch of 49 parts.
(b) Determine the marginal cost of the 50th part in a batch.

4.6 HIGHER-ORDER DERIVATIVES

We have shown that the derivative of any function, when evaluated at a point $x=a$, can be interpreted as the slope of the tangent line drawn to the graph of the function at $x=a$. We abbreviated this by saying $f'(a)$ can be interpreted as the *slope of the graph of the function* $f(x)$ at $x=a$. We also interpreted the derivative of a function as the rate of change of the function. Furthermore, we have found that the derivative of a function is a function. Thus if we are interested in determining the *rate of change of the slope of a function* at $x=a$, we could simply determine the derivative of the function that represents the slope and evaluate it at $x=a$. This would involve differentiating a function that is already a derivative. The derivatives obtained by differentiating functions that are already derivatives are called *higher-order derivatives*, and each higher-order derivative gives information concerning previous derivatives. When considering higher-order derivatives, the derivative of the given function is called the first derivative, the derivative of the first derivative is called the *second derivative*, etc. In the following example, we determine the second derivative of a function and offer a geometrical interpretation of it.

Example 4.47 Rate of Change of Slope Find the rate of change of the slope of the curve given by the function $f(x)=4-x^2$ and interpret the result.

Solution The slope of the curve given by the function $f(x)=4-x^2$ is given by the first derivative $f'(x)=-2x$. The rate of change of the slope of the

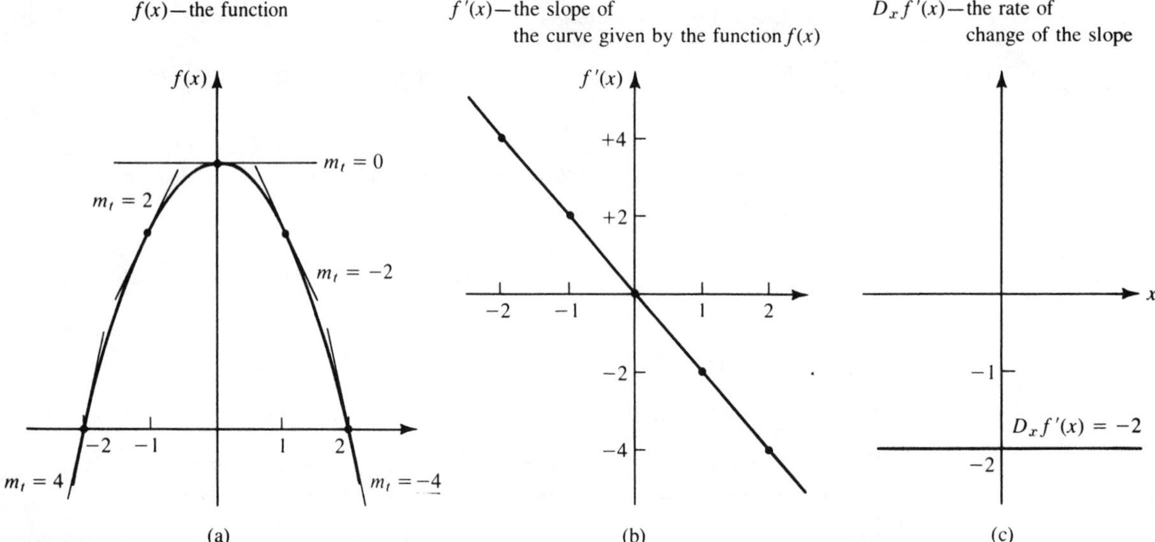

FIGURE 4.16

curve given by the function is the derivative of $f'(x)$; that is, $D_x f'(x) = D_x(-2x) = -2$.

Figure 4.16*a* is the graph of the function $f(x)$ with several tangent lines drawn to the curve. Figure 4.16*b* is the graph of $f'(x)$, and Fig. 4.16*c* is the graph of the rate of change of $f'(x)$ or $D_x f'(x)$. The rate of change of the slope equaling -2 implies that the slope of the tangent lines drawn to the function will decrease by 2 units for each unit increase in x. This can be verified by inspecting the tangent lines drawn to the graph in Fig. 4.16*a*.

If we are going to consider second derivatives, third derivatives, etc., we must agree upon a notation. We have used several symbols to denote the derivative of a function, namely, $f'(x)$, $D_x f(x)$, dy/dx, and y'. If we wish to determine the derivative of $f'(x)$, we will simply differentiate the function given by $f'(x)$ and denote it as $f''(x)$, which is read "f double prime of x." Similarly, if we wish to determine the derivative of $f''(x)$, we will simply differentiate $f''(x)$ and denote it as $f'''(x)$, which is read "f triple prime of x." To denote higher derivatives, it is customary to use natural numbers rather than primes; for example, the fourth derivative would be denoted as $f^{(4)}(x)$ and the nth derivative as $f^{(n)}(x)$.

Extensions of the other symbols can also be used to represent the higher-order derivatives, and a few of these are given in Table 4.3. It should be noted that $f^{(n)}(x)$ represents the nth derivative of $f(x)$ and not $f(x)$ raised to the nth power, which we denote by $[f(x)]^n$. However, when referring to other textbooks, the reader should be careful to verify what notation is being

Table 4.3

Function	*First Derivative*	*Second Derivative*	*Third Derivative*	*nth Derivative*
$f(x)$	$f'(x)$	$f''(x)$	$f'''(x)$	$f^{(n)}(x)$
$f(x)$	$D_x f(x)$	$D_x^2 f(x)$	$D_x^3 f(x)$	$D_x^n f(x)$
$f(x)$	$f'(x)$	$f''(x)$	$f^{(3)}(x)$	$f^{(n)}(x)$
$f(x)$	$\frac{df}{dx}$	$\frac{d^2f}{dx^2}$	$\frac{d^3f}{dx^3}$	$\frac{d^nf}{dx^n}$
y	$\frac{dy}{dx}$	$\frac{d^2y}{dx^2}$	$\frac{d^3y}{dx^3}$	$\frac{d^ny}{dx^n}$
y	$D_x y$	$D_x^2 y$	$D_x^3 y$	$D_x^n y$

used. For example, some authors use $f^n(x)$ to represent the function raised to the nth power, which can easily be confused with $f^{(n)}(x)$ because they are similar in appearance. In this textbook we shall primarily use the notation $f'(x), f''(x), f'''(x), \ldots, f^{(n)}(x)$ and occasionally the notation $D_x^n f(x)$.

Suppose we wish to determine the first four consecutive derivatives of $f(x)=x^5+3x^4+4x^3-10x^2+x-1$. We simply obtain the first derivative in the usual manner; next, we determine the second derivative by differentiating the function that represents the first derivative; etc. Proceeding until we have obtained the fourth derivative of $f(x)$, we obtain the following derivatives:

$$f'(x) = 5x^4 + 12x^3 + 12x^2 - 20x + 1$$
$$f''(x) = 20x^3 + 36x^2 + 24x - 20$$
$$f'''(x) = 60x^2 + 72x + 24$$
$$f^{(4)}(x) = 120x + 72$$

Evaluating the respective derivatives at $x=1$, we obtain the following values:

$$f'(1) = 10$$
$$f''(1) = 60$$
$$f'''(1) = 156$$
$$f^{(4)}(1) = 192$$

Example 4.48 Determine the first four derivatives of $f(x)=x^3+3x^2-4x+7$, and determine the values of the function and the respective derivatives at $x=0$ and $x=2$.

Solution

$$f'(x) = 3x^2 + 6x - 4$$
$$f''(x) = 6x + 6$$
$$f'''(x) = 6$$
$$f^{(4)}(x) = 0$$

The function values are as follows:

$$\begin{array}{llll} f(0) = 7 & & f(2) = 19 \\ f'(0) = -4 & & f'(2) = 20 \\ f''(0) = 6 & & f''(2) = 18 \\ f'''(0) = 6 & & f'''(2) = 6 \\ f^{(4)}(0) = 0 & & f^{(4)}(2) = 0 \end{array}$$

Example 4.49 Determine the first five derivatives of $f(x)=(2x-3)^4$, and determine the values of the function and the respective derivatives at $x=1$, $x=1\frac{1}{2}$, and $x=2$.

Solution

$$\begin{aligned} f(x) &= (2x-3)^4 \\ f'(x) &= D_x[(2x-3)^4] \\ &= 4(2x-3)^3(2) = 8(2x-3)^3 \\ f''(x) &= D_x[8(2x-3)^3] \\ &= 3 \cdot 8(2x-3)^2(2) = 48(2x-3)^2 \\ f'''(x) &= D_x[48(2x-3)^2] \\ &= 2 \cdot 48(2x-3)^1(2) = 192(2x-3) \\ f^{(4)}(x) &= D_x[192(2x-3)] \\ &= 192(2) = 384 \\ f^{(5)}(x) &= 0 \end{aligned}$$

The respective function values are as follows:

$$\begin{array}{lll} f(1)=1 & f(\frac{3}{2})=0 & f(2)=1 \\ f'(1)=-8 & f'(\frac{3}{2})=0 & f'(2)=8 \\ f''(1)=48 & f''(\frac{3}{2})=0 & f''(2)=48 \\ f'''(1)=-192 & f'''(\frac{3}{2})=0 & f'''(2)=192 \\ f^{(4)}(1)=384 & f^{(4)}(\frac{3}{2})=384 & f^{(4)}(2)=384 \\ f^{(5)}(1)=0 & f^{(5)}(\frac{3}{2})=0 & f^{(5)}(2)=0 \end{array}$$

Example 4.50 Determine the first three derivatives of $f(x)=(x^3+2)^2$, and determine the value of the function and the derivatives at $x=1$.

Solution

$$\begin{aligned} f(x) &= (x^3+2)^2 \\ f'(x) &= D_x\left[(x^3+2)^2\right] \\ &= 2(x^3+2)^1(3x^2) = 6x^2(x^3+2)^1 \\ f''(x) &= D_x[6x^2(x^3+2)] \\ &= 6x^2(3x^2)+(x^3+2)(12x) = 30x^4+24x \\ f'''(x) &= D_x[30x^4+24x] \\ &= 120x^3+24 \end{aligned}$$

The respective function values are

$$\begin{aligned} f(1) &= 9 \\ f'(1) &= 18 \\ f''(1) &= 54 \\ f'''(1) &= 144 \end{aligned}$$

EXERCISES

For each of the functions in Exercises 1–8, determine all of the successive derivatives until all remaining derivatives equal zero and evaluate the derivatives for the given value.

1. $f(x)=x^2+3x-4$ (at $x=-2$)

2. $f(x)=4x-x^3$ (at $x=5$)

3. $f(x)=x^4+5x^3-7x^2+3x-10$ (at $x=1$)

4. $f(x)=(x^2+4)(x^3-5)$ (at $x=1$)

5. $f(x)=x(x-1)^3$ (at $x=1$)

6. $s(t)=t^3(t+2)^2$ (at $t=0$)

7. $z(t)=t(t+1)(t+2)$ (at $t=-1$)

8. $g(z)=(z+1)^2(z+2)^2$ (at $z=-1$)

For each of the functions in Exercises 9–16, determine the first and second derivatives.

9. $f(x)=\dfrac{1}{\sqrt{x}}$

10. $f(x)=\dfrac{x-1}{x+1}$

11. $g(x)=\dfrac{(x+2)^2}{x}$

12. $g(x)=\dfrac{x^2}{x+1}$

13. $h(x) = \sqrt{3x - x^3}$

14. $h(x) = (x^2 + 4)^3$

15. $f(x) = (x^2 + 3x + 2)^{-2}$

16. $f(x) = (x^4 - x^2 + 1)^{-3}$

17. Sketch the graph of $f(x) = 4 + x^2$ and

(a) Draw the tangent lines to the curve at the x values: $\{-2, -1, 0, 1, 2\}$.
(b) Determine the derivative $f'(x)$ and sketch the graph of $f'(x)$.
(c) Show that $f''(x)$ is greater than zero and verify by inspection of the tangent lines in part (a) that the rate of change of the slope is positive.

18. If $f(x) = x^n + x^{n-1} + x^{n-2} + \cdots + x^2 + x^1 + a_0$, where n is any positive integer and a_0 is a constant, determine a formula for $f^{(n)}(x)$.

IMPORTANT TERMS AND CONCEPTS

Calculus
Chain rule
  shortcut chain rule
Composite function
Continuous
  at a point
  function
  in the domain
Derivative
Derivative of a
  composite function
  constant
  difference
  function
  product
  quotient
  sum
Difference quotient
Differential calculus
Discontinuous function
Higher-order derivatives
Instantaneous rate of change
Integral calculus
Limit of a function
Limits
  right-hand limit
  left-hand limit
Marginal analysis
  cost
  profit
  revenue
Rate of change
Second derivative
Slope of a tangent line

SUMMARY OF RULES AND FORMULAS

Limit Rules

4.1(a) $\lim_{x \to a} kx = ka$

4.1(b) $\lim_{x \to a} k = k$

4.2 $\lim_{x \to a} [f(x) + g(x)] = \lim_{x \to a} f(x) + \lim_{x \to a} g(x)$

4.3 $\lim_{x \to a} [f(x) \cdot g(x)] = \left[\lim_{x \to a} f(x)\right] \cdot \left[\lim_{x \to a} g(x)\right]$

4.4 $\lim_{x \to a} \left[\dfrac{f(x)}{g(x)}\right] = \dfrac{\lim_{x \to a} f(x)}{\lim_{x \to a} g(x)}$, provided $\lim_{x \to a} g(x) \neq 0$

4.5 $\lim_{x \to a} [f(x)]^n = \left[\lim_{x \to a} f(x) \right]^n$

4.6 $\lim_{x \to a} [f(x)]^{1/n} = \left[\lim_{x \to a} f(x) \right]^{1/n}$

Differentiation Rules

4.7 If $f(x) = k$, a constant, then $f'(x) = 0$

4.8 If $f(x) = x^n$, then $f'(x) = nx^{n-1}$

4.9 $D_x[kf(x)] = kD_x f(x) = kf'(x)$

4.10 If $f(x) = kx^n$, then $f'(x) = nkx^{n-1}$

4.11 $D_x[f(x) + g(x)] = f'(x) + g'(x)$
$D_x[f(x) - g(x)] = f'(x) - g'(x)$

4.12 $D_x[f(x) \cdot s(x)] = f(x)s'(x) + s(x)f'(x)$

4.13 $D_x\left[\frac{f(x)}{g(x)} \right] = \frac{g(x)D_x f(x) - f(x)D_x g(x)}{[g(x)]^2}$

4.14 $D_x[f(u(x))] = D_u f(u) D_x u(x)$

Eq. (4.17) If $f(x) = [u(x)]^n$, then $f'(x) = n[u(x)]^{n-1} D_x u(x)$

Eq. (4.18) $D_x[f(g[u(x)])] = D_g f(g) \cdot D_u g(u) \cdot D_x u(x)$

5 DIFFERENTIAL CALCULUS: APPLICATIONS

5.1 INTRODUCTION

In Chapter 4 we considered the definition and various interpretations of a derivative. We also developed the rules needed to determine the derivatives of the algebraic functions we introduced in Chapters 1 and 2. We will utilize the calculus we have developed as a tool to investigate various mathematical models. It is in this area of analysis that calculus makes one of its most significant contributions.

Specifically, we will use differential calculus to determine on what intervals a function is increasing or decreasing, to determine the extreme values of functions, to continue our study of marginal analysis, and as an aid in sketching the graphs of functions.

5.2 INCREASING AND DECREASING FUNCTIONS

We have noted that once a mathematical model of a real-life problem has been formulated, we can use the model to give us information about the interaction of various components of the problem. For example, a mathematical model of the financial posture of a state may be used to determine whether state revenues will increase or decrease as corporate taxes are increased. (Obviously, there is a point where corporate taxes would become so high that businesses would leave the state, putting people out of work, and the state's revenue would decline.) Additionally, some practical problems require us to determine the intervals on which a function is increasing or decreasing. For example, when we considered the revenue function $R(x) = 100x - 4x^2$ of Example 4.28, we were interested in determining the intervals on which the revenue, $R(x)$, increased. The graph of this function is reproduced for reference in Fig. 5.1.

We concluded from the graph of $R(x)$ that the function is increasing on the interval $(5, 12.5)$ and decreasing on the interval $(12.5, 20)$. We intuitively know what we mean by saying a function is increasing on an interval (a,b);

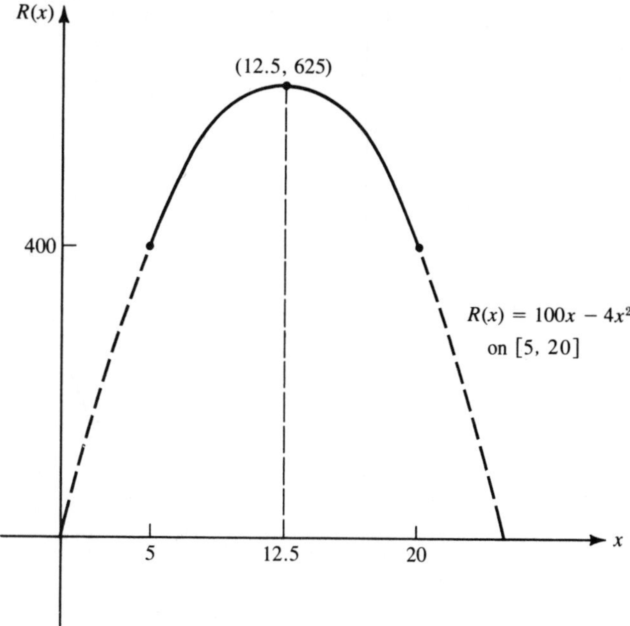

FIGURE 5.1

that is, as x increases in the interval, $f(x)$ increases or geometrically the graph rises as we move to the right in the interval. The following definition based on our intuition is more precise.

Definition 5.1 The function $f(x)$ is increasing on the open interval (a,b) if and only if for any two points x_1, x_2 in the interval, $f(x_1) < f(x_2)$ whenever $x_1 < x_2$. See Fig. 5.2.

In a like manner, a function is decreasing on an interval if as x increases in the interval, $f(x)$ decreases, or geometrically the graph falls as we move to the right in the interval. We state the following definition for a decreasing function.

Definition 5.2 The function $f(x)$ is decreasing on the open interval (a,b) if and only if for any two points x_1, x_2 in the interval, $f(x_1) > f(x_2)$ whenever $x_1 < x_2$. See Fig. 5.3.

The open interval in these definitions can be considered to be the entire real line $(-\infty, \infty)$, or the *open rays* (a, ∞) or $(-\infty, b)$.

Once we have a graph of a function it is relatively easy to determine on what intervals the function is increasing and decreasing. For example, in Fig.

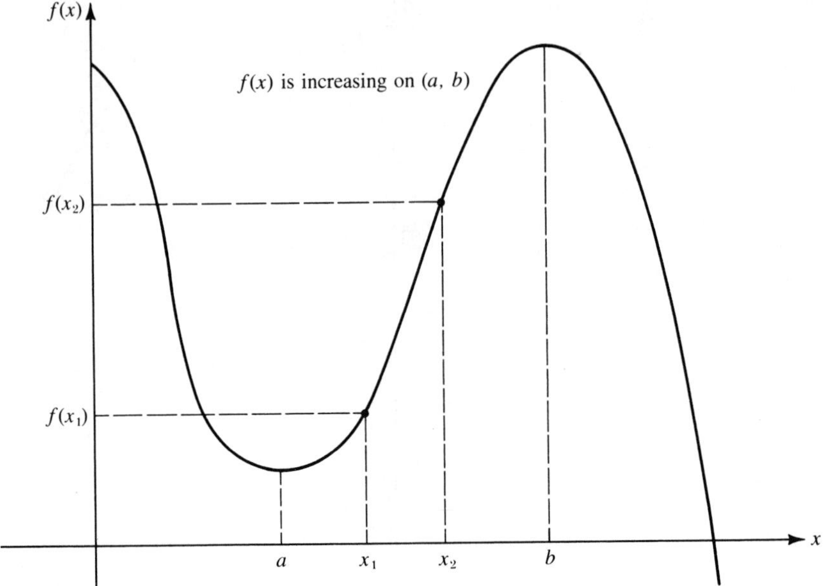

FIGURE 5.2

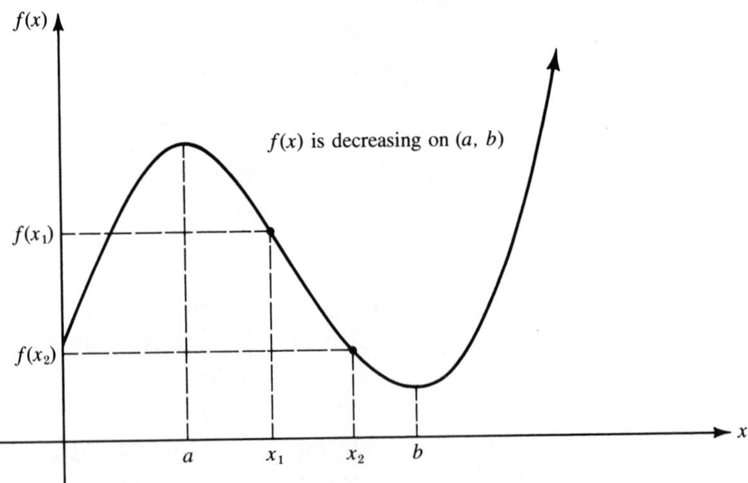

FIGURE 5.3

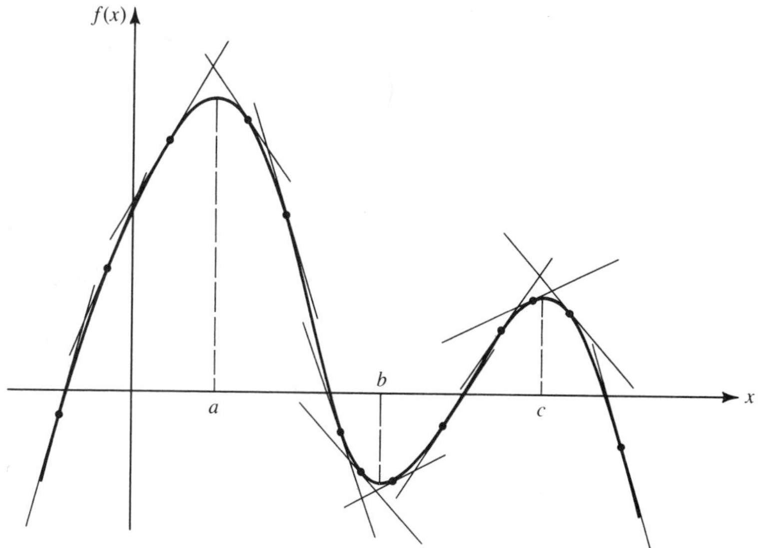

FIGURE 5.4

5.4 we conclude that $f(x)$ is increasing (sometimes denoted by $f(x)\uparrow$) on $(-\infty, a)\cup(b,c)$ and decreasing on $(a,b)\cup(c,\infty)$.

If, however, we do not have the graph of a function, how could we determine the values of a, b, and c? A point-by-point method of graphing the function is not only tedious, but will sometimes lead to only approximate answers. Fortunately, there is a method to determine these values based on the properties of the first derivative of a function.

A study of Fig. 5.4 shows that in each open interval where the function is increasing, a tangent line drawn to the curve would have a positive slope, or in other words, the first derivative of the function at any point in these intervals is positive. Similarly, we note that if a function is decreasing on an interval, the first derivative is negative at any point in the interval. Based on our observations, we state the following rule.

Rule 5.1 Assuming that the function $f(x)$ has a derivative at every point in an open interval (a,b),

(a) If $f'(x)>0$ for all values of x in (a,b), then the function $f(x)$ is increasing on (a,b), and if $f(x)$ is increasing on (a,b), then for all x in (a,b), $f'(x)>0$ or $f'(x)=0$ for a finite number of x's.

(b) If $f'(x)<0$ for all values of x in (a,b), then the function $f(x)$ is decreasing on (a,b), and if $f(x)$ is decreasing on (a,b), then for all x in (a,b), $f'(x)<0$ or $f'(x)=0$ for a finite number of x's.

This rule also applies to the open interval $(-\infty, \infty)$, or the open rays $(-\infty, b)$ or (a, ∞). To apply Rule 5.1 we need only examine the first derivative of a function, determine on what intervals it is positive or negative, and conclude that our function is increasing or decreasing on those respective intervals.

Example 5.1 Assuming $f(x)=4+2x-x^2$ is defined on the set of real numbers, find the intervals on which the function is increasing or decreasing.

Solution To apply Rule 5.1 we take the first derivative:

$$f'(x) = 2 - 2x$$

To find where $f(x)$ is increasing, we determine on what interval $f'(x)>0$ or we solve the inequality

$$2-2x>0$$
$$-2x>-2$$
$$x<1$$

Therefore, $f(x)$ is increasing when $x<1$ or on the interval $(-\infty, 1)$. To find where $f(x)$ is decreasing, we determine on what interval $f'(x)<0$ by solving the inequality

$$2-2x<0$$
$$-2x<-2$$
$$x>1$$

Therefore, $f(x)$ is decreasing when $x>1$ or on the interval $(1, \infty)$. Since our function is quadratic, we can determine the turning point or vertex by setting $f'(x)=0$ and solving for x.

If $\qquad f'(x) = 2 - 2x = 0$

then $\qquad x = 1$

We graph the function $f(x)=4+2x-x^2$ in Fig. 5.5 and by inspection verify our conclusions. Note that the point on the curve where the function changes from being an increasing function to a decreasing function (or vice versa) is a turning point and that the value of the derivative at this point is zero.

As another example consider:

Example 5.2 Find the intervals on which the function $f(x)=x^3-8$ is increasing or decreasing.

Solution To apply Rule 5.1 we take the first derivative:

$$f'(x) = 3x^2$$

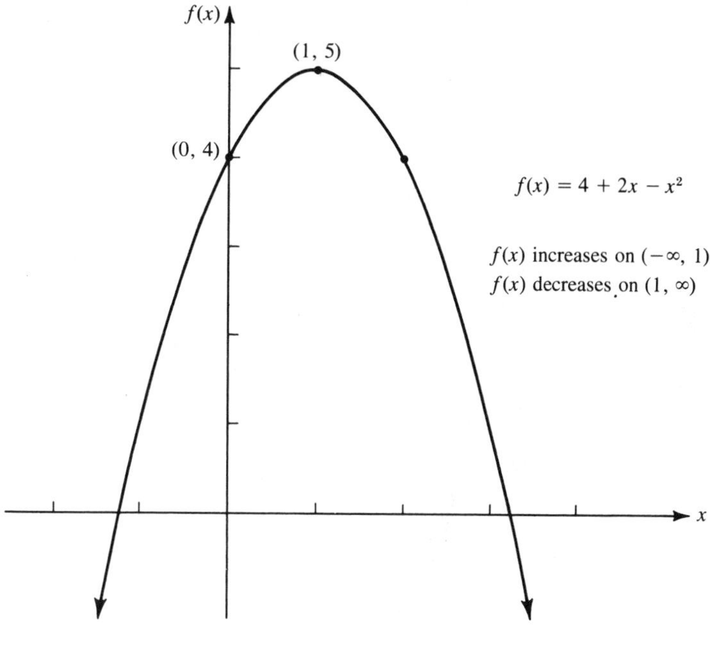

FIGURE 5.5

To find where $f(x)$ is increasing, we determine on what interval $f'(x)>0$ or equivalently we solve the inequality

$$3x^2 > 0$$

from which we find either $x>0$ or $x<0$. This implies that $f(x)$ increases on $(-\infty,0)\cup(0,\infty)$. At $x=0$, $f'(x)=0$ or the tangent line drawn to the curve at the point $(0,f(0))$ is horizontal. Since there exists only one point in the interval $(-\infty,\infty)$, where $f'(x)\not>0$ and at this point $f'(x)=0$, we conclude from Rule 5.1 that $f(x)$ increases on $(-\infty,\infty)$. Figure 5.6 is a sketch of the graph of $f(x)=x^3-8$, and we verify our conclusions by inspection.

In the next example we consider a function that is the model of a physical problem we have previously considered.

Example 5.3 The Box Problem Revisited In Chapter 1 we found that the relation $V(x)=(10-2x)^2(x)$ was a mathematical model for the volume of an open box made from a piece of 10 in.$\times$10 in. cardboard by cutting out squares of dimensions x by x from each corner. The domain of this function is restricted by the physical problem to be $(0,5)$. Use Rule 5.1 to find all intervals in this domain where $V(x)$ is increasing or decreasing.

Solution Since $V(x)=(10-2x)^2(x)=100x-40x^2+4x^3$, the first derivative $V'(x)=100-80x+12x^2$. To find where $V(x)$ is increasing, we find on what

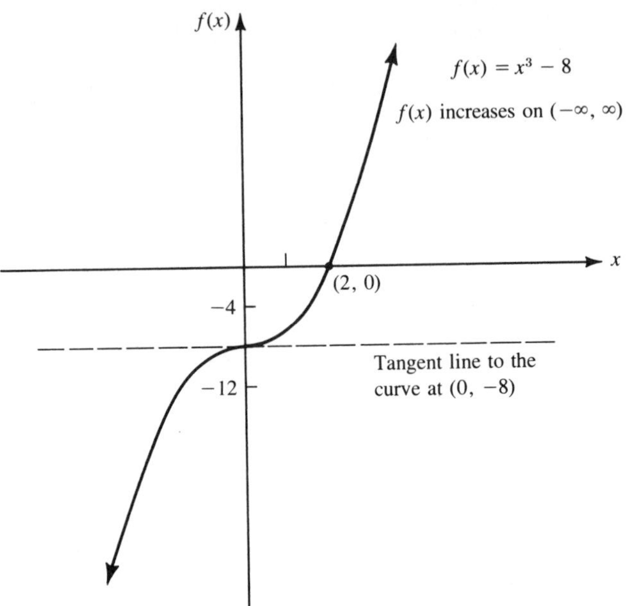

FIGURE 5.6

interval $V'(x)>0$ by solving the quadratic inequality

$$12x^2 - 80x + 100 > 0$$
$$4(3x^2 - 20x + 25) > 0$$
$$4(3x-5)(x-5) > 0$$

which is equivalent to solving

$$(3x-5)(x-5) > 0$$

To solve this quadratic inequality, we use the graphical cut-method discussed in Section 2.4 (see Fig. 5.7). Because of the restricted domain in this problem, we are only interested in $V'(x)$ on the interval (0,5), marked on the sketch by the arrow. From this sketch we conclude that

$$V'(x) > 0 \text{ and } V(x) \text{ is increasing on } \left(0, \tfrac{5}{3}\right)$$
$$V'(x) < 0 \text{ and } V(x) \text{ is decreasing on } \left(\tfrac{5}{3}, 5\right)$$
$$V'(x) = 0 \text{ when } x = \tfrac{5}{3}$$

Information provided by the first derivative of a function can also be used to graph the function. In this example we know that $V(x)$ is increasing

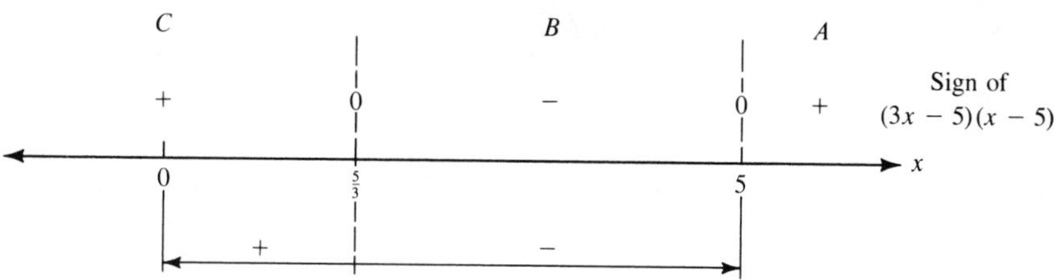

FIGURE 5.7

on $(0, \frac{5}{3})$ and decreasing on $(\frac{5}{3}, 5)$, which would imply that the maximum value must be $V(\frac{5}{3})$; hence the ordered pair $(\frac{5}{3}, V(\frac{5}{3}))$ is a critical point in our graph. Figure 5.8 is a graph of the function $V(x) = (10 - 2x)^2(x)$, and you will note that the ordered pair $(\frac{5}{3}, V(\frac{5}{3}))$ has been plotted.

Perhaps you recall from a discussion of this problem in Chapter 1 that we were only able to find an approximate value of the maximum volume of the box. With the use of the calculus we have found the exact dimensions to maximize the volume. In the next section we will continue our investigation of determining maximum and minimum values of functions.

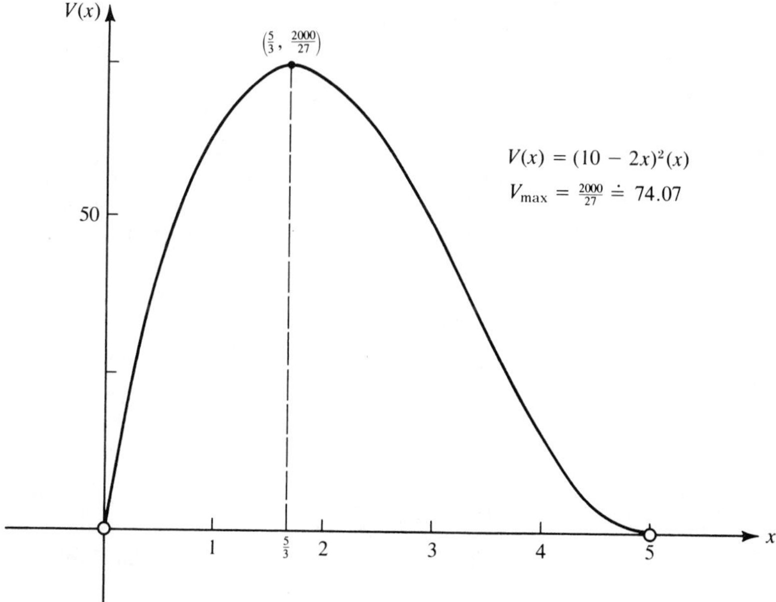

FIGURE 5.8

EXERCISES

In Exercises 1–14 *find the intervals where the functions are increasing or decreasing.*

1. $f(x)=x^2-6x+5$

2. $f(x)=3x^2-12x-1$

3. $g(x)=4x-2$

4. $g(x)=-\frac{1}{2}x+7$

5. $f(x)=2x^3+3x^2-12x-4$

6. $f(x)=2+3x-x^3$

7. $s(x)=\frac{1}{2}x^4-2x^3$

8. $s(x)=x^4-2x^3+4$

9. $h(x)=\dfrac{4x}{x^2+4}$

10. $g(x)=\dfrac{2x+1}{x-1}$

11. $f(x)=x+\dfrac{1}{x}$

12. $g(x)=\dfrac{4}{x-3}$

13. $h(x)=\sqrt{x^2+1}$

14. $f(x)=\sqrt{x+1}$

15. A manufacturer has determined that if he manufactures x items in one week, his profit, P, on the x items is given by

$$P = 0.001x^2(40-x)^3$$

where P is measured in dollars.

(a) Determine the interval on which his profit will be increasing.
(b) At what rate is his profit increasing when he manufactures 10 items?

16. The Togahide Plastics Company manufactures and sells wholesale leather briefcases. Their cost of manufacturing x briefcases is given by $C=30+50x$ where C is in dollars. To encourage their outlets to purchase larger quantities they charge \$40/briefcase for sales over 20 and $\$(60-x)$/briefcase when 20 or fewer are purchased.

(a) Determine their revenue, R, from the sale of x briefcases to Kaufmans Luggage, one of their outlets, which purchases fewer than 20.
(b) Determine the interval on which the revenue function in part (a) is increasing.
(c) Determine the profit, P, from the sale of x briefcases to Kaufmans.
(d) Determine the interval on which the profit function, P, in part (c) increases.

17. **Per Capita Expenditure for Coffee** Harriet Maxwell, an economist, formulated a price-demand function for coffee in the United States as follows: $D(x)=25-4x$, where x = wholesale price in dollars/lb of coffee and $D(x)$ = per capita consumption in pounds of coffee in the United States per year.

(a) Determine the per capita yearly expenditure for coffee as a function of the wholesale price.
(b) Determine the interval on which the per capita expenditure function is increasing.
(c) What is the rate of increase of the per capita yearly expenditure function when the wholesale price of coffee is \$2/lb? Interpret your results.

18. Charles, the owner of the Ritz Cocktail Lounge, knows by experience that on the average his customers spend \$10 each evening when the lounge is not overcrowded. He has noted that for each person over 100 this average decreases by 5¢.

(a) If x represents the number of customers over 100 in the Ritz Lounge on a particular evening, verify that Charles' income for that evening is $I(x)=(100+x)(10-.05x)$ dollars. (Assume there are at least 100 customers in the lounge.)
(b) If the lounge has a maximum capacity of 160 people, determine the interval on which the function $I(x)$ is increasing.

19. **Temperature Pattern Near a Reactor Coolant Outlet** Biologists for S.W. Utility Company concerned with the effect of a proposed nuclear power plant on water temperatures in the Snake River calculated that the water temperature would be increased by varying amounts, depending on the distance from the outlet of the plant's coolant system. They concluded that the temperature of the water would be increased by $T(x)=10/[(x-2)^2+1]$ degrees Celsius, where x is the distance in hundreds of yards measured from the point S shown in the following sketch.

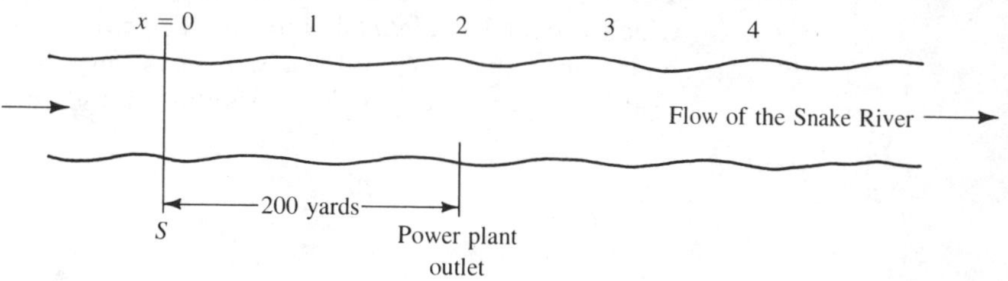

EXERCISE 19

(a) Determine the increase in temperature 200 yards above the plant's outlet. (*Hint*: $x=0$.)
(b) Determine the interval on which the function $T(x)$ is increasing. (assume $x \geqslant 0$.)
(c) Determine the interval on which the function $T(x)$ is decreasing. (assume $x \geqslant 0$.)

20. **Spread of an Epidemic** The percentage of the citizens in Transylvania infected by virus x, t days after the first case was discovered, is given by

$$N(t) = \frac{100t}{t^2+100}$$

(a) Determine the time interval on which the percentage was increasing.
(b) Determine the rate of increase 5 days after the first case was discovered.

5.3 MAXIMA AND MINIMA

At the end of the last section it was pointed out that we are often interested in determining the maximum or minimum value of a function on a given domain. We have encountered several problems of this nature in previous chapters. In Example 4.28 we were interested in finding the maximum revenue from the sale of T-Slot cutters. In Example 5.3 we were looking for the dimensions of a box that would yield a maximum volume, and in Chapter 2 we were interested in finding maximum and minimum values of quadratic functions. In this section we will develop techniques to find the maxima (plural of maximum) and minima (plural of minimum) of a function defined on a given domain, and in the next two sections we will apply these techniques to solving practical problems.

Since the functions that are models for most real-life problems are continuous and defined on a closed interval, we will restrict our discussion to such functions unless otherwise indicated. To learn more about maximum and minimum values of such functions, let us examine the graph of the polynomial function $f(x)$ defined on the closed interval $[a,b]$ in Fig. 5.9.

From an inspection of the graph of $f(x)$, we see that the maximum value of $f(x)$ occurs at the end point $x=b$ and equals $f(b)$. Geometrically, $(b,f(b))$ is the highest point on the graph of the function on the closed interval $[a,b]$. The value $f(b)$ is called the *absolute maximum* in contrast to the value $f(x_0)$, which is called a *relative maximum*. The word relative is used since $f(x_0)$ is a maximum value of $f(x)$ if one considers only those values of $f(x)$ "relatively near" the point $(x_0,f(x_0))$. The value of the function $f(x_2)$ is the *absolute*

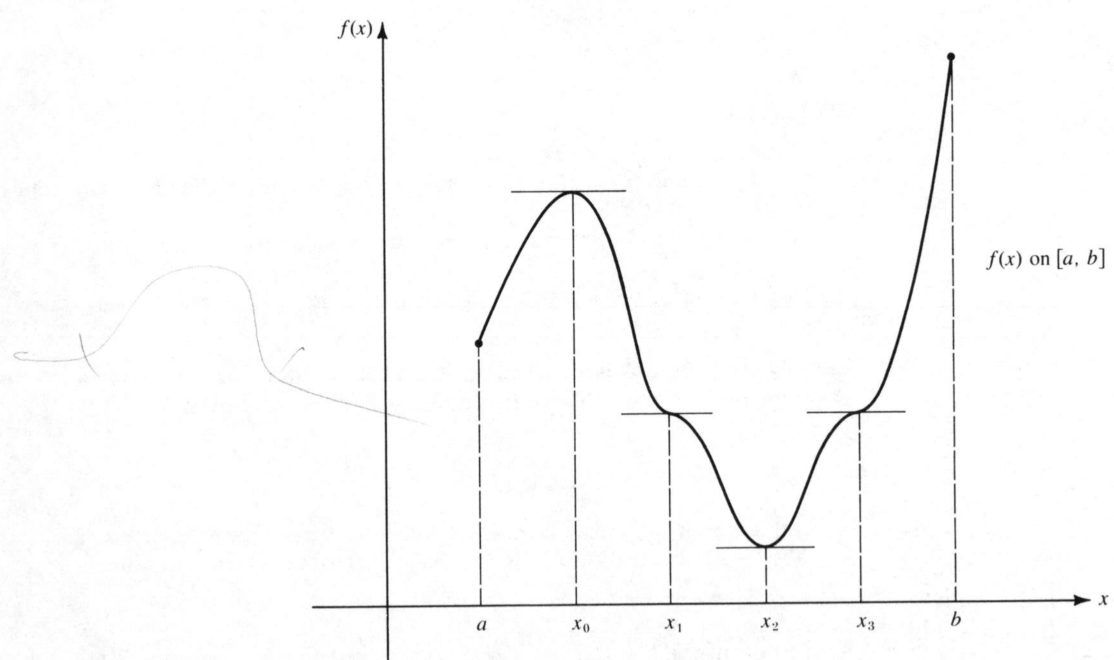

FIGURE 5.9

minimum of the function, as it corresponds to the lowest point on the graph in the interval $[a,b]$. The value $f(x_2)$ is also a *relative minimum*. Although the tangent lines drawn to the curve at $(x_1,f(x_1))$ and $(x_3,f(x_3))$ are horizontal, $f(x_1)$ and $f(x_3)$ are neither maximum nor minimum values. To clarify these concepts, we will make use of the following definitions.

Definition 5.3 The value $f(x^*)$ is the absolute maximum of the function $f(x)$ on an interval if and only if x^* is in the interval and $f(x^*) \geqslant f(x)$ for all x in the interval.

Definition 5.4 The value $f(x^*)$ is a relative maximum of the function $f(x)$ if and only if there exists an open interval containing x^* in the domain of the function where $f(x^*) \geqslant f(x)$ for all x in the open interval.

Definition 5.5 The value $f(x^*)$ is the absolute minimum of the function $f(x)$ on an interval if and only if x^* is in the interval and $f(x^*) \leqslant f(x)$ for all x in the interval.

Definition 5.6 The value $f(x^*)$ is a relative minimum of the function $f(x)$ if and only if there exists an open interval containing x^* in the domain of the function where $f(x^*) \leqslant f(x)$ for all x in the open interval.

Definitions 5.3 and 5.5 are straightforward and agree with our intuitive descriptions of absolute maximum and minimum values. Definitions 5.4 and 5.6, however, need some clarification. The sketch of the graph of the function $g(x)$ in Fig. 5.10 will help us to interpret these definitions.

Based on Definition 5.4, $g(x_0)$ is a relative maximum since we can find an open interval containing x_0 in the domain of the function where $g(x_0) \geqslant g(x)$ for all x in the open interval. Analogously, $g(x_1)$ is a relative minimum. Additionally, $g(x_1)$ is an absolute minimum, since it satisfies Definition 5.5. The value $g(b)$ is an absolute maximum, but it is not a relative maximum since no open interval containing b is in the domain $[a,b]$ of the function. Analogously, $g(a)$ is not a relative minimum. We conclude from this that the value of the function at an end point of a closed interval may be an absolute maximum or absolute minimum but it cannot be a relative maximum or minimum.

From the cases considered, we might conjecture that absolute maximum or minimum values, customarily referred to as *absolute extrema*, occur at either "turning points" of the function or at end points of the function and that relative maxima and minima, or *relative extrema*, occur only at turning points of the function. Let us investigate this conjecture by considering several examples.

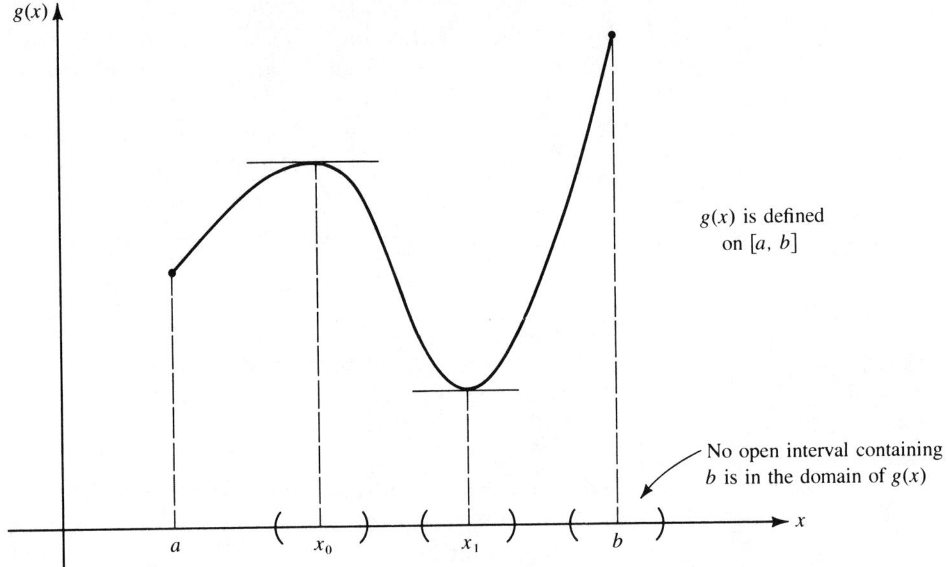

FIGURE 5.10

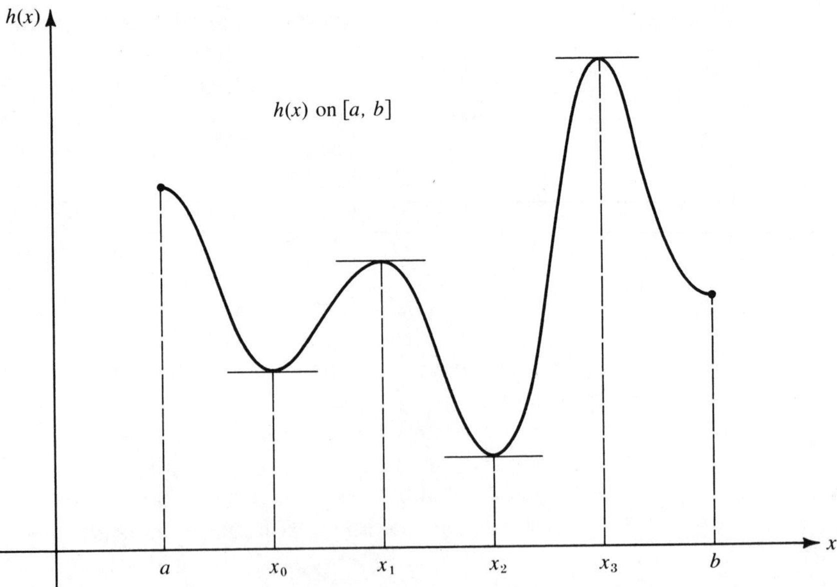

FIGURE 5.11

Example 5.4 From an inspection of the graph of $h(x)$ shown in Fig. 5.11 determine the absolute and relative extrema.

Solution By inspection of the graph of $h(x)$ shown in Fig. 5.11, we conclude that $h(x_0)$ and $h(x_2)$ are relative minima, and $h(x_1)$ and $h(x_3)$ are relative maxima. The absolute maximum is $h(x_3)$ and the absolute minimum is $h(x_2)$. The values of the function at the end points are not extrema in this case.

At each of the turning points in Figs. 5.9–5.11 the tangent line drawn to the curve is horizontal. We might infer from this that relative extrema occur only where the first derivative is zero. This conclusion is invalid as we see in the next example.

Example 5.5 From an inspection of the graph of $f(x)=|x-4|+2$ defined on $[0,6]$, determine the extrema of the function.

Solution The graph of $f(x)=|x-4|+2$ shown in Fig. 5.12 can be sketched from the following table of values.

x	0	1	2	3	4	5	6
$f(x)$	6	5	4	3	2	3	4

By inspection of the graph of $f(x)$ we note that $f(0)=6$ is the absolute maximum and $f(4)=2$ is a relative minimum and an absolute minimum. The point $(4, f(4))$ is a "sharp" turning point, indicating that $f'(4)$ is undefined. If $x<4, f'(x)=-1$ and if $x>4, f'(x)=1$, while at $x=4, f'(x)$ is undefined. You may recall that in Sec. 4.3, Fig. 4.11*c*, we investigated a similar point where the first derivative was not defined. From this example we conclude that a relative extremum can occur where the first derivative is undefined.

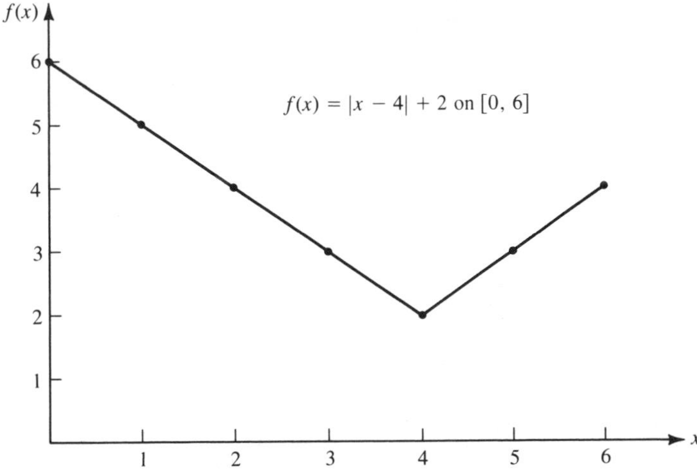

FIGURE 5.12

On the other hand, to show that relative extrema do not necessarily occur where the first derivative is zero, consider the following example.

Example 5.6 From an inspection of the graph of $f(x)=(x-3)^3+4$ defined on $[1,5]$ shown in Fig. 5.13, determine the extrema of the function.

Solution The graph of $f(x)$ can be sketched from the following table of values, noting that $f'(3)=0$.

x	1	2	3	4	5
$f(x)$	-4	3	4	5	12

The absolute minimum, $f(1)$, occurs at the end point where $x=1$ and the absolute maximum, $f(5)$, occurs at the end point where $x=5$. Although we might suspect that $f(3)$ is a relative extremum since $f'(3)=0$, this is not the case. It is not a relative maximum because in every open interval containing $x=3$ there exist values of $f(x)$ (where $x>3$) such that $f(x)>f(3)$, and similarly, it is not a relative minimum because in every open interval containing $x=3$ there exist values of $f(x)$ (where $x<3$) such that $f(x)<f(3)$.

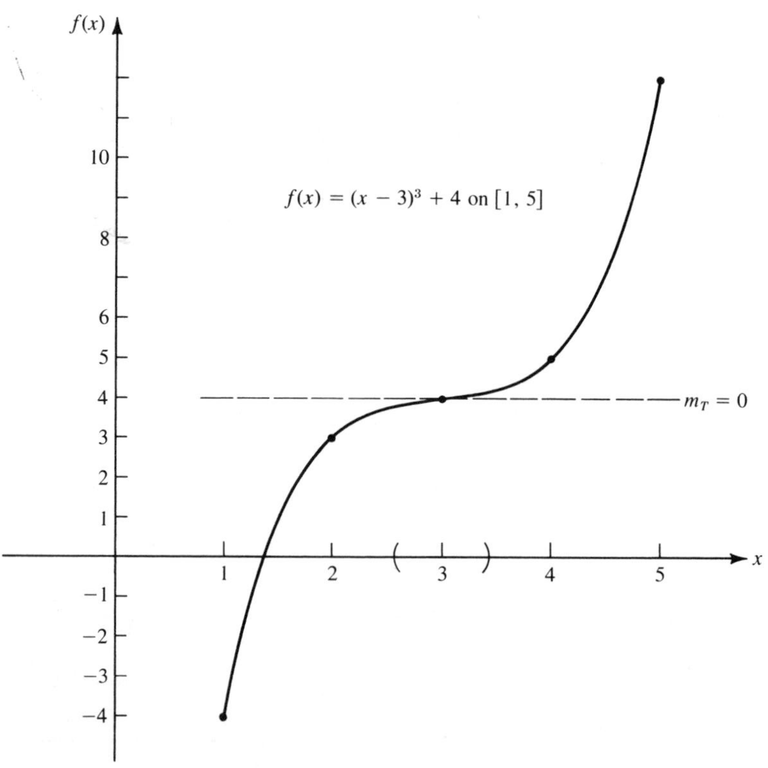

FIGURE 5.13

Hence, even though there exists an x^* in the interval (a,b) where $f'(x^*)=0$ there is no relative extremum in this case.

Let us summarize the conclusions we have drawn from an investigation of the functions considered in Figs. 5.9 through 5.13. Since we are primarily interested in absolute extrema, as we implied at the beginning of this section, let us list those properties first.

Property 5.1 If a continuous function $f(x)$ is defined on a closed interval $[a,b]$, then there exist numbers x_0 and x_1 in $[a,b]$ such that $f(x_0)$ is an absolute maximum, and $f(x_1)$ is an absolute minimum. (Simply stated, the function must have an absolute maximum and an absolute minimum in the closed interval.)

Property 5.2 If a continuous function $f(x)$ is defined on a closed interval $[a,b]$, then either the absolute maximum (or minimum) will occur at a or b, an end point of the interval, or the absolute maximum (or minimum) will be a relative maximum (or minimum) and occur at a point in the open interval (a,b).

This property is very important when looking for absolute extrema, since it implies that we need only consider the end points and points where relative extrema exist.

Property 5.3 If $f(x)$ is continuous on a closed interval $[a,b]$ and $f(x^*)$ is a relative extremum, then x^* is in the open interval (a,b). (In other words, relative extrema do not occur at end points).

Property 5.4 If $f(x)$ is continuous on a closed interval $[a,b]$ and $f(x^*)$ is a relative extremum, then $f'(x^*)=0$ or $f'(x^*)$ is undefined.

These properties can be proved true, but the proofs would be beyond the scope and objectives of this text. *A word of caution*: Property 5.4 does not state that if $f'(x^*)=0$ or is undefined, $f(x^*)$ is a relative extremum. We verified this statement in Example 5.6. Property 5.4, however, does provide us with a method of determining the relative extrema without graphing the function. If we determine all those values of x in the domain of the function that either make $f'(x)=0$ or $f'(x)$ undefined, and denote them by $x_0, x_1, x_2, \ldots, x_n$, then the only possibilities for relative extrema are the corresponding ordinates $f(x_0), f(x_1), \ldots, f(x_n)$. There are rules we can develop to determine which of these values are relative extrema. Those values of x that make $f'(x)=0$ or undefined are so critical in this step, we make the following definitions.

Definition 5.7 The x values, excluding the end points, in the domain of the function $f(x)$ satisfying the equation $f'(x)=0$ or those for which $f'(x)$ is undefined are called *critical values* of the function.

Definition 5.8 If x^* is a critical value of the function, $(x^*, f(x^*))$ is a *critical point*.

Example 5.7 Determine the critical values and the associated critical points of the function $f(x)=x^3-6x^2+9x+4$ defined on the interval $[0, 10]$.

Solution To find the critical values of the function $f(x)$ we determine the derivative $f'(x)=3x^2-12x+9$. Since $f'(x)$ is defined on the interval $[0, 10]$, we know that the critical values will be found by solving

$$f'(x) = 0$$

or

$$3x^2 - 12x + 9 = 0$$

$$3(x^2-4x+3) = 0$$

$$3(x-3)(x-1) = 0$$

$$x = 3 \qquad \text{or} \qquad x = 1$$

Thus the critical values are $x=1$ and $x=3$ and the associated critical points are $(1, f(1))$ or $(1, 8)$ and $(3, f(3))$ or $(3, 4)$.

In Example 5.7 we have determined that $f(1)$ and $f(3)$ are possible relative extrema. We could graph the function $f(x)=x^3-6x^2+9x+4$ between $x=0$ and $x=5$ and by inspection of the graph determine if these values are relative extrema. A shortcut procedure is more desirable. From an inspection of Figs. 5.9 through 5.13 we can conclude that if $f'(x_0)=0$, one of the four situations depicted in Fig. 5.14 exists at the point $(x_0, f(x_0))$.

In order to determine which one of the four cases we have for the function $f(x)$ of Example 5.7 at the point $(1, f(1))$, we compare the value of $f(1)$ with the values of $f(a)$ and $f(b)$, where a is to the left of 1 and b is to the right. The values a and b selected must be between the critical value $x=1$ and the next adjacent critical value on the corresponding side of 1; if none exists, then any value may be selected on that side. Since there are only two critical values, $x=1$ and $x=3$, for this function, we select $a=0$ and $b=2$. Evaluating and comparing $f(x)$, we find $f(0)=4$, $f(1)=8$, and $f(2)=6$. Since both $f(0)$ and $f(2)$ are less than $f(1)$, this value is a relative maximum as depicted in Fig. 5.14*a*.

Similarly, to investigate the critical point $(3, f(3))$, we select $a=2$ and $b=4$. Evaluating and comparing $f(x)$, we find $f(2)=6$, $f(3)=4$, and $f(4)=8$. Since $f(3)$ is less than the two other values, it is a relative minimum as depicted in Fig. 5.14*b*.

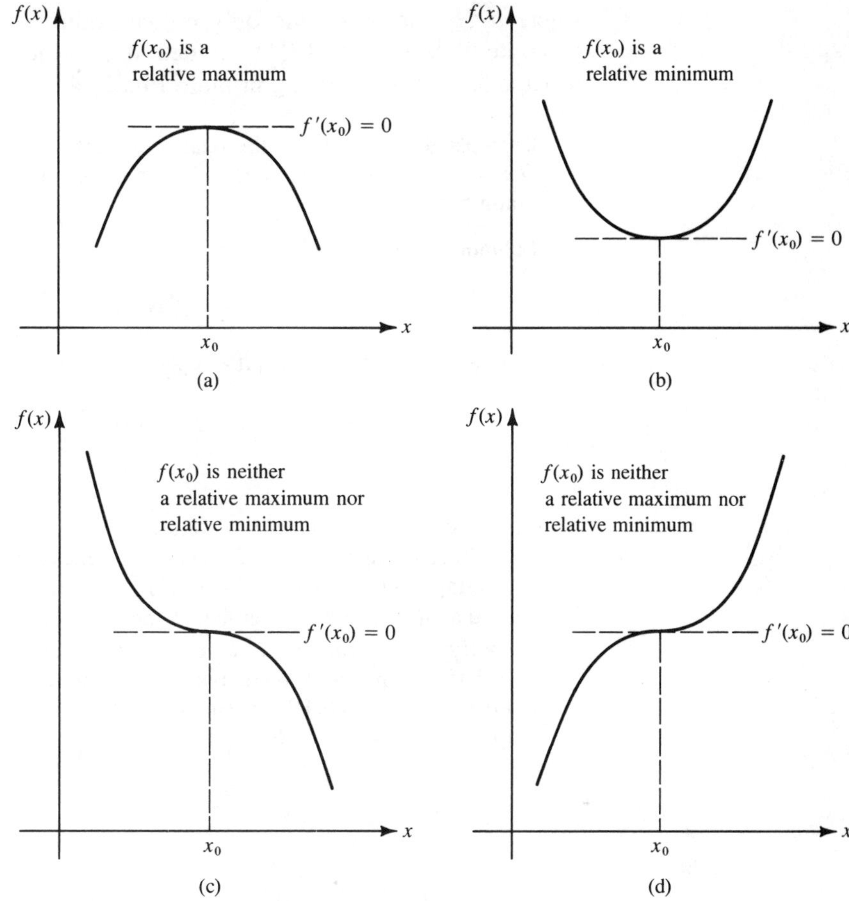

FIGURE 5.14

We summarize this procedure in Rule 5.2, which is commonly known as the *function test*. It determines whether a critical point corresponds to a relative maximum, a relative minimum, or neither.

Rule 5.2 If $f(x)$ is a continuous function for which x^* is the only critical value in the interval $[a,b]$, then

(a) $f(x^*)$ is a relative maximum if $f(x^*)>f(a)$ and $f(x^*)>f(b)$.
(b) $f(x^*)$ is a relative minimum if $f(x^*)<f(a)$ and $f(x^*)<f(b)$.
(c) Otherwise, $f(x^*)$ is not an extremum.

To apply Rule 5.2 we find the critical values of a function in the domain of interest, and for each critical value x^* we select values of $x=a$ and $x=b$

where $a<b$ and x^* is the only critical value between a and b. We then compute $f(a)$, $f(x^*)$, and $f(b)$ and determine which of the three cases we have in Rule 5.2, always keeping in mind Fig. 5.14.

Example 5.8 Determine the relative maxima and minima for the function $f(x)=x^2-5x+6$ on the interval $[1,4]$ and use this information to sketch the graph of the function.

Solution The first derivative is

$$f'(x) = 2x - 5$$

Setting this equal to zero and solving for x, we obtain

$$2x - 5 = 0$$
$$x = \tfrac{5}{2}$$

Thus, we have only one critical value.

Choosing the values $x=2$ and $x=3$ on each side of the critical value and evaluating $f(x)$, we have $f(2)=0$, $f(\frac{5}{2})=-\frac{1}{4}$, and $f(3)=0$. These values are such that $f(\frac{5}{2})$ is less than either of the adjacent values; hence, the critical value $x=\frac{5}{2}$ represents the x value for which $f(x)$ has a relative minimum. In Fig. 5.15 the previous three points and the end points $f(1)=f(4)=2$ are plotted and connected by a smooth curve to give a sketch of the related graph of $f(x)=x^2-5x+6$.

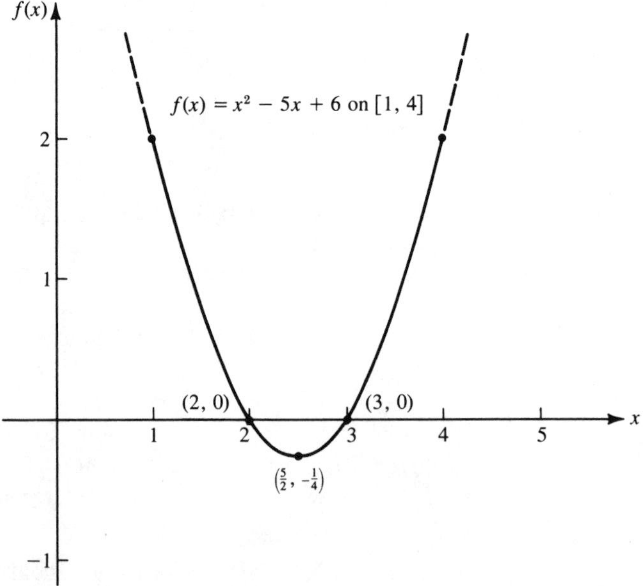

FIGURE 5.15

Since the function in Example 5.8 is a quadratic function, we could employ the information from Chapter 2 and achieve the same results; however, for more complicated functions, generally we shall rely on differential calculus to obtain the desired information.

Example 5.9 Determine the relative maxima and minima for the function $f(x)=x^5-15x^3$ defined on $\mathbb{R}$, the real numbers, and sketch the graph of the function.

Solution To determine the critical values of x, we determine the derivative of $f(x)$:

$$f'(x) = 5x^4 - 45x^2$$

Setting the derivative equal to zero and solving, we have

$$\begin{aligned} 5x^4 - 45x^2 &= 0 \\ 5x^2(x^2-9) &= 0 \\ 5x^2(x+3)(x-3) &= 0 \end{aligned}$$

which yields the critical values

$$x = -3 \qquad x = 0 \qquad x = 3$$

Determining the value of the function at each critical value and two adjacent x values, we obtain

$$\begin{array}{lll} f(-4) = -64 & f(-1) = 14 & f(2) = -88 \\ f(-3) = 162 & f(0) = 0 & f(3) = -162 \\ f(-2) = 88 & f(1) = -14 & f(4) = 64 \end{array}$$

Thus, we see that the curve has a relative maximum at $x=-3$, neither at $x=0$, and a relative minimum at $x=3$. The nine points are plotted and a sketch of the curve is presented in Fig. 5.16.

Examples 5.8 and 5.9 show that the major contribution that differential calculus makes to the graphing of functions is the indication of regions that require further investigation of the shape of the curve. That is, it tells us where the directional changes of the function are occurring, and hence we can concentrate on evaluating the function around these critical values. We will pursue the topic of graphing functions in more detail in Sec. 5.6.

In many situations, it is rather awkward and tedious to evaluate a function; occasionally it is easier to evaluate the first and/or second derivative. Fortunately, there are tests that use the first and second derivatives for determining whether the function has attained a relative maximum or relative

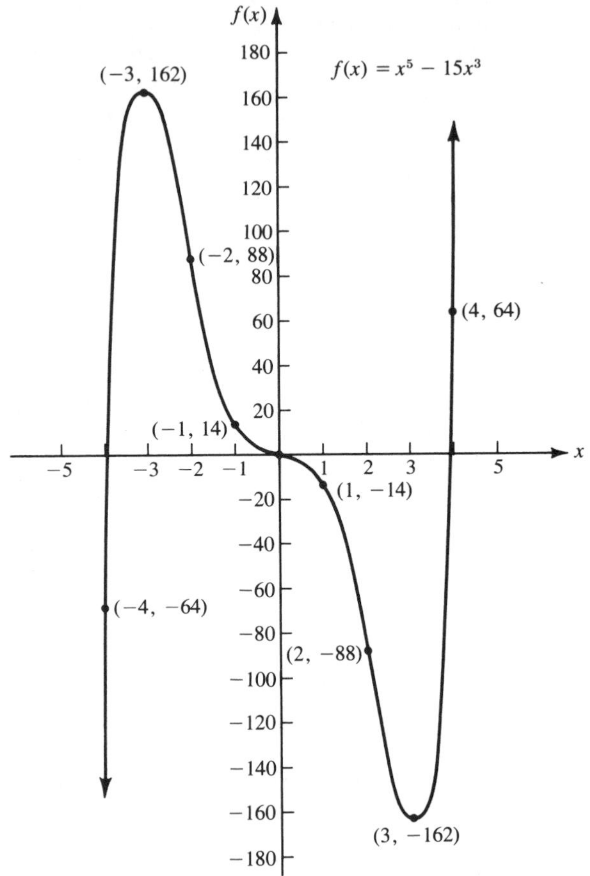

FIGURE 5.16

minimum at the critical value. Two of these tests are stated below without proof.

Rule 5.3 If $f(x)$ is a continuous function on the interval $[a,b]$ and x^* is the only critical value in the interval (a,b), then

(a) $f(x^*)$ is a relative maximum if $f'(a)>0$ and $f'(b)<0$.
(b) $f(x^*)$ is a relative minimum if $f'(a)<0$ and $f'(b)>0$.
(c) Otherwise, $f(x^*)$ is not an extremum.

This rule is commonly known as the *first derivative test* because it uses only the values of the first derivative to determine whether the critical value yields a relative maximum, a relative minimum, or neither.

A graphic analogy of Rule 5.3 is given in Fig. 5.17. If we keep in mind the general relationships given in Fig. 5.17, we should not have any difficulty

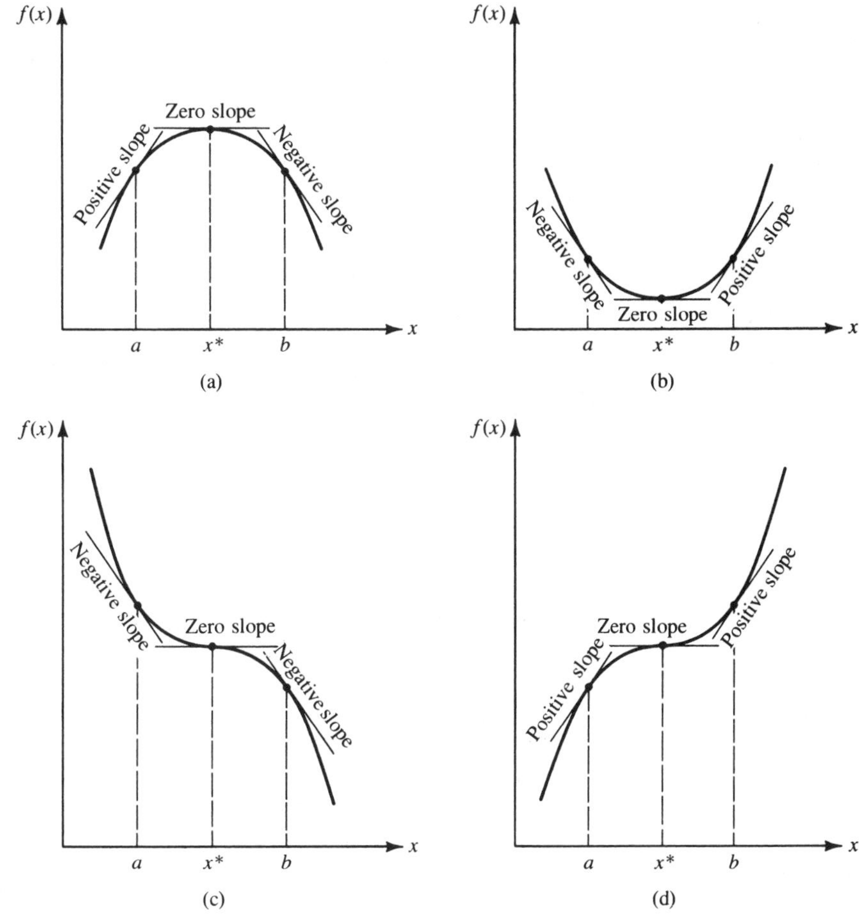

FIGURE 5.17

recalling and applying the first derivative test. Figure 5.17*a* corresponds to part (a) of Rule 5.3; Fig. 5.17*b* corresponds to part (b); and Figs. 5.17*c* and *d* correspond to the situation where $f(x^*)$ is not an extremum. We have assumed that $f'(x^*)=0$ in Figs. 5.17 since by far the majority of critical values occur when this is true. Similar sketches can be drawn to illustrate the case where $f'(x^*)$ does not exist.

Rule 5.4 If the first and second derivatives of $f(x)$ exist for every value of x in the interval $[a,b]$, and x^* is a critical value in the interval (a,b), then

(a) $f(x^*)$ is a relative maximum if $f''(x^*)<0$.
(b) $f(x^*)$ is a relative minimum if $f''(x^*)>0$.
(c) if $f''(x^*)=0$, no conclusion can be drawn.

This rule is commonly known as the *second derivative test* because it uses the second derivative at the critical value to determine whether we have a relative maximum or a relative minimum at the critical point. This rule has a simple graphical interpretation if we think of the second derivative as the rate of change of the first derivative. In Fig. 5.17a, as x increases from a to b the first derivative changes from positive to negative values; hence the first derivative is a decreasing function and its derivative (that is, the second derivative of $f(x)$) must be negative or less than zero, confirming Rule 5.4(a). In Fig. 5.17b, as x increases from a to b the first derivative changes from negative to positive values; hence the first derivative is an increasing function and its derivative (that is, the second derivative of $f(x)$) must be positive or greater than zero, confirming Rule 5.4(b). Note that Rule 5.4(c) states that if $f''(x^*)=0$, the test is not conclusive and you must use either the function test or the first derivative test.

Let us apply Rules 5.3 and 5.4 to the functions in Examples 5.8 and 5.9.

Example 5.10 Apply Rules 5.3 and 5.4 to determine the relative maxima and relative minima for $f(x)=x^2-5x+6$ on $[1,4]$.

Solution From the results of Example 5.8 the first derivative is

$$f'(x) = 2x - 5$$

and the critical value is $x=\frac{5}{2}$. Applying the first derivative test (Rule 5.3) for $a=2$ and $b=3$, we have

$$f'(2) = 2(2) - 5 = -1$$

and $$f'(3) = 2(3) - 5 = 1$$

Thus, $f'(a)<0$ and $f'(b)>0$, which means that $f(x)=x^2-5x+6$ attains a relative minimum at $x=\frac{5}{2}$.

Applying the second derivative test (Rule 5.4), we have

$$f''(x) = 2$$

Thus, for the critical value $x^*=\frac{5}{2}$, we have $f''(\frac{5}{2})=2>0$; therefore, $f(x)=x^2-5x+6$ has a relative minimum at $x=\frac{5}{2}$.

Example 5.11 Use the first and second derivative tests to determine the relative maxima and relative minima for $f(x)=x^5-15x^3$ on $(-\infty,\infty)$.

Solution From the results of Example 5.9 the critical values are $x=-3$, $x=0$, and $x=3$.

Applying the first derivative test for each of the critical values, we choose a value from each side of the critical values and determine the values of $f'(x) = 5x^4 - 45x^2$ as follows:

$x^* = -3$	$x^* = 0$	$x^* = 3$
$f'(-4) = 560$	$f'(-1) = -40$	$f'(2) = -100$
$f'(-2) = -100$	$f'(1) = -40$	$f'(4) = 560$

Thus, we see that $f(-3)$ is a relative maximum, $f(3)$ is a relative minimum, and $f(0)$ is not an extremum.

Rather than trying to memorize Rule 5.3 it is easier to graphically represent the information given by the first derivative. For example, in the three cases above we could draw the sketches in Fig. 5.18 based on the information provided by the first derivative. Representative tangent lines are drawn to the left and right of each of the critical values. From Fig. 5.18 it is obvious that $f(-3)$ is a relative maximum, $f(0)$ is neither, and $f(3)$ is a relative minimum.

Applying the second derivative test, we have

$$f''(x) = 20x^3 - 90x$$

which yields

$$f''(-3) = -270$$
$$f''(0) = 0$$
$$f''(3) = 270$$

Thus, $f''(-3)$ is negative, which indicates that $f(-3)$ is a relative maximum, and $f''(3)$ is positive, which indicates that $f(3)$ is a relative minimum. Since $f''(0)$ is zero, the test is inconclusive at $x = 0$ and we must use either the function test or the first derivative test for this critical point.

The following example shows that if x^* is a critical value and $f''(x^*) = 0$, the value $f(x^*)$ may be a relative maximum, a relative minimum, or neither of these.

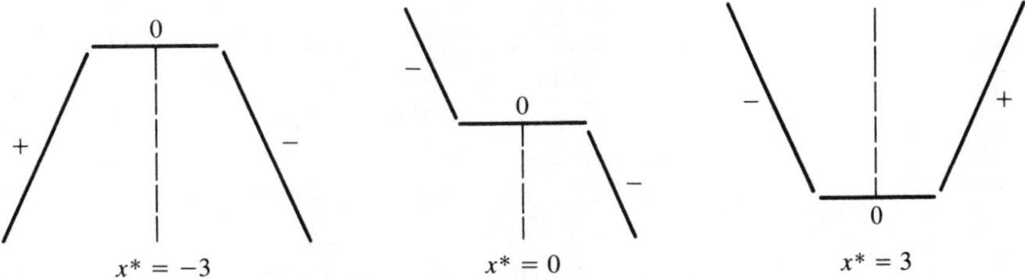

FIGURE 5.18

Example 5.12 Find the relative extrema for the functions $f(x) = -x^4$, $g(x) = x^4$, and $h(x) = x^3$ by applying Rules 5.3 and 5.4.

Solution Determining the critical values by the use of the first derivative, we have

$$f'(x) = -4x^3 \qquad g'(x) = 4x^3 \qquad h'(x) = 3x^2$$

and $x = 0$ is the only critical value for each function. Applying the first derivative test (Rule 5.3) for each function, we have

$$f'(-1) = +4 \qquad g'(-1) = -4 \qquad h'(-1) = +3$$

$$f'(1) = -4 \qquad g'(1) = +4 \qquad h'(1) = +3$$

We also represent graphically the information provided by this test in Fig. 5.19. Thus, we see that $f(0)$ is a relative maximum, $g(0)$ is a relative minimum, and $h(0)$ is neither.

Applying the second derivative test (Rule 5.4), we have $f''(x) = -12x^2$, $g''(x) = 12x^2$, and $h''(x) = 6x$ and $f''(0) = 0$, $g''(0) = 0$, and $h''(0) = 0$. Thus we can see that if the second derivative is zero at the critical value x^*, the test is inconclusive as $f(x^*)$ can be a relative maximum, a relative minimum, or neither.

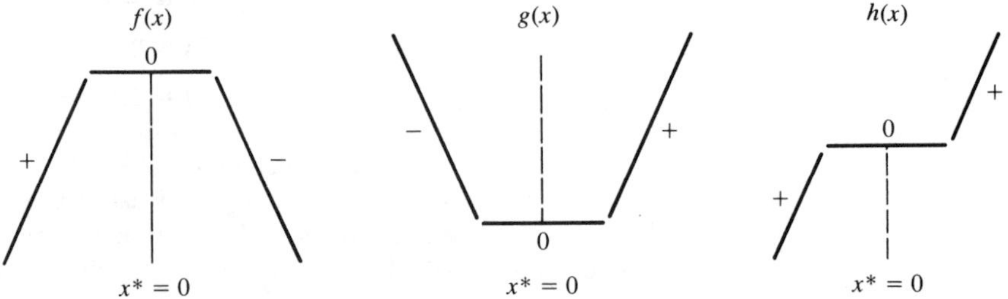

FIGURE 5.19

The choice of which of the three methods to use for determining the extreme values is dependent upon the complexity of the particular function. The second derivative test is generally the easiest to use because it involves determining only one function value per critical value; however, for those situations where the second derivative is very difficult to determine, it may be better to use one of the other methods. The first derivative test is generally better than the method of evaluating the original function because it involves determining only two function values for each critical value, whereas the latter method requires us to determine three function values per critical value. Nevertheless, if the first derivative is difficult to evaluate, it may be best to use the method of evaluating the original function at each critical value and two adjacent values.

If we are only interested in finding the absolute maximum (or minimum) of a function defined on a closed interval $[a,b]$, we know from Property 5.2 that it must occur at an end point of the interval or the absolute extremum will also be a relative extremum. Based on this property, we make the following rule

---

Rule 5.5 To find the absolute maximum (or minimum) value of a continuous function defined on the closed interval $[a,b]$ determine the critical values $x_0, x_1, \ldots, x_n$ in the interval and calculate the values of the function $f(a), f(x_0), f(x_1), \ldots, f(x_n)$ and $f(b)$. The largest of these values will be the absolute maximum and the smallest will be the absolute minimum.

We apply Rule 5.5 in the following two examples.

Example 5.13 Estatic Plastics, Inc., has determined that the cost function for producing a batch of a particular type of unit is given by $C(x)=x^2-14x+100$, where $C(x)$ is the cost in dollars and x is hundreds of units/batch. Their equipment is such that they cannot produce more than 1,500 units in a single batch. Determine the batch size that will yield a *minimum cost*.

Solution First, we determine the critical values by obtaining the first derivative and setting it equal to zero.

$$C(x) = x^2 - 14x + 100$$
$$C'(x) = 2x - 14$$
$$2x - 14 = 0$$
$$x = 7$$

Secondly, we determine from the description of the problem the domain on which the cost function $C(x)$ is defined. It is implied that the number of hundreds of units produced, x, is equal to or greater than zero and equal to or less than 15. Hence $C(x)$ is defined on $[0, 15]$. Applying Rule 5.5 we find

$$C(0) = 0^2 - 14(0) + 100 = \$100$$
$$C(7) = (7)^2 - 14(7) + 100 = \$51$$
$$C(15) = (15)^2 - 14(15) + 100 = \$115$$

Hence, the minimum cost per batch is \$51 and is achieved when 700 units/batch are made.

Example 5.14 Depletion of Oxygen by Organic Wastes Organic matter discharged from a waste disposal plant into a river is oxidized by the dissolved oxygen in the water. This action can deplete the oxygen concentration to a level below that necessary to sustain plant and animal life. Ecologists studying the effects of a city waste disposal system on a large eastern river found that the oxygen content, $C(x)$, was related to the distance, x, measured in meters, from the discharge source according to the equation

$$C(x) = 10 - \frac{500x}{x^2+900}$$

where it is assumed that $0 \leqslant x \leqslant 200$ and $C(x)$ is in parts per million. Determine the distance from the discharge source where the concentration of oxygen will be a minimum.

Solution First we determine the critical value(s) by obtaining the first derivative and setting it equal to zero.

$$C(x) = 10 - \frac{500x}{x^2+900}$$

$$C'(x) = -\left[\frac{(x^2+900)\cdot 500 - 500x(2x)}{(x^2+900)^2}\right]$$

$$0 = -\left[\frac{500x^2+900(500)-1{,}000x^2}{(x^2+900)^2}\right]$$

$$0 = \frac{500x^2-900(500)}{(x^2+900)^2}$$

Multiplying the left- and right-hand side of the equation by $(x^2+900)^2$, we have

$$500x^2 - 900(500) = 0$$

$$x^2 - 900 = 0$$

$$(x+30)(x-30) = 0$$

$$x = -30 \qquad \text{or} \qquad x = 30$$

Hence, $x=30$ is the only critical value in the interval $[0,200]$.

Evaluating $C(x)$ at the end points of the interval $[0,200]$ and at the critical value $x=30$, we have

$$C(0) = 10 - \frac{500(0)}{0^2+900} = 10 \text{ parts per million}$$

$$C(30) = 10 - \frac{500(30)}{(30)^2+900}$$

$$= 10 - \frac{50}{6} = \frac{10}{6} \doteq 1.66 \text{ parts per million}$$

$$C(200) = 10 - \frac{500(200)}{(200)^2+900}$$

$$= 10 - \frac{1{,}000}{409} \doteq 10 - 2.45 \doteq 7.55 \text{ parts per million}$$

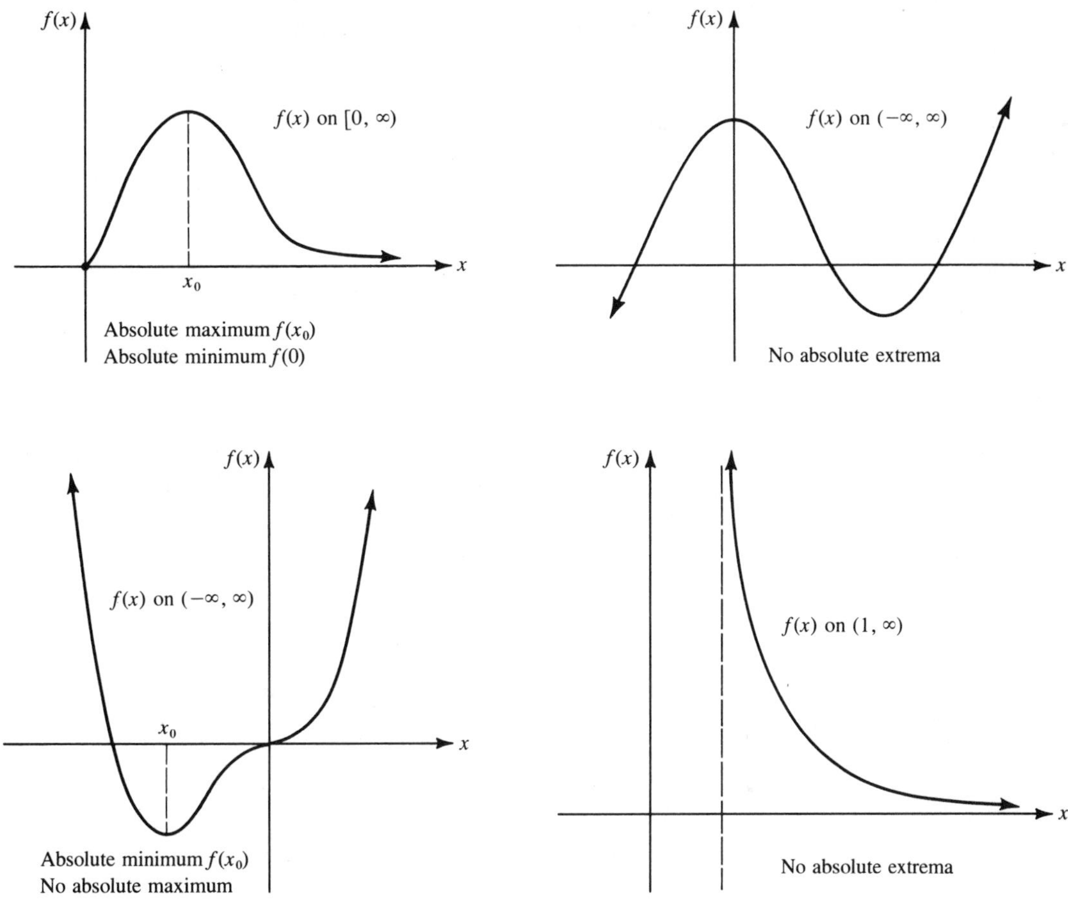

FIGURE 5.20

From this we conclude that the minimum concentration of oxygen, 1.66 parts per million, is found 30 meters downstream from the source.

Although most real-life problems involve continuous functions defined on a closed interval, there are times when we wish to consider such functions defined on the entire set of real numbers, on an open interval, or on a half-open interval. In such cases the function may or may not have absolute extrema. For example, consider the four functions shown in Fig. 5.20.

To determine if such functions have absolute extrema on the defined domain, we first determine the critical values of the function in the domain $x_0, x_1, \ldots, x_n$ and the values of the function $f(x_0), f(x_1), \ldots, f(x_n)$, and then compare these to the values of the function at any end point or "near" the end point if it is not included. By "the value of the function near the end point" we mean the limit of $f(x)$ as x approaches the end point. For example, if the function $f(x)$ is defined on the domain $[a, \infty)$ we must determine $f(a)$ and consider the $\lim_{x\to\infty} f(x)$.

Example 5.15 Determine the absolute extrema for the function $f(x)=x^3-9x^2+24x+3$ defined on $[0, \infty)$.

Solution Determining the critical values by using the first derivative we have

$$
\begin{aligned}
f'(x) &= 3x^2 - 18x + 24 \\
0 &= 3(x^2-6x+8) \\
0 &= 3(x-2)(x-4)
\end{aligned}
$$

Therefore, the critical values are $x=2$ and $x=4$ and the corresponding values of the function are $f(2)=23$ and $f(4)=19$. If we are careless we might conclude that $f(2)=23$ is the absolute maximum and $f(4)=19$ is the absolute minimum. Investigating the end points, we find

$$f(0) = 3$$

and

$$\lim_{x\to\infty} f(x) = +\infty$$

Hence, this function has an absolute minimum of $f(0)=3$, but has no absolute maximum on the given domain since $f(x)$ has no upper limit.

Example 5.16 Determine the absolute extrema for the function $f(x)=x^2/(1-x)$ defined on $(1,5]$.

Solution Determining the critical values of $f(x)$ in the domain $(1,5]$ we have

$$
\begin{aligned}
f'(x) &= \frac{(1-x)(2x)-x^2(-1)}{(1-x)^2} \\
0 &= \frac{2x-x^2}{(1-x)^2}
\end{aligned}
$$

Solving this equation by setting $2x-x^2=0$ we obtain

$$
\begin{gathered}
x(2-x) = 0 \\
x = 0 \quad \text{or} \quad x = 2
\end{gathered}
$$

The only critical value in the domain $(1,5]$ is $x=2$. Evaluating $f(x)$ at the critical value and at the end points we find $f(2)=-4$, $f(5)=\frac{-25}{4}$, and $\lim_{x\to 1^+} f(x)=-\infty$. Therefore, the absolute maximum is $f(2)=-4$, and there is no absolute minimum as $f(x)$ is not bounded below on the interval $(1,5]$.

EXERCISES

In Exercises 1–6 determine the critical points for each function on the specified interval. Determine the relative maxima and relative minima by using the function test. Roughly draw a sketch of the graph of the function from the available data.

1. $f(x)=2x^2+6x-8$: $[-3,5]$

2. $f(x)=x^2-16$: $[-5,8]$

3. $g(x)=x^3-6x^2-15x+4$: $(-\infty,\infty)$

4. $g(x)=x^3-3x$: $[0,4]$

5. $s(x)=(x^2-4)^2$: $(-\infty,\infty)$

6. $s(x)=(x^2-5x+6)^3$: $(-\infty,\infty)$

In Exercises 7–14 determine the critical points for each function on the specified interval. Determine the relative maxima and relative minima by using the first derivative test. Roughly draw a sketch of the graph of the function from the available data.

7. $f(x)=x^2+4x-5$: $[-4,2]$

8. $f(x)=9-3x-2x^2$: $(-\infty,\infty)$

9. $g(x)=\frac{1}{2}x^4-2x^3$: $[-2,4]$

10. $g(x)=x^3+3x^2+3x+1$: $[0,5]$

11. $s(x)=x^4-8x^2$: $[-3,3]$

12. $s(x)=x^3+9x^2$: $[-8,4]$

13. $h(x)=2+3x-x^3$: $(-\infty,\infty)$

14. $h(x)=x^3+3x^2-9x+5$: $(-\infty,\infty)$

In Exercises 15–22 determine the critical points for each function on the specified interval. Determine the relative maxima and relative minima by using the second derivative test. If this test is inconclusive, use the first derivative test.

15. $f(x)=x^3+3x^2+2$: $(-\infty,\infty)$

16. $g(x)=3+8x-x^2$: $[0,10]$

17. $h(x)=\frac{1}{3}x^3-4x+1$: $[-4,4]$

18. $s(x)=\dfrac{4x}{x^2+1}$: $[0,5]$

19. $f(x)=x+\dfrac{1}{x}$: $(0,\infty)$

20. $g(x)=4x^3+7x^2-10x+2$: $(-\infty,\infty)$

21. $h(x)=2x^3-\frac{1}{2}x^4$: $(-\infty,\infty)$

22. $g(x)=\dfrac{10}{x^2+1}$: $[-1,1]$

In Exercises 23–30 determine the absolute maxima and absolute minima of each function on the specified interval.

23. $f(x)=x^2+3x-5$: $[-2,6]$

24. $g(x)=x^3-3x+1$: $[-2,8]$

25. $h(x)=\dfrac{x^2}{x+4}$: $[0,5]$

26. $f(x)=\frac{1}{5}x^5-\frac{1}{3}x^3+2$: $[0,10]$

27. $f(x)=\dfrac{1}{(x+1)^2}$: $[0,10]$

28. $f(x)=(x-2)^3$: $[0,10]$

29. $p(x)=x^2(4{,}000-x)^3$: $[0,4000]$

30. $p(x)=\frac{1}{5}x^5-\frac{4}{3}x^3$: $[-3,3]$

31. The profit in dollars for the manufacture and sale of x units is given by

$$P(x) = 50x - 0.002x^2$$

Determine the number of units that will maximize profit and the associated profit.

32. The cost in dollars for the manufacture of x units is given by

$$C(x) = x^2 - 1{,}000x + 400{,}000$$

Determine the number of units that will minimize cost and the associated cost.

33. The total cost in dollars of producing x units is given by

$$C(x) = 0.04x^2 + 8x + 80$$

The selling price of a unit is \$16. Determine

(a) the profit as a function of x,
(b) the number of units that will maximize profit,
(c) the maximum profit.

34. **Optimum Number of Miles a Truck Should Be Driven** The Dixie Transport Company has determined that the cost/thousand miles for operating a certain type of truck is given by $C(x)=1{,}000{,}000/(100x-x^2)$, where x is thousands of miles driven. Company policy states that no truck will be sold prior to accumulating 20,000 miles and no truck will be retained beyond 90,000 miles. Determine the optimum number of miles that a truck should be driven before disposing of it and the associated cost/thousand miles.

35. The Jurgens Equipment Company determines that its profit is related to the number of products it sells, x, by $P(x)=720+352x-4x^2$. Determine the optimum number of products Jurgens should sell in order to maximize profit.

36. In Exercise 18 in Sec. 5.2, determine the number of customers to maximize Charles' income for any given evening.

37. In Exercise 19 in Sec. 5.2, determine the maximum temperature increase $T(x)$.

38. In Exercise 20 in Sec. 5.2, determine the maximum percent of infected citizens during the height of the epidemic.

5.4 PRACTICAL EXTREMA PROBLEMS

In Sec. 5.3 we developed the tools of differential calculus to find the absolute maximum and minimum values of a function. We also applied these tools to some simple real-life problems to find an optimum (i.e., absolute maximum or minimum) value. In this section we will continue our study of real-life optimization problems to further develop our problem-solving skills.

We will develop a simple strategy that will allow us to solve many kinds of optimization problems even though they may appear quite diverse. In every case the goal is to determine a function where the *dependent variable* is the quantity to be optimized. Once this is accomplished we can apply the tools of differential calculus that we developed in Sec. 5.3 to obtain the optimum value. No amount of skill in differentiating will be of any avail until this function is determined. Of course, if this function is given as in Examples 5.13 and 5.14, we need only apply Rule 5.5 directly. However, in many practical optimization problems we will be required to develop the functional relationship involving the variable to be optimized. In these cases the following steps will help us solve the problem.

Step 1 Read the problem carefully and determine what it is you are asked to maximize or minimize.

Step 2 Choose a letter to represent this quantity. This letter will be the dependent variable for your function. It is often preferable to use a letter associated with the quantity (i.e., C for cost, P for population, V for volume, etc.). Some students prefer to use the standard letter f for all cases since we are looking for a function.

Step 3 Select or introduce the independent variable for your function. We usually let x represent this quantity, although sometimes a letter associated with this variable is preferred. In carrying out this step, we ask ourselves what does the variable in Step 2 depend on. Sometimes the choice is obvious. For example, in business problems the revenue, R, in Step 2 will depend on x, the number of items sold. In a biology problem the number of bacteria, N, will depend on the time, t. A sketch of the important information given in the problem is often very helpful in this step.

Step 4 Express the quantity being optimized, the dependent variable, as a function of the independent variable you have introduced in Step 3. The relationship between these variables will be implied in the statement of the problem. Sometimes it is a geometric relation that will be implied by your sketch.

Step 5 Once you have determined the functional relation in Step 4, determine the domain of the function based on the problem and the form of your function. Usually practical considerations will limit the domain to a closed interval.

This step completes the determination of the mathematical model of the problem. The rest of the work is routine.

Step 6 Apply Rule 5.5 to optimize your function on the domain.

Let us apply this strategy to several problems to firmly establish the procedure.

Example 5.17 Optimum Box Configuration The Southwest Metal Company wishes to manufacture tin boxes from a 20 in.$\times$20 in. piece of tin by cutting squares from each corner and folding up the sides. Determine the size of the square that should be cut from each corner in order to produce a box with maximum volume. Determine the maximum volume.

Solution Although we have considered a problem very similar to this in Chapter 1 and elsewhere in the text, we will find it instructive to follow our step-by-step procedure of analysis.

Step 1 The quantity to be maximized is the Volume of the box.

Step 2 Let V represent the volume of the box and be our dependent variable.

Step 3 Before selecting the independent variable, we draw a sketch or model of this problem (see Fig. 5.21). From the geometry of the figure, the volume, V, of the box is given by $V = l \cdot w \cdot h$. This would imply that the volume depends on the three variables, l, w, and h. However, from the statement of the problem and the sketch we note that these variables are related. Since it is required to write V in terms of one variable in Step 4, we express l, w, and h in terms of one independent variable. From Fig. 5.21 we see that if we let x = the edge of the square being cut out from the 20 in.$\times$20 in. sheet, we have

$$h = x$$
$$l = (20-2x)$$
$$w = (20-2x)$$

Step 4 The volume, V, can be expressed in terms of the independent

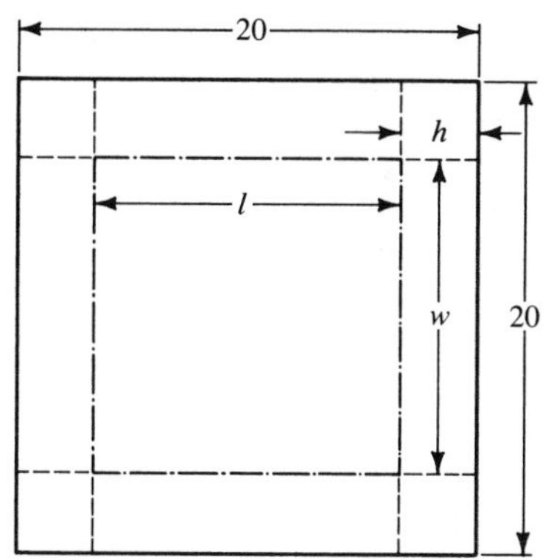

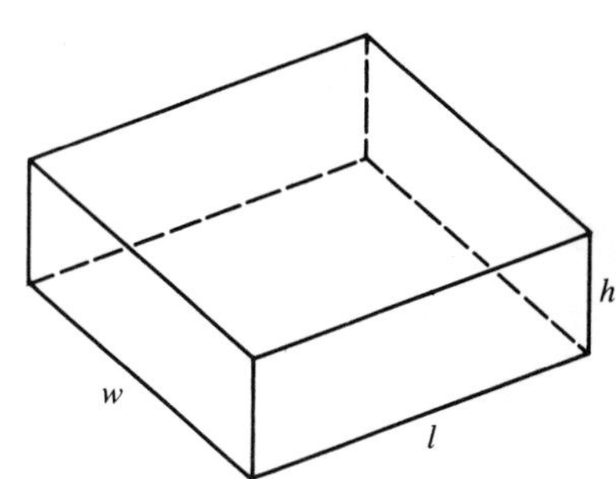

FIGURE 5.21

variable, x, by substitution.

$$\begin{aligned} V &= l \cdot w \cdot h \\ &= (20-2x)(20-2x)x \\ V(x) &= (20-2x)^2(x) \end{aligned}$$

(As you can see, Steps 3 and 4 are closely related and crucial in our procedure.)

Step 5 The domain of the function $V(x)$ is determined from the equation $V(x)=(20-2x)^2(x)$ and the physical restrictions imposed by the problem. To make sense in this problem, x would have to be equal to or greater than zero and equal to or less than 10. Therefore, our domain is $[0,10]$ and we are seeking to maximize $V(x)=(20-2x)^2(x)$ on $[0,10]$.

Step 6 We apply Rule 5.5 by first finding the critical values of $V(x)$.

$$\begin{aligned} V(x) &= (20-2x)^2(x) \\ &= 400x - 80x^2 + 4x^3 \\ V'(x) &= 400 - 160x + 12x^2 \\ 0 &= 4(100-40x+3x^2) \\ 0 &= 4(3x^2-40x+100) \\ 0 &= 4(3x-10)(x-10) \\ x &= \tfrac{10}{3} \qquad \text{or} \qquad x = 10 \end{aligned}$$

Hence, $x=\frac{10}{3}$ is the only critical value. (*Note*: $x=10$ is an end point and hence not a critical value.) Evaluating $V(x)$ at the critical value and the end points of [0, 10], we have

$$V(10) = 0$$

$$V(\tfrac{10}{3}) = (20 - \tfrac{20}{3})^2(\tfrac{10}{3}) = (\tfrac{40}{3})^2(\tfrac{10}{3}) \doteq 592.6$$

$$V(0) = 0$$

The maximum volume is 592.6 cu in. and the dimensions of the box are

$$h = \tfrac{10}{3} \text{ or } 3\tfrac{1}{3} \text{ in.}$$

$$l = \left(20 - \tfrac{20}{3}\right) = \tfrac{40}{3} \text{ or } 13\tfrac{1}{3} \text{ in.}$$

$$w = \left(20 - \tfrac{20}{3}\right) = \tfrac{40}{3} \text{ or } 13\tfrac{1}{3} \text{ in.}$$

Example 5.18 The Johnson Tool Company sells oil well drilling bits for \$300 each and the cost of making x bits a month is given by $C(x) = 100 + 10x + 5x^2$. If production is limited to 40 bits a month, determine the number of bits that the company should make to maximize their profit.

Solution Following our step-by-step procedure, we have

Step 1 We want to maximize the profit.

Step 2 Let $P =$ the profit, the dependent variable.

Step 3 Since the profit, P, will depend on the number of bits sold per month, the obvious choice is to let $x =$ the number of bits sold per month. (No sketch is required for this problem.)

Step 4 The profit can be obtained by subtracting the total cost from the sales (or revenue). The total sales in dollars is $S(x) = \$300x$, and the total cost is $C(x) = 100 + 10x + 5x^2$. The profit function is given by

$$\begin{aligned} P(x) &= S(x) - C(x) \\ &= 300x - 100 - 10x - 5x^2 \\ &= -100 + 290x - 5x^2 \end{aligned}$$

Step 5 Reading the problem we realize the company can produce and sell as few as zero bits and as many as 40 bits, hence the domain is [0, 40].

Step 6 To maximize $P(x)$ on [0, 40] we apply Rule 5.5 by first calculating the critical values:

$$\begin{aligned} P(x) &= -100 + 290x - 5x^2 \\ P'(x) &= 290 - 10x \\ 0 &= 290 - 10x \\ x &= 29 \end{aligned}$$

Evaluating $P(x)$ at the critical value $x = 29$ and the end points $x = 0$ and

$x=40$, we have

$$\begin{aligned} P(0) &= -100 \\ P(29) &= -100+290(29)-5(29)^2 = 4{,}105 \\ P(40) &= -100+290(40)-5(40)^2 = 3{,}500 \end{aligned}$$

Thus, the Johnson Tool Company can make a maximum profit of \$4,105 a month if they make and sell 29 bits a month.

Example 5.19 Super Bowl Charter to New Orleans The Starburst Travel Agency offers a package plan for a flight to New Orleans to watch the Super Bowl. The flight has a fare of \$200/person plus \$4/person for each unsold seat on the plane. If the plane holds 100 passengers, find the number of passengers necessary for the company to maximize their revenue. Assume that the flight will be canceled if there are fewer than 40 passengers.

Solution

Steps 1 and 2 Since the revenue (or income) is to be maximized we let $R=$the dependent variable.

Steps 3 and 4 To help us select the independent variable we should determine what the revenue, R, depends on. The revenue, R, equals the product of the number of passengers on the flight and the fare per passenger. The number of passengers and the fare/passenger both depend on the number of empty seats. If we let x equal the number of empty seats, we have:

$$\begin{aligned} \text{the number of passengers} &= (100-x) \\ \text{fare per passenger} &= (200+4x) \end{aligned}$$

Hence, the revenue, $R(x)=(100-x)(200+4x)$.

Step 5 Since we could fill the plane and have no empty seats, or have at most 60 empty seats, the domain of $R(x)$ is $[0,60]$.

Step 6 To maximize $R(x)=(100-x)(200+4x)$ on $[0,60]$ we apply Rule 5.5 by first calculating the critical values.

$$\begin{aligned} R(x) &= (100-x)(200+4x) \\ &= 20{,}000+200x-4x^2 \\ R'(x) &= 200-8x \\ x &= \tfrac{200}{8} = 25 \text{ empty seats} \end{aligned}$$

Evaluating $R(x)$ at the end points $x=0$ and $x=60$ and at the critical value $x=25$, we find

$$\begin{aligned} R(0) &= [100-0][200+4(0)] = 20{,}000 \\ R(25) &= (100-25)[200+4(25)] = (75)(300) = 22{,}500 \\ R(60) &= (100-60)(200+4(60)) = (40)(440) = 17{,}600 \end{aligned}$$

To maximize R we select $x=25$ and conclude that the travel agency will receive a maximum revenue of $22,500 when they fly $(100-25)$, i.e., 75, passengers to the Super Bowl.

Example 5.20 Optimum Mix of Hydrogen and Hydroxyl Ions Chemists know that in water the product of the concentration of hydrogen ions $[H^+]$ and hydroxyl ions $[OH^-]$ is very close to 10^{-14}. Determine the value of $[H^+]$ that will minimize the sum $[H^+]+[OH^-]$.

Solution

Steps 1 and 2 Since we wish to minimize the sum of $[H^+]$ and $[OH^-]$, select S as the dependent variable.

Steps 3 and 4 If we let $x=$ the concentration of $[H^+]$ and y the concentration of $[OH^{-1}]$, the sum S is given by

$$S = x + y$$

Unfortunately, S is now written in terms of two variables rather than one as is required in Step 4. Rereading the problem, we note that there is a *side condition* given in the problem relating x and y; that is, $x \cdot y = 10^{-14}$. Using this side condition to express y in terms of x, we have

$$y = \frac{10^{-14}}{x}$$

and substituting this expression for y into the equation for S, we can write

$$S = x + \frac{10^{-14}}{x}$$

Step 5 Since the concentration of $[H^+]$ ions must be greater than zero and no upper limit is given, the domain for $S(x)$ is $(0, \infty)$.

Step 6 To minimize $S(x)$ on $(0, \infty)$, we first calculate the critical values of $S(x)$.

$$\begin{aligned} S(x) &= x + \frac{10^{-14}}{x} \\ &= x + 10^{-14}(x^{-1}) \\ S'(x) &= 1 + -10^{-14}(x^{-2}) \\ 0 &= 1 - \frac{10^{-14}}{x^2} \\ 0 &= x^2 - 10^{-14} \\ x &= 10^{-7} \qquad \text{or} \quad x = -10^{-7} \end{aligned}$$

The only critical value in the domain is $x=10^{-7}$. Since S is defined on an open interval $(0, \infty)$, we must investigate the values of $S(x)$ as $x \to 0$ from the right and as $x \to \infty$. We have

$$\lim_{x \to 0^+} S(x) = \lim_{x \to 0^+} \left(x + \frac{10^{-14}}{x} \right) = +\infty$$

$$\lim_{x \to \infty} S(x) = \lim_{x \to \infty} \left(x + \frac{10^{-14}}{x} \right) = +\infty$$

Both these limits imply that $S(x)$ does not have a minimum near the end points. The minimum value of $S(x)$ must occur at the critical value $x = 10^{-7}$. Calculating $S(10^{-7})$, we find

$$S(10^{-7}) = 10^{-7} + \left(\frac{10^{-14}}{10^{-7}} \right)$$
$$= 2 \times 10^{-7}$$

We can conclude from this that the sum of $[H^+]$ and $[OH^-]$ will be a minimum when the concentration of each of these ions is the same.

Example 5.21 Selecting the Dimensions of a Cattle Pasture The owners of Rocking W Cattle Ranch wish to pasture some cattle on a field of winter wheat located on the Snake River. They have 1,500 ft of wire for building a fence. Using the Snake River as one boundary, determine the dimensions of the largest rectangular field that can be enclosed with the available wire. (Assume that the section of the Snake River being used as a boundary is straight.)

Solution

Steps 1 and 2 We are asked to find the largest rectangular field. This is equivalent to maximizing the area of the field. Therefore, we select A as the dependent variable.

Steps 3 and 4 Figure 5.22 will help us determine our independent variable. The problem implies that we will use the fence material on three sides of the field, the fourth side being the Snake River. The area of the rectangular field

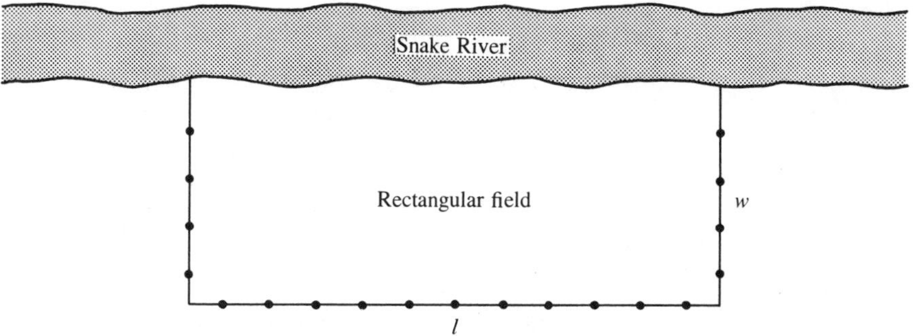

FIGURE 5.22

can be written as

$$A = w \cdot l$$

Since our goal is to express A in terms of one variable, we must consider our problem to determine if there is a side condition that will relate the variables l and w. Knowing that the total amount of wire available is 1500 ft, we can state

$$l + 2w = 1{,}500$$

or

$$l = 1{,}500 - 2w$$

With this information, A can be expressed in terms of the one variable w.

$$A = l \cdot w$$
$$A = (1{,}500 - 2w)(w)$$
$$A(w) = (1{,}500 - 2w)(w)$$

Step 5 Based on the physical restrictions of the problem and the equation $A(w) = (1{,}500 - 2w)(w)$, w must be equal to or greater than zero and less than or equal to 750. Hence, we are seeking a maximum value of $A(w)$ on $[0, 750]$.

Step 6 To maximize $A(w)$ on $[0, 750]$, we apply Rule 5.5 by first finding the critical values.

$$A(w) = (1{,}500 - 2w)(w)$$
$$A(w) = 1{,}500w - 2w^2$$
$$A'(w) = 1{,}500 - 4w$$
$$0 = 4(375 - w)$$
$$w = 375$$

Evaluating $A(w)$ at the end points of $[0, 750]$ and at the critical value $w = 375$, we have

$$A(0) = 0$$
$$A(750) = [1{,}500 - 2(750)](750) = 0$$
$$A(375) = [1{,}500 - 2(375)]375 = 281{,}250$$

To enclose a field with this maximum area of 281,250 sq ft the ranch owners must erect their fencing with a length of 750 ft, parallel to the river, and "two lengths" of 375 ft perpendicular to the river.

As our last example we will consider a typical inventory problem and show how calculus can be used to solve it.

Example 5.22 Minimizing Inventory Costs A biologist who purchases 1,000 mice a year for laboratory use is trying to determine how many times a year he should order the mice. If he orders them all at once or infrequently, each delivery will be large and the cost of feeding and housing the mice will be high. On the other hand, if he orders too frequently the ordering and delivery fees will mount up.

Suppose that the ordering and shipping fees are fixed at \$4/order, independent of the size of the order, that the cost of feeding and housing a mouse is \$5/year, and that each mouse costs 50¢. Assuming that the mice are used at a constant rate throughout the year, the biologist knows when each new order should arrive. In return for ordering the same number of mice and at least 10 each time, his suppliers guarantee they will deliver a new order the same day he runs out of mice.

Solution

Steps 1 and 2 Let C be the dependent variable representing the total yearly cost of the mice.

Steps 3 and 4 Before selecting the independent variable, let us investigate the factors that make up the total cost, C.

$$C = \text{cost of the mice} + \text{feeding and housing costs} \\ + \text{ordering and shipping costs}$$

The cost of the mice is fixed at 50¢ /mouse. The total cost for 1,000 mice is \$500.

Since the feeding and housing costs per mouse are \$5/year, these costs depend on the average number of mice being housed during the year. Suppose there are x mice in each shipment. As stated, each shipment arrives as the previous one runs out. If the mice are used at a constant rate, an inventory graph for the mice for the year would appear as in Fig. 5.23. From Fig. 5.23 we conclude that the average number of mice being housed and fed *between shipments* is $(x/2)$. Since the mice are being used at a constant rate and each shipment size is the same, the time between shipments is equal. Based on these facts we conclude that the average number of mice being *housed and fed during the year* is $(x/2)$. If there are $(x/2)$ mice being housed

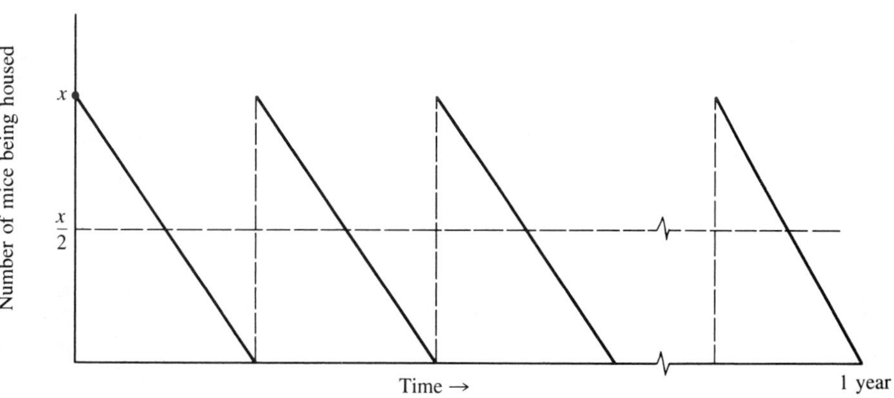

FIGURE 5.23

and fed during the year at a cost of \$5/mouse, the yearly feeding and housing costs are

$$\left(\frac{x}{2}\right)\cdot \$5 = \$2.5x$$

The ordering and shipping costs equal the number of shipments times the cost/shipment. Since the number of mice/shipment is x, there are $(1{,}000/x)$ shipments per year. The cost/shipment was given as \$4; hence, the ordering and shipping cost equals

$$\left(\frac{1{,}000}{x}\right)\cdot \$4 = \frac{\$4{,}000}{x}$$

Now the total cost, $C(x)$, is given by

$$C(x) = 500 + 2.5x + \frac{4{,}000}{x}$$

Step 5 Since each shipment must contain at least 10 mice and no more than 1,000 mice, the domain of $C(x)$ is $[10, 1{,}000]$.

Step 6 To minimize $C(x)=500+2.5x+(4{,}000/x)$ on $[10, 1{,}000]$, we apply Rule 5.5 by first calculating the critical values.

$$\begin{aligned}
C(x) &= 500 + 2.5x + \frac{4{,}000}{x} \\
&= 500 + 2.5x + 4{,}000x^{-1} \\
C'(x) &= 2.5 - 4{,}000x^{-2} \\
0 &= 2.5 - \frac{4{,}000}{x^2} \\
2.5x^2 &= 4{,}000 \\
x^2 &= 1{,}600 \\
x &= 40 \text{ or } x = -40
\end{aligned}$$

Since $x=-40$ is not in the interval $[10, 1{,}000]$, the only critical value is $x=40$. Evaluating $C(x)$ at $x=10$, $x=40$, and $x=1{,}000$, we find

$$C(10) = 500 + 2.5(10) + \frac{4{,}000}{(10)} = 925$$

$$C(40) = 500 + 2.5(40) + \frac{4{,}000}{40} = 700$$

$$C(1{,}000) = 500 + 2.5(1{,}000) + \frac{4{,}000}{1{,}000} = 3{,}004$$

The minimum total cost of \$700 occurs when each order contains 40 mice. So the biologist should order 40 mice each time he orders. During the year there will be a total of 25, (1,000/40), orders delivered, or approximately 2 orders/month.

EXERCISES

1. Find two numbers x and y whose sum is 100 and whose product is a maximum.

2. Find two positive numbers x and y that have a sum of 50 such that the product x^2y is a maximum.

3. The product of two positive numbers is 64. The square of the first number is added to twice the second. Determine the minimum value of this sum.

4. A farmer has available 2,000 ft of fencing to enclose a rectangular field. Find the dimensions of the field that will produce a maximum area. Find the maximum area.

5. **The Lifeguard Dilemma Revisited** The chief lifeguard at a public swimming beach has available 400 ft of rope to lay out a rectangular restricted swimming area.

   (a) What will be the dimensions of the rectangle if he maximizes the swimming area?
   (b) To ensure the safety of swimmers he decides that no one should be more than 50 ft from shore. With this added restriction, what should be the dimensions of the swimming area?

6. The Southwest Laundry Company manufactures automatic coin-operated washers. The total cost of manufacturing x washers is $\$2x^2-270x+200$, and the selling price for each unit is $\$525-3x$. Determine the number of washers they should produce to maximize profit.

7. **Profit from the Sale of Dune Buggies** The Leisure-Time Manufacturing Company has determined that the function that represents manufacturing cost for dune buggies is given by $C(x)=x^2-250x+20{,}000$, where x is the number of buggies produced. Their revenue function was determined to be $R(x)=650x-x^2$. Determine their profit function, the optimum number of dune buggies to maximize profits, and the associated profit.

8. International Travel Agency is offering a plane tour of Europe and the Middle East. Air East has agreed to charter a DC-10 to the travel agency if they are guaranteed at least 60 passengers. The air fare is to be \$1,000/person if 60 people go, and it will decrease by \$2/person for every person above the minimum number. A maximum of 300 spaces is available.

   (a) Determine the number of tourists that will give Air East the maximum revenue.
   (b) Will they be able to get maximum revenue?
   (c) Determine the revenue if all available spaces are occupied.

9. **Cornering the Wheat Market** Green Acres Farm grows wheat and has harvested 25,000 bushels this season. When the wheat is harvested, the price per bushel is \$3.50. The cost of storing the wheat is \$15 plus a daily charge of 0.0003 dollars for each bushel stored. The price per bushel is expected to increase by \$3.50 $(0.02t-0.002t^2)$, where t is time in weeks. Determine how long they should hold the wheat before selling in order to obtain maximum profit.

10. A dealer of Brazilian coffee can ship a cargo of 200 sacks of coffee now at a profit of \$10/sack. However, by waiting he can add 10 sacks to the shipment for every week he waits, but the profit on the entire shipment will be reduced by 20¢ /sack for each week he waits. When should he ship the coffee to realize maximum profit?

11. The rental manager of an 80-unit apartment building is trying to decide on the rent to charge. When the rent of each unit is \$160/month all the units are occupied. However, for each \$5 increase in the rent each month, one of the units becomes vacant. What rent should she charge to maximize the rental income? Find the maximum income.

12. If each occupied apartment in Exercise 11 requires on the average \$20/month for repairs and maintenance, what rent should she charge to maximize the profits. (*Hint*: profit = revenue − expenses.)

13. A Florida orange grower wishes to ship as early as possible in the season to catch higher prices. At this time he can ship 6 tons at a profit of \$400/ton. By waiting he estimates he can add 3 tons/week to his shipment, but the profit will be reduced by $\$33\frac{1}{3}$/ton each week. Assuming that he will ship within the next 8 weeks, how many weeks should he wait to maximize his profit? Find the maximum profit.

14. An orchard manager gets 400 lb of apples/tree when he plants 46 trees/acre. If each additional tree planted per acre decreases the yield per tree by 5 lbs, how many trees/acre should the farmer plant to provide a maximum yield?

15. A manufacturer of cardboard boxes has an order for boxes that are to have a square base and no top, with a volume of 32 cu ft. Find the dimensions of the box that will minimize the material if no edge of the box can be less than 1 ft.

16. A rancher wants to enclose two rectangular areas along a straight section of the Snake River, as shown in the following sketch. If he uses no fencing along the river, what is the largest total area he can enclose if he has 240 yards of fencing? What are the dimensions of the field?

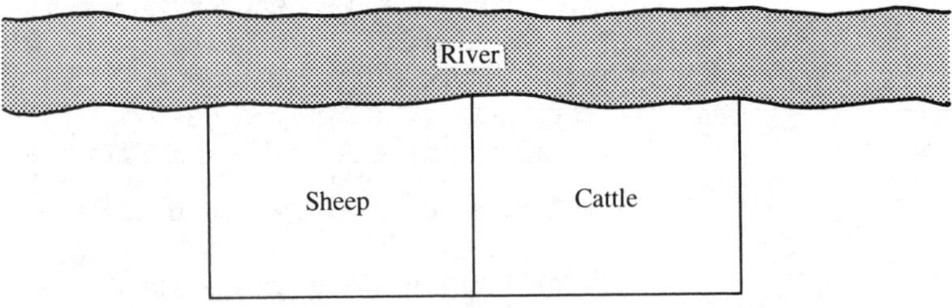

EXERCISE 16

17. **Picking Peaches for Profit** The manager of a peach field is trying to decide when to arrange for the picking of his peaches within the next fifteen days. If they are picked now, the average yield per tree will be 100 lb, which can be sold for 60¢ /lb. Past experience shows that the yield per tree will increase about 5

lb/day, while the price will decrease about 2¢/lb each day. When should the peaches be picked in order to produce maximum income? What is the maximum income?

18. A chicken farmer wants to enclose a rectangular area of four sides, and then run two additional fences to give him three separate runs, as shown in the following sketch. The outside fencing costs \$4/linear ft and the inside fencing costs \$2/linear ft. Find the maximum area the farmer can enclose if he has \$2,400 to spend on fencing. What are the dimensions of the enclosure?

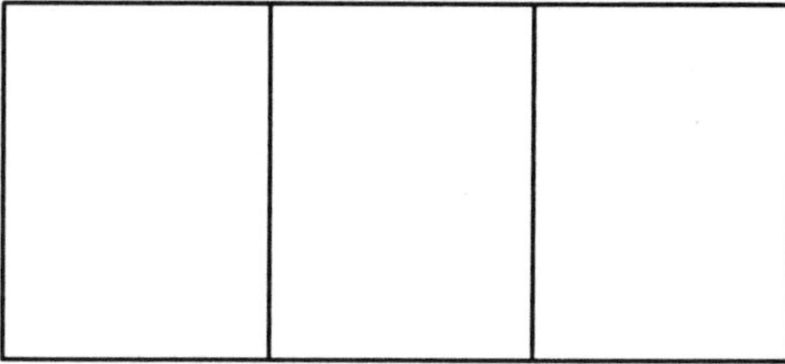

EXERCISE 18

19. **World Series Charter** A travel agency advertises all-expenses-paid trips to the World Series for special groups. Transportation is by chartered bus, which seats 48 passengers, and the charge per person is \$80 plus an additional \$2 for each empty seat. If the travel agency is interested in making the total receipts as large as possible, how many passengers should they take?

20. A manufacturer wishes to construct a box with a square base and an open top to hold 40.5 cu ft. He wants to choose dimensions that will enable him to minimize the cost of the box. The bottom of the box costs \$3/sq ft and the sides cost \$1/sq ft. What are the dimensions of the box if no edge is less than 6 in.?

21. **Forestry** The Maine Forest Service finds that if 16 trees are planted on an acre of land each tree will grow approximately 3 ft/year. For each additional tree planted, the growth will be reduced by $\frac{1}{2}$ in/yr. Find the number of trees per acre that will yield the largest amount of timber.

22. **An Inventory Problem** A specialty paint manufacturer uses 5,000 cases of linseed oil each year. The cost of storing each case for a year is \$2. To accommodate their suppliers they have agreed to make each order of equal size, and in return their delivery charge is fixed at \$50/order. Assuming that the oil is used at a uniform rate throughout the year and that each order is delivered when the previous one runs out, how frequently should the company place orders to minimize their costs?

23. A rectangular field of area 5,000 sq ft borders a stream on one side. Find the *least* amount of fencing required for the other three sides.

24. A closed box with a square base is to have a volume of 360 cu ft. The material for the top of the box costs \$2/sq ft, the material for the sides costs \$1.50/ sq ft, while the material for the bottom of the box costs \$3/sq ft. If each edge must be at least 5 ft in length, find the dimensions of the box that will minimize the total cost. Find the minimum cost.

25. A fabric manufacturer must supply a customer with 12,000 yd of a certain cloth during the coming year. There is a constant demand for the cloth during the year and the manager wants to produce the same number of yards at each production run. We have the following information:

(a) Set-up cost per production run is \$300.
(b) Inventory costs are 80¢ times the average number of yards in stock during the year.
(c) Production costs are \$3/yd.
(d) Each production run must be at least 500 yd.

Determine the number of yards in each production run to minimize the total cost of this business activity. What is the minimum total yearly cost for this operation?

26. **Winning Political Office by Proper Allocation of Funds** Two candidates compete for office in a district composed of two wards, with v_1 and v_2 voters in the first and second wards, respectively. The Democrat has a total campaign fund of D dollars, and the Republican has a total campaign fund of R dollars. The two candidates can divide their funds however they wish between the two wards, and it is assumed that the vote for each candidate in a ward is directly proportional to the share of the total money spent by the candidates in that ward. Thus, if the Democrat spends d dollars in ward 1, and $d' = D - d$ dollars in ward 2, and the Republican spends r and $r' = R - r$ dollars in wards 1 and 2, respectively, then the total vote for the Democratic candidate is

$$\begin{aligned} V_D &= \left[\frac{d}{d+r}\right]v_1 + \left[\frac{d'}{d'+r'}\right]v_2 \\ &= \left[\frac{d}{d+r}\right]v_1 + \left[\frac{(D-d)}{(D-d)+(R-r)}\right]v_2 \end{aligned}$$

What is the optimal amount that the Democrat should spend in each ward if there are 100 voters in each ward, both candidates have the same total campaign fund of \$1,300, and the Republican allocates his funds by spending \$400 in ward 1 and \$900 in ward 2? What will be the outcome of the election if the Democrat makes his optimal expenditure?

27. If the budget maximizing agency of Exercise 21 in Sec. 2.3 were not constrained from running a deficit, what level of output would maximize the budget? What would be its budget and deficit at this level of output?

28. If the agency of Exercise 21 in Sec. 2.3 were a profit maximizing firm whose revenue function is given by the budget function (i.e., $R = 1{,}200Q - \frac{1}{2}Q^2$), determine the optimal level of output, and the associated revenue and profit at this level of output.

29. **Scheduling Production** The Miro-Muffler Company experiences a yearly demand for 100,000 of their Life-Time Mufflers. This demand is spread uniformly throughout the year. The set-up cost for a production run of mufflers is \$800. If the cost of carrying one muffler in inventory is \$10/year, how large should each equal size production run be and how many should they schedule each year to minimize the sum of the set-up and inventory costs? Assume each production run yields at least 500 mufflers. [*Hint*: If x represents the number of mufflers in each production run, the average inventory is $(x/2)$ mufflers during the year.]

30. **Beer Consumption at the College Pub** The manager of the College Pub expects to sell 500 kegs of beer during the 10-month college year. To store the kegs of beer he loses valuable space. He estimates that it costs him \$10/year/keg. The manager and his supplier agree that if equal-size orders are placed during the college year, a fixed fee of \$25/order will be charged. Assuming that beer consumption is uniform during the 10 months and that each shipment is made as he runs out, how many shipments should be made to minimize his delivery and storage costs?

5.5 MARGINAL ANALYSIS

In the preceding sections we were concerned with determining the extrema of certain functions. Several of the examples considered were cost, revenue, and profit functions. Another consideration in management decision making involves the derivatives of these functions: the marginal cost, marginal revenue, and marginal profit that we introduced in Sec. 4.4. From that discussion we learned that if the total cost, $C(x)$, is defined in terms of x, the number of items produced or manufactured, the marginal cost, $C'(x)$, is the approximate cost of producing one more item. Analogously, the marginal revenue is the approximate additional revenue generated from the sale of an additional item and the marginal profit is the approximate additional profit generated from the sale of an additional item.

In many situations, the information about marginal values is more important and useful than information about total values. For example, generally management is concerned with questions such as whether to produce a larger or smaller batch of units, to increase or decrease the number of machines for production, to increase or decrease staff for a certain task, or to increase or decrease advertising expenditures. If a company wished to maximize total profit, they would certainly be willing to spend more on advertising only if the return would be at least sufficient to offset the expenditure; similarly, if they wished to minimize total cost, they would be willing to add another machine if the reduction in production cost would be sufficient to offset the cost of using the additional machine. The analysis of problems that consider the effect of a decision on the total function is termed *marginal analysis*.

For those situations where it is of interest to determine the marginal cost for a specific value of the independent variable and for which the total cost function is known, the analysis is performed by differentiating the total cost function to obtain the marginal cost function. For example, if the total cost function is given by $C(x)=4x^3-30x^2+200x+100$, the marginal cost function is given by $C'(x)=12x^2-60x+200$. (The graphic relationship for

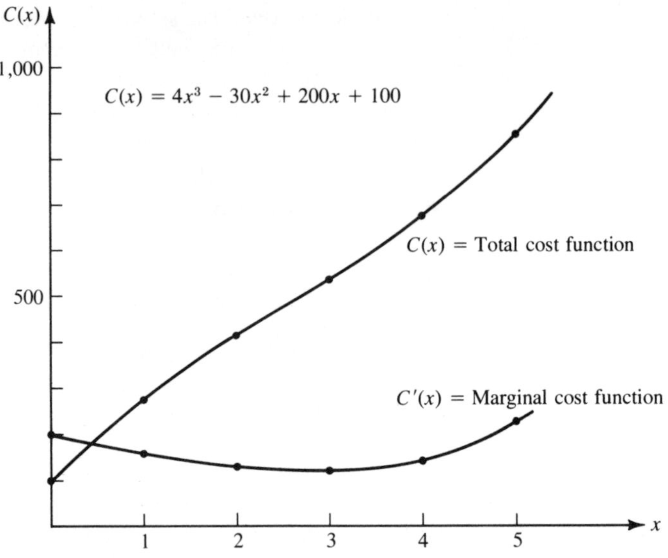

FIGURE 5.24

these two functions is illustrated in Fig. 5.24.) Furthermore, if we wish to determine the value of x that renders the minimum *marginal* cost, we simply differentiate the marginal cost function (i.e., find the second derivative of the total cost function) to determine the critical values and then proceed in the usual manner. The same procedure applies for marginal revenue and marginal profit.

Example 5.23 The total cost in dollars for a particular operation is given by $C(x)=x^3-21x^2+360x+3{,}025$, where x represents the number of units made. Determine

(a) the marginal cost when the tenth unit is produced,
(b) the number of units for which the marginal cost is a minimum,
(c) the minimum marginal cost,
(d) the total cost and average cost for the number of units that minimizes the marginal cost,
(e) the total cost and average cost for 10 units.

Solution The marginal cost function $c(x)$ is the first derivative of the total cost function; that is,

$$c(x) = C'(x) = 3x^2 - 42x + 360$$

(a) The marginal cost at the time the tenth unit is produced is

$$c(10) = 3(10)^2 - 42(10) + 360 = \$240$$

(b) The first derivative of the marginal cost function (i.e., the second derivative of the total cost function) is

$$c'(x) = C''(x) = 6x - 42$$

Setting this equal to zero and solving, we obtain the critical value of the marginal cost function.

$$6x - 42 = 0$$
$$x = 7$$

Applying the second derivative test, we have $c''(x)=6$; thus $c''(7)=6>0$, which means that $c(x)$ attains a minimum at $x=7$.

(c) The minimum marginal cost when $x=7$ is

$$c(7) = 3(7)^2 - 42(7) + 360 = \$213$$

Hence, the approximate cost of manufacturing the 8th item is \$213.

(d) The total cost for $x=7$ is

$$C(7) = (7)^3 - 21(7)^2 + 360(7) + 3{,}025 = \$4{,}859$$

The average cost of manufacturing 7 items is

$$\frac{C(7)}{7} = \frac{\$4{,}859}{7} = \$694.14/\text{unit}$$

(e) The total cost for $x=10$ is

$$C(10) = (10)^3 - 21(10)^2 + 360(10) + 3{,}025 = \$5{,}525$$

The average cost/item when manufacturing 10 items is:

$$\frac{C(10)}{10} = \frac{\$5{,}525}{10} = \$552.50/\text{unit}$$

From Example 5.23 we see that the lowest marginal cost does not necessarily give the lowest average cost.

Example 5.24 The total profit function in hundreds of dollars for a particular achievement motivation course is given by $P(x) = -x^3 + 30x^2 + 600x - 1{,}000$, where x represents the dollars in thousands spent on advertising. Determine

(a) the marginal profit function and the marginal profit for spending \$9,000 and \$11,000, respectively, on advertising,

(b) the advertising expenditure that yields a maximum marginal profit and the associated maximum marginal profit,
(c) the average profit/thousand dollars spent on advertising for the advertising expenditure obtained in part (b).

Solution

(a) The marginal profit function is $p(x)=P'(x)=-3x^2+60x+600$.

$$p(9) = -3(9)^2 + 60(9) + 600 = 897 \qquad \text{or } \$89{,}700$$

which means an additional profit of approximately \$89,700 is made by increasing the advertising expenditures from \$9,000 to \$10,000.

$$p(11) = -3(11)^2 + 60(11) + 600 = 897 \qquad \text{or } \$89{,}700$$

which means an additional profit of approximately \$89,700 is made by increasing the advertising expenditures from \$11,000 to \$12,000.

(b) The first derivative of the marginal profit function is $p'(x)=-6x+60$, which, when set equal to zero, yields $x=10$. Applying the second derivative test, we have

$$p''(x) = -6$$

Thus, $p''(10)=-6<0$, and therefore the marginal profit function is a maximum at $x=10$. The associated marginal profit is

$$p(10) = -3(10)^2 + 60(10) + 600 = 900$$

or the marginal profit is \$90,000.

(c) The average profit/thousand dollars spent is

$$\frac{P(10)}{10} = \frac{1}{10}\left[-(10)^3+30(10)^2+600(10)-1{,}000\right] = \frac{7{,}000}{10} = 700$$

or \$70,000/\$1,000 spent on advertising.

In Example 5.18, we considered the total profit function as a difference between the sales (or revenue) function $S(x)$ and the cost function $C(x)$; that is,

$$P(x) = S(x) - C(x)$$

To determine the x value which yields maximum profit, we applied an appropriate test to $P(x)$. As an alternative method for solving for the maximum profit, we note that

$$P'(x) = S'(x) - C'(x)$$

In terms of marginal analysis, this simply means that marginal profit equals marginal sales (or revenue) minus marginal cost. Furthermore, to obtain the critical values, we have

$$S'(x) - C'(x) = 0$$

which is satisfied when

$$S'(x) = C'(x)$$

Again, in terms of marginal analysis, this says that maximum profit occurs only when marginal sales (or revenue) equal marginal cost. When this occurs, we say that we have satisfied the *optimality condition of marginal analysis*. Furthermore, applying the second derivative test, the profit will be a maximum if the second derivative of $P(x)$ is negative at the critical values. Thus,

$$P''(x^*) = S''(x^*) - C''(x^*) < 0$$

or

$$S''(x^*) < C''(x^*)$$

which means that the marginal sales (or revenue) function is increasing at a slower rate than the marginal cost function. In summary, the profit will attain a maxima at those points for which the marginal sales equals marginal cost and for which the rate of increase of marginal sales is less than the rate of increase of marginal costs. This is a basic law of economics.

Example 5.25 Use the information given in Example 5.18 and the marginal analysis approach to determine the number of drilling bits that will maximize profit.

Solution Essentially we are given the sales and cost functions

$$S(x) = 300x$$

and

$$C(x) = 100 + 10x + 5x^2$$

respectively. Thus, the marginal sales function is

$$S'(x) = 300$$

and the marginal cost function is

$$C'(x) = 10 + 10x$$

The maximum will occur when

$$S'(x) = C'(x)$$

or

$$300 = 10 + 10x$$

Solving, $x=29$ is the number of bits that makes marginal sales equal to marginal cost.

To ensure that we have a maximum, we compare the rates of change of the two marginal functions at the critical value. Since $S''(x)=0$ and $C''(x)=10$, for any value of x, the rate of change of marginal sales is less than the corresponding rate of change of marginal costs. Specifically, this is satisfied at $x=29$, and hence a relative maximum profit is attained when the Johnson Tool Company produces 29 bits/month. To verify that it is an absolute maximum, we need only evaluate the profit function $P(x)$ at the end points as we did in Example 5.18 and show that $P(29)$ is greater than $P(x)$ at either end point.

EXERCISES

1. **Cost of Housing in a Penal Institution** The total weekly cost in dollars of housing x inmates in a penal institution is given by

$$C(x) = 0.0002x^3 - 0.3x^2 + 50x + 100{,}000$$

Determine

(a) the marginal cost of housing when 1,000 inmates are housed,
(b) the total cost of housing 1,000 inmates.

2. If the cost in dollars of making x units of a critical drug for coronary patients is

$$C(x) = 0.0009x^3 - 0.54x^2 + 108x + 2{,}000$$

determine the marginal cost function, and plot the marginal cost function for the following x values: 100, 150, 200, 250, and 300. Use differential calculus to determine the number of units that yields a minimum marginal cost.

3. The total cost for manufacturing x units of a certain type of radio equipment is given by the function

$$C(x) = 50{,}000 + 2{,}000x - 15x^2 + x^3$$

Determine

(a) the marginal cost function,
(b) the marginal cost when 10 units are being manufactured,
(c) the total cost of manufacturing 10 units,
(d) the minimum marginal cost,
(e) the total cost for producing the number of units that minimizes marginal cost.

4. The total profit function in hundreds of dollars for a building contractor is given by the function

$$P(x) = 400x - 40x^2 + x^3$$

where x represents the number of four-unit apartment houses constructed. Determine

(a) the marginal profit function,
(b) the marginal profit when five apartment houses have been constructed,
(c) the total profit for five apartment houses.

5. **Sales Force Size and Profits** The Excellent Encyclopedia Company is considering increasing their sales force from the current level of 225. They have established that their profit function in dollars is $P(x)=60x^{3/2}-200x$, where x represents the number of sales representatives.

(a) Determine the total profit for the current staff.
(b) Determine the marginal profit for the current staff.

6. The manufacturer of Excello Widgets has determined that the cost of manufacturing x widgets is given by

$$C(x) = 2x^2 - 10x + 100$$

and the sales function is given by

$$S(x) = x^2 + 40x - 200$$

(a) Determine the profit function and use calculus to determine the optimal number of units and the associated profit.
(b) Verify the results of part (a) by applying the optimality condition of marginal analysis.

7. The Freehoff Trailer Company has determined that the total cost (in hundreds of dollars) of making x trailers is given by

$$C(x) = 12x - 0.06x^2 + 0.0001x^3$$

Determine

(a) the marginal cost function,
(b) the marginal cost when 300 trailers have been made,
(c) the total cost of manufacturing 200 trailers,
(d) the minimum marginal cost,
(e) the total cost of making the number of trailers that minimizes marginal cost.

8. A manufacturer has determined that if he manufactures x items in one week, his profit, P, on the x items is given by (see Exercise 15, Sec. 5.2)

$$P(x) = 0.001x^2(40-x)^3$$

Determine

(a) the marginal profit function,
(b) the number to manufacture to maximize the profit.

9. The Slick Pen Company has determined that the total cost in dollars for manufacturing x cartons of ball point pens is given by

$$C(x) = .002x^3 - .03x^2 + 4x + 5$$

To encourage larger purchases by their outlets, they charge \$8/carton for purchases of 20 or more cartons and $\$(12-.2x)$/carton when fewer than 20 cartons are purchased. One of their outlets is Ace Stationers. Determine

(a) the marginal profit function from the sale of x cartons of pens to Ace Stationers if $x<20$,

(b) the marginal profit when 10 cartons of pens have been sold to Ace Stationers.

10. Mallory Cut Rate is one of Slick Pen Company's largest outlets (see Exercise 9). Determine

(a) the marginal profit function from the sale of x cartons of pens to Mallory Cut Rate if $x \geqslant 20$,

(b) the marginal profit when 21 cartons of pens have been sold to Mallory Cut Rate.

5.6 CURVE SKETCHING

It has been said many times that a picture is worth a thousand words. Throughout our study of the characteristics of functions we immediately turned to the graph of the function for a concrete picture of the abstract concept we were studying. Earlier in the course we determined the graph of a function by the point-by-point plotting technique. As we progressed, we found this tedious and time-consuming method could be replaced or refined by new tools that we developed in the area of algebra and calculus. In this section we will summarize the noncalculus and calculus techniques we have developed to graph functions. While studying the calculus tools we will introduce two new concepts that will make curve sketching easier.

The noncalculus tools we have developed to aid us in sketching the graph of a function play a significant role and should never be neglected. Table 5.1 summarizes the most important ones we have considered in this text. The characteristics we are considering are shown in column 1. The second column lists those functions that display these characteristics. The third column lists the test used to investigate the characteristics, and the fourth column gives references to sections in the text where this material was introduced.

The following example illustrates the many features or characteristics that can be determined by noncalculus tools.

Example 5.26 Draw a sketch of the graph of $f(x)=(2x+1)/(x-1)$ by using tools 1–5 listed in Table 5.1.

Solution As we perform each of the tests as they appear in Table 5.1 we will geometrically show the results in Fig. 5.25.

1. *Extent:*

$$\lim_{x\to\infty} f(x) = \lim_{x\to\infty}\left(\frac{2x+1}{x-1}\right) = 2$$

$$\lim_{x\to-\infty} f(x) = \lim_{x\to-\infty}\left(\frac{2x+1}{x-1}\right) = 2$$

Table 5.1 Noncalculus Graphing Tools

	Characteristic		Function(s)	Test(s)	Refer to Section
1.	Extent		All $f(x)$	(a) $\lim_{x\to\infty} f(x)$ (b) $\lim_{x\to-\infty} f(x)$	2.5
2.	Zeros of the function		All $f(x)$	Solve $f(x)=0$	2.3 2.4
3.	y-intercept		All $f(x)$	$y=f(0)$	
4.	Vertical asymptote(s)	(a)	Rational functions $\dfrac{p(x)}{q(x)}$	$x=a$ where $q(a)=0$ and $p(a)\neq 0$	2.5
		(b)	Logarithmic functions $y=\log_b x$	$x=0$	3.5
5.	Horizontal asymptote	(a)	Rational functions $\dfrac{p(x)}{q(x)}$	$y=\lim_{x\to\infty}\dfrac{p(x)}{q(x)}$	2.5
		(b)	Exponential functions $y=ab^x$	$y=0$	3.3

As $x\to+\infty$, $f(x)$ approaches 2 from above as can be found by evaluating $f(10)$ and $f(100)$.

$$f(10) = \tfrac{21}{9} \doteq 2.33$$

$$f(100) = \tfrac{201}{99} \doteq 2.03$$

As $x\to-\infty$, $f(x)$ approaches 2 from below as can be found by evaluating $f(-10)$ and $f(-100)$.

$$f(-10) = \frac{-19}{-11} = \frac{19}{11} \doteq 1.73$$

$$f(-100) = \frac{-199}{-101} = \frac{199}{101} \doteq 1.97$$

Since $\lim_{x\to\infty} f(x) = \lim_{x\to-\infty} f(x) = 2$, we can conclude that there is a horizontal asymptote whose equation is $y=2$.

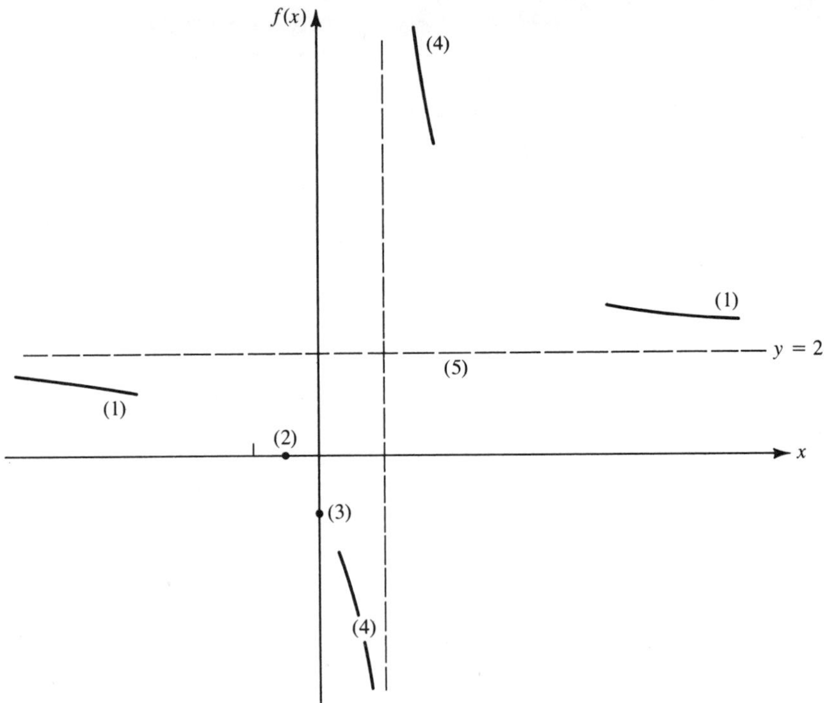

FIGURE 5.25

2. *Zeros of the function:* We find by solving the equation $f(x)=0$ or,

$$\frac{2x+1}{x-1} = 0$$
$$2x+1 = 0$$
$$x = -\frac{1}{2}$$

3. *y-Intercept:* We find by solving $y=f(0)$,

$$f(x) = \frac{2x+1}{x-1}$$
$$y = f(0) = -1$$

4. *Vertical asymptote(s):* Since $f(x)=(2x+1)/(x-1)$, the vertical asymptote is found by solving $x-1=0$ and checking to see that $2x+1 \neq 0$ for this value of x. Thus, we find

$$x-1 = 0 \text{ or}$$
$$x = 1$$

is a vertical asymptote since $2(1)+1 \neq 0$. To determine the behavior of the function $f(x)$ near the asymptote $x=1$ we investigate and find

$$\lim_{x\to1^+}\left(\frac{2x+1}{x-1}\right) = +\infty$$

[To evaluate this limit we have chosen for values of x: 1.1, 1.01, and 1.001. Note that $f(x)$ increases without bound.]

$$\lim_{x\to1^-}\left(\frac{2x+1}{x-1}\right) = -\infty$$

[To evaluate this limit we have chosen for values of x: .9, .99, and .999. Note that $f(x)$ decreases without bound.]

5. *Horizontal Asymptote:* Considered in step 1 above.

A composite picture of all this information is shown in Fig. 5.25. By inspecting the information provided in this figure we sketch the graph of $f(x)$ in Fig. 5.26.

The calculus tools we have to aid us in sketching the graph of a function $f(x)$ are summarized in Table 5.2. Although we discussed increasing and

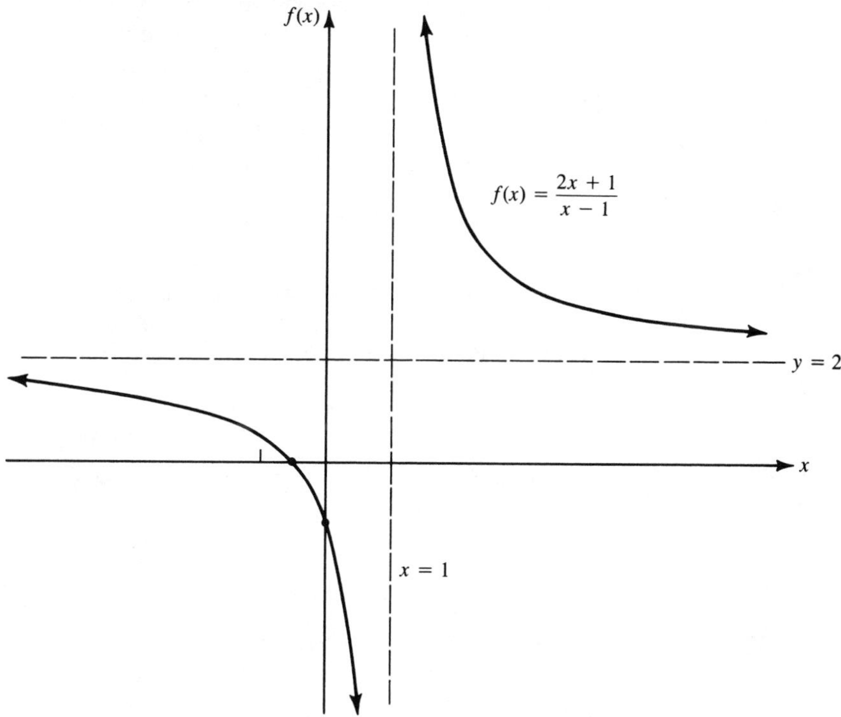

FIGURE 5.26

Table 5.2 Calculus Graphing Tools

	Characteristic	*Functions*	*Test(s)*	*Refer to Section*
6.	Critical values	All	Those values of x where $f'(x)=0$ or $f'(x)$ is undefined	5.3
7.	Critical points	All	Points $(x^*, f(x^*))$ where x^* is a critical value	5.3
8.	Interval(s) where the function is increasing	All	Domain where $f'(x)>0$	5.2
9.	Interval(s) where the function is decreasing	All	Domain where $f'(x)<0$	5.2

decreasing functions before we discussed critical values and critical points, the latter are more important in curve sketching as they give us information on where the curve may be turning. With the use of the function test, the first derivative test, or the second derivative test, we can determine if our critical points correspond to relative maxima, relative minima, or neither. Very often these tests are not required if we combine our noncalculus and calculus curve sketching tools, as we will see in the following examples.

Example 5.27 Draw a sketch of the graph of the polynomial function $f(x)=x^4-4x^3$.

Solution Since every polynomial function is continuous on the set of real numbers there are no holes or jumps in the curve. Performing our tests as they appear in Tables 5.1 and 5.2 and showing the results geometrically in Fig. 5.27, we have:

Noncalculus Tests

1. *Extent:*

$$\lim_{x\to\infty} f(x) = +\infty$$

$$\lim_{x\to-\infty} f(x) = +\infty$$

(*Note*: The extent is indicated in Fig. 5.27 by arrows.)

2. *Zeros of the function:* Solve $f(x)=0$

$$x^4-4x^3=0$$

$$x^3(x-4)=0$$

$$x=0, \quad x=4$$

are zeros of the function.

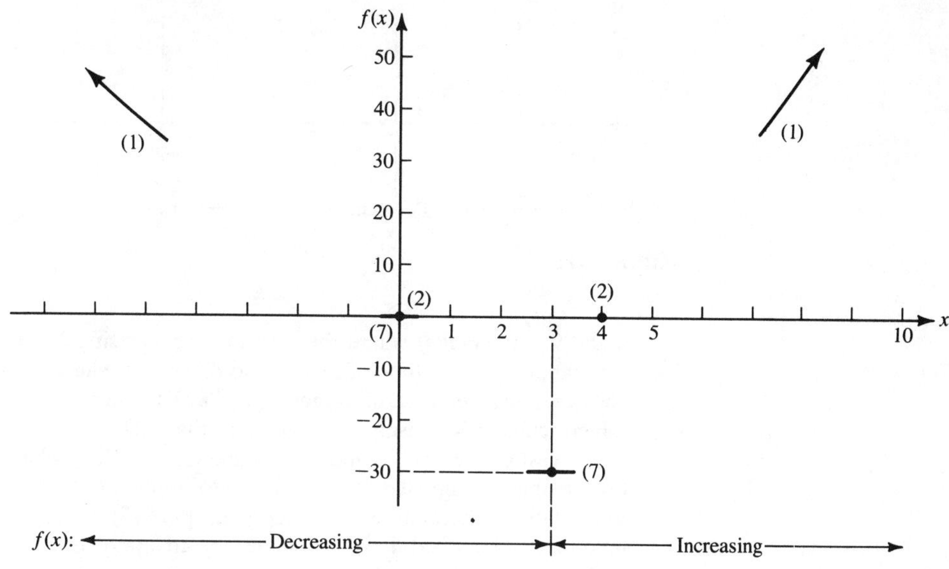

FIGURE 5.27

3. *y-Intercept:*

$$y = f(0) = 0$$

Since we have the point (0,0) from Step 2, this does not provide us with any new information.

4 and 5. *Asymptotes:* There are none when considering polynomial functions.

Calculus Tests

6. *Critical values:* Solve $f'(x) = 0$

$$f(x) = x^4 - 4x^3$$

$$f'(x) = 4x^3 - 12x^2$$

$$0 = 4x^2(x-3)$$

$$x = 0, \quad x = 3$$

are critical values.

7. *Critical points:* $(0, f(0))$ and $(3, f(3))$. Solving for $f(0)$ and $f(3)$, we have

$$f(0) = 0$$

and

$$f(3) = (3)^4 - 4(3)^3 = -27$$

The critical points (0,0) and (3, −27) are shown in Fig. 5.27 as $-\!\!\bullet\!\!-$ to indicate that the tangent lines drawn to the curve at these points are horizontal.

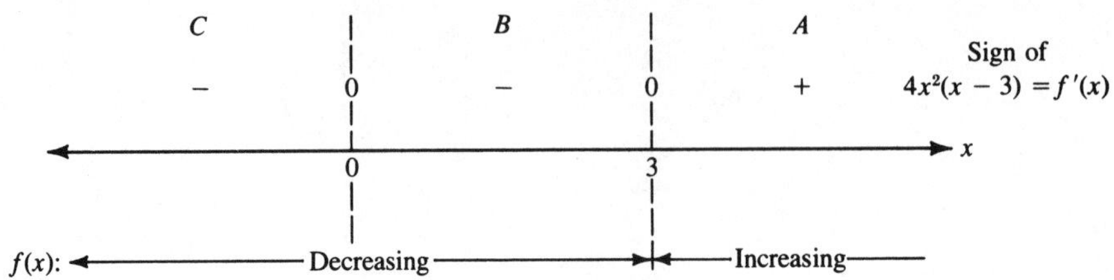

FIGURE 5.28

8 and 9. Interval(s) where the function is increasing or decreasing: Using the graphical cut-method of Sec. 2.4 to determine where $f'(x)=4x^2(x-3)$ is positive, negative, or zero, we get Fig. 5.28. The intervals where the function is increasing or decreasing are denoted in Fig. 5.27.

Investigating the composite picture in Fig. 5.27, which illustrates the tests we have completed, we can see that the point (0,0) will not correspond to a relative extremum, whereas the point $(3,-27)$ corresponds to a relative minimum. Note that it is not necessary to apply the function test, first derivative test, or the second derivative test at the critical points to arrive at this conclusion. However, we can easily verify by these tests that our conclusions are correct.

Several smooth curves can be drawn meeting all of the requirements that are described by Fig. 5.27. Two of the possible graphs are shown in Figs. 5.29*a* and *b*. Although these graphs show certain similarities, the shape of the curves on the domains $(-\infty,0)$ and $(4,\infty)$ are quite different. We shall shortly develop a calculus tool to predict which curves are the correct ones on these domains. For now, we complete the following table, and conclude that the correct graph of $f(x)$ is given by Fig. 5.29*a*.

x	-3	-2	-1	4.5	5
$f(x)$	189	48	5	45.5625	125

As a second example showing the use of our curve sketching tools, consider the following.

Example 5.28 Draw a sketch of the graph of the function $f(x)=2x^3+3x^2-36x+1$.

Solution Performing the tests as shown in Tables 5.1 and 5.2, and geometrically indicating their results in Fig. 5.30, we have

Noncalculus Tests

1. *Extent:*

$$\lim_{x\to\infty} f(x) = +\infty$$

$$\lim_{x\to-\infty} f(x) = -\infty$$

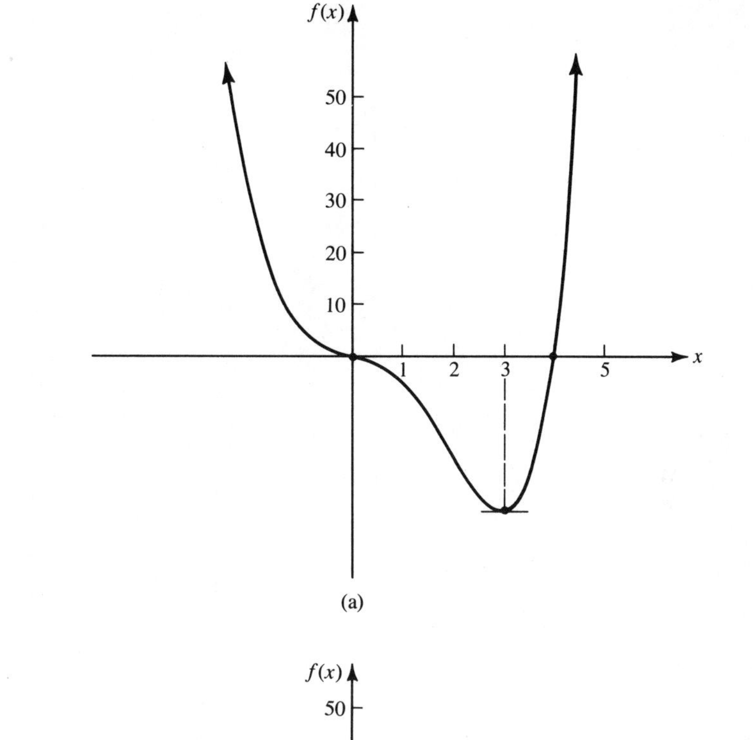

(a)

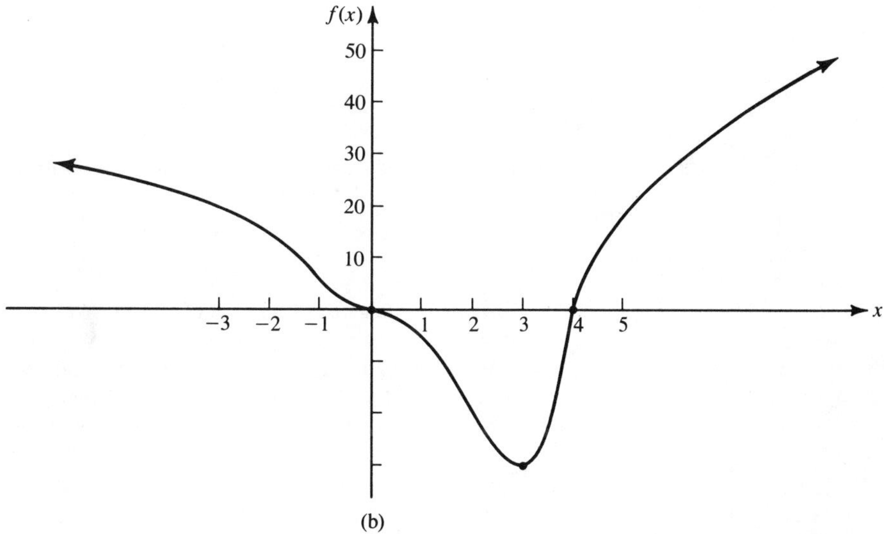

(b)

FIGURE 5.29

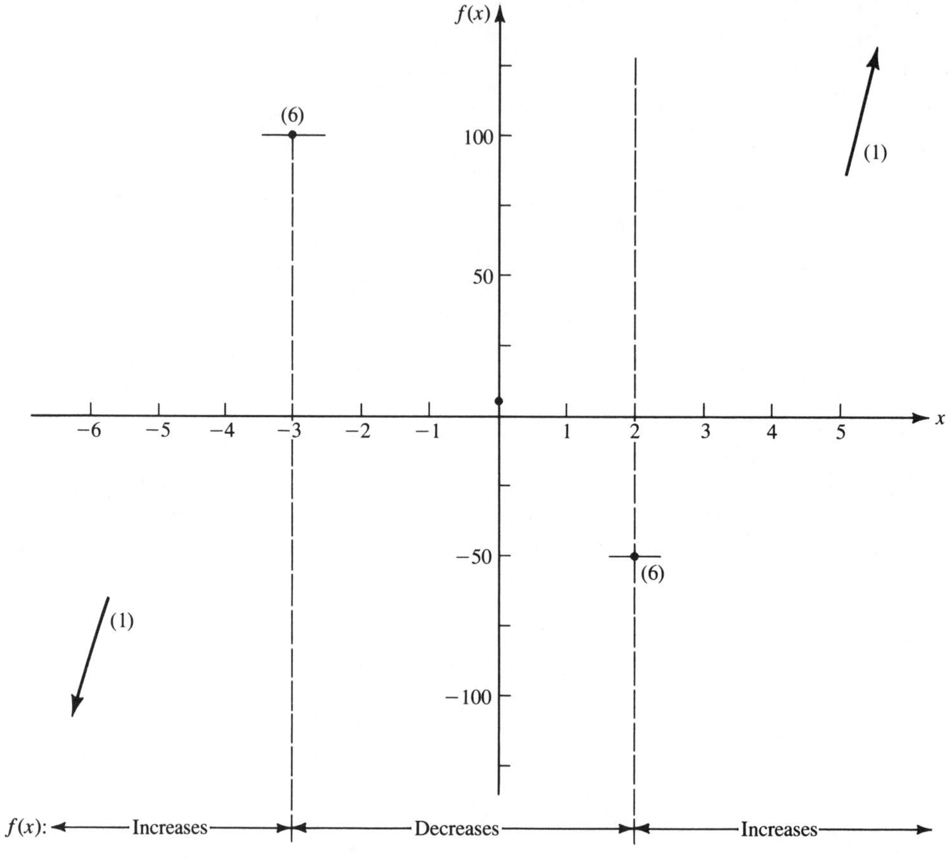

FIGURE 5.30

2. *Zeros of the function:* Solve $2x^3+3x^2-36x+1=0$. The only possible rational roots of this polynomial equation (see Sec. 2.4) are $\pm 1, \pm \frac{1}{2}$. A check shows that none of these possibilities are roots and we conclude that if there are any zeros of this function, they are irrational numbers. (In Sec. 5.7 we will develop a method to obtain approximate values of these roots.)

3. *y-Intercept:* $y=f(0)=1$.

4 and 5. *Asymptotes:* None.

Calculus Tests

6. *Critical values:*

$$f(x) = 2x^3+3x^2-36x+1$$
$$f'(x) = 6x^2+6x-36$$
$$0 = 6(x^2+x-6)$$
$$0 = 6(x+3)(x-2)$$
$$x = -3 \quad \text{and} \quad x = 2$$

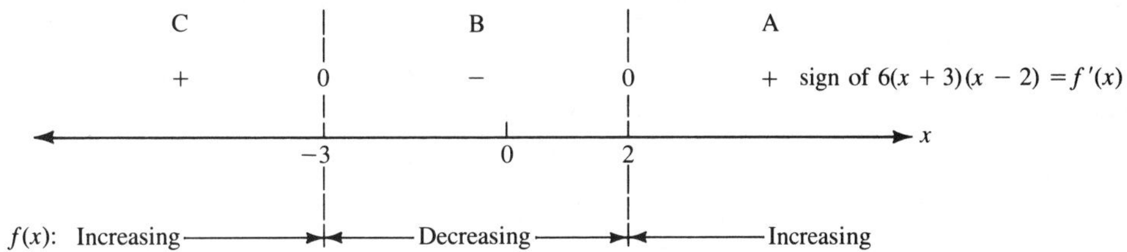

FIGURE 5.31

7. *Critical points:* $(-3, f(-3))$ and $(2, f(2))$

$$f(-3) = 2(-3)^3 + 3(-3)^2 - 36(-3) + 1 = 82$$

$$f(2) = 2(2)^3 + 3(2)^2 - 36(2) + 1 = -43$$

8 and 9. *Intervals where $f(x)$ is increasing or decreasing*: Using the cut-method to determine where $f'(x) = 6(x+3)(x-2)$ is positive, negative, or zero, we get Fig. 5.31. By considering the composite picture in Fig. 5.30 of the results of tests 1–5 of Table 5.1 and 6–9 of Table 5.2, and determining the ordered pairs in the following table

x	-6	-4	3	4
$f(x)$	-107	65	-26	33

we can draw a sketch of the graph of $f(x) = 2x^3 + 3x^2 - 36x + 1$ as shown in Fig. 5.32. The several tangent lines drawn to the graph of $f(x)$ will be discussed below.

By inspection of the tangent lines drawn to the graph of the function $f(x) = 2x^3 + 3x^2 - 36x + 1$ shown in Fig. 5.32 we note the following:

(a) The tangent lines drawn to the curve in the interval $(-\infty, x_0)$ turn clockwise as the point of tangency moves from left to right in this interval.
(b) The tangent lines drawn to the curve in the interval (x_0, ∞) turn counterclockwise as the point of tangency moves from left to right in this interval.
(c) The tangent line at the point $(x_0, f(x_0))$ crosses the curve at this point.

We describe the shape of the curve on the intervals in (a) and (b) above by the following definitions.

Definition 5.9 A curve is *concave downward* on an open interval (a, b) if and only if the tangent lines drawn to the curve in the open

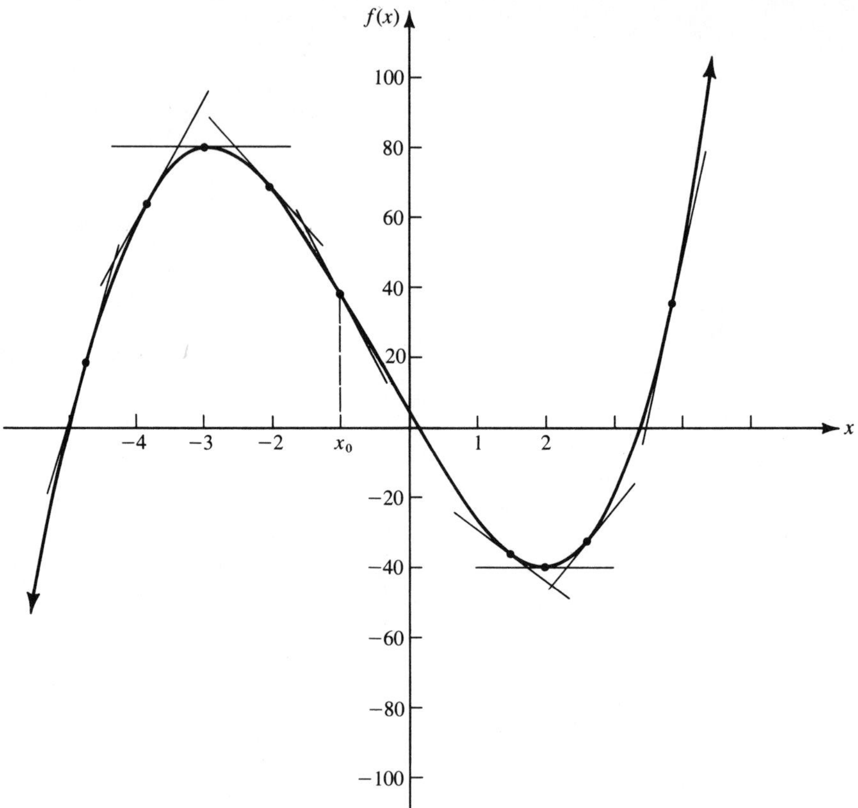

FIGURE 5.32

interval turn clockwise as the point of tangency moves from left to right.

Definition 5.10 A curve is *concave upward* on an open interval (a, b) if and only if the tangent lines drawn to the curve in the open interval turn counterclockwise as the point of tangency moves from left to right.

The point $(x_0, f(x_0))$ is defined by Definition 5.11.

Definition 5.11 A point on the graph of a function where the curve changes from concave downward on one side of the point to concave upward on the other side of the point (or from concave upward to concave downward) is called an *inflection point*.

Based on Definition 5.9 we can conclude that the following sections of a curve are concave downward. If a section of a curve is concave downward,

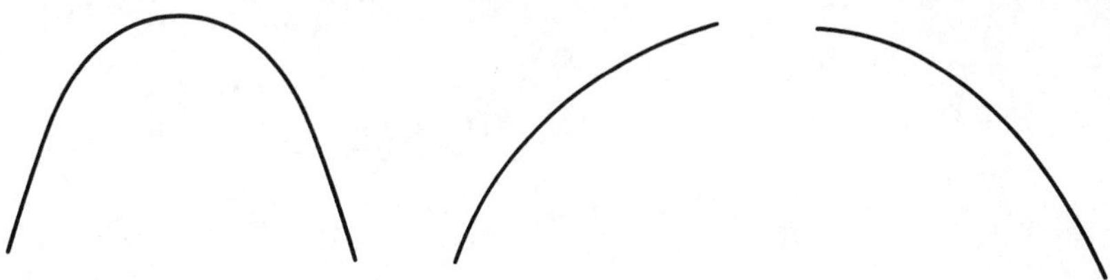

we can think of it as "not holding water." Based on Definition 5.10 we can conclude that the following sections of a curve are concave upward. If a section of a curve is concave upward, we can think of it as "holding water."

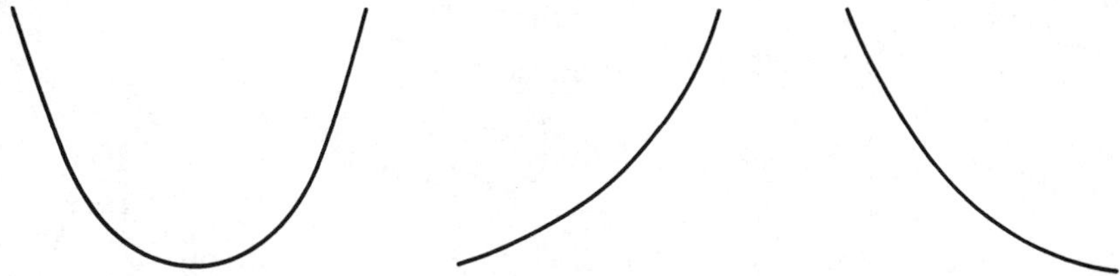

From Definition 5.9 and an inspection of the sections of curves that are concave downward, we see that the slopes of the tangent lines decrease as the point of tangency moves from left to right. This implies that the rate of change of the slope or the rate of change of the first derivative is negative. Since the rate of change of the first derivative equals the second derivative, a negative second derivative will be associated with a curve that is concave downward. Analogously, a positive second derivative will be associated with a curve that is concave upward. This is summarized in the following *concavity test* or rule.

Rule 5.6

(a) If $f''(x)<0$ for each value of x in an interval (a,b), then the graph of $f(x)$ is concave downward in the interval (a,b).
(b) If $f''(x)>0$ for each value of x in an interval (a,b), then the graph of $f(x)$ is concave upward in the interval (a,b).
(c) If $f''(x^*)=0$ or is undefined and the graph of the function changes from concave upward on one side of x^* to concave downward on the other side (or from concave downward to concave upward), the point $(x^*,f(x^*))$ is an inflection point.

Note the agreement of Rule 5.6(a) with our second derivative test for a relative maximum, i.e., Rule 5.4(a), and the agreement of Rule 5.6(b) with our second derivative test for a relative minimum, i.e., Rule 5.4(b) (see Fig. 5.33). Figure 5.34 shows four common cases of inflection points. Note that in Figs. 5.34 *a* and *b* the tangent line to the curve at the inflection point is horizontal and in each of the four cases it crosses the curve.

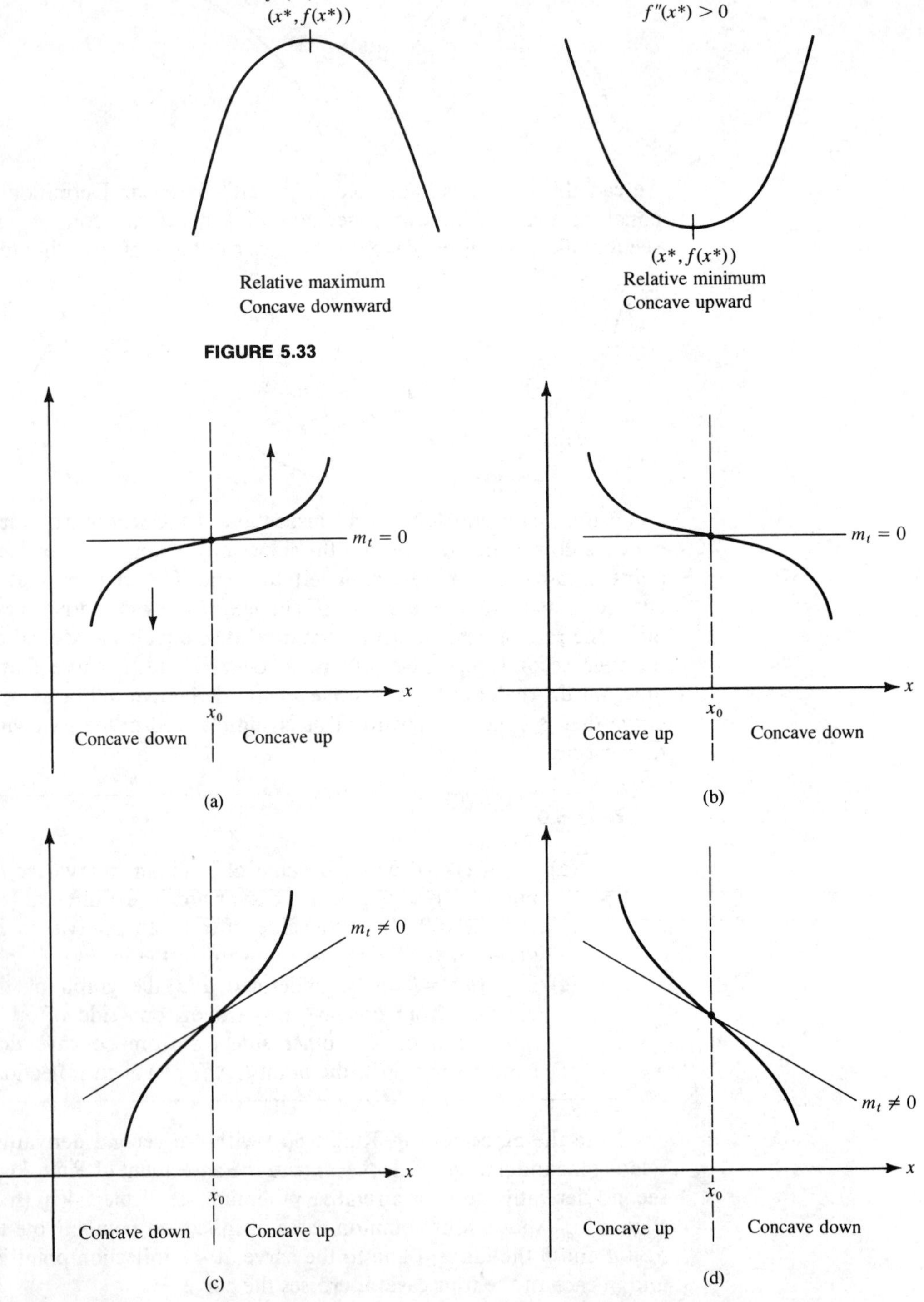

FIGURE 5.33

FIGURE 5.34

From Rule 5.6 we conclude that inflection points occur only at values of x where $f''(x)=0$ or where $f''(x)$ does not exist. The solutions of the equation $f''(x)=0$ or values of x that make $f''(x)$ undefined are called *hypercritical values*. If $x_1, x_2, \ldots, x_n$ are hypercritical values, the points $(x_1, f(x_1))$, $(x_2, f(x_2)), \ldots, (x_n, f(x_n))$ are possible inflection points. If the concavity of the curve changes from one side of the given point to the other side, it is an inflection point.

In the following example we will consider the concavity of a function whose graph we have already sketched (Example 5.27).

Example 5.29 Determine where the graph of the function $f(x)=x^4-4x^3$ is concave upward or concave downward. Also find any inflection points.

Solution To apply Rule 5.6 we determine the second derivative and use the graphical cut-method to determine where it is positive, negative, or zero.

$$\begin{aligned} f(x) &= x^4 - 4x^3 \\ f'(x) &= 4x^3 - 12x^2 \\ f''(x) &= 12x^2 - 24x \\ &= 12x(x-2) \end{aligned}$$

From Rule 5.6 and Fig. 5.35 we conclude that $f(x)$ is concave upward on $(-\infty, 0) \cup (2, \infty)$ and concave downward on $(0, 2)$. The hypercritical values found by solving $f''(x)=0$; are:

$$x = 0, x = 2$$

Since the concavity of the curve changes at $x=0$ and at $x=2$ as noted in Fig. 5.35, we conclude that $(0, f(0))$ and $(2, f(2))$ are inflection points. The graph of $f(x)=x^4-4x^3$ is reproduced in Fig. 5.36, where we denote the inflection points and concavity.

We can see that with the aid of the concavity test we were able to sketch the graph of $f(x)=x^4-4x^3$ without having to plot several additional points of the graph as we did in Example 5.27. Hence, the concavity test is an important tool in curve sketching and we add it as Steps 10 through 12 in Table 5.3.

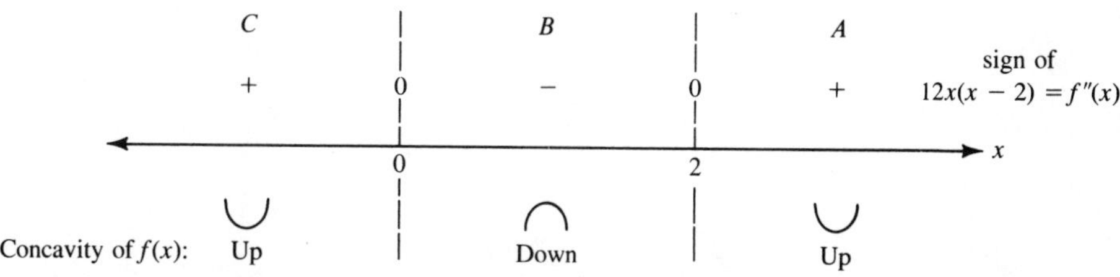

FIGURE 5.35

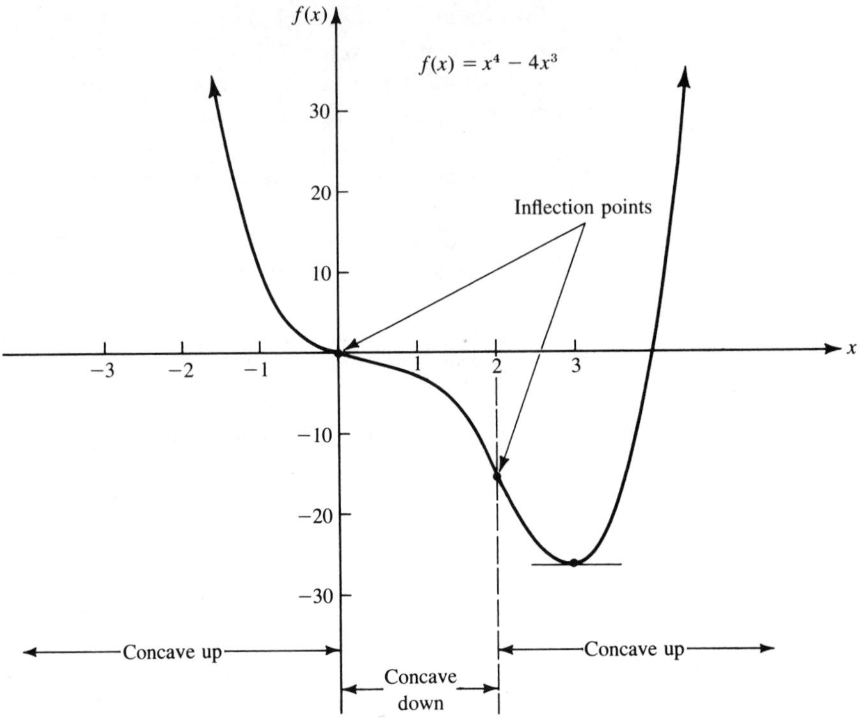

FIGURE 5.36

Table 5.3 Calculus Graphing Tools (cont.)

	Characteristic	*Functions*	*Test(s)*	*Refer to Section*
10.	Interval(s) where curve is concave downward	All	$f''(x) < 0$ in (a, b)	5.6
11.	Interval(s) where curve is concave upward	All	$f''(x) > 0$ in (a, b)	5.6
12.	Inflection point(s) $(x^*, f(x^*))$	All	$f''(x^*) = 0$ or is undefined and curve changes concavity at x^*	5.6

We conclude our study of concavity by considering one more example.

Example 5.30 Determine the intervals on which the function $f(x) = (2x+1)/(x-1)$ is concave upward or downward. Also, find any inflection points.

Solution To apply Rule 5.6 we determine the second derivative and use the graphical cut-method to determine where it is positive, negative, or zero.

$$\begin{aligned} f(x) &= \frac{2x+1}{x-1} \\ f'(x) &= \frac{(x-1)(2)-(2x+1)(1)}{(x-1)^2} \\ &= -3(x-1)^{-2} \\ f''(x) &= 6(x-1)^{-3}(1) \\ &= \frac{6}{(x-1)^3} \end{aligned}$$

From Rule 5.6 and Fig. 5.37, we conclude that $f(x)$ is concave downward on $(-\infty, 1)$ and concave upward on $(1, \infty)$. To determine if there are any hypercritical values, we solve $f''(x)=0$ and also find these values where $f''(x)$ is undefined. Since $f''(x)=6/(x-1)^3$, there are no solutions of $f''(x)=0$, and the only value of x where $f''(x)$ is undefined is $x=1$. The value of $x=1$ is not in the domain of the function since the function is not defined at $x=1$; hence, there are no hypercritical values and no inflection points.

The above conclusions can be verified by considering the graph of $f(x)$ shown in Fig. 5.26.

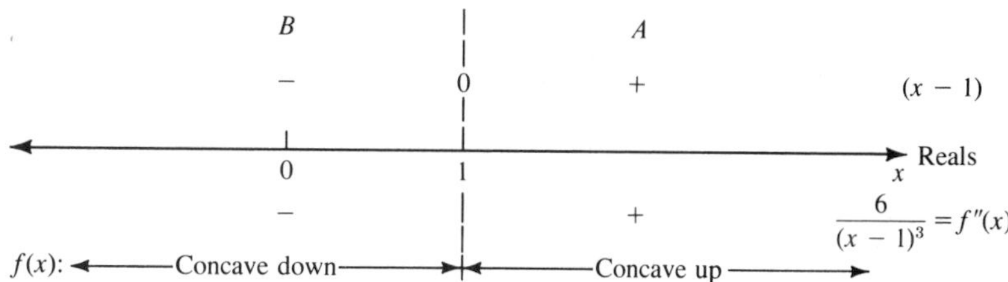

FIGURE 5.37

EXERCISES

Using the tools we have summarized in Tables 5.1–5.3, sketch the graph of each of the following functions. Determine the intervals on which the function is increasing or decreasing and determine the intervals on which the function is concave upward or concave downward. Find the coordinates of any inflection points.

1. $f(x)=2+5x-x^2$
2. $f(x)=3x^2+12x-6$
3. $g(x)=2-x^3$
4. $g(x)=2+3x-x^3$
5. $h(x)=4x^{2/3}$
6. $h(x)=\sqrt{x+1}$
7. $f(x)=\frac{1}{3}x^3-4x+1$
8. $f(x)=x^3-3x+1$
9. $g(x)=\dfrac{4}{x^2+1}$
10. $g(x)=\dfrac{4x}{x^2+1}$
11. $s(x)=\dfrac{2x-1}{x+1}$
12. $s(x)=\dfrac{1}{(x+1)^2}$
13. $h(x)=\dfrac{x^2}{x+4}$
14. $p(x)=\frac{1}{5}x^5-\frac{4}{3}x^3$
15. $f(x)=x+\dfrac{1}{x}$
16. $f(x)=x-\dfrac{1}{x}$
17. $g(x)=x^4-x^3+1$
18. $g(x)=4-x^4$
19. $f(x)=(x-3)^3$
20. $f(x)=\sqrt[3]{x+1}$
21. $f(x)=\dfrac{1}{x^2-1}$
22. $f(x)=\dfrac{4}{x^2-2x+1}$
23. If the derivative of the function $f(x)$ is $f'(x)=(x-4)(x+3)$, determine

    (a) the intervals on which $f(x)$ is increasing or decreasing,
    (b) the intervals on which $f(x)$ is concave upward or concave downward,
    (c) the x coordinate of any possible inflection point.

24. If the derivative of the function $g(x)$ is $g'(x)=(x-1)(x-2)(x+3)$, determine

    (a) the intervals on which $g(x)$ is increasing or decreasing,
    (b) the intervals on which $g(x)$ is concave upward or concave downward,
    (c) the x coordinate of any possible inflection point.

25. If the derivative of the function $h(x)$ is $h'(x)=(x-1)/x^2$, determine

    (a) the intervals on which $h(x)$ is increasing or decreasing,
    (b) the intervals on which $h(x)$ is concave upward or concave downward,
    (c) the x coordinate of any possible inflection point.

5.7 NEWTON'S METHOD OF ROOT APPROXIMATION

In many problems we need to determine the zeros of a function $f(x)$. That is, we need to determine the values of x that will satisfy an equation of the form

$$f(x) = 0 \tag{5.1}$$

For example, when we are determining the relative extrema of $f(x)$, we need to determine the *root* (or roots) of the equation $f'(x)=0$. In most situations that we have considered previously, the determination of the roots was relatively easy in that the equation was generally a quadratic equation or at least a polynomial that could be factored with little effort. At worst if the equation was neither quadratic nor factorable, we were able to determine the roots by applying the rational roots theorem of Sec. 2.4. Unfortunately, these types of equations are not very common in real-life problems, and for this reason we need to have available a technique that will assist us in solving for the roots in a wide range of situations. Thus, we must settle for a technique that will yield a good *approximation* of the root, be it a rational or irrational number. The development and use of such a method is the concern of this section.

The method we shall consider is developed from reasoning associated with Fig. 5.38*a* and *b*, which will assist the reader in following the logical development of the technique. Again, the reader is encouraged to understand the *development of the technique* instead of *memorizing the result*. The following reasoning can be applied to both parts of Fig. 5.38 with equal validity and the same results.

First, we must make an estimate x_0 of the desired root x^*. Next, the *iteration procedure*, called *Newton's method*, can be followed on either part of Fig. 5.38. From the point $(x_0, 0)$, go up to the graph of the function. At the point $(x_0, f(x_0))$ construct the line that is tangent to the function. Follow the tangent line to the point where it intersects the x axis and call this point $(x_1, 0)$. Thus, x_1 is the second estimate of x^*, and we see that x_1 is closer to x^* than is x_0. By repeating this procedure, we can obtain successive values of x_i that will converge to x^*.

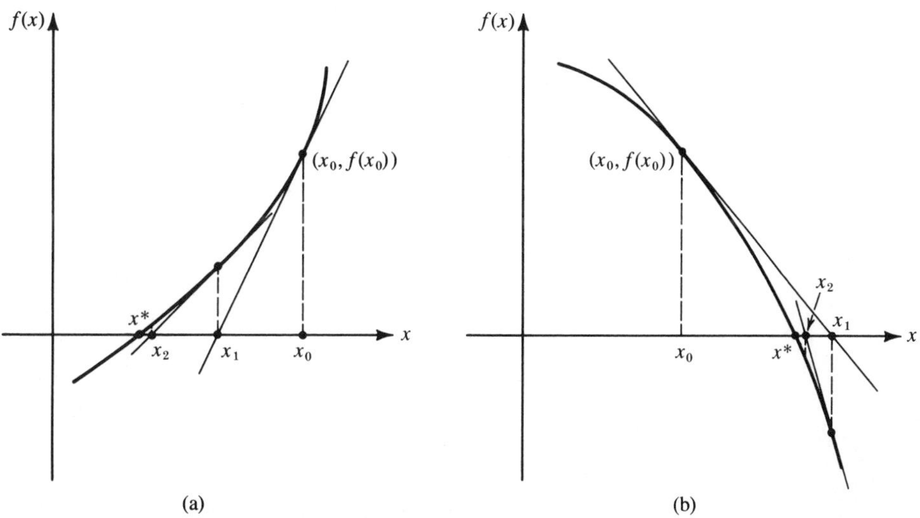

FIGURE 5.38

The computational procedure can be developed from the relationship between the function and its derivative. For example, the slope of the line that is tangent to $f(x)$ at $x=x_0$ is given by

$$\text{Slope} = \frac{f(x_0)-0}{x_0-x_1} = \frac{f(x_0)}{x_0-x_1}$$

Furthermore, the slope of the function at the point $x=x_0$ is the same as the slope of the tangent line, and it is given by the first derivative evaluated at $x=x_0$; that is, $f'(x_0)$. Thus, we have

$$f'(x_0) = \frac{f(x_0)}{x_0-x_1} \tag{5.2}$$

Since x_0 is the original estimate of a root of $f(x)=0$, we wish to determine a second estimate; that is, we wish to solve for x_1. Solving Eq. (5.2) for x_1, we obtain

$$x_1 = x_0 - \frac{f(x_0)}{f'(x_0)} \tag{5.3}$$

which yields an estimate of x^* based on values determined from x_0.

For closer estimates of x^*, we can repeat the above process; and the next approximation of x^* based on values obtained for x_i, the ith approximation after x_0, is given by

$$x_{i+1} = x_i - \frac{f(x_i)}{f'(x_i)} \tag{5.4}$$

Eventually successive values of x_i will converge to the value of x^*. The number of iterations required to obtain an acceptable approximation will depend on the behavior of the function and the desired accuracy of the approximation. In some situations the curve is so shaped that the method will not converge if x_0 is not *sufficiently* close to x^*; however, this is not a serious problem for most functions with which we are concerned. Hence, we see that Newton's method is applicable to a very wide range of problems.

Example 5.31 For the function $f(x)=x^2-5$, apply Newton's method to obtain the real root (accurate to three decimal places) that results from beginning with $x_0=2$.

Solution In order to obtain the next approximation of x^*, that is, x_1, we must evaluate $f(x_0)$ and $f'(x_0)$ and substitute their values into Eq. (5.4). Thus, $f'(x)=2x$, which yields $f'(2)=4$; also, $f(2)=(2)^2-5=-1$. Substituting into

Eq. (5.4),

$$\begin{aligned} x_1 &= 2 - \frac{-1}{4} \\ &= 2 + \frac{1}{4} \\ &= 2.25 \end{aligned}$$

We now have x_1 with two decimal places in the answer, but it is not necessarily accurate to two decimal places. In order to establish accuracy to three decimal places, we continue obtaining new estimates of x^* until two consecutive estimates, when rounded to three decimal places, have the same values; thus, we need to obtain the value for x_2.

To determine the value for x_2, we have $f(2.25)=(2.25)^5-5=0.0625$ and $f'(2.25)=2(2.25)=4.50$. Therefore,

$$\begin{aligned} x_2 &= 2.25 - \frac{0.0625}{4.50} \\ &= 2.25 - 0.014 \\ &= 2.236 \end{aligned}$$

Using $x_2=2.236$, we determine $f(2.236)=(2.236)^2-5=-0.000304$ and $f'(2.236)=4.472$. Substituting into Eq. (5.4), we obtain

$$\begin{aligned} x_3 &= 2.236 - \frac{-0.000304}{4.472} \\ &= 2.236 + 0.000 \\ &= 2.236 \end{aligned}$$

Therefore, one zero of $f(x)=x^2-5$ accurate to three decimal places is $x^*=2.236$; furthermore, the value of $f(2.236)=-0.000304$ is nearly zero, as we would expect, because we are seeking a root of $f(x)=0$.

Generally it is more convenient and orderly to place the estimates and their associated values in a table for easy reference as the computations progress. Table 5.4 contains the computed values for this example, and we can observe what happens to $f(x)$ with each new approximation. We note that $f(x)$ started at -1 for the original estimate and became 0.0625 for the next estimate. Finally, it became -0.000304 for the final estimate, which is accurate to three decimal places. Thus, we do not have the exact value of the root of the equation; however, we do have a very good approximation, which is generally adequate for most problems.

Table 5.4

i	x_i	$f(x_i)$	$f'(x_i)$	x_{i+1}
0	2	-1	4	2.25
1	2.25	0.0625	4.50	2.236
2	2.236	-0.000304	4.472	2.236

In Example 5.31, we were required only to determine *one zero* of the function $f(x)=x^2-5$; however, from our knowledge of quadratic equations, we know that if there is one root, then either it is a double root or there is another root. We can factor the given equation $x^2-5=0$ into $(x+\sqrt{5})(x-\sqrt{5})=0$, and hence we know both roots exactly. However, we could determine the estimate of the root $x^*=-\sqrt{5}$ in the same manner as we did for the estimate of the root $x^*=\sqrt{5}$, except that we would begin with a different value of x_0 near $-\sqrt{5}$; for example, $x_0=-2$. Unfortunately, most of our problems are not such that we can compare our estimate with the actual root; if this were the case, there would be no need to approximate the root by Newton's method. However, Example 5.31 does illustrate a method for determining the decimal representation of the square root of 5.

If a continuous function is such that $f(a)$ is positive and $f(b)$ is negative (or vice versa) for any two values a and b, we know by Property 4.3 that there is at least one value x^*, as depicted in Fig. 5.39, between a and b such that $f(x^*)=0$. This property of continuous functions is used when seeking the root of an equation. That is, if we can find values a and b for which $f(a)$ and $f(b)$ are opposite in sign, we know that there is at least one root of $f(x)=0$ between a and b; thus, we can choose for the value of x_0 either a or b or some value in between.

Example 5.32 Use Newton's method to determine the real positive zero of $f(x)=x^3-2x-5$ accurate to three decimal places.

Solution We determine that $f(2)=-1$ and $f(3)=16$, which indicates that the root is between 2 and 3 and is considerably nearer 2. Thus, we shall take $x_0=2$ as our original estimate of x^*. Determining the first derivative of $f(x)$, we have

$$f'(x) = 3x^2 - 2$$

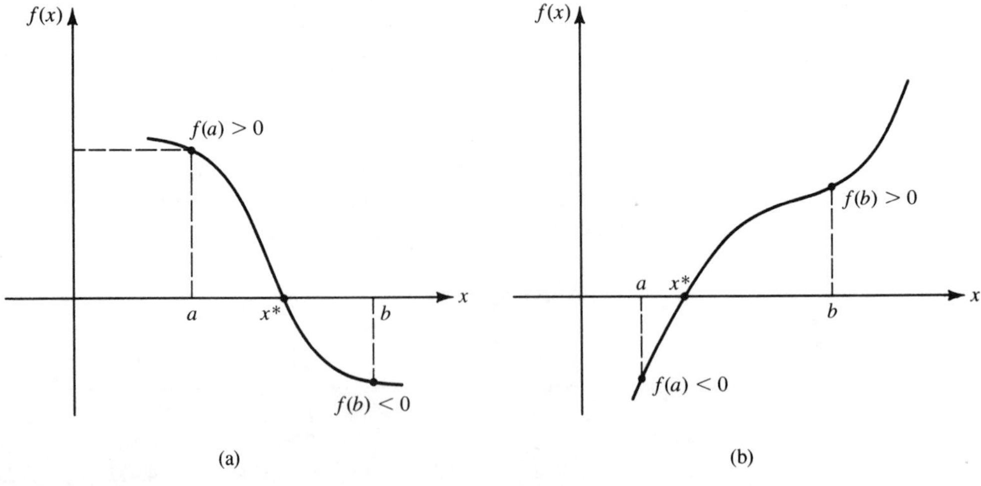

FIGURE 5.39

Furthermore, $f'(2)=3(2)^2-2=10$; substituting into (5.4), we have

$$x_1 = 2 - \frac{-1}{10} = 2.1$$

We enter these values in the first row of Table 5.5.

Table 5.5

i	x_i	$f(x_i)$	$f'(x_i)$	x_{i+1}
0	2	-1	10	2.1
1	2.1	0.061	11.23	2.095
2	2.095	0.00501	11.167	2.095

Next, we compute $f(2.1)=(2.1)^3-2(2.1)-5=0.061$ and $f'(2.1)=3(2.1)^2-2=11.23$. Substituting into (5.4), we obtain

$$\begin{aligned} x_2 &= 2.1 - \frac{0.061}{11.23} \\ &= 2.1 - 0.005 \\ &= 2.095 \end{aligned}$$

We enter these values in the second row of Table 5.5.

In order to determine x_3, we compute $f(2.095)=(2.095)^3-2(2.095)-5=0.00501$ and $f'(2.095)=3(2.095)^2-2=11.167$. Substituting into (5.4), we obtain

$$\begin{aligned} x_3 &= 2.095 - \frac{0.00501}{11.167} \\ &= 2.095 - 0.000 \\ &= 2.095 \end{aligned}$$

We enter these values in the third row of Table 5.5 and note that x_2 and x_3 are the same to three decimal places; furthermore, we note that $f(2.095)=0.00501$, which is near zero. Thus, the estimate of x^* to three decimal places is 2.095.

Newton's method is a particularly beneficial method to use when seeking relative maxima and minima.

Example 5.33 Determine the relative extrema of the function $f(x)=x^4+x^3-13x^2-x+12$ defined on $[0, \infty)$ by finding the critical values accurate to two decimal places.

Solution Using the first derivative to determine the critical values of the function $f(x)$, we have

$$f'(x) = 4x^3 + 3x^2 - 26x - 1$$

$$0 = 4x^3 + 3x^2 - 26x - 1 \tag{5.5}$$

Since Eq. (5.5) is not readily factorable, we check to determine if it has any rational roots. The possible rational roots, p/q, by the rational roots theorem are $\{\pm 1, \pm \frac{1}{4}, \pm \frac{1}{2}\}$. Since we are interested in critical values in $[0, \infty)$, we only need to check the positive roots: 1, $\frac{1}{4}$, and $\frac{1}{2}$. Substituting these values into $4x^3 + 3x^2 - 26x - 1$, we have

$$4(1)^3 + 3(1)^2 - 26(1) - 1 \neq 0$$

$$4(\tfrac{1}{4})^3 + 3(\tfrac{1}{4})^2 - 26(\tfrac{1}{4}) - 1 \neq 0$$

$$4(\tfrac{1}{2})^3 + 3(\tfrac{1}{2})^2 - 26(\tfrac{1}{2}) - 1 \neq 0$$

Hence, we conclude that if there are any solutions of Eq. (5.5) they must be irrational numbers. To determine if there are any solutions we let $g(x) = 4x^3 + 3x^2 - 26x - 1$ and draw a sketch of $g(x)$. By inspecting the graph of $g(x)$ shown in Fig. 5.40, we conclude that there is a zero of the function $g(x)$ or a root of the equation $4x^3 + 3x^2 - 26x - 1 = 0$ between $x = 2$ and $x = 3$.

Selecting $x_0 = 2$ as our first approximation in Newton's method, we determine x_1:

$$\begin{aligned} g(x) &= 4x^3 + 3x^2 - 26x - 1 \\ g(2) &= 4(2)^3 + 3(2)^2 - 26(2) - 1 \\ &= -9 \\ g'(x) &= 12x^2 + 6x - 26 \\ g'(2) &= 12(2)^2 + 6(2) - 26 \\ &= 34 \\ x_1 &= 2 - \frac{-9}{34} = 2.26 \end{aligned}$$

We enter these values in the first row of Table 5.6.

Using x_1 as our second approximation, we determine x_2.

$$\begin{aligned} g(2.26) &= 4(2.26)^3 + 3(2.26)^2 - 26(2.26) - 1 \\ &= 1.736 \\ g'(2.26) &= 12(2.26)^2 + 6(2.26) - 26 \\ &= 48.85 \\ x_2 &= 2.26 - \frac{1.736}{48.85} \\ &= 2.22 \end{aligned}$$

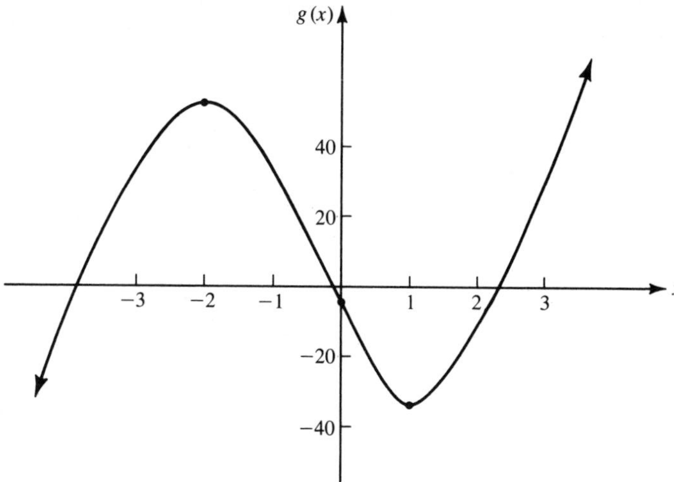

FIGURE 5.40

We enter these values in the second row of Table 5.6 and compute x_3.

$$\begin{aligned} g(2.22) &= 4(2.22)^3 + 3(2.22)^2 - 26(2.22) - 1 \\ &= -1.71 \\ g'(2.22) &= 12(2.22)^2 + 6(2.22) - 26 \\ &= 46.46 \\ x_3 &= 2.22 - \frac{-.171}{46.46} = 2.22 + .004 = 2.224 \end{aligned}$$

The approximations x_2 and x_3 are the same (accurate to two decimal places); hence, the critical value is $x = 2.22$ and the relative extrema is

$$\begin{aligned} f(2.22) &= (2.22)^4 + (2.22)^3 - 13(2.22)^2 - (2.22) + 12 \\ &\doteq -19.06 \end{aligned}$$

We use the second derivative test to determine if $f(2.22)$ is a relative

Table 5.6

	x_i	$g(x_i)$	$g'(x_i)$	x_{i+1}
0	2	−9	34	2.26
1	2.26	1.736	48.85	2.22
2	2.22	−.171	46.46	2.22

maximum or minimum:

$$f''(x) = 12x^2 + 6x - 26$$
$$f''(2.22) > 0$$

We conclude that $f(2.22) = -19.06$ is a relative minimum.

EXERCISES

For each of the functions in Exercises 1–8 determine the zeros indicated correct to three decimal places.

1. $f(x)=x^3-4x+1$ (zero between 1 and 2)
2. $f(x)=x^3+3x-5$ (zero between 1 and 2)
3. $f(x)=2x^3+5x-50$ (zero between 2 and 3)
4. $f(x)=x^3-7x-15$ (zero between 3 and 4)
5. $f(x)=2x^3+2x^2-4x-8$ (zero between 1 and 2)
6. $f(x)=2x^3-3x^2-2x+2$ (zero between 0 and 1)
7. $f(x)=x^4-2x^3+15x-23$ (zero between 1 and 2)
8. $f(x)=x^4-2x^3+x-1$ (zero between 1 and 2)
9. Determine whether the function $f(x)=x^4-7x^2+6x+5$ has a relative maximum or relative minimum between 0 and 1. If it does, determine its value.
10. Determine whether the function in Exercise 9 has a relative maximum or relative minimum between 1 and 2. If it does, determine its value.
11. Determine whether the function in Exercise 9 has a relative maximum or relative minimum between -3 and -2. If it does, determine its value.
12. Use the information obtained in Exercises 9–11 and graph the function in Exercise 9.

IMPORTANT TERMS AND CONCEPTS

Absolute maximum
Absolute minimum
Approximation
Concave downward
Concave upward
Concavity
Critical point
Critical value
Curve sketching
Extrema
First derivative test
Function test
Hypercritical value
Inflection point
Iteration procedure
Marginal analysis
Newton's method
Optimality condition of marginal analysis
Relative maximum
Relative minimum
Roots
Second derivative test
Zeros of a function

SUMMARY OF RULES AND FORMULAS

Rule 5.1: Increasing and Decreasing Functions

Assuming that the function $f(x)$ has a derivative at every point in an open interval (a,b),

(a) If $f'(x)>0$ for all x in (a,b), then the function $f(x)$ is increasing on (a,b), and if $f(x)$ is increasing on (a,b), then for all x in (a,b), $f'(x)>0$ or $f'(x)=0$ for a finite number of x's.
(b) If $f'(x)<0$ for all x in (a,b), then the function $f(x)$ is decreasing on (a,b), and if $f(x)$ is decreasing on (a,b), then for all x in (a,b), $f'(x)<0$ or $f'(x)=0$ for a finite number of x's.

Rule 5.2: The Function Test

If $f(x)$ is a continuous function for which x^* is the only critical value in the interval $[a,b]$, then

(a) $f(x^*)$ is a relative maximum if $f(x^*)>f(a)$ and $f(x^*)>f(b)$.
(b) $f(x^*)$ is a relative minimum if $f(x^*)<f(a)$ and $f(x^*)<f(b)$.
(c) Otherwise, $f(x^*)$ is not an extremum.

Rule 5.3: The First Derivative Test

If $f(x)$ is a continuous function for every value of x in the interval $[a,b]$ and x^* is the only critical value in the interval (a,b), then

(a) $f(x^*)$ is a relative maximum if $f'(a)>0$ and $f'(b)<0$.
(b) $f(x^*)$ is a relative minimum if $f'(a)<0$ and $f'(b)>0$.
(c) Otherwise, $f(x^*)$ is not an extremum.

Rule 5.4: The Second Derivative Test

If the first and second derivatives of $f(x)$ exist for every value of x in the interval $[a,b]$ and x^* is a critical value in the interval (a,b), then

(a) $f(x^*)$ is a relative maximum if $f''(x^*)<0$.
(b) $f(x^*)$ is a relative minimum if $f''(x^*)>0$.
(c) if $f''(x^*)=0$, no conclusion can be drawn.

Rule 5.5: Absolute Extrema

To find the absolute maximum (or minimum) value of a continuous function defined on a closed interval $[a,b]$, determine the critical values $x_0,x_1,\ldots,x_n$ in the interval and calculate the values of the function $f(a),f(x_0),f(x_1),\ldots,f(x_n)$ and $f(b)$. The largest of these values will be the absolute maximum and the smallest will be the absolute minimum.

Rule 5.6: Concavity

(a) If $f''(x)<0$ for each value of x in an interval (a,b), then the graph of $f(x)$ is concave downward in the interval (a,b).
(b) If $f''(x)>0$ for each value of x in an interval (a,b), then the graph of $f(x)$ is concave upward in the interval (a,b).
(c) If $f''(x^*)=0$ or is undefined and the graph of the function changes from concave upward on one side of x^* to concave downward on the other side (or from concave downward to concave upward), the point $(x^*,f(x^*))$ is an inflection point.

6 DIFFERENTIAL CALCULUS: ADVANCED METHODOLOGY

6.1 INTRODUCTION

In Chapters 4 and 5 we learned to determine the derivatives of algebraic functions and to apply the differential calculus to solve problems involving these functions.

In Chapter 3 we introduced two nonalgebraic or *transcendental functions*: exponential and logarithmic, and found that these functions are often used in management, social science, and biology to describe growth and decay. In this chapter we will develop the necessary rules to find the derivatives of these functions, and as we did in Chapter 5, apply the differential calculus to solve real-world problems. Only in this case the mathematical models for these problems will be exponential or logarithmic functions, either alone or combined with algebraic functions.

6.2 THE DERIVATIVES OF EXPONENTIAL FUNCTIONS

In this section we will develop the rules necessary to find the *derivatives of exponential functions*. You will recall from Sec. 3.5 that we demonstrated that an exponential function with base b could be converted into one with base e by the relationship

$$b^x = e^{x \ln b} \tag{6.1}$$

One of the reasons for converting an exponential function to one with base e is that certain results in calculus are simpler for exponential functions with this base. The simplest exponential function with base e is $f(x)=e^x$.

We found in Chapter 3 that the graph of $f(x)=e^x$ is drawn as shown in Fig. 6.1 where we have also indicated the domain and range of the function.

To find the derivative of an exponential function, we begin by considering the function $f(x)=e^x$. Since the derivative can be interpreted as the slope of the curve, we can, by inspection of Fig. 6.1, draw some conclusions about the derivative of $f(x)=e^x$ before we actually have a formula for $f'(x)$.

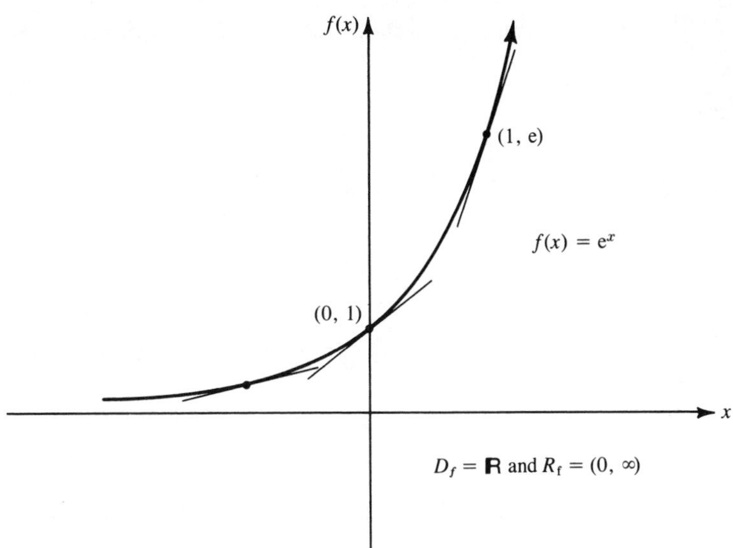

FIGURE 6.1

By inspection of the graph we can conclude that a tangent line can be drawn to the curve at every real number x and that each of the tangent lines so drawn will have a positive slope. This implies that $f'(x)$ will be defined for every real number x and that $f'(x)>0$ for all x. Furthermore, we can conclude from Fig. 6.1 that the slopes of the tangent lines increase as x increases. This implies that the derivative $f'(x)$ must be an increasing function. Now, let us find a formula for $f'(x)$ and verify these conclusions.

Unfortunately, we cannot find the derivative of $f(x)=e^x$ by applying Rules 4.1–4.13. These rules apply only to algebraic functions where the variable x is a base and not an exponent as in Eq. (6.1). To find the derivative of $f(x)=e^x$, we must begin with the basic definition of a derivative, Definition 4.3. To apply Definition 4.3 we begin with the general statement that if $f(x)$ is differentiable, then

$$f'(x) = \lim_{h \to 0}\left[\frac{f(x+h)-f(x)}{h}\right] \tag{6.2}$$

If $f(x)=e^x$, then $f(x+h)=e^{x+h}$ and

$$\begin{aligned} f'(x) &= \lim_{h \to 0}\left[\frac{e^{x+h}-e^x}{h}\right] \\ &= \lim_{h \to 0}\left[\frac{e^x \cdot e^h - e^x}{h}\right] \\ &= \lim_{h \to 0}\left[\frac{e^x(e^h-1)}{h}\right] \end{aligned} \tag{6.3}$$

Since e^x is independent of h, we can rewrite the right-hand side of Eq. (6.3) as

$$f'(x) = e^x \cdot \lim_{h \to 0} \left(\frac{e^h - 1}{h} \right) \tag{6.4}$$

It can be shown that

$$\lim_{h \to 0} \left(\frac{e^h - 1}{h} \right) = 1$$

(The proof of this statement is beyond the scope of this textbook.) Therefore

$$\begin{aligned} f'(x) &= e^x \cdot (1) \\ &= e^x \end{aligned}$$

Thus we now have a rule for differentiating $f(x) = e^x$.

Rule 6.1 If $f(x) = e^x$, then $f'(x) = e^x$, or $D_x(e^x) = e^x$.

This result is startling! The function $f(x) = e^x$ is such that its derivative is the same as the original function. Let us determine some of the properties of the derivative by considering the formula $f'(x) = e^x$. First, since e^x is defined for all real values of x, we can conclude that $f'(x)$ is defined for every real number x. Secondly, since e^x is an increasing function, $f'(x)$ is increasing. You will note that these properties agree exactly with our conclusions based on the graph of $f(x) = e^x$ in Fig. 6.1.

At first glance, Rule 6.1 does not appear to be very important because there is only one function of the form $f(x) = e^x$; however, we can use Rule 6.1 in conjunction with the chain rule (Rule 4.14) to determine the derivative of a rather large class of functions. For example, suppose $f(x) = e^{5x}$. Treating this as a composite function for which the variable is $u = u(x) = 5x$, we can write $f(x) = g(u) = e^u$. Then employing Rule 6.1, we determine the derivative with respect to u of $g(u)$ to be

$$g'(u) = e^u$$

Applying the chain rule, we obtain the derivative with respect to x of $f(x)$ to be

$$\begin{aligned} f'(x) &= D_x[\, g(u(x))] \\ &= D_u g(u) \cdot D_x u(x) \\ &= e^u D_x(5x) \\ &= e^u \cdot 5 \end{aligned}$$

Substituting $u(x)=5x$ in the exponent for u we have

$$f'(x) = 5e^{5x}$$

Similarly, if $f(x)=e^{u(x)}$ and the derivative with respect to x of $u(x)$ exists, we apply Rule 6.1 and the chain rule to obtain

$$\begin{aligned} f'(x) &= D_x(e^{u(x)}) = D_u e^u \cdot D_x u(x) = e^{u(x)} D_x u(x) \\ &= e^{u(x)} \cdot u'(x) \end{aligned}$$

Thus we have a rule for determining the derivative of a function of the form $f(x)=e^{u(x)}$.

Rule 6.2 The derivative with respect to x of the exponential function $f(x)=e^{u(x)}$, provided $u'(x)$ exists, equals $e^{u(x)}$ times the derivative of $u(x)$; that is, if $f(x)=e^{u(x)}$ where $u'(x)$ exists, then

$$\begin{aligned} f'(x) &= D_x(e^{u(x)}) = e^{u(x)} \cdot D_x u(x) \\ &= e^{u(x)} \cdot u'(x) \end{aligned}$$

Example 6.1 Determine the derivative with respect to x of each of the following functions.

(a) $f(x)=e^{7x}$ (b) $f(x)=e^{x^3}$
(c) $g(x)=e^{x^2+3x+1}$ (d) $s(x)=e^{6x^3+4\sqrt{x}}$

Solution Employing Rule 6.2 and the rules of differentiation from Chapter 4, we obtain

(a) $f'(x)=D_x(e^{7x})=e^{7x}D_x(7x)=e^{7x}\cdot 7=7e^{7x}$
(b) $f'(x)=D_x(e^{x^3})=e^{x^3}D_x(x^3)=e^{x^3}\cdot 3x^2=3x^2e^{x^3}$
(c) $g'(x)=D_x(e^{x^2+3x+1})=(e^{x^2+3x+1})D_x(x^2+3x+1)$
$=(2x+3)e^{x^2+3x+1}$
(d) $s'(x)=D_x(e^{6x^3+4x^{1/2}})=e^{6x^3+4x^{1/2}}D_x(6x^3+4x^{1/2})$
$=(18x^2+2x^{-1/2})e^{6x^3+4x^{1/2}}$

On several occasions we have stated that the advantage of exponential functions with base e is that they yield relatively simple results. Rules 6.1 and 6.2 verify the simplicity of the results for exponential functions with base e. Let us now consider the derivative of an exponential function for which the base is b. To determine $D_x b^x$, we use Eq. (6.1) and Rule 6.2 to obtain

$$D_x b^x = D_x(e^{x\ln b}) = e^{x\ln b}\ln b$$

However, since $b^x=e^{x\ln b}$, we make this substitution on the right side and obtain the following rule.

Rule 6.3 If b is a positive number that does not equal 1 and $f(x)=b^x$, then $f'(x)=D_x(b^x)=b^x\ln b$

Example 6.2 Determine the derivative with respect to x for each of the following functions. Use Table A.2 in the Appendix to determine the numerical values of the natural logarithms.

(a) $f(x)=6^x$ (b) $g(x)=10^x$
(c) $s(x)=2.51^x$ (d) $h(x)=11^x$

Solution

(a) $f'(x)=D_x(6^x)=6^x\ln 6=(1.79176)\cdot 6^x$
(b) $g'(x)=D_x(10^x)=10^x\ln 10=(2.30259)\cdot 10^x$
(c) $s'(x)=D_x(2.51^x)=2.51^x\ln 2.51=(0.92028)\cdot 2.51^x$
(d) $h'(x)=D_x(11^x)=11^x\ln 11=(2.39790)\cdot 11^x$

From Example 6.2 it is apparent that the derivatives of exponential functions with base e are simpler than with any other base. Since there exist tables of e^x, it is also easy to evaluate these derivatives at a particular value of x.

To determine the derivative of a composite function of the form $f(x)=b^{u(x)}$, we simply employ Rule 6.3 and the chain rule, which yields the following rule.

Rule 6.4 If $u'(x)$ exists and $f(x)=b^{u(x)}$, then

$$f'(x)=D_x(b^{u(x)})=b^{u(x)}\ln b\cdot D_xu(x)$$
$$=b^{u(x)}\ln b\cdot u'(x)$$

Example 6.3 Determine the derivative with respect to x of each of the following functions.

(a) $f(x)=10^{x^2}$ (b) $g(x)=5^{3x^2+2x}$
(c) $s(x)=3^{\sqrt{x}}$ (d) $h(x)=6^{3e^{x^2}}$

Solution

(a) $f'(x)=D_x(10^{x^2})=10^{x^2}\ln 10(2x)=(4.60518)x\cdot 10^{x^2}$
(b) $g'(x)=D_x(5^{3x^2+2x})=5^{3x^2+2x}\ln 5(6x+2)$
$=(1.60944)(6x+2)5^{3x^2+2x}$
(c) $s'(x)=D_x(3^{x^{1/2}})=3^{x^{1/2}}\ln 3(\frac{1}{2}x^{-1/2})=\dfrac{.54931}{\sqrt{x}}3^{\sqrt{x}}$
(d) $h'(x)=D_x(6^{3e^{x^2}})=6^{3e^{x^2}}\ln 6D_x(3e^{x^2})=6^{3e^{x^2}}\ln 6(3e^{x^2}\cdot 2x)$
$=(10.75056)x\cdot 6^{3e^{x^2}}\cdot e^{x^2}$

Example 6.4 Determine the first four derivatives of $f(x)=e^x$.

Solution

$$f'(x) = D_x(e^x) = e^x$$
$$f''(x) = D_x(e^x) = e^x$$
$$f'''(x) = D_x(e^x) = e^x$$
$$f^{(4)}(x) = D_x(e^x) = e^x$$

The exponential functions $e^{u(x)}$ and $b^{u(x)}$ can be combined under the operations of addition, subtraction, multiplication, and division with various algebraic functions. To determine the derivatives of these functions, we employ, as needed, the product rule, the quotient rule, and the chain rule. As in most complicated problems, our task is to decide which rule will be employed first. The following example will help to establish the proper procedure.

Example 6.5 Determine the derivative of each of the following functions.

(a) $f(x)=5e^{x^2}$ (b) $f(x)=2x^3e^{4x}$

(c) $g(x)=\dfrac{x^2}{e^{3x}}$ (d) $s(x)=(x+e^x)^{1/3}$

(e) $h(x)=(x+1)^2 3^x$ (f) $h(x)=(x+10^x)^2$

Solution

(a) $f(x)=5e^{x^2}$ is of the form $f(x)=k\cdot g(x)$. Therefore,

$$f'(x) = k\cdot g'(x) \quad \text{or} \quad f'(x) = 5\cdot D_x(e^{x^2}) = 5e^{x^2}(2x) = 10xe^{x^2}$$

(b) $f(x)=2x^3\cdot e^{4x}$ is in the form of a product. Therefore,

$$\begin{aligned} f'(x) &= (2x^3)D_x(e^{4x}) + e^{4x}D_x(2x^3) \\ &= 2x^3e^{4x}(4) + e^{4x}(6x^2) \\ &= 8x^3e^{4x} + 6x^2e^{4x} \end{aligned}$$

(c) $g(x)=x^2/e^{3x}$ is in the form of a quotient. Therefore,

$$\begin{aligned} g'(x) &= \frac{e^{3x}D_x(x^2)-x^2D_x(e^{3x})}{(e^{3x})^2} = \frac{e^{3x}(2x)-x^2e^{3x}\cdot 3}{e^{6x}} \\ &= \frac{e^{3x}(2x-3x^2)}{e^{6x}} = \frac{2x-3x^2}{e^{3x}} \end{aligned}$$

(*Note*: We could have written $g(x)$ as $g(x)=x^2e^{-3x}$ and used the product rule.)

(d) $s(x)=(x+e^x)^{1/3}$ is a composite where $u(x)=(x+e^x)$. Therefore,

$$\begin{aligned} s'(x) &= \tfrac{1}{3}(x+e^x)^{-2/3}D_x(x+e^x) \\ &= \tfrac{1}{3}(x+e^x)^{-2/3}(1+e^x) \end{aligned}$$

(e) $h(x)=(x+1)^2 3^x$ is in the form of a product. Therefore,

$$\begin{aligned} h'(x) &= (x+1)^2 D_x(3^x) + 3^x D_x(x+1)^2 \\ &= (x+1)^2 3^x \ln 3 + 3^x 2(x+1)^1(1) \\ &= (x+1)3^x[(x+1)\ln 3+2] \end{aligned}$$

(f) $h(x)=(x+10^x)^2$ is in the form of a composite where $u(x)=(x+10^x)$. Therefore,

$$\begin{aligned} h'(x) &= 2(x+10^x)^1 D_x(x+10^x) \\ &= 2(x+10^x)\ (1+10^x \ln 10) \end{aligned}$$

EXERCISES

Determine the derivative for each of the functions in Exercises 1–24. (Use Table A.2 to determine numerical values for natural logarithms.)

1. $f(x)=e^{3x}$

2. $g(x)=e^{1/2x}$

3. $g(x)=e^{x^2}$

4. $f(t)=e^{-5t+2}$

5. $y=e^{x^3-1/2x^2}$

6. $y=e^{x^2-3x+1}$

7. $S(t)=te^t$

8. $S(x)=x^2e^x$

9. $f(x)=(x^2-1)e^{-x}$

10. $h(x)=\dfrac{e^x}{1+x}$

11. $f(x)=\dfrac{e^x}{1+x^2}$

12. $g(x)=\dfrac{x^2-1}{e^x}$

13. $g(x)=\sqrt{(2x+e^x)}$

14. $f(x)=\sqrt[3]{x^3+e^{-x}+5}$

15. $f(x)=2e^{1/x}$

16. $g(t)=4^t$

17. $y=10^{3x}$

18. $y=10^{\sqrt{x}}$

19. $f(x)=8^{x^2+x}$

20. $S(x)=(1+e^{-x})^2$

21. $f(t)=100-80e^{-.5t}$

22. $g(t)=\dfrac{e^t-e^{-t}}{2}$

23. $f(x)=\dfrac{1{,}000}{10+5e^{-.01x}}$

24. $S(t)=\dfrac{200}{5+10e^{-.5t}}$

Determine the first and second derivatives for each function in Exercises 25–28.

25. $f(x)=e^{x^2}$

26. $g(x)=3e^{-x^3}$

27. $g(x)=x^2e^x$

28. $h(x)=10^{x^2+x}$

In Exercises 29–36 determine where each function is increasing, decreasing, concave downward, and concave upward.

29. $f(x)=e^{3x}$

30. $g(x)=e^{-2x}$

31. $g(x)=xe^x$

32. $f(x)=xe^{-x}$

33. $h(x)=e^x-e^{-x}$

34. $g(x)=x^2e^x$

35. $f(x)=100-80e^{-.5x}$

36. $s(x)=80-100e^{-.01x}$

In Exercises 37–40 determine the slope of the curve at the given x values.

37. $f(x)=e^x$: $x=-1,0,1$

38. $f(x)=xe^x$: $x=-1,0,1$

39. $g(x)=e^{x^2}$: $x=-1,0,1$

40. $g(x)=xe^{x^2}$: $x=-1,0,1$

6.3 THE DERIVATIVES OF LOGARITHMIC FUNCTIONS

Having determined the rules for differentiating exponential functions in the last section, let us turn our attention to developing rules for *differentiating logarithmic functions*.

Knowing that certain results in calculus are simpler if we work with a base of e, and that the function $f(x)=\ln x$ is the inverse of $g(x)=e^x$, we will begin our considerations with the natural logarithmic function, $f(x)=\ln x$. We know the graph of the logarithmic function $f(x)=\ln x$ can be determined by reflecting the graph of $g(x)=e^x$ across the line $y=x$ as shown in Fig. 6.2. The domain and range of the logarithmic function are also shown in Fig. 6.2.

Let us inspect the graph of the natural logarithmic function $f(x)=\ln x$ and predict some of the characteristics of its derivative before we derive a formula for $f'(x)$. By inspection of Fig. 6.2 we can conclude that a tangent line can be drawn to the curve at every real number x in the domain $(0,\infty)$ and that each of these tangent lines has a positive slope. This implies that $f'(x)$ exists for every real number x in the domain of the function and that $f'(x)>0$ for all such x. Furthermore, we can conclude from inspection of Fig. 6.2 that the slopes of the tangent lines decrease as x increases, implying that the derivative $f'(x)$ will be a decreasing function. Let us now find a formula for $f'(x)$ and verify these conclusions about the derivative.

To determine the derivative of the natural logarithmic function $f(x)=\ln x$ we could begin with the basic definition of the derivative; however, it is much easier to use the results of the previous section. By Definition 3.1, $f(x)=\ln x$

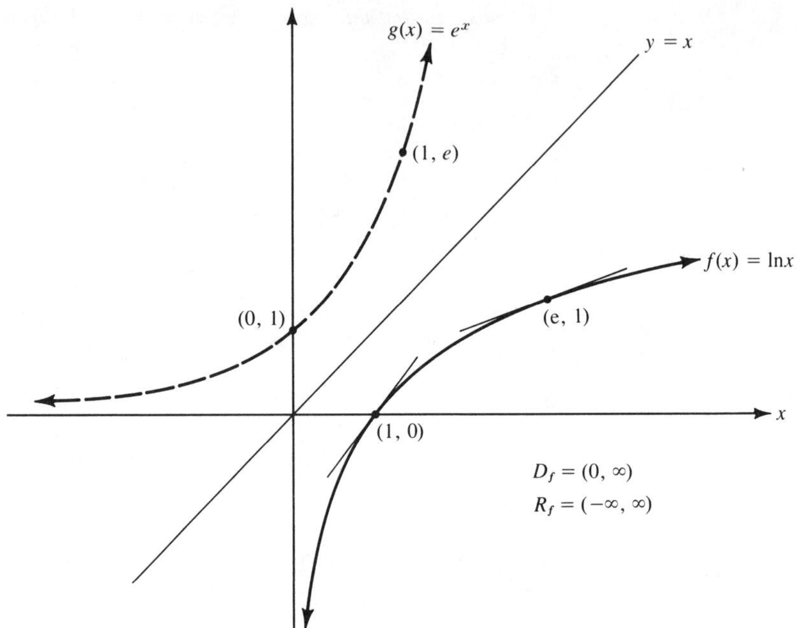

FIGURE 6.2

is equivalent to

$$x = e^{f(x)} \qquad \text{where} \quad x > 0 \tag{6.5}$$

Differentiating the left and right sides of Eq. (6.5) with respect to x, we find

$$D_x(x) = D_x(e^{f(x)})$$

or

$$1 = D_x(e^{f(x)}) \tag{6.6}$$

Using Rule 6.2 and assuming from our previous geometric consideration that $f'(x)$ exists for all $x > 0$, we conclude that

$$\begin{aligned} D_x(e^{f(x)}) &= e^{f(x)} \cdot D_x f(x) \\ &= e^{f(x)} \cdot f'(x) \end{aligned}$$

Substituting this expression for the right side of Eq. (6.6), we have

$$1 = e^{f(x)} \cdot f'(x)$$

Solving for $f'(x)$ and substituting x for $e^{f(x)}$ by Eq. (6.5), we obtain

$$f'(x) = \frac{1}{e^{f(x)}}$$

or

$$f'(x) = \frac{1}{x}$$

whenever $x > 0$, which justifies the following rule.

Rule 6.5 For all real $x > 0$, the derivative of $f(x) = \ln x$ is

$$f'(x) = \frac{1}{x} \qquad \text{or} \qquad D_x(\ln x) = \frac{1}{x}$$

Let us determine some of the properties of the derivative of $f(x) = \ln x$ by examining the formula $f'(x) = 1/x$. First, since $1/x$ is defined for all real $x > 0$, we can conclude that $f'(x)$ is defined for all x in the domain of the function $f(x)$. Secondly, since $(1/x) > 0$ for all x values in this domain, $f'(x) > 0$, and lastly since $1/x$ is a decreasing function, $f'(x)$ is a decreasing function. You will note that these properties agree with our conclusions based on the graph of the function $f(x) = \ln x$ shown in Fig. 6.2.

To determine the derivative for the more general function $f(x) = \ln[u(x)]$ we apply Rule 6.5 and the chain rule as follows. If we let $g(u) = \ln u$ where $u = u(x)$, then $f(x) = \ln[u(x)]$ can be written as

$$f(x) = g(u(x))$$

By Rule 4.14

$$\begin{aligned} f'(x) &= D_u g(u) \cdot D_x u(x) \\ &= D_u(\ln u) D_x u(x) \\ &= \frac{1}{u} \cdot D_x u(x) \\ &= \frac{1}{u(x)} \cdot u'(x) \end{aligned}$$

This gives us the following rule:

Rule 6.6 For a differentiable function $u(x) > 0$ the derivative of $f(x) = \ln u(x)$ is

$$f'(x) = \frac{1}{u(x)} \cdot u'(x) \qquad \text{or} \qquad D_x[\ln u(x)] = \frac{1}{u(x)} \cdot D_x u(x)$$

Example 6.6 For positive values of x, determine the derivative with respect to x of each of the following functions.

(a) $f(x)=\ln x^2$

(b) $h(x)=\ln(x^2+3x+4)$

(c) $s(x)=\ln\sqrt{x}$

(d) $q(x)=\ln\dfrac{2x+1}{3x+2}$

Solution Applying Rule 6.6 we obtain

$$\text{(a)}\quad f'(x) = D_x(\ln x^2) = \frac{1}{x^2}\cdot D_x(x^2) = \frac{2x}{x^2} = \frac{2}{x}$$

$$\text{(b)}\quad h'(x) = D_x[\ln(x^2+3x+4)] = \frac{1}{x^2+3x+4}\cdot D_x(x^2+3x+4)$$

$$= \frac{2x+3}{x^2+3x+4}$$

$$\text{(c)}\quad s'(x) = D_x(\ln\sqrt{x}\,) = \frac{1}{\sqrt{x}}\cdot D_x(\sqrt{x}\,) = \frac{1}{\sqrt{x}}\left(\frac{1}{2}x^{-1/2}\right) = \frac{1}{2x}$$

$$\text{(d)}\quad q'(x) = D_x\left[\ln\left(\frac{2x+1}{3x+2}\right)\right] = \frac{1}{\left(\dfrac{2x+1}{3x+2}\right)}\cdot D_x\left(\frac{2x+1}{3x+2}\right)$$

$$= \left(\frac{3x+2}{2x+1}\right)\left[\frac{(3x+2)(2)-(2x+1)(3)}{(3x+2)^2}\right]$$

$$= \frac{1}{(2x+1)(3x+2)}$$

Alternative ways of determining the derivatives for parts (a), (c), and (d) involve the use of rules for logarithms of powers and quotients, i.e., the rules given in Eqs. (3.16) and (3.17), respectively. Applying these rules, we differentiate in the following manner:

$$\text{(a)}\qquad D_x(\ln x^2) = D_x(2\ln x) = 2D_x(\ln x) = \frac{2}{x}$$

$$\text{(c)}\qquad D_x(\ln\sqrt{x}\,) = D_x(\ln x^{1/2}) = D_x\left(\frac{1}{2}\ln x\right) = \frac{1}{2}D_x(\ln x) = \frac{1}{2x}$$

$$\text{(d)}\quad D_x\left(\ln\frac{2x+1}{3x+2}\right) = D_x[\ln(2x+1)-\ln(3x+2)]$$

$$= D_x[\ln(2x+1)] - D_x[\ln(3x+2)]$$

$$= \frac{1}{2x+1}\cdot 2 - \frac{1}{3x+2}\cdot 3$$

$$= \frac{1}{(2x+1)(3x+2)}$$

Example 6.6 illustrates the advantage of using the rules for logarithms; usually, it is more convenient to apply the rules given in Eqs. (3.15)–(3.17) before differentiating.

As in the case for exponential functions, first we determined the derivatives for logarithmic functions with base e. Now, to determine the derivative of $f(x)=\log_b x$ where $b>0$ and $\neq 1$, we follow the same procedure used to determine Rule 6.5. By Definition 3.1, $f(x)=\log_b x$ is equivalent to

$$x = b^{f(x)} \tag{6.7}$$

By Eq. (6.1) we can write $b^{f(x)}$ as $e^{f(x)\ln b}$ and Eq. (6.7) becomes

$$x = e^{f(x)\ln b} \tag{6.8}$$

Differentiating both left and right sides of Eq. (6.8) with respect to x and solving for $f'(x)$ as before, we have

$$1 = e^{f(x)\ln b} \cdot f'(x)\ln b$$

or

$$f'(x) = \frac{1}{e^{f(x)\ln b} \cdot \ln b}$$

and

$$f'(x) = \frac{1}{x\ln b}$$

which justifies the following rule.

Rule 6.7 For positive x the derivative of $f(x)=\log_b x$ is

$$f'(x) = \frac{1}{x\ln b} \quad \text{or} \quad D_x(\log_b x) = \frac{1}{x\ln b}$$

Using Rule 6.7 to determine the derivative of the common logarithmic function $f(x)=\log_{10} x$, we obtain

$$f'(x) = \frac{1}{x\ln 10}$$

The numerical value of $\ln 10$ to five decimal places from Table A.2 is 2.30259 and the derivative of $f(x)=\log_{10} x$ is approximated by

$$f'(x) \doteq \frac{1}{2.30259x} \quad \text{or} \quad \frac{.434294}{x}$$

(At this point it is obvious, as we stated, that certain results in calculus are simpler if e is used as the base of exponential and logarithmic functions.)

Example 6.7 Determine derivatives for the following functions and evaluate the constant $\ln b$.

(a) $f(x)=\log_{10}x$ (b) $f(x)=\log_2 x$
(c) $s(x)=\log_5 x$ (d) $g(x)=\log_7 x$

Solution Using Rule 6.7 and Table A.2 we obtain

(a) $f'(x)=\dfrac{1}{x\ln 10}=\dfrac{1}{2.30259x}$

(b) $f'(x)=\dfrac{1}{x\ln 2}=\dfrac{1}{.69315x}$

(c) $s'(x)=\dfrac{1}{x\ln 5}=\dfrac{1}{1.60944x}$

(d) $g'(x)=\dfrac{1}{x\ln 7}=\dfrac{1}{1.94591x}$

By using Rule 6.7 and the chain rule, we obtain the rule for differentiating $f(x)=\log_b u(x)$.

Rule 6.8 For $u(x)>0$ the derivative of $f(x)=\log_b u(x)$ is

$$f'(x) = \frac{1}{u(x)\ln b}\cdot u'(x)$$

or

$$D_x[\log_b u(x)] = \frac{1}{u(x)\ln b}\cdot D_x u(x)$$

Example 6.8 For positive x, determine the derivative with respect to x of each of the following functions. Do not evaluate $\ln b$.

(a) $f(x)=\log_{10}x^2$ (b) $g(x)=\log_5 x^4$
(c) $q(x)=\log_3(\dfrac{x^2}{x+1})$ (d) $p(x)=\log_{11}(x+2)(x^2+1)$

Solution

(a) $f'(x) = D_x(\log_{10}x^2) = D_x(2\log_{10}x) = \dfrac{2}{x\ln 10}$

(b) $g'(x) = D_x(\log_5 x^4) = D_x(4\log_5 x) = \dfrac{4}{x\ln 5}$

(c) $q'(x) = D_x\left[\log_3\left(\dfrac{x^2}{x+1}\right)\right] = D_x[2\log_3 x-\log_3(x+1)]$

$$= \frac{2}{x\ln 3}-\frac{1}{(x+1)\ln 3}$$

$$= \frac{2x+2-x}{x(x+1)\ln 3}$$

$$= \frac{x+2}{x(x+1)\ln 3}$$

(d)

$$p'(x) = D_x\left[\log_{11}(x+2)(x^2+1)\right] = D_x\left[\log_{11}(x+2)+\log_{11}(x^2+1)\right]$$

$$= \frac{1}{(x+2)\ln 11} + \frac{1}{(x^2+1)\ln 11}\cdot(2x)$$

$$= \frac{x^2+1+2x^2+4x}{(x+2)(x^2+1)\ln 11}$$

$$= \frac{3x^2+4x+1}{(x+2)(x^2+1)\ln 11}$$

Example 6.9 Determine the first four derivatives of $f(x)=\ln x$.

Solution

$$f'(x) = D_x(\ln x) = \frac{1}{x}$$

$$f''(x) = D_x f'(x) = D_x(x^{-1}) = -x^{-2}$$

$$f'''(x) = D_x f''(x) = D_x(-x^{-2}) = 2x^{-3}$$

$$f^{(4)}(x) = D_x f'''(x) = D_x(2x^{-3}) = -6x^{-4}$$

Example 6.10 Determine the first and second derivatives of

$$f(x) = \ln\left(\frac{x^2+2}{x^3+4x}\right)$$

Solution

$$f(x) = \ln\left(\frac{x^2+2}{x^3+4x}\right) = \ln(x^2+2) - \ln(x^3+4x)$$

$$f'(x) = \frac{2x}{x^2+2} - \frac{(3x^2+4)}{x^3+4x}$$

$$f''(x) = \frac{(x^2+2)(2)-2x(2x)}{(x^2+2)^2} - \frac{(x^3+4x)(6x)-(3x^2+4)(3x^2+4)}{(x^3+4x)^2}$$

$$= \frac{-2x^2+4}{(x^2+2)^2} + \frac{3x^4+16}{(x^3+4x)^2}$$

EXERCISES

Determine the derivative for each of the functions in Exercises 1–20. (*Use Table A.2 to determine numerical values for natural logarithms.*)

1. $f(x)=3\ln x$

2. $f(x)=\ln(x^2+1)$

3. $g(x)=\ln(3x^2)$

4. $g(x)=\ln\left(\frac{3x+1}{x}\right)$

5. $s(x)=2\ln(x^3-1)$

6. $p(x)=5\ln(x^2+2x+3)$

7. $s(t)=t\ln t$

8. $y=\ln\sqrt{x^2+5}$

9. $f(x)=\log_{10}(2x+5)$

10. $f(x)=\log_{10}\left(\frac{x}{x^2+1}\right)$

11. $h(x)=e^x\ln x$

12. $s(x)=(x+\ln x)^2$

13. $s(t)=t^2\ln(t-1)$

14. $y=\ln(x^2+3)^3$

15. $y=[\ln(x^2+3)]^3$

16. $y=\ln(2+e^x)$

17. $y=\log_5(x+1)$

18. $f(x)=\log_2(x^2+5x)$

19. $g(x)=\ln(3x^2+4)^{1/2}$

20. $g(x)=[\ln(3x^2+4)]^{1/2}$

Determine the first and second derivatives for each of the functions in Exercises 21–26.

21. $f(x)=\ln x^2$

22. $f(x)=x\ln x$

23. $g(x)=\ln\left(\frac{x^2+2}{x}\right)$

24. $s(x)=x^2\ln(x+1)$

25. $f(x)=[\ln(x+1)]^2$

26. $f(x)=\ln(x+1)^2$

In Exercises 27–30 *specify the domain of the function and determine where each function is increasing, decreasing, concave upward, and concave downward.*

27. $f(x)=\ln x$

28. $g(x)=\ln(x-5)$

29. $f(x)=x\ln x$

30. $h(x)=\ln(x+1)^2$

Determine the slope of each of the curves in Exercises 31–35 *at the given x values.*

31. $f(x)=\ln x$: $x=\frac{1}{2},1,2$

32. $g(x)=x\ln x$: $x=\frac{1}{2},1,2$

33. $f(x)=\ln x^2$: $x=-1,1,-2,2$

34. $y=[\ln(x+1)]^2$: $x=0,1,2$

35. $y=\ln(x+1)^2$: $x=0,1,2$

6.4 APPLICATIONS TO GROWTH AND DECAY MODELS

Now that we have rules for finding the derivatives of exponential and logarithmic functions, in this section we will consider real-world problems where the mathematical models are exponential or logarithmic functions. We will be primarily concerned with the exponential functions since they have many more practical applications. In the next section we will find additional uses for the derivatives of logarithmic functions.

Initially, let us consider a problem where we will use the derivative as a tool to sketch the graph of an exponential function.

Example 6.11 Sketch the graph of the function $f(x)=(x-1)^2 \cdot e^x$.

Solution To graph this function, which is the product of an algebraic function and an exponential function, let us follow the procedure developed in Chapter 5 and summarized in Tables 5.1, 5.2, and 5.3.

1. *Extent:* Since both $(x-1)^2$ and $e^x \to \infty$ as $x \to \infty$, the product $(x-1)^2 \cdot e^x \to \infty$; therefore, we state $\lim_{x\to\infty} f(x) = +\infty$. Now since $e^x \to 0$ and $(x-1)^2 \to \infty$ as $x \to -\infty$, there is some question about the product. As an aid to finding this limit, let us determine the value of $(x-1)^2 \cdot e^x$ when $x=-100$. Evaluating $f(-100)$, we find

$$f(-100) = (-100-1)^2 \cdot e^{-100} = \frac{(-101)^2}{e^{100}} \doteq \frac{10^4}{e^{100}}$$

By computation we can show that $e^{10} \doteq 2 \times 10^4$. Therefore, $e^{100} = (e^{10})^{10} \doteq (2 \times 10^4)^{10} = 2^{10} \cdot 10^{40}$ and the ratio $10^4 / e^{100} \doteq 1/(2^{10} \cdot 10^{36}) \doteq 0$. Based on this calculation, we can conclude that $(x-1)^2 \cdot e^{-x} \to 0$ as $x \to -\infty$ or

$$\lim_{x \to -\infty} f(x) = 0$$

2. *Zeros of the function:* To find those values of x such that $f(x)=0$ we solve the equation

$$(x-1)^2 \cdot e^x = 0$$

Equating each of these factors to zero, we have

$$e^x = 0 \qquad \text{or} \qquad (x-1)^2 = 0$$

Since e^x is always positive there is no solution of $e^x = 0$. The solution of $(x-1)^2 = 0$ is $x=1$. The zero of the function is $x=1$ and the point $(1,0)$ is on the graph.

3. *y-Intercept:* To find the y-intercept we find $f(0)$. Evaluating $f(0)$ we obtain

$$f(0) = (0-1)^2 \cdot e^0 = 1$$

and conclude that the point $(0,1)$ is on the graph.

4. *Vertical asymptotes:* None. (See Table 5.1 for conditions for vertical asymptotes.)

5. *Horizontal asymptotes:* From our consideration of extent, we note that as $x \to -\infty$, $f(x) \to 0$; hence, the "negative" x-axis is a horizontal asymptote. Since $f(x) = (x-1)^2 \cdot e^x$ is the product of two positive numbers as $x \to -\infty$, we conclude the graph approaches the line $y=0$ from above. The "positive" x axis is not an asymptote since $f(x) \to +\infty$ as $x \to \infty$.

6. *Critical values:* To find the critical values we differentiate $f(x)=(x-1)^2 \cdot e^x$ by the product rule and equate $f'(x)=0$. Thus

$$\begin{aligned} f'(x) &= (x-1)^2 D_x e^x + e^x D_x (x-1)^2 \\ &= (x-1)^2 e^x + e^x 2(x-1)^1(1) \end{aligned}$$

Factoring the right-hand side we obtain

$$\begin{aligned} f'(x) &= (x-1)(e^x)[(x-1)+2] \\ &= (x-1)e^x(x+1) \end{aligned}$$

Solving $(x-1)(x+1)e^x=0$, we obtain the critical values $x=1$ and $x=-1$. (Remember, $e^x \neq 0$.)

7. *Critical points:* The critical points are

$$(1,f(1)) \text{ and } (-1,f(-1))$$

Evaluating $f(1)$ and $f(-1)$ using Table A.3, we find that $f(1)=(1-1)^2e^1=0$ and $f(-1)=(-1-1)^2e^{-1}=4/e \doteq 1.5$. Therefore, the critical points are $(-1,1.5)$ and $(1,0)$. (We denote the critical points in Fig. 6.4 by —•— .)

8. To determine the intervals on which $f(x)$ increases or decreases, we employ the graphical cut-method to determine where $f'(x)$ is positive or negative. Since $f'(x)=(x-1)(x+1)e^x$, the cuts are at $x=1$ and $x=-1$ as marked on the real line in Fig. 6.3.

Evaluating $f'(x)$ in the three regions, we find that $f'(x)$ is positive in regions I and III and negative in region II. Therefore, $f(x)$ increases on $(-\infty,-1)\cup(1,\infty)$ and decreases on $(-1,1)$.

Assimilating the information from Steps 1–8 above and geometrically representing this data, we obtain Fig. 6.4. Based on the plotted points and the information provided in this figure, we sketch the completed graph in Fig. 6.5.

Note that in this problem it has not been necessary to determine and plot additional points that belong to the graph. As a check you might complete the following table and plot the points on the graph in Fig. 6.5.

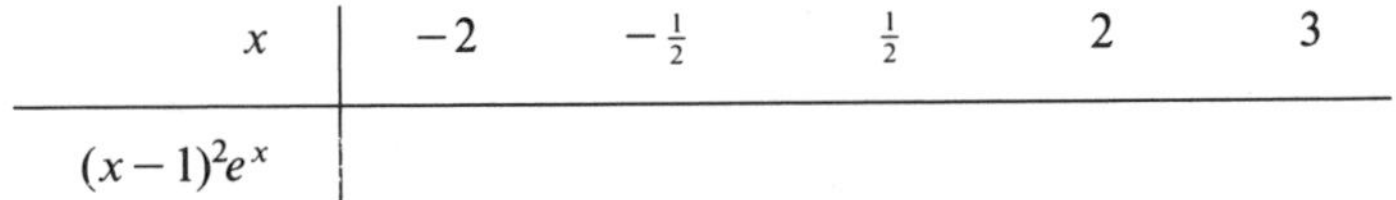

x	-2	$-\frac{1}{2}$	$\frac{1}{2}$	2	3
$(x-1)^2e^x$					

(Values for e^x are found in Table A.3 in the Appendix.)

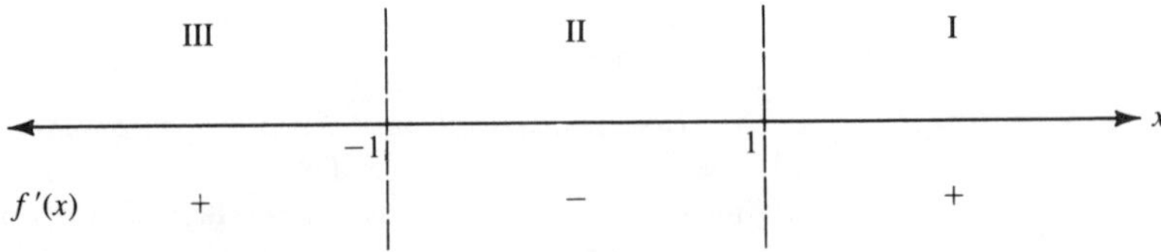

FIGURE 6.3

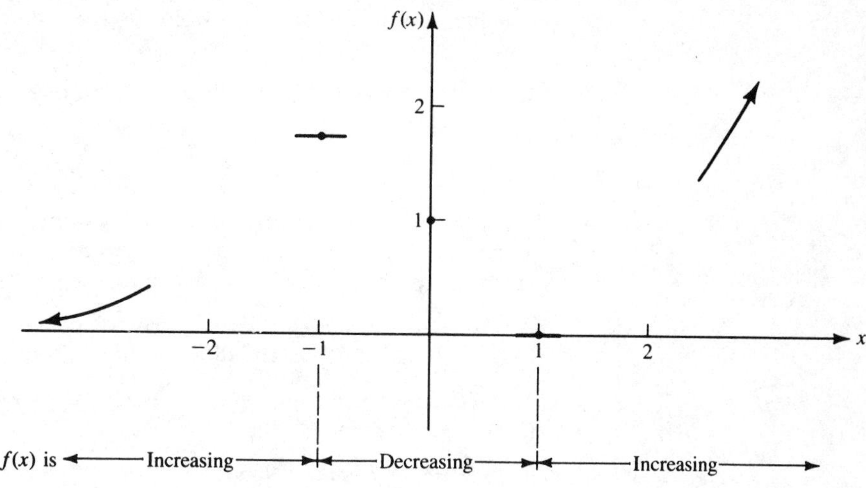

FIGURE 6.4

An important exponential function in probability and statistics is

$$P(x) = ae^{-b(x-c)^2}$$

The values of a, b, and c are positive constants and for particular values of these constants the function is called the *normal* or *Gaussian probability density function*. The graph of this function is referred to as the *normal curve* or the *bell-shaped curve*. The graph of the function $f(x) = ae^{-bx^2}$ has the same shape as the graph of the function $P(x)$.

Example 6.12 Assuming a and b are positive constants, sketch the graph of $f(x) = ae^{-bx^2}$ and determine the coordinates of any inflection points.

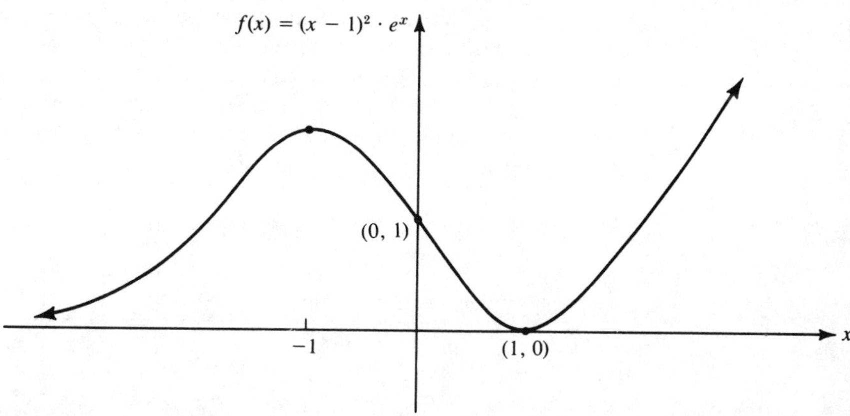

FIGURE 6.5

Solution Again following the eight-step procedure of Tables 5.1–5.3, we find:

1. *Extent:* Since $b>0$, $e^{-bx^2}\to 0$ as $x\to+\infty$ or as $x\to-\infty$. Therefore,

$$\lim_{x\to\infty} f(x) = 0 \text{ and } \lim_{x\to-\infty} f(x) = 0$$

From this we also conclude that the line $y=0$ or the x axis is a horizontal asymptote. Since $a>0$, the function $f(x)=ae^{-bx^2}$ is always positive and its graph approaches the axis from above as $x\to+\infty$ or as $x\to-\infty$.

2. *Zeros of the function:* There is no solution of the equation $ae^{-bx^2}=0$ since $a>0$ and $e^{-bx^2}>0$ for all values of x. Therefore, there are no zeros.

3. *y-Intercept:* Evaluating $f(0)$ we obtain $f(0)=ae^0=a$; hence, the point $(0,a)$ is on the graph of $f(x)$.

4. *Vertical asymptotes:* None.

5. *Horizontal asymptotes:* Discussed in Step 1.

6. *Critical values:* If $f(x)=ae^{-bx^2}$,

$$f'(x) = ae^{-bx^2}D_x(-bx^2) = ae^{-bx^2}(-2bx)$$

or

$$f'(x) = (-2ab)xe^{-bx^2}$$

Equating $f'(x)=0$, we have

$$(-2ab)xe^{-bx^2} = 0$$

Since $(-2ab)$ is a constant and $e^{-bx^2}>0$, the only solution is $x=0$. Therefore, there is one critical value, $x=0$.

7. *Critical points:* The critical point is $(0,f(0))$ or $(0,a)$, which in this case is the same as the y-intercept.

8. To determine the intervals on which the function is increasing or decreasing we employ the graphical cut-method and determine where $f'(x)$ is positive or negative. Since $f'(x)=0$ at $x=0$ only, there is just this one cut. Graphing $x=0$ and the real line and evaluating $f'(x)$ in regions I and II below, we get Fig. 6.6. Therefore, $f(x)$ increases on $(-\infty,0)$ and decreases on $(0,\infty)$.

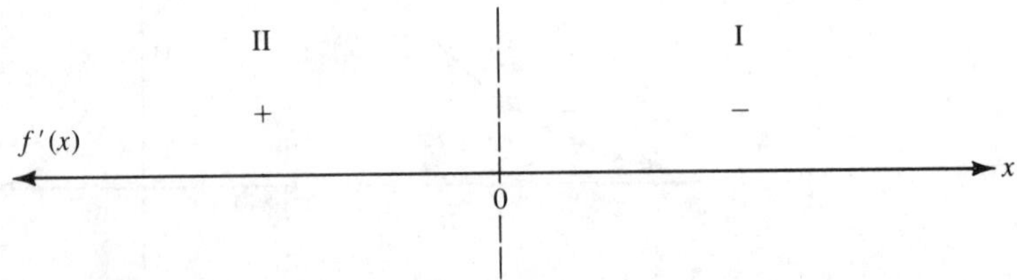

FIGURE 6.6

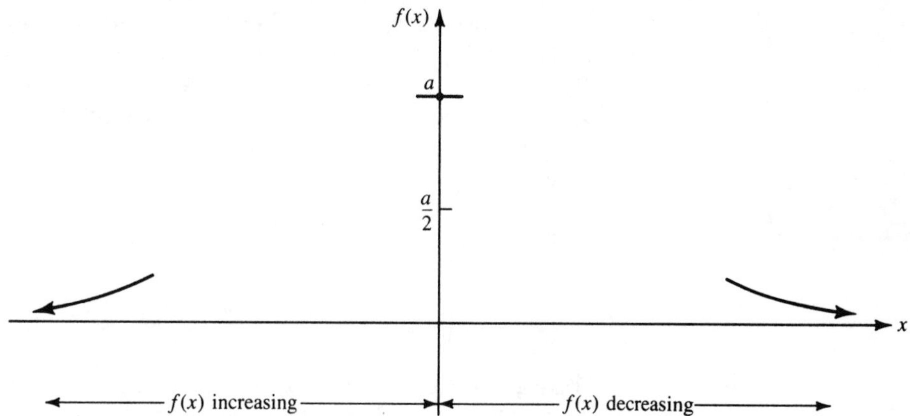

FIGURE 6.7

Graphing the information provided by steps 1–8, we obtain Fig. 6.7. (*Note*: The scale of the vertical axis is in terms of the constant a.)

Before completing the graph of $f(x)$ we find the second derivative to determine the coordinates of any inflection points. To find the derivative of $f'(x)=(-2ab)xe^{-bx^2}$, we use the product rule and obtain

$$\begin{aligned} f''(x) &= (-2ab)xD_x(e^{-bx^2}) + (e^{-bx^2})D_x(-2abx) \\ &= (-2ab)xe^{-bx^2}(-2bx) + e^{-bx^2}(-2ab) \end{aligned}$$

Factoring the right-hand side, we have

$$f''(x) = 2abe^{-bx^2}(2bx^2-1)$$

Using the cut-method to determine where $f''(x)$ is positive or negative we find cuts where $(2bx^2-1)=0$ or where

$$x = \frac{1}{\sqrt{2b}} \quad \text{or} \quad x = \frac{-1}{\sqrt{2b}}$$

Graphing the cuts on the real line and evaluating $f''(x)$ in regions I, II, and III, we get Fig. 6.8.

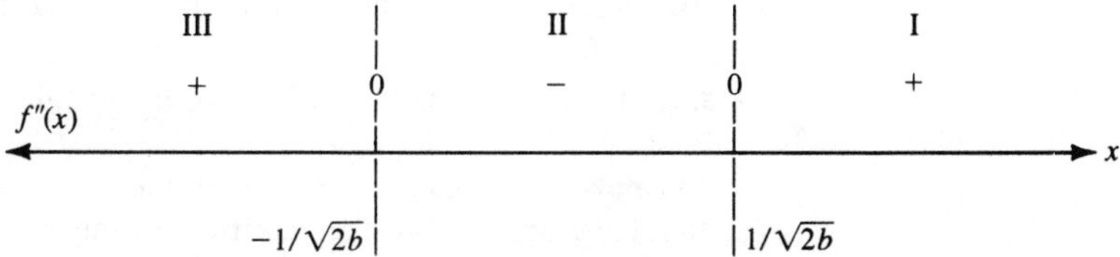

FIGURE 6.8

By Rule 5.6 the function $f(x)$ is concave downward on the interval $(-1/\sqrt{2b}\,, 1/\sqrt{2b}\,)$ and concave upward on $(-\infty, -1/\sqrt{2b}\,) \cup (1/\sqrt{2b}\,, \infty)$. The inflection points by Rule 5.6 occur at $x = -1/\sqrt{2b}$ and $x = 1/\sqrt{2b}$. To find the coordinates of the inflection points we evaluate $f\left(\dfrac{-1}{\sqrt{2b}}\right)$ and $f\left(\dfrac{1}{\sqrt{2b}}\right)$ as follows:

$$f\left(\frac{-1}{\sqrt{2b}}\right) = ae^{-b(-1/\sqrt{2b}\,)^2} = ae^{-1/2} = \frac{a}{\sqrt{e}} \doteq .6a$$

$$f\left(\frac{1}{\sqrt{2b}}\right) = ae^{-b(1/\sqrt{2b}\,)^2} = ae^{-1/2} = \frac{a}{\sqrt{e}} \doteq .6a$$

Taking into consideration concavity, inflection points, and the information provided by Fig. 6.7, we sketch the graph of $f(x) = ae^{-bx^2}$ in Fig. 6.9. This graph has the same shape as the graph of the normal curve shown in Fig. 8.10.

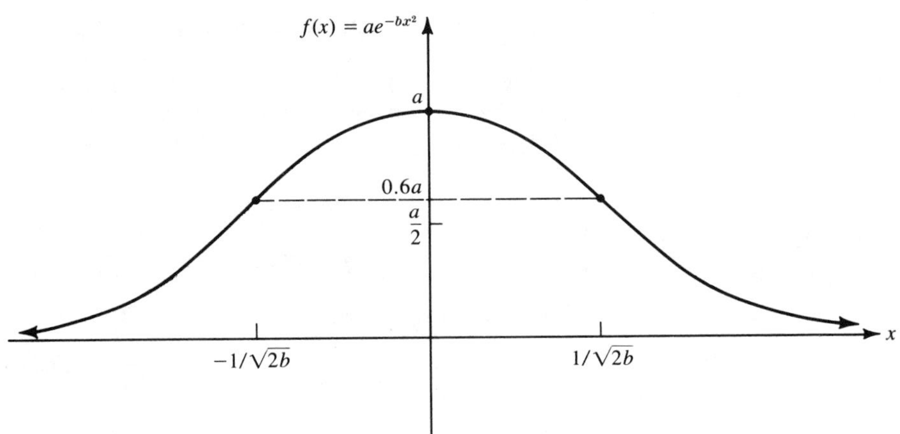

FIGURE 6.9

Many real-world problems involve curves that increase rapidly at first and then taper off and approach a horizontal asymptote as shown in Fig. 6.10. For example, the sales, $S(x)$, of a new product often grow rapidly at first and then taper off with time and reach a saturation level. The value of A in Fig. 6.10 is often referred to as the *saturation value*. A function whose graph has the shape of the curve illustrated in Fig. 6.10 has the general form $f(x) = A - Be^{-Cx}$, where A, B, and C are positive constants. The constant A is the saturation value as we will show in the following example.

Example 6.13 Sketch the graph of the function $f(x) = A - Be^{-Cx}$ and verify that the graph has the characteristics of the curve shown in Fig. 6.10. Assume A, B, and C are positive constants, that $A > B$, and the domain is $[0, \infty)$.

Solution We will still follow the eight-step procedure of curve sketching in this problem, but will not identify each of the steps.

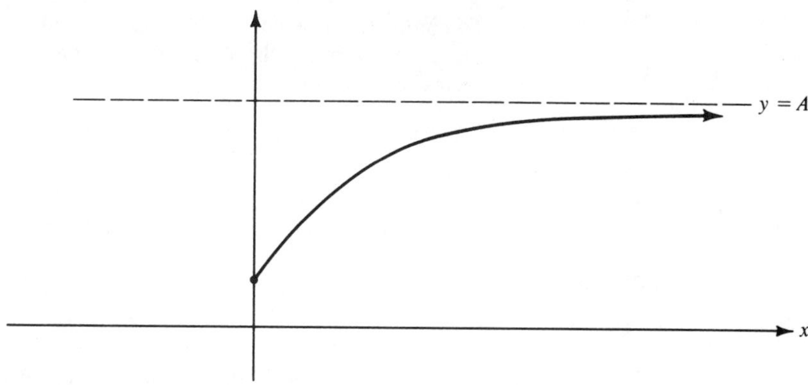

FIGURE 6.10

Since $f(x)$ is defined on $[0, \infty)$, we evaluate $f(0)$ and $\lim_{x\to\infty} f(x)$ as follows:

$$f(0) = A - B = A - Be^0$$

Since $A > B$ by assumption, the point $(0, A - B)$ is on the positive y axis.

As $x \to +\infty$, e^{-Cx} or $(1/e^{Cx}) \to 0$; therefore, $Be^{-Cx} \to 0$ as $x \to \infty$ and we conclude:

$$\lim_{x\to\infty} f(x) = \lim_{x\to\infty} (A - Be^{-Cx}) = A$$

Hence, the function $f(x)$ has a horizontal asymptote whose equation is $y = A$. Since the term Be^{-Cx} is always positive and $f(x) = A - Be^{-Cx}$, we conclude that $f(x) < A$ and the graph of $f(x)$ approaches the horizontal asymptote from below as illustrated in Fig. 6.11.

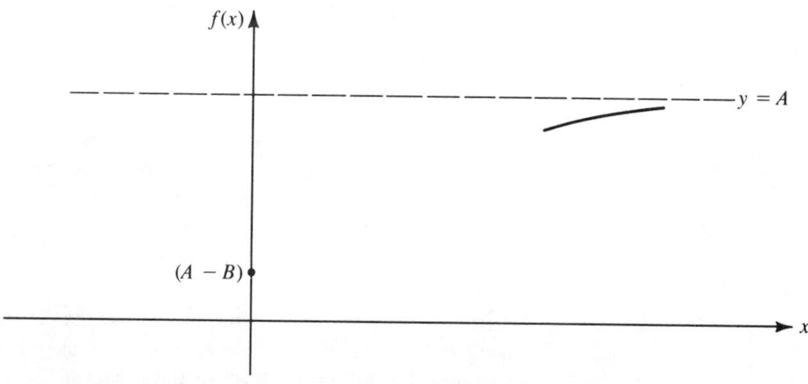

FIGURE 6.11

To complete the sketch of the graph of $f(x)$, we use the first and second derivatives to determine the shape of the curve. Taking the first and second derivatives of $f(x)$, we have

$$\begin{aligned} f'(x) &= D_x(A) - D_x(Be^{-Cx}) \\ &= 0 - Be^{-Cx}(-C) = BCe^{-Cx} \\ f''(x) &= D_x(BCe^{-Cx}) \\ &= BCe^{-Cx}(-C) \\ &= -BC^2e^{-Cx} \end{aligned}$$

Since e^{-Cx} is always positive, we conclude that $f'(x) = BCe^{-Cx} > 0$ for all x and $f''(x) = -BC^2e^{-Cx} < 0$ for all x. From this we note that the slope is positive and the curve is concave downward on the interval $[0, \infty)$. These facts together with the information provided by Fig. 6.11 lead us to sketch the graph of $f(x)$ as shown in Fig. 6.12, which has all the characteristics of the curve drawn in Fig. 6.10. You can easily verify that the slope of the graph drawn in Fig. 6.12 is a decreasing function and that the larger the value of C, the faster the curve will taper off.

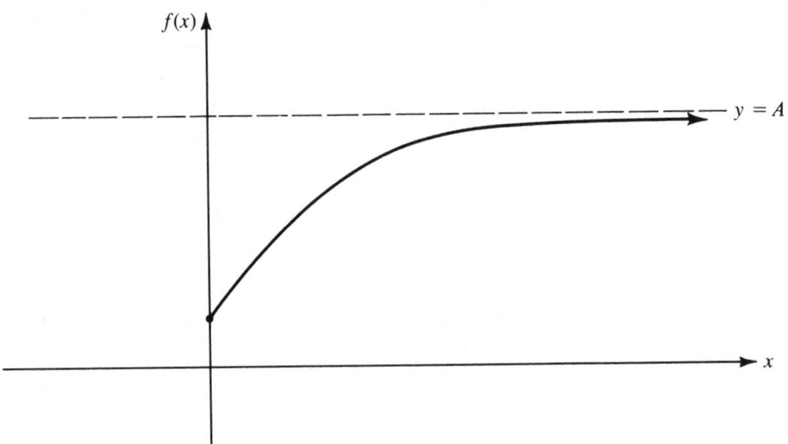

FIGURE 6.12

Example 6.14 Sales Saturation Level for a Breakfast Cereal The daily sales, $S(x)$, in thousands of boxes from the marketing of a new breakfast cereal is approximated by the function

$$S(t) = 100 - 80e^{-.2t}$$

where t represents the time in months that the cereal has been on the market. Using the results of Example 6.13, draw a sketch of the graph of $S(t)$ and determine the saturation level of the sales.

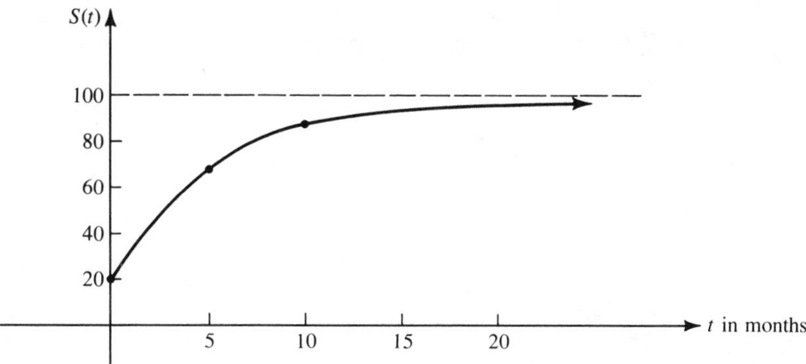

FIGURE 6.13

Solution Since $S(t)$ has the same form as $f(x)=A-Be^{-Cx}$, we conclude that its graph will be similar to Fig. 6.12. At time $t=0$, we find $s(0)=100-80=20$. We determine the horizontal asymptote and the saturation value by determining $\lim_{t\to\infty} S(t)$:

$$\lim_{t\to\infty} S(t) = 100$$

To complete the graph we find two additional points on the curve by selecting t values of 5 and 10 months. The associated $S(t)$ values are

$$S(5) = 100 - 80e^{-1} \doteq 100 - 80(.368) \doteq 70.56$$

$$S(10) = 100 - 80e^{-2} \doteq 100 - 80(.135) \doteq 89.2$$

Graphing these points, the y-intercept, and the horizontal asymptote, we obtain the graph of $S(t)$ as shown in Fig. 6.13.

In learning a particular skill such as typing or golf, a person progresses faster at the beginning and then levels off. A graph of performance against time for such tasks would be similar to the curve of Fig. 6.12. Psychologists call this a *learning curve*. Learning curves have applications in psychology, education, and industry.

Example 6.15 The Graph of a Learning Curve It has been found that on the average a typing student's performance is given by the function

$$N(t) = 80 - 80e^{-.05t}$$

where $N(t)$ is the number of words typed per minute after t weeks of instruction. Sketch the graph of $N(t)$.

Solution Since $N(t)$ has the form of $f(x)=A-Be^{-Cx}$, we need only determine the end point $(0, N(0))$, the saturation level A, and one or two points

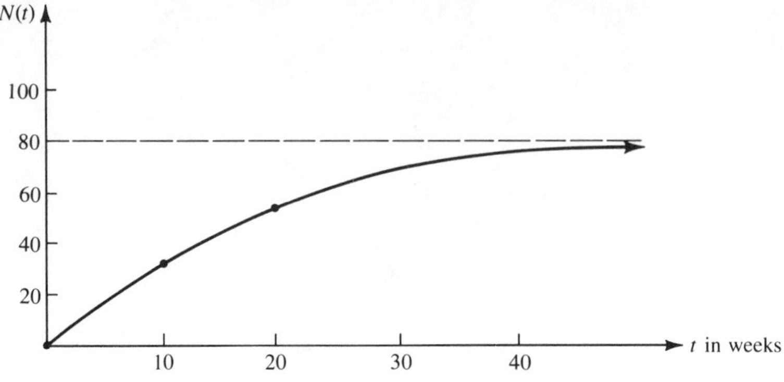

FIGURE 6.14

to draw a good graph. Doing this, we find

$$N(0) = 80 - 80e^0 = 0$$

$$\lim_{t \to \infty} N(t) = 80$$

and evaluating $N(t)$ at $t = 10$ and 20 (note that since C is small we select fairly large values of t) we get

$$N(10) = 80 - 80e^{-.5} \doteq 80 - 80(.606) \doteq 31.5$$

$$N(20) = 80 - 80e^{-1} \doteq 80 - 80(.368) \doteq 50.6$$

Graphing these points and the horizontal asymptote, we obtain the graph of $N(t)$ as shown in Fig. 6.14.

We have seen in previous chapters that the first derivative interpreted as the rate of change of a function has many applications. Since growth and decay are often described by exponential and logarithmic functions, the *rate of growth* and the *rate of decay* will involve the derivative of these functions. For example, in Sec. 3.4 we found that if P dollars is invested at a rate r compounded continuously and allowed to "grow" for t years, the maturity value given by Eq. (3.14) is

$$A(t) = Pe^{rt}$$

Example 6.16 The Rate of Growth of an Investment If we invest \$5,000 at 9% compounded continuously, what is the rate of growth of our investment 10 years after the original investment?

Solution By substitution, the maturity value of our investment at any time t is given by

$$A(t) = 5{,}000e^{.09t}$$

The rate of growth at any time t is given by $A'(t)$, which we find by differentiating the exponential function $A(t)$ as follows:

$$A'(t) = D_t(5{,}000e^{.09t})$$
$$= 5{,}000e^{.09t}(.09)$$

or
$$A'(t) = 450e^{.09t}$$

To find the rate of growth after 10 years, we substitute $t = 10$ into $A'(t)$ and find by use of Table A.3.

$$A'(10) = 450(e^{.9}) = 450(2.4596) = \$1{,}106.82/\text{year}$$

We interpret this to mean that after 10 years our investment is growing at the instantaneous rate of \$1,106.82/year. The actual interest earned during the 11th year, i.e., from $t = 10$ to $t = 11$, will be greater than $A'(10)$ and can be found by evaluating the difference $A(11) - A(10)$. The reason for this difference is depicted in Fig. 6.15. We find the actual interest earned in the 11th year by evaluating $A(11) - A(10)$ as follows:

$$A(11) = 5{,}000e^{.09(11)} = 5{,}000e^{.99}$$
$$A(10) = 5{,}000e^{.09(10)} = 5{,}000e^{.90}$$
$$A(11) - A(10) = 5{,}000(e^{.99} - e^{.90})$$
$$= 5{,}000(2.6912 - 2.4596)$$
$$= 5{,}000(.2316) = \$1{,}158.00$$

Note that the actual interest earned is greater than $A'(10)$, which is in agreement with Fig. 6.15.

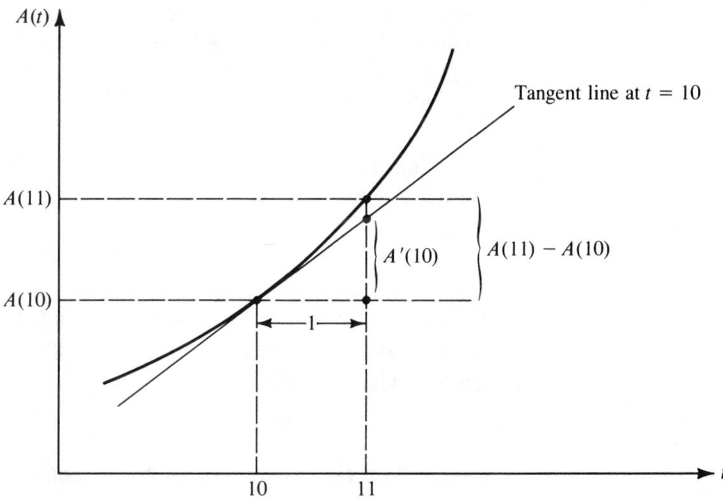

FIGURE 6.15

Example 6.17 The concentration of the pollutant P.C.B. in mg/liter in the Housatonic River is approximated by

$$P(x) = .01e^{-.5x}$$

where x is the number of miles downstream from the source. (See Exercise 20, Sec. 3.4.) Determine $P(1)$ and $P'(1)$ and interpret the results.

Solution By substituting $x=1$ into the function $P(x)$ and evaluating by using Table A.3, we find

$$\begin{aligned} P(1) &= .01e^{-.5(1)} = .01e^{-.5} \\ &= (.01)(.606531) \\ &\doteq .00607 \text{ (rounded to 3 significant digits)} \end{aligned}$$

This means that there are .00607 mgs of P.C.B./liter of water 1 mile downstream from the source.

To find $P'(1)$ we find $P'(x)$ and then substitute $x=1$ as follows

$$\begin{aligned} P'(x) &= (.01)e^{-.5x}(-.5) \\ &= -.005e^{-.5x} \end{aligned}$$

Therefore

$$\begin{aligned} P'(1) &= -.005e^{-.5(1)} \\ &= -(.005)(.607) \\ &= -.00304 \text{ (rounded to 3 significant figures)} \end{aligned}$$

We interpret this to mean that the concentration of pollutant in the water 1 mile downstream is *decreasing* at the rate of .00304 mg/liter/mile. This implies that since we have .00607 mg/liter of P.C.B. 1 mile downstream, 2 miles downstream there will be *approximately* $.00607-.00304$ or .00303 mg/liter of P.C.B. in the water.

In Chapter 3 we stated that experiments show that radioactive decay follows the rule

$$Q(t) = Pe^{-rt} \tag{6.9}$$

(see Exercise 14, Sec. 3.4) where P is the quantity of the radioactive isotope present at time $t=0$, r is the decay constant, and $Q(t)$ is the amount present at time t. The radioactive isotope iodine, I^{131}, used in the study of thyroid disorders, has a decay constant of 8.6%/day.

Example 6.18 Availability of the Tracer Iodine Isotope I^{131} If 2 g of I^{131} are delivered to a local hospital for use as an isotopic tracer,

(a) determine the amount of I^{131} available for experimentation 16 days after delivery, and

(b) determine the rate of change in the amount of I^{131} present at this time.

Solution

(a) If we assume $t=0$ at the time of delivery, then we can determine the amount of I^{131} present 16 days later by substituting the values $P=2$, $r=.086/\text{day}$, and $t=16$ into Eq. (6.9). Thus we find

$$Q(16) = 2e^{-.086(16)} = 2e^{-1.376}$$

Rounding 1.376 to 1.38 to use Table A.3, we find

$$\begin{aligned} Q(16) &\doteq 2(.25) \\ &\doteq \tfrac{1}{2}\text{ g} \end{aligned}$$

(b) To determine the rate of change in the amount of I^{131} present when $t=16$ we find $Q'(t)$ and evaluate by substituting $t=16$ as follows:

$$\begin{aligned} Q'(t) &= D_t(2e^{-.086t}) \\ &= 2e^{-.086t}(-.086) \\ &= (-.086)(2e^{-.086t}) \end{aligned}$$

Evaluating $Q'(16)$ we find

$$Q'(16) = (-.086)(2e^{-.086(16)})$$

The second factor from part (a) above is $\frac{1}{2}$. Therefore,

$$\begin{aligned} Q'(16) &\doteq (-.086)(\tfrac{1}{2}) \\ &= -.043\text{ g/day} \end{aligned}$$

This means that the iodine isotope I^{131} after 16 days is decreasing at the rate of .043 g/day.

We can conclude from the above example that in general if $Q(t)=Pe^{-rt}$, then $Q'(t)=(Pe^{-rt})(-r)$ and since $Pe^{-rt}=Q(t)$, we can by substitution state

$$\begin{aligned} Q'(t) &= Q(t)(-r) \\ &= -rQ(t) \end{aligned}$$

This equation implies that the rate of change of the quantity present is proportional to the amount present and the *proportionality constant* is the decay constant r.

We conclude this section with an application of logarithmic functions.

It is typical for a company to find that when money is first spent on advertising its sales increase very rapidly with the number of dollars spent. As

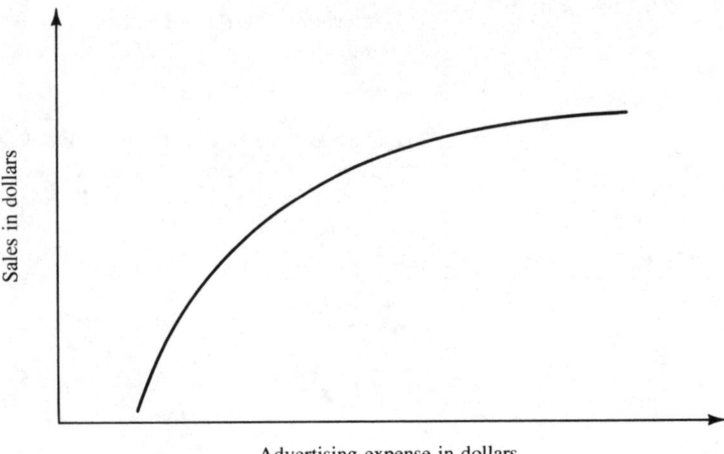

FIGURE 6.16

more money is put into advertising, sales continue to increase, but at a much slower rate. We illustrate this in Fig. 6.16.

Since the graphs of logarithmic functions (Fig. 6.2 for example) have this general shape these functions may be used as models to approximate situations where the rate of growth is always positive but decreasing. As an illustration we consider the following example.

Example 6.19 Sales Growth per Dollar Spent on Advertising A company finds that its annual sales, $S(x)$, in millions of dollars is related to the yearly advertising budget, x, by the equation

$$S(x) = \ln(1+x)$$

where x is in tens of thousands of dollars.

(a) Sketch the graph of $S(x)$ assuming that the maximum advertising budget is \$500,000, i.e., $x=50$.
(b) Determine the rate of growth of the sales when the advertising budget is \$100,000, i.e., $x=10$.
(c) Determine the size of the advertising budget when each additional dollar in sales is costing 50¢ in advertising expense.

Solution

(a) The graph of $S(x)$ shown in Fig. 6.17 is found by selecting x values of 5, 10, 20, 30, 40, and 50 and evaluating $\ln(1+x)$ by using Table A.2. [*Note*: When $x=0$, $S(0)=\ln 1=0$.]
(b) The rate of growth when $x=10$ is found by determining $S'(x)$ and evaluating $S'(10)$ as follows

$$S'(x) = \frac{1}{1+x}$$

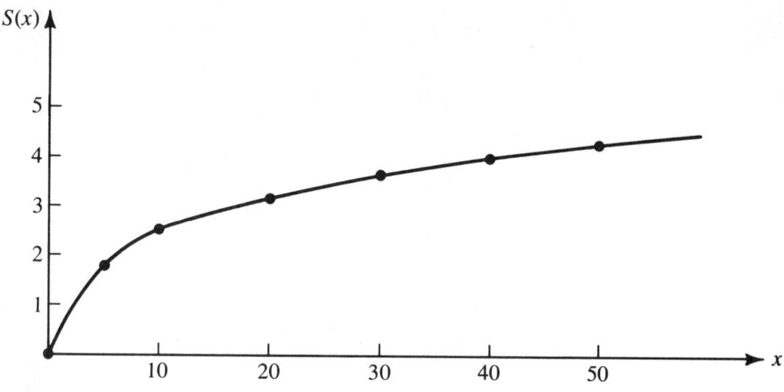

FIGURE 6.17

and
$$S'(10) = \frac{1}{10+1} \doteq .091$$

Hence, the sales are increasing at the rate of approximately .091 millions of dollars/year for each $10,000 spent on advertising or at the rate of $91,000/year/$10,000.

(c) If each additional dollar in sales costs 50¢, then for an increase in the advertising expenditure of $10,000 the sales will increase by $20,000, i.e., 2 : 1. Hence, since $S'(x)$ is the increase in millions of dollars in sales for a $10,000 increase in advertising, we need to determine when $S'(x) = .020$. (*Note*: $20,000 = .020 million dollars.)

From part (b) we know that $S'(x) = 1/(1+x)$; hence, we solve

$$\frac{1}{1+x} = .020$$
$$.02 + .02x = 1$$
$$.02x = .98$$
$$x = 49$$

Therefore, when the annual advertising expenditure is 49 ten thousand dollars, i.e., $490,000, the sales will increase by $1.00 for every additional 50¢ spent on advertising.

EXERCISES

In Exercises 1–8 determine where each function is increasing, decreasing, concave upward, and concave downward. Sketch the graph of the function and indicate on the sketch relative extrema and inflection points.

1. $f(x) = xe^x$

2. $g(x)=xe^{-x}$

3. $h(x)=x^2e^x$

4. $f(x)=x^2e^{-x}$

5. $g(x)=(x+4)^2e^x$

6. $f(x)=\dfrac{1}{1+e^{-2x}}$

7. $g(x)=\ln\left(\dfrac{1}{x}\right), x>0$

8. $f(x)=x\ln x, x>0$

9. Find the absolute minimum value of the function $f(x)=e^x+e^{-x}$.

10. Learning Shorthand The average performance of a student studying shorthand is given by the function

$$N(t) = 90 - 90e^{-.04t}$$

where $N(t)$ is the number of words taken/minute after t weeks of instruction.

(a) Determine the saturation level.
(b) Sketch the graph of $N(t)$.
(c) Find $N'(10)$ and give the meaning of this value.

11. Production Level and On-the-Job Training A time-study engineer has found that production workers performing a certain task improve their performance according to the function

$$N(t) = 100 - 50e^{-.2t}$$

where $N(t)$ is the number of parts assembled per hour after t weeks of on-the-job training.

(a) Determine the saturation level.
(b) Sketch the graph of $N(t)$.
(c) Find $N'(5)$ and give the meaning of this value.

12. Increasing Yield with Fertilizer The percentage increase in produce yield as a function of the number of pounds of fertilizer used in a certain acreage is given by

$$f(x) = 200(1-e^{-.01x})$$

where $f(x)$ is the percentage increase in yield and x is the number of pounds of fertilizer.

(a) Determine the saturation level.
(b) Sketch the graph of $f(x)$.
(c) Find $f'(100)$ and give the meaning of this value.

13. Determine the rate of growth in dollars/year from \$10,000 invested at 10%, compounded continuously, after 5 years; 10 years; and 20 years.

14. A college professor invests $20,000 at 8% compounded continuously.

(a) Determine the rate of growth of her investment in dollars/year after 10 years.
(b) When will the rate of growth equal $2,000/year?

15. Inflation and the Worth of a Dollar Assume that inflation is such that the value of a dollar is decreasing at the rate of 8%/year compounded continuously.

(a) Determine the value of a dollar 10 years from now.
(b) Find the rate of change of the value of a dollar 10 years from now.

16. Half-life of the Iodine Isotope I^{131} The half-life of a radioactive substance is defined as being the time it takes for 50% of any given amount to decay.

(a) Show in general that if a radioactive substance has a decay constant of r the half-life, T, is given by $T = \ln 2/r$ and that Eq. (6.9) can be rewritten as $Q(t) = Pe^{-(\ln 2/T)t}$.
(b) Determine the half-life of the iodine isotope I^{131}. (See Example 6.18.)
(c) If the original amount of I^{131} is 10 g, determine $Q'(T)$ and give the meaning of this value.

17. Medical Research and Potassium K^{42} The potassium isotope K^{42} is used as a tracer in medical research. The half-life of K^{42} is 12.5 hr. (See Exercise 16.)

(a) If the original amount available for experimentation is 10 g, how much will be available in 25 hr?
(b) Determine the rate of change of the amount in 25 hr.

18. To study thyroid disease 0.1 milligram of the iodine isotope I^{131} was injected into a patient.

(a) Assuming total absorption of the isotope, determine the amount of I^{131} in the patient 24 days after injection.
(b) What is the rate of change of the isotope in the patient in mgs/day 24 days after injection?

19. The population of a bacteria culture is given by the formula

$$N(t) = 100e^{.02t}$$

where $N(t)$ is the number of cells present at time t where t is measured in minutes.

(a) Determine the population size when $t = 100$ min.
(b) Determine $N'(100)$ and give the meaning of this value.

20. Destroying Tobacco Virus with X-rays As the result of exposure to X-rays certain strains of tobacco virus found on the leaves of tobacco plants are decreased exponentially as a function of the dosage applied. The amount present after a dosage of r roentgens of X-ray is given by

$$N(r) = N_0 e^{-r}$$

where N_0 represents the initial amount present.

(a) Determine the value of r such that 90% of the virus is destroyed. (*Hint*: $N(r) = 10\% N_0$.)

(b) Find the value of $N'(r)/N_0$ when 90% of the virus has been destroyed. Explain the meaning of this value of $N'(r)/N_0$.

21. **Per-Pupil Expenditure in Education** In the State of Connecticut recent data show that the per-pupil annual public expenditure for education has increased exponentially at 9%/year, compounded continuously, since 1960. That year the annual dollar expenditure per pupil was \$350.

(a) Express the annual per-pupil expenditure, $E(t)$, as a function of the time t where t is the number of years since 1960.
(b) Find the number of dollars spent annually per pupil in 1965, 1970, 1975, and currently.
(c) Find the rate of growth (in dollars/year) of the annual per-pupil expenditure in 1975.
(d) Find the current rate of growth (in dollars/year) of the annual per-pupil expenditure.

22. Reconsider Exercise 21 assuming that the rate of growth had been held to 4%.

23. **Minimum Income and the Effect on Inflation** Inflation has been such that it required \$202/week as minimum for a family of four to make ends meet in 1977, whereas in 1967 this figure was \$101. Data show that the living costs increased exponentially during this period according to the formula

$$C(t) = C_0 e^{rt}$$

where t is the number of years measured from 1967.

(a) Determine C_0 and r in the above formula.
(b) Determine the current minimum income required for a family of four based on the above formula.
(c) Find $C'(13)$ and give the meaning of this value.

24. **Spread of Infection—the Logistic Model** A mathematical model* that is used to predict the spread of infection under certain conditions is given by the formula

$$N(t) = \frac{n+1}{1+ne^{-\beta(n+1)t}}$$

where

$N(t)$ = the number of persons infected at time t

n = the total size of the population to which one infected person is introduced

t = the time

β = a positive constant called the specific infection rate

*This function is generally known as the logistic function and its graph is called the logistic curve.

This model assumes an infected individual remains infected during the process, that no one is removed from the population, and that everyone is likely to be infected. Assuming $n=999$ and that $\beta=.0005$ when t is measured in days

(a) Sketch the graph of this function.
(b) Determine $N'(16)$ and give the meaning of this value.

25. Bounded World Population Growth The logistic function of Exercise 24 may be used as a model to predict bounded population growth. For example, assuming that the world resources cannot support more than 30 billion people, the formula

$$N(t) = \frac{30}{1+14e^{-.20t}}$$

may be used to predict the population size $N(t)$ at any time t measured in years since 1940.

(a) Sketch the graph of $N(t)$.
(b) Determine the time t when the rate of change of the population size will be a maximum. [*Hint*: Find the maximum value of $N'(t)$ by solving $N''(t)=0$.]

26. The B.M.I. Corporation finds that its annual sales, $S(x)$, in millions of dollars is related to the yearly advertising budget, x, by the equation

$$S(x) = 2\ln(1+x)$$

where x is in tens of thousands of dollars.

(a) Sketch the graph of $S(x)$ assuming that the maximum advertising budget is \$500,000; i.e., $x=50$.
(b) Determine the rate of growth of sales when the advertising budget is \$100,000; i.e., $x=10$.
(c) Determine the size of the advertising budget when each additional dollar in sales is costing 10¢ in advertising expense.

6.5 IMPLICIT DIFFERENTIATION

All the functions that we have considered are of the form where y or $f(x)$ is expressed in terms of x; that is, the functions are of a form that *explicitly* states that the dependent variable is a function of the independent variable x. For example, the function $y=x^3+x$ explicitly defines y as a function of x. These functions are called *explicit functions*. Generally there is no difficulty in determining the derivative of functions of this form by applying the appropriate rules from Chapter 4 or Secs. 6.2 and 6.3.

A function can also be stated *implicitly*; that is, the function is stated so that it is not apparent which variable is the dependent variable. Such functions are called *implicit functions*. For example, the equation $3x-6y=11$ implicitly defines either y in terms of x or x in terms of y. Since in explicit functions we define y in terms of x, we will continue to let x be the independent variable in our consideration of implicit functions. Thus, if x is permitted to assume a given value, then y is implicity defined by the equation.

In this case, the equation can be solved such that y is explicitly defined in terms of x; that is, $y=\frac{1}{2}x-\frac{11}{6}$.

However, there are many implicit functions that cannot conveniently be converted to explicit functions of the form $y=f(x)$. For example, the implicit function $y^3+3x^2y+x^3=25$ cannot be converted easily into an explicit function by solving for y in terms of x. Hence, we could not apply any of our differentiation rules to determine the derivative D_xy. Fortunately, there is a method of differentiation called *implicit differentiation* that can be applied to an implicit function. It is based on the assumption that the implicit function is such that y is a differentiable function of x, and involves differentiating each term in the function with respect to the variable x and applying the chain rule where necessary.

To determine D_xy for the previous implicit function, we simply differentiate each term with respect to x as follows:

$$D_x(y^3)+D_x(3x^2y)+D_x(x^3)=D_x(25) \tag{6.10}$$

Assuming that y is the differentiable function $u(x)$, we can substitute $u(x)$ for y and write Eq. (6.10) as

$$D_x[u(x)]^3+D_x[3x^2\cdot u(x)]+D_x(x^3)=D_x(25) \tag{6.11}$$

Differentiating Eq. (6.11) term-by-term (*Note*: we apply the short-cut chain rule to the first term and the product rule to the second term), we obtain:

$$3[u(x)]^2D_xu(x)+3x^2\cdot D_xu(x)+u(x)D_x(3x^2)+D_x(x^3)=D_x(25)$$

or

$$3[u(x)]^2\cdot u'(x)+3x^2\cdot u'(x)+u(x)\cdot 6x+3x^2=0 \tag{6.12}$$

Now replacing $u(x)$ by y and $u'(x)$ by D_xy in Eq. (6.12), we obtain

$$3y^2D_xy+3x^2D_xy+y\cdot 6x+3x^2=0$$

We can also obtain this result directly from Eq. (6.10) without substitution by treating y as a function of x. Differentiating Eq. (6.10) term-by-term we obtain

$$3y^2D_xy+3x^2D_xy+yD_x(3x^2)+D_x(x^3)=D_x(25)$$

or

$$3y^2D_xy+3x^2D_xy+6xy+3x^2=0$$

which quickly gives us the same result as Eq. (6.12). Solving for D_xy we have

$$D_xy(3y^2+3x^2)=-3x^2-6xy$$

or
$$D_xy = \frac{-(3x^2+6xy)}{3y^2+3x^2}$$
$$= \frac{-(x^2+2xy)}{y^2+x^2}$$

When using implicit differentiation, it is quite common for the resulting derivative to contain both variables. In these cases we need both the value of x and y to evaluate D_xy.

Example 6.20 Assuming that $y \geqslant 0$, use the method of implicit differentiation to determine D_xy for $x^2+xy+y^2=16$, and determine D_xy when $x=4$.

Solution Differentiating each term with respect to x and applying the chain rule where necessary, we obtain

$$D_x(x^2) + D_x(xy) + D_x(y^2) = D_x(16)$$

or
$$2x + x \cdot D_xy + y + 2yD_xy = 0$$

Solving for D_xy, we have

$$D_xy(x+2y) = -2x - y$$

or
$$D_xy = \frac{-(2x+y)}{x+2y}$$

From this expression we note that to determine D_xy when $x=4$ we also need the value of y. To determine y we substitute $x=4$ into the original equation $x^2+xy+y^2=16$ and solve for y as follows:

$$16 + 4y + y^2 = 16$$
$$y^2 + 4y = 0$$
$$y(y+4) = 0$$
$$y = 0 \quad \text{or} \quad y = -4$$

Since we have assumed $y \geqslant 0$ we select the value $y=0$. Therefore, by substituting $x=4$ and $y=0$ into

$$D_xy = -\frac{2x+y}{x+2y}$$

we obtain

$$D_xy = -2$$

Example 6.21 Use the method of implicit differentiation to determine the slope of the curve whose equation is $x^2+y^2=25$ at $(4,3)$.

Solution Differentiating each term with respect to x, we obtain

$$D_x(x^2) + D_x(y^2) = D_x(25)$$
$$2x + 2yD_xy = 0$$

therefore
$$D_xy = \frac{-x}{y}$$

Substituting $x=4$ and $y=3$ we obtain the slope at (4,3):

$$D_xy = \frac{-4}{3}$$

We have considered rules for differentiating x^n (a variable to a constant power) and b^x (a constant to a variable power), but we have not considered a variable to a variable power; for example, the function $f(x)=x^x$. In order to determine the derivative of such a function, we shall use a process called *logarithmic differentiation*, which is an application of the properties of logarithms and implicit differentiation.

Suppose we have a function of the form $y=f(x)^{g(x)}$. If $f(x)>0$, we can take the natural logarithm of both sides of this equation and obtain

$$\ln y = \ln\left[f(x)^{g(x)}\right] = g(x)\ln f(x) \tag{6.13}$$

We could have used the logarithm to any base b; however, when differentiating, the results would be more cumbersome. Using implicit differentiation for the left and right sides of Eq. (6.13), we obtain

$$D_x[\ln y] = D_x[g(x)\ln f(x)]$$

or
$$\frac{1}{y}D_xy = g(x)\cdot\frac{1}{f(x)}D_xf(x) + \ln f(x)\cdot D_xg(x)$$

Since we wish to determine D_xy, we need only solve the last equation for D_xy:

$$D_xy = y\left[\frac{g(x)}{f(x)}\cdot D_xf(x) + D_xg(x)\cdot\ln f(x)\right]$$

or
$$D_xy = y\left[\frac{g(x)}{f(x)}\cdot f'(x) + g'(x)\ln f(x)\right] \tag{6.14}$$

Eq. (6.14) could now be used to determine the derivative with respect to x of any function of the form $y=f(x)^{g(x)}$. However, the reader is advised to use the process of logarithmic differentiation instead of the resulting formula. This advice is given for two reasons; the formula is too complex to memorize and it is usually easier to differentiate a function using the logarithmic differentiation process than to substitute into Eq. (6.14).

Example 6.22 Use the process of logarithmic differentiation to determine $D_x y$ for $y = x^x$. (Assume $x > 0$.)

Solution Take the natural logarithm of both sides:

$$\ln y = x \ln x$$

Differentiating the implicit function with respect to x, we obtain

$$D_x(\ln y) = D_x(x \ln x)$$

$$\frac{1}{y} D_x y = x \cdot \frac{1}{x} + \ln x$$

or

$$D_x y = y[1 + \ln x]$$

Since $y = x^x$, we can rewrite this last equation in terms of the variable x:

$$D_x y = x^x[1 + \ln x]$$

Example 6.23 Determine $D_x y$ for $y = (2x^3)^{5x+4}$.

Solution For $x > 0$, we obtain,by taking the natural logarithm of both sides,

$$\ln y = (5x+4)\ln(2x^3)$$

Differentiating the implicit function with respect to x, we obtain

$$\frac{1}{y} D_x y = (5x+4) D_x \ln(2x^3) + [\ln(2x^3)] \cdot D_x(5x+4)$$

$$\frac{1}{y} D_x y = (5x+4)\frac{1}{2x^3} \cdot 6x^2 + [\ln(2x^3)] \cdot (5)$$

or

$$D_x y = y\left[\frac{3(5x+4)}{x} + 5\ln(2x^3)\right]$$

Replacing y by $(2x^3)^{5x+4}$, we have

$$D_x y = (2x^3)^{5x+4}\left[\frac{3(5x+4)}{x} + 5\ln(2x^3)\right]$$

EXERCISES

Determine $D_x y$ for each of the equations in Exercises 1–10.

1. $2xy = x - 5$

2. $xy = x + 1$

3. $3xy = 2x + 4$

4. $x^2 + y^3 = 25$

5. $xy - y^2 = 2x^2y - 3$

6. $y^3 + 3y = 4x^2 + 2x + 1$

7. $x^2y = x^2y^2 + y^2x$

8. $3x^2y^3 + 6xy = 4x^3y^2 + 2x$

9. $3xy = \dfrac{1}{x^2 + y}$

10. $3xy^3 = 5x^2 - 6y + 4x$

Determine the slope of each of the curves in Exercises 11–14 *at the given points.*

11. $x^2 + xy + y^2 = 12$, at $(2, 2)$

12. $xy + 2x^2 = 9$, at $(3, -3)$

13. $y^3x + 3x^2y = 14$, at $(1, 2)$

14. $y^2 + 3y = 4x^2 + 11x + 1$, at $(-3, 1)$

In Exercises 15–24 *use the process of logarithmic differentiation to determine* D_xy.

15. $y = x^{x+1}$

16. $y = (3x - 2)^{2x}$

17. $y = x^{x^2 + 4x + 1}$

18. $y = (10x + 3)^x$

19. $y = (5x + 4)^{5x+4}$

20. $y = (3x^2 + 6x + 1)^{2x+3}$

21. $y = 10^x \cdot x^{10}$

22. $y = (e^{3x^2})^{2x}$

23. $y = x^{4x+5}$

24. $y = (2x)^{x^2}$

In Exercises 25–36 *determine the derivative* D_xy *by the rules that are normally applicable to the respective function and verify the results through use of logarithmic differentiation.*

25. $y = x^7$

26. $y = x^n$

27. $y = (x^2 + 3x + 2)^{-1/2}$

28. $y = e^x$

29. $y = b^x$

30. $y = 10^{4x+3}$

31. $y = e^{\ln x}$

32. $y = 9^{x^2 + 3x + 7}$

33. $y = (6x + 5)^4(x + 3)^{-2}$

34. $y = \dfrac{x^2}{(x^3 + 2)^3}$

35. $y = \sqrt{x^2 + 2x - 3}$

36. $y = \dfrac{1}{\sqrt{x^2 + 5x - 4}}$

IMPORTANT TERMS AND CONCEPTS

Bell-shaped curve
Derivatives of:
- exponential functions
- implicit functions
- logarithmic functions

Explicit function
Implicit differentiation
Implicit function
Learning curve
Logarithmic differentiation
Normal curve
Normal probability density function
Rate of decay
Rate of growth
Saturation value
Transcendental function

SUMMARY OF RULES AND FORMULAS

Differentiation Rules

Rule 6.1: If $f(x)=e^x$, then $f'(x)=e^x$

Rule 6.2: If $f(x)=e^{u(x)}$, then $f'(x)=e^{u(x)}\cdot u'(x)$

Rule 6.3: If $f(x)=b^x$, then $f'(x)=b^x\ln b$

Rule 6.4: If $f(x)=b^{u(x)}$, then $f'(x)=b^{u(x)}\ln b\cdot u'(x)$

Rule 6.5: If $f(x)=\ln x, x>0$, then $f'(x)=\dfrac{1}{x}$

Rule 6.6: If $f(x)=\ln u(x), u(x)>0$, then $f'(x)=\dfrac{1}{u(x)}\cdot u'(x)$

Rule 6.7: If $f(x)=\log_b x, x>0$, then $f'(x)=\dfrac{1}{x\ln b}$

Rule 6.8: If $f(x)=\log_b u(x), u(x)>0$, then $f'(x)=\dfrac{1}{u(x)\ln b}\cdot u'(x)$

7 INTEGRAL CALCULUS: BASIC METHODOLOGY

7.1 INTRODUCTION

In the study of differential calculus we are concerned with finding the derivative of a function to determine the rate of change of that function. For example, we are interested in finding the rate of growth of a colony of bacteria when we are given a function that represents the number of bacteria present at any time t, or we are interested in finding the marginal profit for a product when we are given a function that represents the total profit as a function of the number of units sold.

On the other hand, in integral calculus we are concerned with finding a function knowing its derivative. For example, we may have a function that represents the rate of growth of a colony of bacteria, and we may wish to determine the function that represents the number of bacteria present at any time t, or we may have a function that represents the marginal profit function for a product, and we may wish to determine the function that represents the total profit for the product.

In this chapter we first consider finding a function when we know its derivative. This process is called *integration* and is much like the reverse of differentiation. We also develop some of the elementary rules of integration and apply these rules to some real-world problems.

7.2 INDEFINITE INTEGRAL

Since the concept of integration involves the determination of a function whose derivative is given, we want to reverse the differentiation process, thus determining an *antiderivative* of the given function. We define an antiderivative of the function $f(x)$ in the following way.

Definition 7.1 An antiderivative of the function $f(x)$ is any function, $F(x)$, whose derivative $F'(x)=f(x)$.

For example, an antiderivative of the function $f(x)=4x^3$ is $F(x)=x^4$ since $F'(x)=4x^3=f(x)$. To determine an antiderivative of most simple functions we need only reverse the differentiation process that led to that function. This can be accomplished by inspection and trial and error. Let us begin by considering an example and establishing some basic terminology.

Example 7.1 Determine an antiderivative of $f(x)=3x^2$.

Solution To determine an antiderivative of $3x^2$, all we need to do is determine a function $F(x)$ such that $F'(x)=3x^2$. Based on our familiarity with finding derivatives by inspection, we note $F(x)=x^3$. Once we have found what we believe is a correct antiderivative, we can always check our answer by differentiating it. In this case $F(x)=x^3$ obviously is a correct antiderivative of $3x^2$.

From Example 7.1 we should also note that $F(x)=x^3+1$, $F(x)=x^3+10$, and $F(x)=x^3+40$ are all antiderivatives of $f(x)=3x^2$ since the derivative of any constant is zero. The *general form of an antiderivative* of $3x^2$ is $F(x)=x^3+C$, where C is a constant.

The process by which a function is determined when its derivative is given is called *antidifferentiation*. We denote the process of determining the general antiderivative by the symbol $\int f(x)\,dx$, so that

$$\int f(x)\,dx = F(x)+C \qquad \text{where } F'(x)=f(x) \tag{7.1}$$

In Eq. (7.1) the function $f(x)$ is called the *integrand*; the symbol $\int$ is called the *integral sign*; the term dx is called the *differential*, which indicates the variable with respect to which the integration is performed; and the term C is called the *constant of integration*. The general form of the antiderivative of $f(x)$, that is, $F(x)+C$, is called the *indefinite integral* of $f(x)$. The symbol $\int f(x)\,dx$ is read "the integral of $f(x)$ with respect to x" or "the integral of $f(x)$'dee'x."

In most problems involving integral calculus, after determining the function $f(x)$ to be integrated, the next step is to determine the indefinite integral of $f(x)$. In those situations involving relatively simple functions, we can determine the indefinite integral by obtaining an antiderivative by "inspection" and adding the constant C to it. In order to determine the value for the constant of integration C, we must have additional information. (We shall discuss this later.)

Example 7.2 Determine

(a) $\int 5x^4\,dx$ (b) $\int 10x^9\,dx$

Solution

(a) By inspection an antiderivative of $5x^4$ is x^5; therefore,

$$\int 5x^4\,dx = x^5 + C$$

(b) An antiderivative of $10x^9$ is x^{10}; therefore,

$$\int 10x^9\,dx = x^{10} + C$$

Example 7.3 Determine $\int e^x\,dx$.

Solution Recalling that $D_x e^x = e^x$, we conclude that an antiderivative of e^x is e^x; therefore,

$$\int e^x\,dx = e^x + C$$

Example 7.4 If $x>0$, determine $\int (1/x)\,dx$.

Solution From Sec. 6.3 we recall that if $x>0$, then $D_x \ln x = 1/x$. Since we are seeking a function whose derivative is $1/x$ we conclude that an antiderivative of $1/x$ is $\ln x$ and

$$\int \frac{1}{x}\,dx = \ln x + C \qquad \text{where } x > 0$$

We found antiderivatives of the integrands in Example 7.1 through 7.4 with little effort. The next two examples require more trial and error.

Example 7.5 Determine $\int x^2\,dx$.

Solution We know that since $F'(x)=x^2$, $F(x)$ must be of the form x^3; however, $D_x(x^3)=3x^2$. To obtain x^2 we must divide $3x^2$ by 3 and hence x^3 by 3. Therefore, $F(x)=\frac{1}{3}x^3$ and

$$\int x^2\,dx = \tfrac{1}{3}x^3 + C$$

Example 7.6 Determine $\int x^6\,dx$.

Solution In a manner analogous to our procedure in Example 7.5, we conclude

$$\int x^6 = \tfrac{1}{7}x^7 + C$$

From Examples 7.5 and 7.6 we conjecture that an antiderivative of x^n is $[1/(n+1)]x^{n+1}$. This is true for all $n \neq -1$ since $D_x\{[1/(n+1)]x^{n+1}\} = x^n$.

(*Note*: When $n=-1$, $1/(n+1)$ is not defined.) Our conjecture provides us with our first rule of integration.

Rule 7.1 If n is any constant not equal to -1, the indefinite integral of x^n is obtained by increasing the exponent of x by 1 and then dividing this expression by the new exponent:

$$\int x^n\,dx = \frac{x^{n+1}}{n+1} + C, \qquad n \neq -1 \tag{7.2}$$

(*Note*: We considered the special case when $n=-1$ in Example 7.4.)

Rule 7.1 is known as the power rule for integration. Additional rules of integration will be considered in the next section.

Example 7.7 Determine $\int x^{-1/2}dx$.

Solution Applying Rule 7.1, we have

$$\int x^{-1/2}dx = \frac{x^{-(1/2)+1}}{\left(-\frac{1}{2}+1\right)} + C$$

$$= \frac{x^{1/2}}{\frac{1}{2}} + C = 2x^{1/2} + C$$

or

$$\int \frac{1}{\sqrt{x}}\,dx = 2\sqrt{x} + C$$

We conclude this section by emphasizing that to determine whether a function $F(x)$ is the indefinite integral of $f(x)$, we need only determine the derivative of $F(x)$ and verify that it is equal to $f(x)$. Thus, we have a fairly easy check of our answer, and we should never leave a problem without making certain that our solution is correct.

EXERCISES

Determine the indefinite integral in each of Exercises 1–20 and check the answer by differentiation.

1. $\int 8x^7\,dx$

2. $\int 4x^3\,dx$

3. $\int x^2\,dx$

4. $\int -e^{-x}\,dx$

5. $\int 2e^{2x}\,dx$

6. $\int 7\,dx$

7. $\int dx$

8. $\int x^4 dx$

9. $\int -\frac{1}{x^2} dx$

10. $\int \frac{-2}{x^3} dx$

11. $\int 2x^5 dx$

12. $\int 2x^9 dx$

13. $\int x^{1/2} dx$

14. $\int 3\sqrt{x}\, dx$

15. $\int \frac{1}{\sqrt[3]{x}} dx$

16. $\int x^{-4} dx$

17. $\int e^{4x} dx$

18. $\int \sqrt[4]{x}\, dx$

19. $\int \frac{2}{x} dx, x>0$

20. $\int 2x^{3/2} dx$

7.3 RULES OF INTEGRATION

Since differentiation and integration are the reverse of each other, many of the rules of integration are analogous to the rules of differentiation. This is especially true for some of the related algebraic rules. In this section we shall develop a few general rules of integration comparable to Rule 7.1 of the last section. We will also establish the important algebraic and general rules for integrals.

Rule 7.1, by itself, is very limited in its applications. For example, we cannot find $\int 3x^6 dx$ with it because the integrand contains an extra factor of 3. If we could factor out the 3 and place it in front of the integral sign and write

$$\int 3x^6 dx = 3 \cdot \int x^6 dx$$

then we could apply Rule 7.1 to determine that $\int x^6 dx = \frac{1}{7}x^7$ and conclude

$$\int 3x^6 dx = 3 \cdot \int x^6 dx = 3\left(\tfrac{1}{7}x^7\right) + C$$

or

$$\int 3x^6 dx = \tfrac{3}{7}x^7 + C$$

Differentiating $(\frac{3}{7}x^7 + C)$ we obtain $3x^6$, which implies this factoring out process may be justified. It is in fact the basis for the following rule.

Rule 7.2 The integral of a constant times a function is equal to the constant times the integral of the function:

$$\int kf(x)\,dx = k\int f(x)\,dx \qquad (7.3)$$

where k is a given constant.

This rule states that a *constant factor* in the integrand may be placed before the integral sign or a constant factor times an integral can be placed in the integrand. (*A word of caution*: A variable cannot be moved in this way.)

Example 7.8 Determine $\int 7x^{1/2}dx$.

Solution By Rule 7.2

$$\int 7x^{1/2}dx = 7\int x^{1/2}dx$$

and by Rule 7.1

$$7\int x^{1/2}dx = 7\cdot\frac{x^{(1/2)+1}}{\frac{1}{2}+1} + C = 7\cdot\tfrac{2}{3}x^{3/2} + C = \tfrac{14}{3}x^{3/2} + C$$

Analogous to the rule for finding the derivative of the sum of two functions is the following rule, which determines the integral of the sum of two functions.

Rule 7.3 The integral of the sum of two functions is equal to the sum of their respective integrals:

$$\int [f(x)+g(x)]\,dx = \int f(x)\,dx + \int g(x)\,dx \tag{7.4}$$

This rule can be extended to include the sum of three or more functions; furthermore, Rules 7.2 and 7.3 can be combined to prove that the integral of the difference of two functions is equal to the difference of the respective integrals.

Example 7.9 Determine $\int (5x^{-2} + \sqrt{x}\,)dx$.

Solution Rewriting both terms of the integrand with rational exponents and using Rule 7.3, we have

$$\int (5x^{-2}+x^{1/2})\,dx = \int 5x^{-2}dx + \int x^{1/2}dx$$

Applying Rules 7.2 and 7.1 to the right side, we have

$$\int (5x^{-2}+x^{1/2})\,dx = 5\int x^{-2}dx + \int x^{1/2}dx$$

$$= \frac{5x^{-2+1}}{-2+1} + \frac{x^{1/2+1}}{\frac{1}{2}+1} + C$$

or
$$\int (5x^{-2}+\sqrt{x}\,)\,dx = -5x^{-1}+\tfrac{2}{3}x^{3/2}+C$$

Rule 7.1 can be combined with the shortcut chain rule for differentiating composite functions to obtain the following rule.

Rule 7.4 If n is any constant that does not equal -1, the integral of $[u(x)]^n$ times the derivative of $u(x)$ with respect to x is obtained by increasing the exponent of $[u(x)]$ by 1 and dividing the result by the new exponent:

$$\int [u(x)]^n \cdot u'(x)\,dx = \frac{[u(x)]^{n+1}}{n+1} + C, \qquad n \neq -1 \tag{7.5}$$

This rule and the two following are justified by the fact that the derivative of the right side of each equation is equal to the integrand on the left side of the equation.

Rule 7.4 implies that whenever an integrand can be written as the product of two functions we should check to determine if one of the factors is of the form $[u(x)]^n, n\neq -1$, and the other is $u'(x)$. If this is the case, we can apply this rule directly.

Example 7.10 Determine $\int (x^3+5)^{1/2}3x^2\,dx$.

Solution Although $(x^3+5)^{1/2}3x^2$ might appear to be the product of two unrelated functions, we see that if we let $u(x)=(x^3+5)$, then $u'(x)=3x^2$ and this integral is of the form $\int [u(x)]^{1/2}u'(x)\,dx$. By Rule 7.4

$$\int (x^3+5)^{1/2}3x^2\,dx = \frac{(x^3+5)^{(1/2)+1}}{\frac{1}{2}+1} + C$$
$$= \tfrac{2}{3}(x^3+5)^{3/2} + C$$

(It is instructive to take the derivative of the right side of this equation and show it is equal to the integrand on the left side.)

For emphasis we state again that the major task in applying Rule 7.4 is to identify the $u(x)$ and $u'(x)$ in the integrand. Note that $[u(x)]^n u'(x) = u'(x)[u(x)]^n$; hence, the $u'(x)$ may be the first or second factor in the product.

We have already shown that $\int e^x\,dx = e^x + C$. This result combined with the chain rule for differentiation yields the following rule.

Rule 7.5 The integral of $e^{u(x)}$ times the derivative of $u(x)$ with respect to x is equal to $e^{u(x)}+C$:

$$\int e^{u(x)}u'(x)\,dx = e^{u(x)} + C \tag{7.6}$$

Whenever there is an exponential function in the integrand we should let $u(x)$ equal the exponent of the base e, determine $u'(x)$, and check to see if the integral is of the form shown in Eq. (7.6).

Example 7.11 Determine $\int e^{x^2} 2x\,dx$.

Solution Letting $u(x) = x^2$, the exponent of the base e, and taking the derivative of $u(x)$, we find $u'(x) = 2x$. Hence, the integrand $e^{x^2}2x$ is of the form $e^{u(x)}u'(x)$. Therefore, by Rule 7.5 we have

$$\int e^{x^2} 2x\,dx = e^{x^2} + C$$

Our last rule of this section involves the other transcendental function studied in Chapter 6, the logarithmic function.

Rule 7.6 The integral of the derivative of $u(x)$ with respect to x divided by $u(x)$ equals the natural logarithm of the absolute value of $u(x)$.

$$\int \frac{u'(x)}{u(x)}\,dx = \ln|u(x)| + C \tag{7.7}$$

(The absolute value is necessary to insure we are considering the logarithm of a positive quantity.)

This rule implies that whenever an integrand is the ratio of two functions we should check to determine if the numerator is the derivative of the denominator. If this is the case, we can apply this rule directly.

Example 7.12 Determine $\int \frac{3}{3x+1}\,dx$.

Solution Since the numerator, 3, equals the derivative of the denominator, $(3x+1)$, we can apply Rule 7.6 where $u(x) = 3x+1$ and $u'(x) = 3$; that is,

$$\int \frac{3}{3x+1}\,dx = \ln|(3x+1)| + C$$

where $x \neq -\frac{1}{3}$.

In applying Rules 7.4, 7.5, and 7.6 we frequently encounter cases where after identifying the $u(x)$ function, we find that the other function in the integrand differs from $u'(x)$ by a constant factor. This problem can be resolved as shown in the next two examples.

Example 7.13 Determine $\int (x^3+5)^{1/2} x^2\,dx$.

Solution The integrand $(x^3+5)^{1/2}x^2$ appears to be of the form $[u(x)]^n u'(x)$; however, if we let $u(x)=(x^3+5)$, then $u'(x)=3x^2$, which differs from our remaining function x^2 in the integrand by the constant factor of 3. To rectify this we multiply the integrand by $\frac{1}{3}\cdot 3$, i.e., 1, and apply Rule 7.2 to obtain

$$\int (x^3+5)^{1/2}x^2\,dx = \int (x^3+5)^{1/2}\cdot\tfrac{1}{3}\cdot 3x^2\,dx = \tfrac{1}{3}\cdot\int (x^3+5)^{1/2}3x^2\,dx$$

Now by Rule 7.4,

$$\tfrac{1}{3}\int (x^3+5)^{1/2}3x^2\,dx = \tfrac{1}{3}\left[\tfrac{2}{3}(x^3+5)^{3/2}\right]$$

Therefore, $\int (x^3+5)^{1/2}x^2\,dx = \frac{2}{9}(x^3+5)^{3/2}+C$.

Example 7.14 Determine $\int xe^{-x^2}\,dx$.

Solution Since the integrand contains an exponential function $e^{u(x)}$ where $u(x)=-x^2$, we determine $u'(x)=-2x$ and check to see if we have an integral of the form $\int u'(x)e^{u(x)}\,dx$. Doing this, we note we need a factor of (-2) in the integrand. Therefore, we multiply the integrand by $(-\frac{1}{2})(-2)$, and obtain by Rule 7.2:

$$\int xe^{-x^2}\,dx = \int \left(-\tfrac{1}{2}\right)(-2)xe^{-x^2}\,dx = -\tfrac{1}{2}\int -2xe^{-x^2}\,dx$$

Now by Rule 7.5, $-\frac{1}{2}\int -2xe^{-x^2}\,dx = -\frac{1}{2}e^{-x^2}+C$. Therefore, $\int xe^{-x^2}\,dx = -\frac{1}{2}e^{-x^2}+C$.

Table 7.1

Rule	*Integration Formula*
7.1	$\int x^n\,dx = \dfrac{x^{n+1}}{n+1}+C, \qquad n\neq -1$
7.2	$\int kf(x)\,dx = k\int f(x)\,dx, \qquad k = \text{a constant}$
7.3	$\int [f(x)+g(x)]\,dx = \int f(x)\,dx + \int g(x)\,dx$
7.4	$\int [u(x)]^n\cdot u'(x)\,dx = \dfrac{[u(x)]^{n+1}}{n+1}+C, \qquad n\neq -1$
7.5	$\int e^{u(x)}u'(x)\,dx = e^{u(x)}+C$
7.6	$\int \dfrac{u'(x)}{u(x)}\,dx = \ln\lvert u(x)\rvert + C$

The six integration rules summarized in Table 7.1 provide us with the tools to find the indefinite integral of many functions; however, there are others that require more sophisticated methods. For example, we cannot determine $\int xe^x\,dx$ by the above rules. (*Note*: If $u(x)=x$, $u'(x)=1$ and the integrand contains an "extra factor" of x.) We will discuss these more advanced methods in Chapter 8.

EXERCISES

In Exercises 1–50 *determine the indefinite integral and check the answer by differentiation.*

1. $\int 3x^4\,dx$

2. $\int 7x^9\,dx$

3. $\int 2x^{-5}\,dx$

4. $\int 4y^{-7}\,dy$

5. $\int 6u^{3/2}\,du$

6. $\int 5v^{-2/3}\,dv$

7. $\int 5\sqrt{z}\,dz$

8. $\int \frac{1}{\sqrt{x}}\,dx$

9. $\int 2\sqrt[3]{x}\,dx$

10. $\int (3x^3+2x-4)\,dx$

11. $\int (2y^2-4y+3)\,dy$

12. $\int (3x^{-2}+\sqrt{x}+6)\,dx$

13. $\int \left(\frac{3}{y^4}-\frac{4}{y^5}\right)dy$

14. $\int \left(x^2-\frac{3}{x^3}+4\right)dx$

15. $\int \left(\frac{1}{\sqrt[4]{y}}-5y^{-2}\right)dy$

16. $\int (2\sqrt[3]{x}-3\sqrt{x})\,dx$

17. $\int \left(\frac{1+x}{x^4}\right)dx$

18. $\int \left(\frac{2+y}{\sqrt{y}}\right)dy$

19. $\int (1-y^3)^2\,dy$

20. $\int t(2t^2-3)^2\,dt$

21. $\int (e^x-5x^{1/3}+2)\,dx$

22. $\int (e^{3x}-4x)\,dx$

23. $\int (x^2-2x+1)^5(2x-2)\,dx$

24. $\int 6y^2(2y^3+1)^{-6}\,dy$

25. $\int (x^2+1)^{1/2}2x\,dx$

26. $\int (u^3+4)^{-1/2}3u^2\,du$

27. $\int \frac{6x}{(3x^2+4)^{1/2}}\,dx$

28. $\int (t^2+3t)^5(2t+3)\,dt$

29. $\int (3t^2-1)^{10} 6t\,dt$

30. $\int \sqrt{(1-u^3)}\,(-3u^2)du$

31. $\int (x+1)^5 dx$

32. $\int \frac{1}{(x+1)^4} dx$

33. $\int (x^2+1)^{-1/2} x\,dx$

34. $\int (x^3+4)^{1/3} x^2 dx$

35. $\int \frac{(x+1)}{(x^2+2x+1)^2} dx$

36. $\int \frac{(2u^3+u)}{\sqrt{u^4+u^2+1}} du$

37. $\int e^{2x} 2\,dx$

38. $\int e^{-3x} dx$

39. $\int e^{x^2} 2x\,dx$

40. $\int e^{ax} dx, \quad a \neq 0$

41. $\int e^{-\frac{1}{2}x} dx$

42. $\int e^{x^3} x^2 dx$

43. $\int \frac{2x}{x^2+1} dx$

44. $\int \frac{3u^2}{u^3+6} du$

45. $\int \frac{t}{4t^2+1} dt$

46. $\int \frac{x}{(x^2+3)^3} dx$

47. $\int \frac{dx}{3x+1}$

48. $\int x^2 e^{-x^3} dx$

49. $\int \frac{e^x}{e^x+1} dx$

50. $\int \frac{t}{t^2+1} dt$

7.4 MARGINAL ANALYSIS

In Sec. 5.5 we noted that the marginal cost function was obtained by determining the derivative of the total cost function; the same holds true for the marginal profit and marginal revenue functions. Now that we have the reverse process of differentiation, we can determine the total cost (or profit or revenue) function from the marginal cost (or profit or revenue) function. For example, if the marginal cost function is given by $c(x)=2x^2+6x+13$ where x represents the number of units produced, the total cost function $C(x)$ is given by

$$C(x) = \int c(x)\,dx$$

$$= \int (2x^2+6x+13)\,dx$$

or

$$C(x) = \tfrac{2}{3}x^3 + 3x^2 + 13x + C_0 \tag{7.8}$$

where C_0 is the constant of integration. [C_0 is used in place of C to avoid confusion with $C(x)$.] The question now arises as to how the constant of integration may be determined.

From Eq. (7.8) it is evident that we can determine the value of C_0 if we know the value of $C(x)$ for some particular value of x. For example, when no units are produced, i.e., $x=0$, we have

$$C(0) = \tfrac{2}{3}(0)^3 + 3(0)^2 + 13(0) + C_0$$

or

$$C(0) = C_0$$

The cost associated with producing 0 units, $C(0)$, is called the *fixed cost* (FC). If the fixed cost $C(0)$ for the above problem is \$4,000, then $C_0=\$4{,}000$ and the total cost function is given by

$$C(x) = \tfrac{2}{3}x^3 + 3x^2 + 13x + 4{,}000$$

Furthermore, the *variable cost* of producing x units, $VC(x)$, is given by

$$VC(x) = \tfrac{2}{3}x^3 + 3x^2 + 13x$$

Example 7.15 Fixed Cost, Marginal Cost, and Total Cost The fixed cost of manufacturing a product is \$5,000 and the marginal cost $c(x)$ is \$3/unit. Determine the total cost function and the cost for manufacturing 6,000 units.

Solution The total cost function is given by

$$C(x) = \int 3\,dx = 3x + C_0$$

Since $C(0)=\$5{,}000$, we find by substitution

$$5{,}000 = 3 \cdot 0 + C_0 \qquad \text{or} \qquad C_0 = 5{,}000$$

Therefore,

$$C(x) = 3x + 5{,}000$$

The cost of manufacturing 6,000 units is

$$C(6{,}000) = 3(6{,}000) + 5{,}000 = \$23{,}000$$

Example 7.16 If the marginal cost function in dollars for x tons of raw material is given by $c(x)=3\sqrt{x+100}$ and the fixed cost is \$4,000, determine the total cost function. Also determine the total cost of 21 tons of raw material.

Solution The total cost function is given by

$$C(x) = \int 3\sqrt{x+100}\ dx = \int 3(x+100)^{1/2} dx$$
$$= 2(x+100)^{3/2} + C_0$$

Since $C(0) = \$4{,}000$, we find by substitution

$$4{,}000 = 2(0+100)^{3/2} + C_0$$

or

$$2{,}000 = C_0$$

Hence, $C(x) = 2(x+100)^{3/2} + 2{,}000$. Then,

$$C(21) = 2(21+100)^{3/2} + 2{,}000$$
$$= 4{,}662$$

Hence, the cost of 21 tons of raw material is \$4,662.

Example 7.17 Given the marginal output function $MO(x) = 100x - x^2$ in units of finished product where x is the number of input units of raw material, determine the total output function and the total output when input is 45 units.

Solution The total output function $TO(x)$ is given by

$$TO(x) = \int (100x - x^2)\, dx = 50x^2 - \frac{1}{3}x^3 + C$$

Since there is no output unless there is some input, $TO(0) = 0$. Therefore, substituting $x = 0$ into the above equation shows $C = 0$ and

$$TO(x) = 50x^2 - \tfrac{1}{3}x^3$$

Therefore,

$$TO(45) = 50(45)^2 - \frac{(45)^3}{3}$$
$$= 70{,}875$$

Hence, the total output given 45 input units is 70,875 units.

EXERCISES

1. The marginal cost in dollars for a particular operation is given by

$$c(x) = 3x^2 - 42x + 360$$

where x represents the number of units made. The fixed cost is \$3,025. Determine the total cost function and the total cost for 10 units.

2. The marginal profit function in hundreds of dollars for a particular product is given by

$$p(x) = -3x^2 + 60x + 600$$

where x represents the dollars in thousands spent on advertising. A cost of \$1,000 is incurred from miscellaneous expenses. Determine the total profit function and the total profit for spending \$10,000 on advertising.

3. The marginal cost of making x units of a product is given by

$$c(x) = 0.0006x^2 - 0.6x + 50$$

The fixed cost is \$100,000. Determine the total cost function and the total cost of producing 1,000 units.

4. If the marginal cost in dollars of making x units is $c(x)=0.0027x^2-1.08x+108$ and the fixed cost is \$2,000, determine the total cost function and the cost of producing 200 units.

5. The marginal cost of manufacturing x units of a certain type of radio equipment for military aircraft is given by

$$c(x) = 3x^2 - 30x + 2{,}000$$

and the fixed cost is \$4,000. Determine the total cost function and the cost of manufacturing 10 units.

6. The marginal profit in hundreds of dollars for a building contractor is given by

$$p(x) = 400 - 80x - 50e^{-x}$$

where x represents the number of four-unit apartment houses constructed. The contractor signed a contract which guarantees him a profit of \$5,000 if he is not permitted to build any apartments. Determine the total payment function and the profit for building six apartment houses.

7. The Krebbie Vacuum Cleaner Company has determined that their marginal profit in hundreds of dollars is given by

$$p(x) = 90x^{1/2} - 200$$

where x represents the number of sales representatives. When they have no sales reps, there is no profit (or loss). Determine the total profit function and the profit for 250 sales reps.

8. The manufacturer of Gungho Widgets has determined that the marginal cost in dollars of manufacturing x widgets is given by

$$c(x) = 4x - 10$$

and the marginal sales revenue in dollars is given by

$$r(x) = 2x + 40$$

A loss of \$300 is incurred if no widgets are sold. Determine the total profit function and the profit for producing and selling 20 widgets.

9. An insurance salesperson has determined that her marginal cost per month is given by

$$c(x) = 0.03x^2 + 0.2x + 0.05$$

where x represents the number of clients she sees per week. Her fixed costs are \$150. Determine the total cost function and the total cost if she sees 50 clients in a week.

10. A theater has determined that its marginal profit for a show is given by

$$p(x) = 0.5x^{-1/2} + 0.3x^{1/2} + 0.4$$

where x is the number of customers. The theater has fixed costs of \$50 per show. Determine the total profit function and the total profit for 225 customers.

11. The Have-All Department Store has been told by the ABC television network that the marginal cost for TV advertising on their network is given by

$$c(x) = 10x^2 + 50x - 180$$

where x is the number of minutes of advertising per week. There is not a fixed charge for advertising on the ABC network. Determine the total cost function and the total cost for 18 minutes of advertising per week.

12. A manufacturing company has determined that the marginal cost of manufacturing x bolts of cloth is given by

$$c(x) = 1.5x + 2$$

and the marginal sales revenue in dollars is given by

$$r(x) = 2x$$

Fixed costs are \$5500. Determine the total profit function and the total profit for producing and selling 500 bolts of cloth.

13. The regional manager of the World Encyclopedia Company has determined that the marginal revenue in thousands of dollars is given by

$$r(x) = \frac{1}{\sqrt[3]{x}} + .1$$

where x represents the number of district sales managers. When there are no managers, income is \$10,000 from direct mail orders. Determine the revenue function and the total revenue when there are 27 district managers.

7.5 DEFINITE INTEGRAL

Mechanically speaking there is little difference between determining the value for a *definite integral* and the function for an indefinite integral; however, the development of each of the concepts is quite different. We have established that the indefinite integral of $f(x)$ is the most general form of the antiderivative of $f(x)$ and is thus a function. The definite integral is defined as a number. First we shall consider the definition of a definite integral and then relate the definition to the concept of an indefinite integral.

In order to define a definite integral, let $f(x)$ be a function that is continuous in the interval $[a,b]$ as shown in Fig. 7.1. Although it is not necessary for the function to be continuous, we will consider only definite integrals of continuous functions.

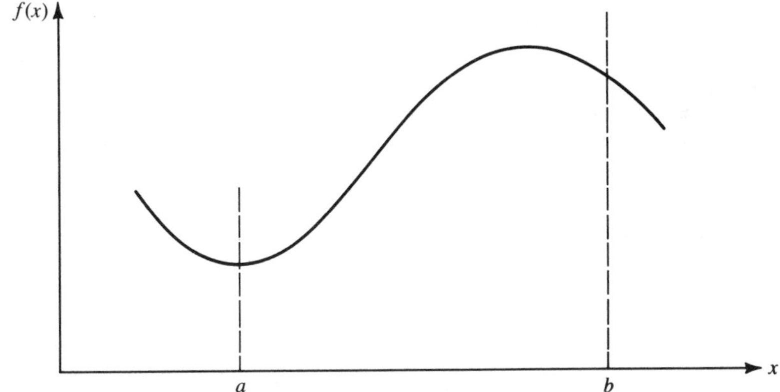

FIGURE 7.1

Next, we divide the interval $[a,b]$ into n subintervals as shown in Fig. 7.2 and let the n points $x_1, x_2, \ldots, x_n$, where $a<x_1<x_2<\cdots<x_n<b$ represent the midpoints of the respective subintervals. If the subintervals are of equal

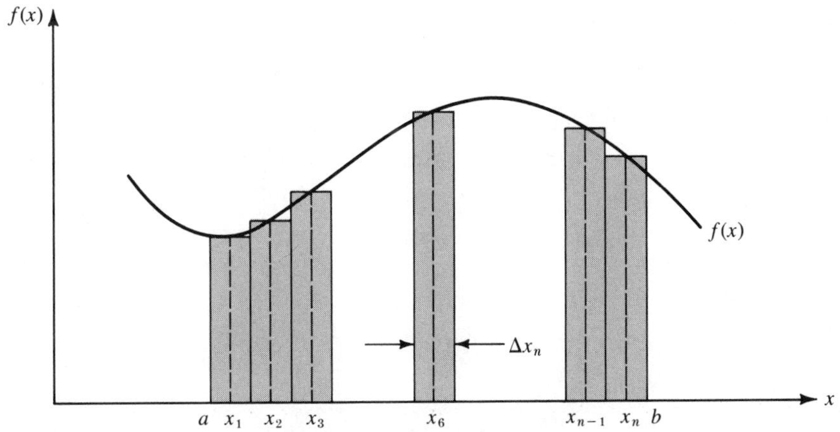

FIGURE 7.2

width, then each of them will have width Δx_n, where

$$\Delta x_n = \frac{b-a}{n}$$

Thus, we see that as n increases, Δx_n becomes smaller. Figure 7.2 demonstrates this *partitioning* of the interval from a to b and the division of the area between the curve and the x axis into rectangles for which the base is Δx_n and the height is given by the value of the function at the midpoint x_i of the subinterval. That is, for the rectangle in the ith subinterval, the base is Δx_n and the height is $f(x_i)$.

For the partitioning in Fig. 7.2, we form the sum of the areas of the rectangles

$$S_n = f(x_1)\Delta x_n + f(x_2)\Delta x_n + \cdots + f(x_n)\Delta x_n$$

which can be represented in summation notation as

$$S_n = \sum_{i=1}^{n} f(x_i)\Delta x_n$$

If we permit the division to become finer by letting n increase, it is likely that we shall obtain different values of S_n. If we permit n to increase without bound, that is, $n \to \infty$, the value of $S_n \to$ a number L. This number L is called the *definite integral* of $f(x)$ from a to b and is denoted by the symbol

$$\int_a^b f(x)\,dx$$

that is,

$$\int_a^b f(x)\,dx = \lim_{n\to\infty} \sum_{i=1}^{n} f(x_i)\Delta x_n \tag{7.9}$$

In the definite integral $\int_a^b f(x)\,dx$, the function $f(x)$ is called the *integrand* and the constants a and b are called, respectively, the *lower limit* and the *upper limit* of the integral. The definite integral is read "the integral from a to b of $f(x)\,dx$." The evaluation of $\int_a^b f(x)\,dx$ from the definitional formula (7.9) is not a very workable method of obtaining the value of a definite integral. However, because there exists a relationship between the definite integral and the indefinite integral, there is a simple and direct method for this evaluation. This relationship is now given without proof because it is beyond the scope of this book and can be found in almost any advanced calculus text.

Rule 7.7 *Fundamental theorem of integral calculus*: If $f(x)$ is a continuous function in the interval $[a,b]$ and $F(x) = \int f(x)\,dx$ is the indefinite integral of $f(x)$, the value of the definite integral $\int_a^b f(x)\,dx$ is given by

$$\int_a^b f(x)\,dx = F(b) - F(a) \tag{7.10}$$

This result connects the two fundamental concepts of integral calculus: the definite and the indefinite integrals.

Another way of denoting (7.10) is

$$\int_a^b f(x)\,dx = F(x)\Big|_a^b \tag{7.11}$$

where $F(x)|_a^b = F(b) - F(a)$. (This notation is usually more convenient because it is more compact.)

Example 7.18 Determine the value of $\int_0^2 x\,dx$.

Solution First we determine the indefinite integral to be

$$F(x) = \int x\,dx = \frac{x^2}{2} + C$$

Thus,

$$\int_0^2 x\,dx = \frac{x^2}{2} + C\Big|_0^2$$

$$= \left[\frac{(2)^2}{2} + C\right] - \left[\frac{(0)^2}{2} + C\right]$$

$$= 2$$

From Example 7.18 we see that the value of the constant of integration is irrelevant because it will always cancel itself. Thus, in (7.10) and (7.11), when we write the indefinite integral, we shall omit the constant of integration. Furthermore, since the value of a definite integral is determined through use of indefinite integrals, we note that any rule that is applicable for indefinite integrals is also applicable for definite integrals.

Example 7.19 Determine the value of $\int_1^5 (3x^2 + 4x + 3)\,dx$.

Solution

$$\int_1^5 (3x^2 + 4x + 3)\,dx = x^3 + 2x^2 + 3x\Big|_1^5$$

$$= \left[(5)^3 + 2(5)^2 + 3(5)\right] - \left[(1)^3 + 2(1)^2 + 3(1)\right]$$

$$= 190 - 6$$

$$= 184$$

Example 7.20 Evaluate $\int_{-2}^{1}(x^3+2x)\,dx$.

Solution

$$\begin{aligned}\int_{-2}^{1}(x^3+2x)\,dx &= \left.\frac{x^4}{4}+x^2\right|_{-2}^{1}\\ &=\left[\frac{(1)^4}{4}+(1)^2\right]-\left[\frac{(-2)^4}{4}+(-2)^2\right]\\ &=\frac{5}{4}-8\\ &=-\frac{27}{4}\end{aligned}$$

There are many applications of definite integrals and we shall discuss several of these in the next two sections and in Chapter 8. However, first let us consider some fundamental properties of definite integrals. Each of the following rules can be developed by referring to Eq. (7.10) and the rules for indefinite integrals.

Rule 7.8 If c is any point such that $a \leqslant c \leqslant b$, we can express the definite integral of $f(x)$ from a to b as two separate definite integrals of $f(x)$, one from a to c and the second from c to b; that is,

$$\int_a^b f(x)\,dx = \int_a^c f(x)\,dx + \int_c^b f(x)\,dx \tag{7.12}$$

This rule can be verified by applying Eq. (7.10) to each term in (7.12). Since the integrand is $f(x)$ in each case, we have

$$F(b) - F(a) = F(c) - F(a) + F(b) - F(c)$$

Example 7.21 Verify that $\int_2^4 x^2\,dx = \int_2^3 x^2\,dx + \int_3^4 x^2\,dx$.

Solution Evaluating, we have

$$\int_2^4 x^2\,dx = \left.\frac{1}{3}x^3\right|_2^4 = \frac{1}{3}(4)^3 - \frac{1}{3}(2)^3 = \frac{64}{3} - \frac{8}{3} = \frac{56}{3}$$

$$\int_2^3 x^2\,dx = \left.\frac{1}{3}x^3\right|_2^3 = \frac{1}{3}(3)^3 - \frac{1}{3}(2)^3 = \frac{27}{3} - \frac{8}{3} = \frac{19}{3}$$

$$\int_3^4 x^2\,dx = \left.\frac{1}{3}x^3\right|_3^4 = \frac{1}{3}(4)^3 - \frac{1}{3}(3)^3 = \frac{64}{3} - \frac{27}{3} = \frac{37}{3}$$

Therefore, $\int_2^4 x^2\,dx = \int_2^3 x^2\,dx + \int_3^4 x^2\,dx$ since $\frac{56}{3} = \frac{19}{3} + \frac{37}{3}$.

The following rule is a direct result of Rule 7.2 and Eq. (7.3).

Rule 7.9 If k is any constant

$$\int_a^b kf(x)\,dx = k\int_a^b f(x)\,dx \tag{7.13}$$

Another rule that is quite useful at times and which results from Rule 7.3 and Eq. (7.4) is the following.

Rule 7.10 If $f(x)$ and $g(x)$ are continuous functions in the interval $[a,b]$

$$\int_a^b [f(x)+g(x)]\,dx = \int_a^b f(x)\,dx + \int_a^b g(x)\,dx \tag{7.14}$$

This rule can be extended to include the sum of three or more functions, and Rules 7.9 and 7.10 can be combined to prove that the definite integral of the difference of two functions is equal to the difference of the respective definite integrals.

Example 7.22 Verify that $\int_0^1 (e^x+x)\,dx = \int_0^1 e^x\,dx + \int_0^1 x\,dx$.

Solution Evaluating, we have

$$\int_0^1 (e^x+x)\,dx = e^x + \frac{1}{2}x^2\Big|_0^1 = \left(e^1+\frac{1}{2}\right)-(e^0)$$

$$= e+\frac{1}{2}-1 = e-\frac{1}{2}$$

$$\int_0^1 e^x\,dx = e^x\Big|_0^1 = e^1-e^0 = e-1$$

$$\int_0^1 x\,dx = \frac{1}{2}x^2\Big|_0^1 = \frac{1}{2}$$

Therefore, $\int_0^1 (e^x+x)\,dx = \int_0^1 e^x\,dx + \int_0^1 x\,dx$ since $e-\frac{1}{2}=e-1+\frac{1}{2}$.

Another property of definite integrals that is worthy of mention is that, like derivatives, we may use any symbol for the independent variable because the definite integral depends only on the integrand and the limits a and b. Thus, for example, we have

$$\int_a^b f(x)\,dx = \int_a^b f(t)\,dt = \int_a^b f(u)\,du = \int_a^b f(z)\,dz$$

EXERCISES

In Exercises 1–20 evaluate the given definite integral.

1. $\int_1^4 3x^2\,dx$
2. $\int_0^3 5x^4\,dx$
3. $\int_0^3 (x^2-x)\,dx$
4. $\int_{-2}^0 (3y^2-2y)\,dy$
5. $\int_1^4 (z-1)^3\,dz$
6. $\int_0^2 t(t^2-4)^3\,dt$
7. $\int_1^{13} \frac{1}{\sqrt{2y-1}}\,dy$
8. $\int_{-2}^2 \sqrt{2-t}\,dt$
9. $\int_0^4 (2-\sqrt{x})^2\,dx$
10. $\int_{-1}^1 t(t^2+1)^3\,dt$
11. $\int_1^2 (2t^2+3)^2\,dt$
12. $\int_{-1}^1 (t^3-1)^2\,dt$
13. $\int_{-1}^1 y\sqrt{4-3y^2}\,dy$
14. $\int_0^4 3y\sqrt{y^2+9}\,dy$
15. $\int_0^3 \frac{z}{\sqrt{16+z^2}}\,dz$
16. $\int_1^4 \frac{3x+2}{\sqrt{x}}\,dx$
17. $\int_1^6 \frac{2}{\sqrt{x+3}}\,dx$
18. $\int_4^{16} \frac{2x^2+2x+1}{x^{3/2}}\,dx$
19. $\int_2^8 \frac{2}{x}\,dx$
20. $\int_1^3 2e^{3x}\,dx$

In Exercises 21–25 verify by evaluation that the given equation is a true statement.

21. $\int_0^4 (x^2-\tfrac{1}{2}x)\,dx = \int_0^2 (x^2-\tfrac{1}{2}x)\,dx + \int_2^4 (x^2-\tfrac{1}{2}x)\,dx$
22. $\int_{-1}^1 e^x\,dx = \int_{-1}^0 e^x\,dx + \int_0^1 e^x\,dx$
23. $\int_0^1 (x+e^{-x})\,dx = \int_0^1 x\,dx + \int_0^1 e^{-x}\,dx$
24. $\int_0^4 (\sqrt{x}+e^x)\,dx = \int_0^4 \sqrt{x}\,dx + \int_0^4 e^x\,dx$
25. $\int_0^1 \frac{u\,du}{\sqrt{u^2+1}} = \int_0^1 \frac{x\,dx}{\sqrt{x^2+1}}$

7.6 AREA BY INTEGRATION

Based on Fig. 7.2 it seems reasonable to conclude that as the width, Δx_n, of each rectangle gets smaller and smaller, the sum of the areas of these n rectangles gets closer and closer to the area of the region between the graph of the function $f(x)$ and the x axis bounded by $x=a$ and $x=b$. Since we have defined the limit of this sum as n increases without bound as the definite integral from a to b of the function $f(x)$, the following rule seems justifiable.

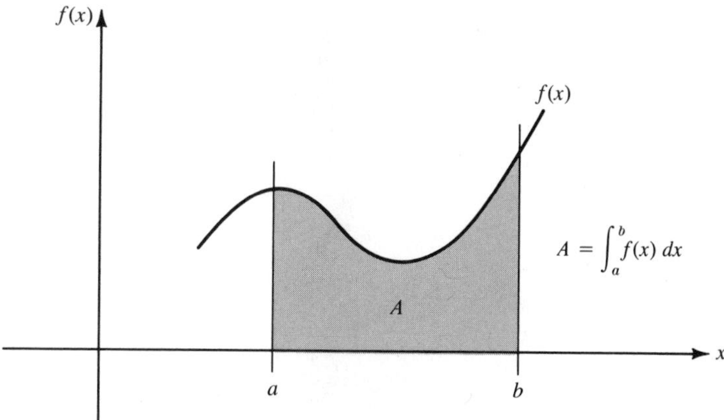

FIGURE 7.3

Rule 7.11 The *area of the region* between the continuous nonnegative function $f(x)$ and the x axis bounded by $x=a$ and $x=b$ (Fig. 7.3) is given by the equation

$$A = \int_a^b f(x)\,dx \tag{7.15}$$

To support this conclusion, let us examine two cases where we know the area of the regions being considered. The areas of the regions shown in Figs. 7.4*a* and *b* can be determined by using formulas from elementary geometry. The area of the rectangular region in Fig. 7.4*a* is 32 square units ($A=\text{base}\times\text{height}$) and the area of the triangular region in Fig. 7.4*b* is 8 square units ($A=\frac{1}{2}\text{base}\times\text{height}$).

The area of the shaded region in Fig. 7.4*a* by Eq. (7.15) is

$$\int_a^b f(x)\,dx$$

where $a=0$, $b=4$, and $f(x)=8$. Evaluating the definite integral, we have

$$\int_0^4 8\,dx = 8x\Big|_0^4 = 8(4) - 8(0) = 32 \text{ square units}$$

which is the correct answer. Similarly, the area of the shaded region in Fig. 7.4*b* by Eq. (7.15) is

$$\int_a^b f(x)\,dx = \int_0^4 x\,dx = \frac{1}{2}x^2\Big|_0^4 = \frac{1}{2}(4)^2 - \frac{1}{2}(0)^2 = 8 \text{ square units}$$

which is also the correct answer.

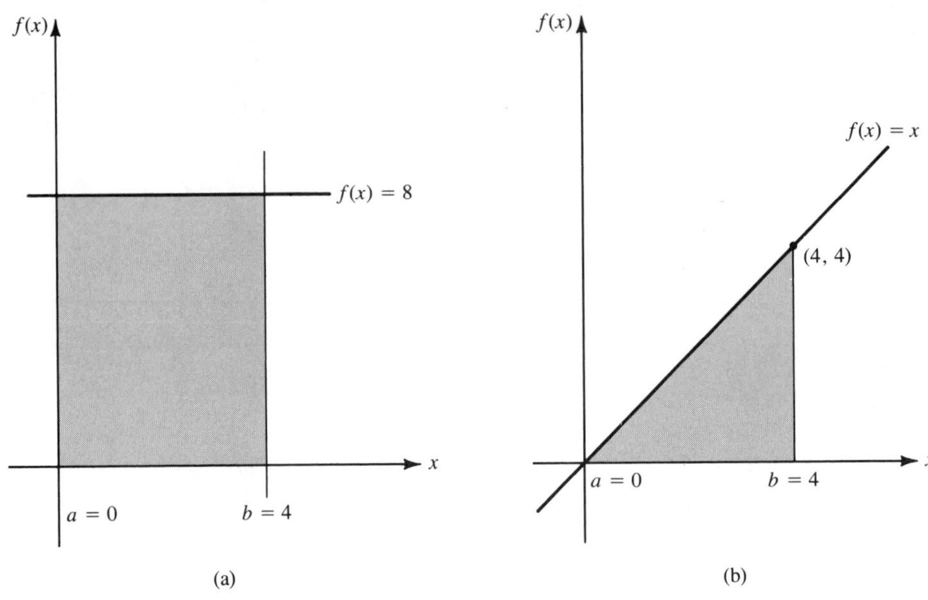

FIGURE 7.4

Having demonstrated that Rule 7.11 is true for these two special cases there seems little doubt of its validity. To remove any doubt, consider the following. If we let A be the area between the curve $f(x)$ and the x axis bounded by the lines $x=a$ and $x=x$, as shown in Fig. 7.5, then A is a function of x, $A(x)$. The area $A(x+h)$ is the area between $f(x)$ and the x axis

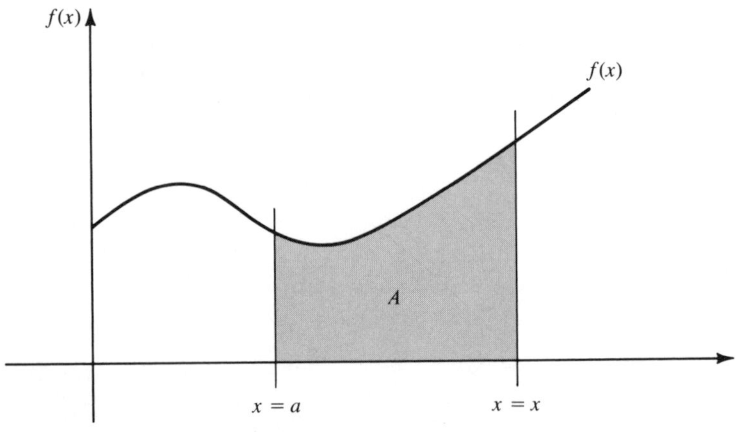

FIGURE 7.5

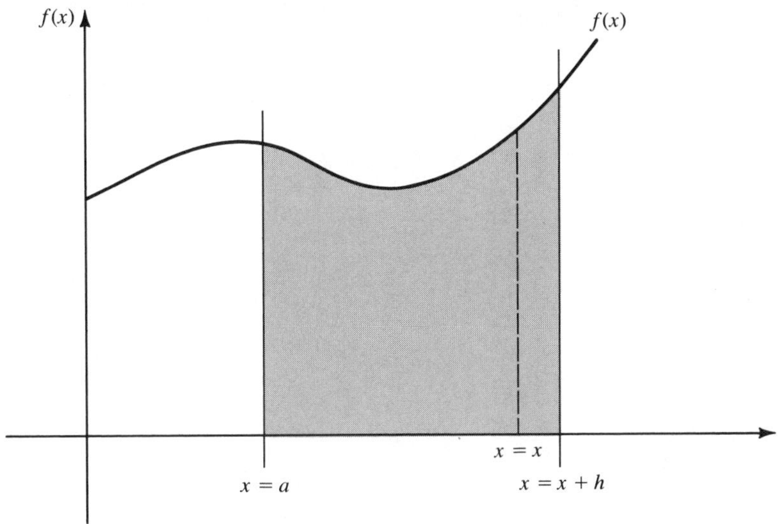

FIGURE 7.6

bounded by $x=a$ and $x=x+h$ as shown in Fig. 7.6. The difference $A(x+h) - A(x)$ is the area of the heavy shaded region in Fig. 7.7. The area $A(x+h)-A(x)$ is closely approximated by the width of the region, h, times the height, $f(x)$; that is,

$$A(x+h) - A(x) \doteq h \cdot f(x)$$

Therefore, the ratio $[A(x+h)-A(x)]/h \doteq f(x)$. Now as h gets smaller and

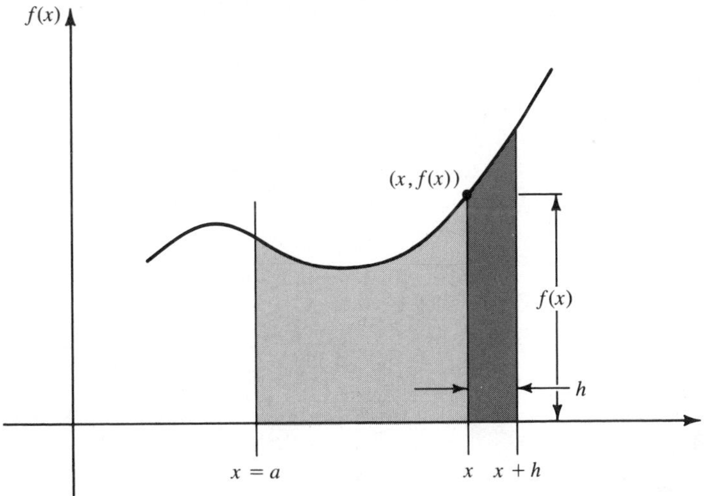

FIGURE 7.7

smaller, this approximation gets better and we have as $h \to 0$:

$$\lim_{h \to 0} \left[\frac{A(x+h) - A(x)}{h} \right] = f(x) \tag{7.16}$$

But the left side of Eq. (7.16) by Definition 4.3 is $A'(x)$; hence, we can conclude that

$$A'(x) = f(x)$$

or $A(x)$ is an antiderivative of $f(x)$. Therefore, the indefinite integral $\int f(x)\,dx = A(x) + C$, and the definite integral from a to b is

$$\int_a^b f(x)\,dx = A(x) + C \Big|_a^b = [A(b) + C] - [A(a) + C]$$

or

$$\int_a^b f(x)\,dx = A(b) - A(a) \tag{7.17}$$

By the definition of $A(x)$, $A(a)$ is the area between $f(x)$ and the x axis bounded by $x = a$ and $x = a$; hence, $A(a) = 0$. Therefore, $A(b)$, the area of the region between the graph of $f(x)$ and the x axis bounded by $x = a$ and $x = b$ given by Eq. (7.17) is

$$A(b) = \int_a^b f(x)\,dx$$

which verifies Rule 7.11.

Since the only limitation on the function $f(x)$ in Rule 7.11 is that it be continuous and nonnegative, we can use Eq. (7.15) to determine the areas of many regions that previously we could at best approximate.

Example 7.23 Area of a Region Determine the area of the region bounded by the function $f(x) = x^2$, the x axis, and the vertical lines $x = 2$ and $x = 5$.

Solution Graphing the function $f(x) = x^2$ and the lines $x = 2$ and $x = 5$, we depict the desired region in Fig. 7.8. The area by Eq. (7.15) is

$$\begin{aligned} A &= \int_2^5 x^2\,dx = \frac{1}{3}x^3 \Big|_2^5 = \frac{1}{3}(5)^3 - \frac{1}{3}(2)^3 \\ &= \frac{125}{3} - \frac{8}{3} \\ &= \frac{117}{3} = 39 \text{ square units} \end{aligned}$$

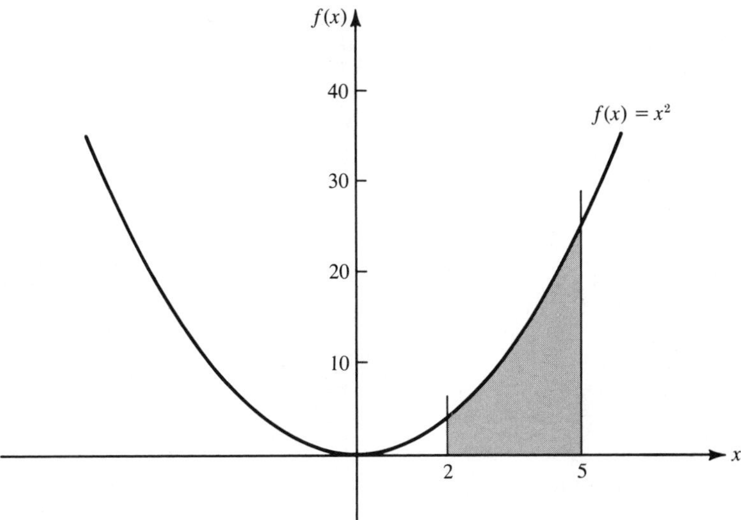

FIGURE 7.8

If we were asked to determine the area of the region between the graph of the function $f(x)=x^2-3x+2$ and the x axis bounded by $x=0$ and $x=3$, we might apply Eq. (7.15) directly and find

$$A = \int_0^3 (x^2-3x+2)\,dx = \frac{x^3}{3} - \frac{3}{2}x^2+2x\Big|_0^3$$

$$= \frac{27}{3} - \frac{27}{2} + 6$$

$$= \frac{9}{6} \text{ or } \frac{3}{2} \text{ square units}$$

But, before we conclude that this is the correct solution, let us identify graphically the region of concern. The graph of the quadratic function $f(x)=x^2-3x+2$ is shown in Fig. 7.9. [$f(x)$ has zeros at $x=1$ and $x=2$ and the coordinates of the vertex are $(\frac{3}{2}, -\frac{1}{4})$]. From the graph of $f(x)$ in Fig. 7.9 we note that $f(x) \ngtr 0$ on the entire interval defined by the limits of integration, which is one of the conditions of Rule 7.11. Then, what is the effect of the interval (1,2) where $f(x)<0$ on the definite integral $\int_0^3 f(x)\,dx$?

By Rule 7.8 we can write

$$\int_0^3 f(x)\,dx = \int_0^1 f(x)\,dx + \int_1^2 f(x)\,dx + \int_2^3 f(x)\,dx$$

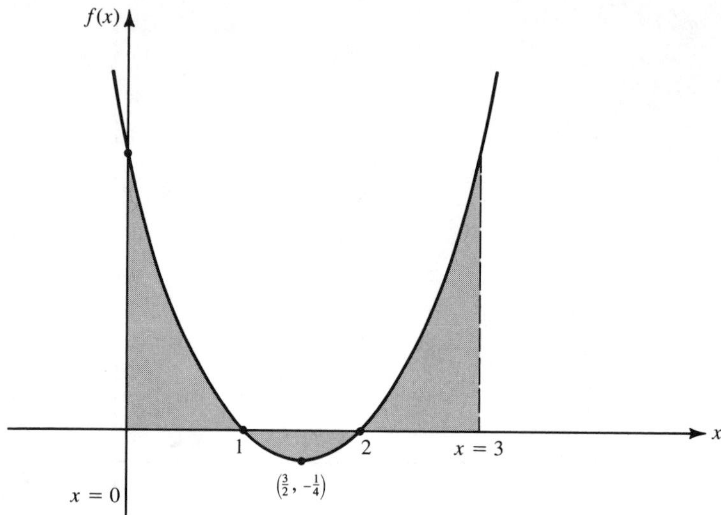

FIGURE 7.9

Evaluating each definite integral we find

$$\int_0^1 (x^2-3x+2)\,dx = \frac{1}{3}x^3 - \frac{3}{2}x^2 + 2x\Big|_0^1 = \frac{1}{3} - \frac{3}{2} + 2 = \frac{5}{6}$$

$$\int_1^2 (x^2-3x+2)\,dx = \left(\frac{1}{3}x^3 - \frac{3}{2}x^2 + 2x\right)\Big|_1^2 = \left(\frac{8}{3} - \frac{12}{2} + 4\right)$$
$$- \left(\frac{1}{3} - \frac{3}{2} + 2\right) = -\frac{1}{6}$$

$$\int_2^3 (x^2-3x+2)\,dx = \left(\frac{1}{3}x^3 - \frac{3}{2}x^2 + 2x\right)\Big|_2^3 = \left(\frac{27}{3} - \frac{27}{2} + 6\right)$$
$$- \left(\frac{8}{3} - \frac{12}{2} + 4\right) = \frac{5}{6}$$

and $$\int_0^3 (x^2-3x+2)\,dx = \frac{5}{6} + -\frac{1}{6} + \frac{5}{6} = \frac{3}{2}$$

which is not the sum of areas of the three regions shown in Fig. 7.9, but is a "net area" where the integral of $f(x)$ over the interval between $x=1$ and $x=2$ has a negative value as depicted in Fig. 7.10.

The correct solution to our original problem is found when we consider the area of each of the three regions shown in Fig. 7.10 to be positive. Hence, the area of the region between the graph of $f(x)$ and the x axis bounded by

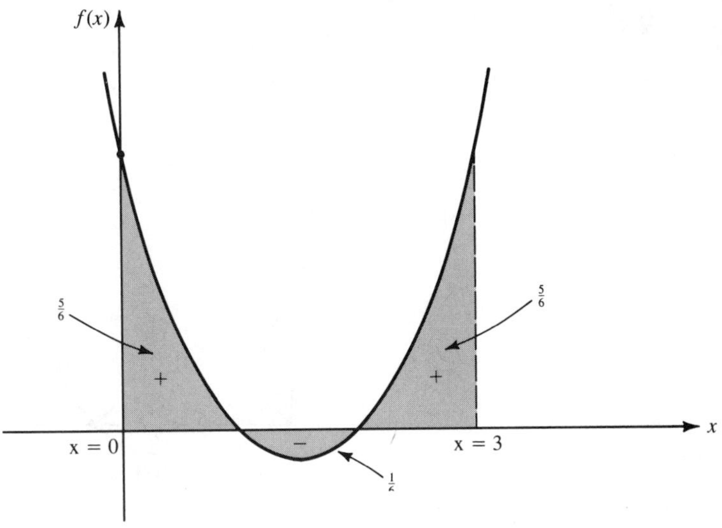

FIGURE 7.10

$x=0$ and $x=3$ is given by

$$A = \frac{5}{6} + \frac{1}{6} + \frac{5}{6}$$
$$= \frac{11}{6} \text{ square units}$$

and not $\frac{3}{2}$ sq. units as we at first calculated.

To prevent such an error, it is necessary to identify the region of interest by drawing a careful sketch of the function or functions involved. Once this is done, we must insure that the integrand is nonnegative over the interval defined by the limits of integration. In the previous example the area is correctly given by

$$A = \int_0^1 f(x)\,dx + \int_1^2 -f(x)\,dx + \int_2^3 f(x)\,dx \tag{7.18}$$

Note that we have changed the integrand in the second integral from $f(x)$, which is negative on the interval $(1,2)$, to $-f(x)$, which is positive on this interval. A graph of the integrands of Eq. (7.18) is shown in Fig. 7.11.

The graph of $-f(x)$ on the interval $[1,2]$ is a reflection across the x axis of the graph of $f(x)$ on this interval. Thus, the area of regions A and B depicted in Fig. 7.11 are equal. The definite integral $\int_1^2 -f(x)\,dx$ can be

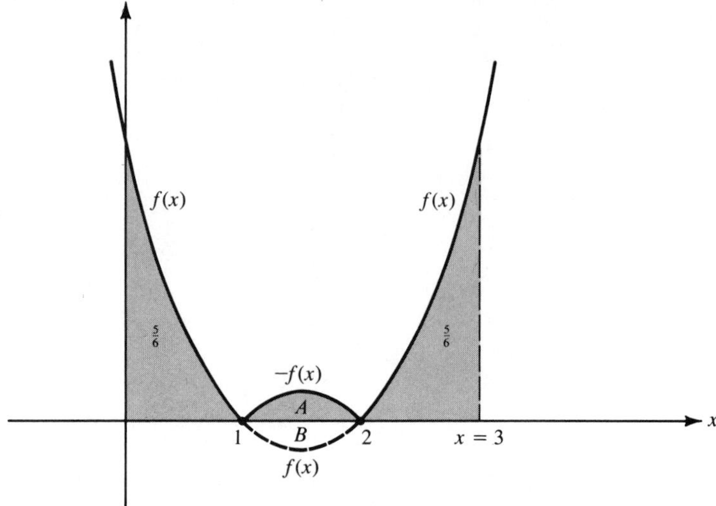

FIGURE 7.11

interpreted as the area of region A by Rule 7.11 since $-f(x) \geqslant 0$ on $[1,2]$. Evaluating this integral we have

$$\int_1^2 -f(x)\,dx = \int_1^2 -(x^2-3x+2)\,dx = \int_1^2 (3x-2-x^2)\,dx$$

$$= \frac{3}{2}x^2 - 2x - \frac{1}{3}x^3 \Big|_1^2 = \left(\frac{12}{2} - 4 - \frac{8}{3}\right) - \left(\frac{3}{2} - 2 - \frac{1}{3}\right)$$

$$= -\frac{4}{6} + \frac{5}{6} = \frac{1}{6} \text{ square units}$$

Hence, the total area A, given by Eq. (7.18) is

$$A = \frac{5}{6} + \frac{1}{6} + \frac{5}{6} = \frac{11}{6} \text{ square units}$$

which verifies our earlier statement.

Example 7.24 Find the area of the region below the x axis and above the graph of $f(x) = x^2 - 6x + 5$.

Solution Graphing the quadratic function $f(x) = x^2 - 6x + 5$ by finding the vertex $(3, -4)$ and the zeros of the function, ($x = 1$ and $x = 5$), we obtain Fig. 7.12. The region of interest is between $x = 1$ and $x = 5$. Since $f(x)$ is negative

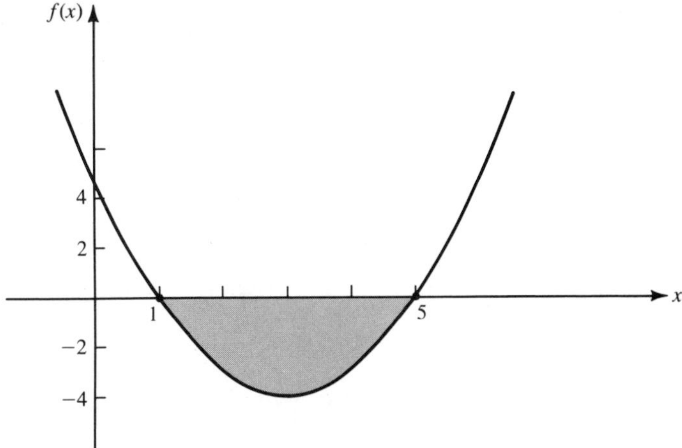

FIGURE 7.12

on this interval, the area A is given by

$$A = \int_1^5 -f(x)\,dx = \int_1^5 -(x^2-6x+5)\,dx$$
$$= \int_1^5 (6x-x^2-5)\,dx$$

Evaluating the definite integral, we have

$$A = 3x^2 - \frac{1}{3}x^3 - 5x\Big|_1^5 = \frac{25}{3} - \left(\frac{-7}{3}\right)$$
$$= \frac{32}{3} \text{ square units}$$

The definite integral can also be used to find the area between curves. For example, consider the shaded region bounded by the graphs of the functions $f(x)$ and $g(x)$ shown in Fig. 7.13a. Since $\int_a^b f(x)\,dx$ is equal to the area bounded by $f(x)$, the x axis and the lines $x=a$ and $x=b$ (see Fig. 7.13b), and $\int_a^b g(x)\,dx$ is equal to the area bounded by $g(x)$, the x axis and the lines $x=a$ and $x=b$ (see Fig. 7.13c), the area of the shaded region in Fig. 7.13a equals $\int_a^b f(x)\,dx - \int_a^b g(x)\,dx$. By Rules 7.9 and 7.10, the desired area is

$$A = \int_a^b [f(x)-g(x)]\,dx$$

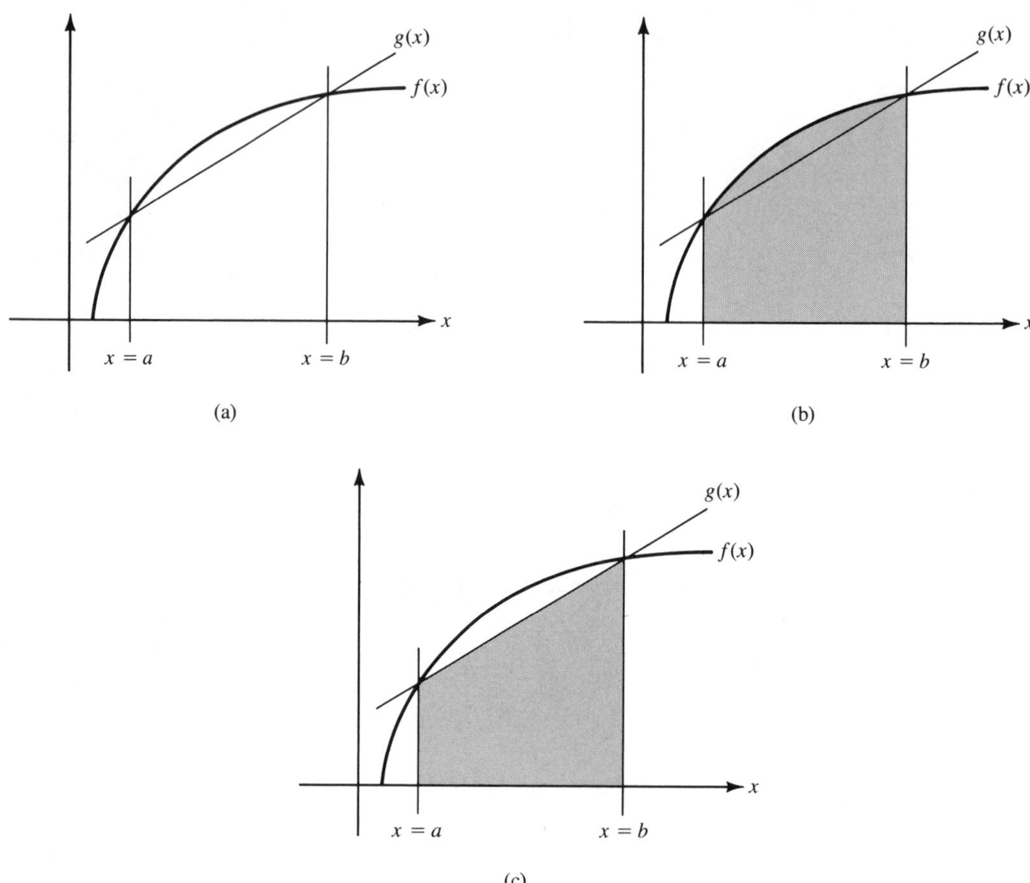

FIGURE 7.13

Example 7.25 Area of a Region Between Two Curves Find the area of the region bounded by the function $f(x)=\sqrt{x}$ and the function $g(x)=x$.

Solution From the graphs of $f(x)$ and $g(x)$ shown in Fig. 7.14, we can determine the region of interest. We see from this graph that the lower limit $a=0$. To determine the value of b we note that $g(b)=f(b)$ or $b=\sqrt{b}$. Solving this equation for the variable b, we find that by squaring both left and right sides of the equation $b=\sqrt{b}$ we have

$$b^2 = b$$
$$b^2 - b = 0$$
$$b(b-1) = 0$$
$$b = 0 \qquad \text{or} \qquad b = 1$$

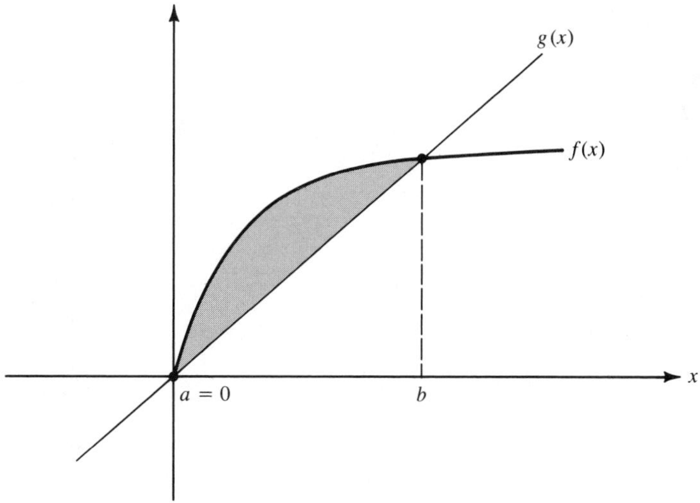

FIGURE 7.14

Since $b=0$ corresponds to the intersection of $f(x)$ and $g(x)$ at the origin, the desired value is $b=1$. Hence, the area is given by

$$\begin{aligned}
A &= \int_0^1 [f(x)-g(x)]\,dx \\
&= \int_0^1 (\sqrt{x}-x)\,dx = \int_0^1 (x^{1/2}-x)\,dx \\
&= \frac{2}{3}x^{3/2}-\frac{1}{2}x^2\Big|_0^1 = \left(\frac{2}{3}-\frac{1}{2}\right)-(0) = \frac{4}{6}-\frac{3}{6} \\
&= \frac{1}{6} \text{ square units}
\end{aligned}$$

Example 7.26 Determine the area of the region bounded by the graphs of the functions $f(x)=1+x^2$ and $g(x)=-x^2-2x+5$.

Solution Graphing these quadratic functions to determine the region of interest we arrive at Fig. 7.15. To find the x coordinates of the points of intersection, P_1 and P_2, we equate $f(x)$ and $g(x)$:

$$\begin{gathered}
1+x^2 = -x^2-2x+5 \\
2x^2+2x-4 = 0 \\
2(x+2)(x-1) = 0 \\
x = -2 \qquad \text{and} \qquad x = 1
\end{gathered}$$

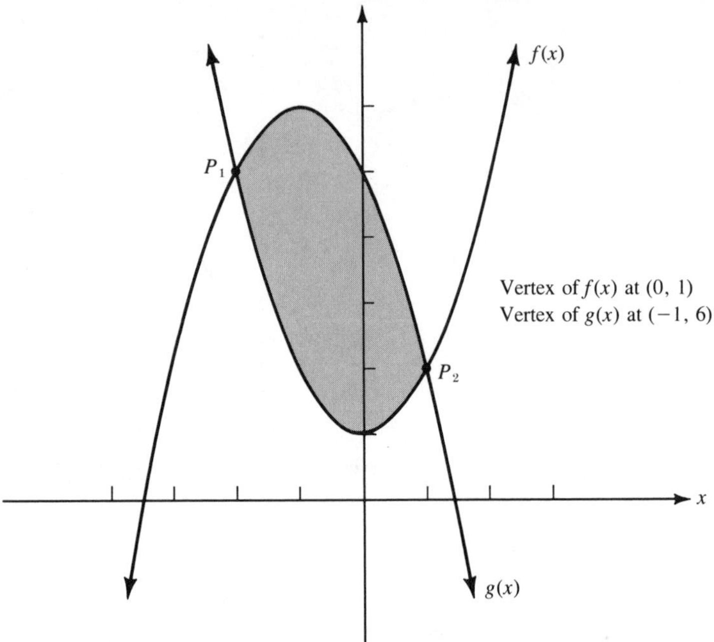

FIGURE 7.15

Therefore, the area of the shaded region shown in Fig. 7.15 is

$$
\begin{aligned}
A &= \int_{-2}^{1} g(x)\,dx - \int_{-2}^{1} f(x)\,dx \\
&= \int_{-2}^{1} [g(x) - f(x)]\,dx \\
&= \int_{-2}^{1} [(-x^2 - 2x + 5) - (1 + x^2)]\,dx \\
&= \int_{-2}^{1} (-2x^2 - 2x + 4)\,dx \\
&= -\frac{2}{3}x^3 - x^2 + 4x\Big|_{-2}^{1} \\
&= \left(-\frac{2}{3} - 1 + 4\right) - \left(\frac{16}{3} - 4 - 8\right) \\
&= 9 \text{ square units}
\end{aligned}
$$

EXERCISES

In Exercises 1–24 determine the area of the region between the graph of the given function and the x axis bounded by the limits of the given interval. In each case draw a sketch and shade the region of interest.

1. $f(x)=2x+3 \quad (0 \leqslant x \leqslant 4)$

2. $f(x)=x \quad (0 \leqslant x \leqslant 6)$

3. $f(x)=x-1 \quad (1 \leqslant x \leqslant 3)$

4. $f(x)=2x-4 \quad (0 \leqslant x \leqslant 4)$

5. $f(x)=-4x+5 \quad (0 \leqslant x \leqslant 3)$

6. $f(x)=-x \quad (-1 \leqslant x \leqslant 1)$

7. $f(x)=x^2 \quad (0 \leqslant x \leqslant 2)$

8. $f(x)=3x^2-4x \quad (2 \leqslant x \leqslant 3)$

9. $f(x)=x^2-4x+5 \quad (0 \leqslant x \leqslant 3)$

10. $f(x)=6x^2+2x+1 \quad (0 \leqslant x \leqslant 2)$

11. $f(x)=-3x^2+4x+4 \quad (0 \leqslant x \leqslant 3)$

12. $f(x)=x^2-6x+8 \quad (0 \leqslant x \leqslant 6)$

13. $f(x)=x^3 \quad (0 \leqslant x \leqslant 1)$

14. $f(x)=x^3 \quad (-1 \leqslant x \leqslant 1)$

15. $f(x)=x^3-2x^2-x \quad (0 \leqslant x \leqslant 1)$

16. $f(x)=x^3-2x^2+x \quad (-1 \leqslant x \leqslant 1)$

17. $f(x)=e^{2x} \quad (0 \leqslant x \leqslant 3)$

18. $f(x)=2e^{-2x} \quad (0 \leqslant x \leqslant 2)$

19. $f(x)=\dfrac{2}{x} \quad (1 \leqslant x \leqslant 2)$

20. $f(x)=\dfrac{3}{(x+4)} \quad (0 \leqslant x \leqslant 1)$

21. $f(x)=x^2-8x+12 \quad (2 \leqslant x \leqslant 6)$

22. $f(x)=x^3+8 \quad (-1 \leqslant x \leqslant 1)$

23. $f(x)=x^2-3x+2 \quad (0 \leqslant x \leqslant 2)$

24. $f(x)=-\sqrt{x+4} \quad (0 \leqslant x \leqslant 5)$

In Exercises 25–35 *draw a sketch and shade the region of interest.*

25. Determine the area of the region bounded by the graphs of the functions $f(x)=x^2$ and $g(x)=x^3$.

26. Determine the area of the region bounded by the graphs of the functions $f(x)=7-x$, $g(x)=7+x$, and the lines $x=0$ and $x=7$.

27. Determine the area of the region bounded by the graphs of the functions $f(x)=\sqrt{x}$ and $g(x)=x^2$.

28. Determine the area of the region in the first quadrant bounded by the graphs of the function $f(x)=4/x^2$ and $g(x)=7-3x$.

29. Determine the area of the region in the first quadrant that lies below the graph of $f(x)=6-x^2$, above the line $y=x$, and is bounded by the line $x=0$.

30. Determine the area of the region that lies below the graph of the function $f(x)=4-x^2$ and above the line $y=x+2$.

31. Determine the area of the region bounded by the graphs of the functions $f(x)=x^2$ and $g(x)=-x^2+6x$.

32. Determine the area of the region bounded by the graphs of the functions $f(x)=e^x$, $y=ex$, and $x=0$.

33. Determine the area of the region bounded by the graphs of the functions $f(x)=\sqrt{x}$ and $g(x)=x^3$.

34. Determine the area of the region bounded by the graphs of the functions $f(x)=x^2-3x+2$, $g(x)=x^2-3x+10$, and the lines $x=0$ and $x=3$.

35. Determine the area of the region bounded by the graphs of the functions $g(x)=x^2-9$ and $f(x)=9-x^2$.

7.7 APPLICATIONS OF THE DEFINITE INTEGRAL

In the last section we used the definite integral to determine areas of regions between curves in a plane. This interpretation of the definite integral arises naturally from Eq. (7.9). In this section and in Sec. 8.7 we will extend this interpretation to consider a few of the many other applications of the definite integral.

The Definite Integral as a Sum or Total

There are many functions of time, t, that represent the rate at which a given quantity changes per unit of time. These functions are often referred to as *rate functions*. The definite integral of such functions with respect to time can be used to express the total change of the quantity over a given period of time. For example, let us determine the total sales in the following situation.

The owner of a small business reports that his annual sales have been increasing every year since he went into business five years ago and that they can be closely approximated by the continuous function

$$S(t) = 25{,}000 + 20{,}000t^{1/2} \tag{7.19}$$

where $S(t)$ = the sales rate in dollars per year and t represents the number of years, or fraction thereof, the owner has been in business.

Eq. (7.19) implies that at the beginning of the first year of business, i.e. $t=0$, the annual sales rate is \$25,000/year, and at the end of the first year, i.e. $t=1$, the sales rate has increased to \$45,000/year. The graph of the sales function $S(t)$ is shown in Fig. 7.16. We can estimate the total sales for the five-year period by taking some average value for the sales in each year as shown in Fig. 7.17. If $S(t_1)$, $S(t_2)$, etc. are estimates of the annual sales in the first year, second year, etc., then the total sales S is approximated by

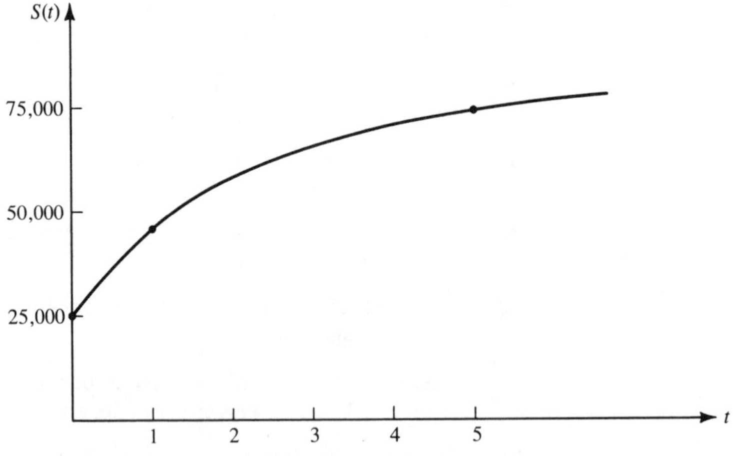

FIGURE 7.16

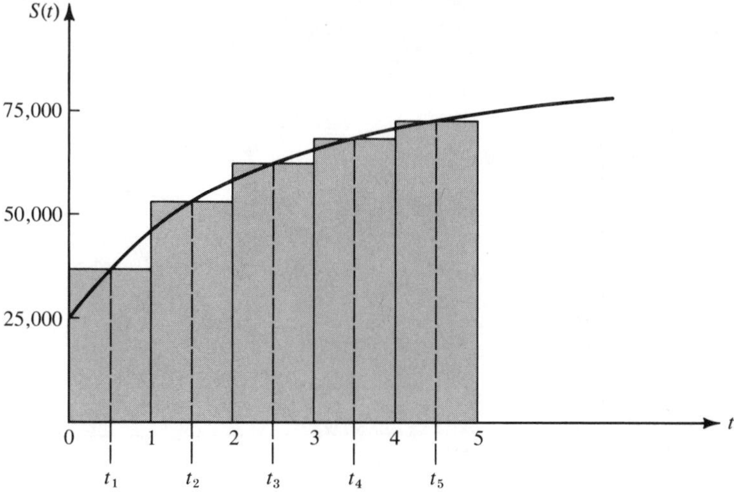

FIGURE 7.17

the sum

$$S = S(t_1) + S(t_2) + S(t_3) + S(t_4) + S(t_5)$$

This sum is equal to the sum of the areas of the rectangles shown in Fig. 7.17 and hence is an approximation of $\int_0^5 S(t)\,dt$.

To improve our estimate of the total sales, we could divide each year into 12 subintervals, i.e. monthly periods. Since $S(t)$ represents the annual sales, the approximate sales for any month would be given by $\frac{1}{12}S(t)$ where $S(t)$ is an approximation of the annual sales in that month. Therefore, a better approximation of our total sales over the five-year period would be given by

$$S = S(t_1)\cdot\tfrac{1}{12} + S(t_2)\cdot\tfrac{1}{12} + S(t_3)\cdot\tfrac{1}{12} + \cdots + S(t_{60})\cdot\tfrac{1}{12}$$

where $S(t_1)$, $S(t_2)$, etc. represent an approximation of the annual sales in the first month, second month, etc., as depicted in Fig. 7.18.

Obviously, a better approximation of the total sales would be obtained if daily intervals were chosen and we found the sum

$$S = S(t_1)\cdot\tfrac{1}{365} + S(t_2)\cdot\tfrac{1}{365} + \cdots + S(t_{1825})\cdot\tfrac{1}{365}$$

Ultimately we could select n intervals of width Δt and find the sum

$$S = S(t_1)\,\Delta t + S(t_2)\,\Delta t + \cdots + S(t_n)\,\Delta t$$

as $n \to \infty$. But this sum is equivalent to the definition of the definite integral of

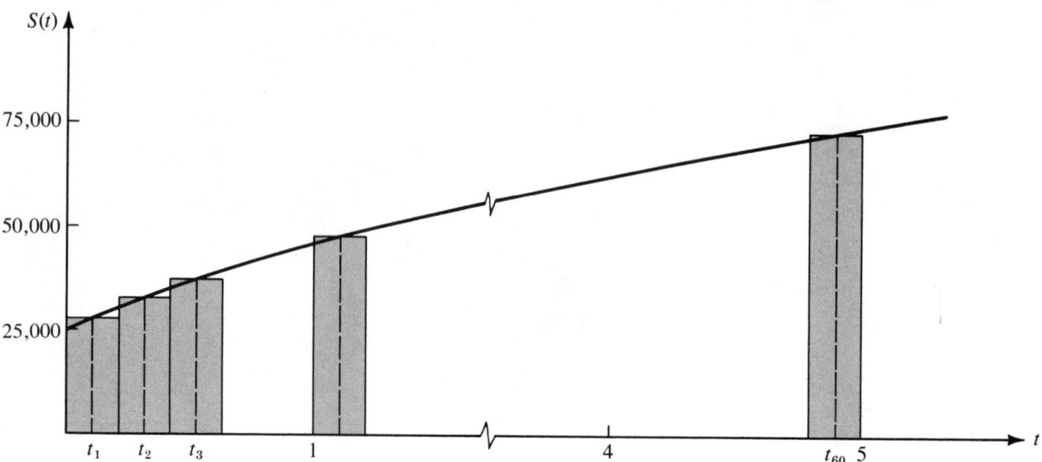

FIGURE 7.18

$S(t)$ from $t=0$ to $t=5$; hence, the total sales, S, is given by

$$S = \int_0^5 S(t)\,dt$$

Evaluating this definite integral we obtain

$$\begin{aligned} S &= \int_0^5 (25{,}000 + 20{,}000t^{1/2})\,dt \\ &= 25{,}000t + \frac{2}{3}(20{,}000)t^{3/2}\Big|_0^5 \\ &= (25{,}000)(5) + \frac{40{,}000}{3}(5)^{3/2} \\ &= 274{,}071 \end{aligned}$$

Hence, the total sales over the five-year period is \$274,071.

As another illustration that a definite integral can be used to determine a total or sum let us find the total interest income from \$10,000 invested at 8% compounded continuously for a period of 10 years. From Sec. 6.4 we learned that the growth rate, $R(t)$, from an investment of P dollars compounded continuously at the rate of r percent is given by

$$R(t) = rPe^{rt} \tag{7.20}$$

where $R(t)$ = the growth rate in dollars/year and t is the number of years since the initial investment. Substituting the appropriate values for P and r

for this problem into Eq. (7.20), we obtain

$$\begin{aligned} R(t) &= .08(10{,}000)e^{.08t} \\ &= 800e^{.08t} \text{ dollars/year} \end{aligned}$$

The graph of the continuous exponential function $R(t)$ is shown in Fig. 7.19.

To determine the total interest income, I, for a period of 10 years, we evaluate the definite integral

$$\begin{aligned} I = \int_0^{10} R(t)\,dt &= \int_0^{10} 800e^{.08t}\,dt \\ &= \frac{800}{.08} e^{.08t} \Big|_0^{10} \\ &= 10{,}000e^{.8} - 10{,}000 \\ &\doteq 10{,}000(2.2255) - 10{,}000 \\ &\doteq 22{,}255 - 10{,}000 \\ &= \$12{,}255 \text{ (rounded to nearest dollar)} \end{aligned}$$

Hence, the total interest income earned from the initial investment of \$10,000 over a period of 10 years is \$12,255.

We can verify this result since we know from Eq. (3.14) that the maturity value $A(t)$ from an initial investment of P dollars compounded continuously at an interest rate of r is given by

$$A(t) = Pe^{rt}$$

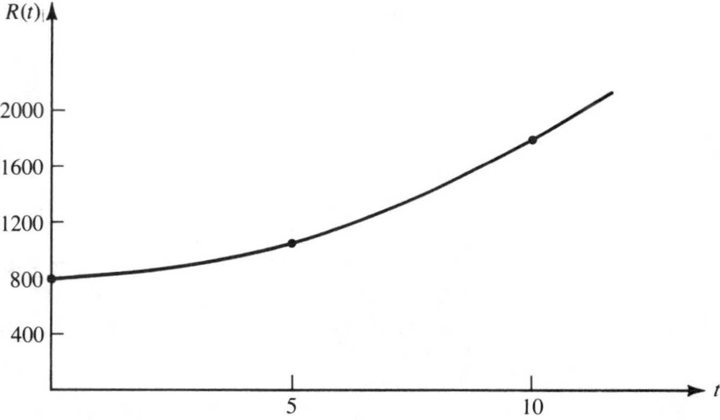

FIGURE 7.19

Therefore, the maturity value in this illustration where $P = 10{,}000$, $r = .08$, and $t = 10$ is

$$A(10) = 10{,}000e^{.08(10)}$$
$$= 10{,}000e^{.8} \doteq 22{,}255$$

Since the total interest income, I, is the difference between the maturity value and the initial investment, the value of I is given by

$$I = \$22{,}255 - \$10{,}000$$
$$= \$12{,}255$$

which is exactly the same results we obtained above by integrating the rate function $R(t)$.

Example 7.27 Car Maintenance Expense A car rental agency has determined that its monthly maintenance expense per car can be approximated by the continuous function

$$M(t) = 20 + .02t^2$$

where $M(t)$ is in dollars and t represents the age of the car in months. Determine the total number of dollars that has been spent on the maintenance of a three-year-old car.

Solution Since $M(t) = 20 + .02t^2$ represents the number of dollars spent per month, we determine the total dollars spent by evaluating the definite integral $\int_0^{36} M(t)\,dt$. Note that the upper limit 36 represents the total number of months in the period. Evaluating $\int_0^{36} M(t)\,dt$, we find

$$\int_0^{36} (20 + .02t^2)\,dt = 20t + \frac{.02t^3}{3}\Bigg|_0^{36}$$
$$= 720 + 311.04$$
$$= 1{,}031.04$$

Therefore, the total expense to maintain a new car for 3 years is $1,031 (rounded to the nearest dollar).

Example 7.28 Total Depreciation from the Depreciation Rate A certain machine depreciates over a 15-year period at a rate that depends on the age of the machine. When the machine is t years old the rate of depreciation, $D(t)$, is given by

$$D(t) = 200(15 - t)$$

where $D(t)$ is in dollars/year. Determine the total depreciation in 10 years and calculate the depreciated value of this machine if the original cost was \$30,000.

Solution Given the rate function $D(t)$ in dollars/year, we determine the total depreciation in 10 years by evaluating the definite integral $\int_0^{10} D(t)\,dt$.

$$\begin{aligned}\int_0^{10} 200(15-t)\,dt &= \int (3{,}000-200t)\,dt\\ &= 3{,}000t - 100t^2\Big|_0^{10}\\ &= 30{,}000 - 10{,}000\\ &= 20{,}000\end{aligned}$$

Therefore, the total depreciation in 10 years is \$20,000 and the depreciated value of the machine is (\$30,000 − \$20,000) or \$10,000.

There are numerous examples that involve rate functions. Many of these functions are similar to the ones we have discussed; however, there are others where the rate is not expressed in dollars per unit of time. The graphs of several different kinds of rate functions are shown in Fig. 7.20, along with the interpretations of each definite integral.

Example 7.29 Divorce Rate A study of the divorce rate in a certain community over the past 10 years indicates that the number of divorces per year is increasing and will be given by the function

$$D(t) = 100e^{.05t}$$

where $D(t)$ = the number of divorces per year t years from now. Determine the total number of divorces that will occur in the next 5 years.

Solution Since $D(t) = 100e^{.05t}$ is a rate function, the total number of divorces during the next 5 years is given by $\int_0^5 D(t)\,dt$. Evaluating this definite integral we have

$$\begin{aligned}\int_0^5 100e^{.05t}\,dt &= \frac{100}{.05}\,e^{.05t}\Big|_0^5\\ &= 2{,}000e^{.25} - 2{,}000e^0\\ &= 2{,}000(1.2840-1)\\ &= 568\end{aligned}$$

Hence, in this particular community there will be an additional 568 divorces in the next 5 years.

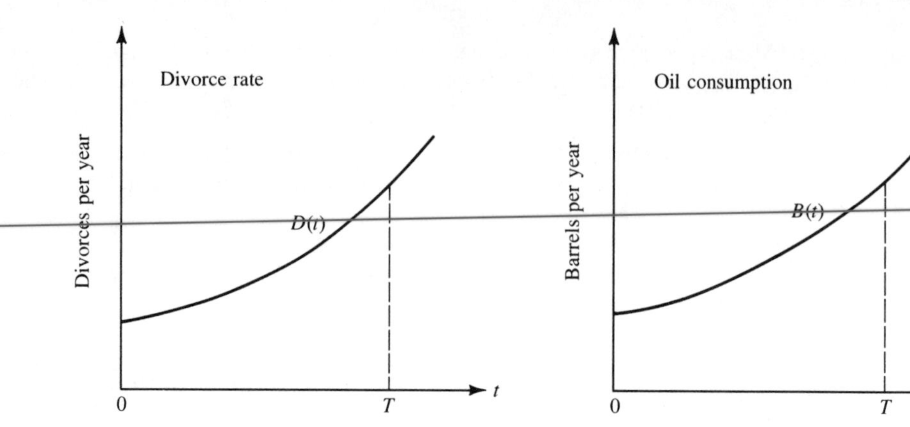

(a) $\int_0^T D(t)\,dt$ represents the number of divorces between $t = 0$ and $t = T$.

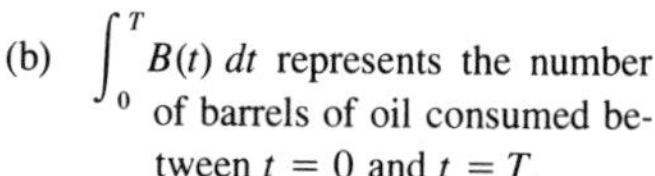

(b) $\int_0^T B(t)\,dt$ represents the number of barrels of oil consumed between $t = 0$ and $t = T$.

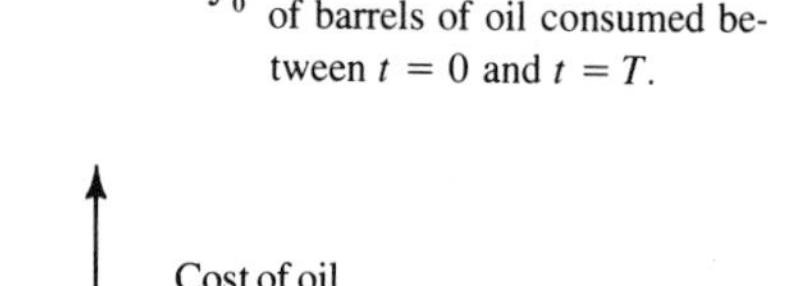

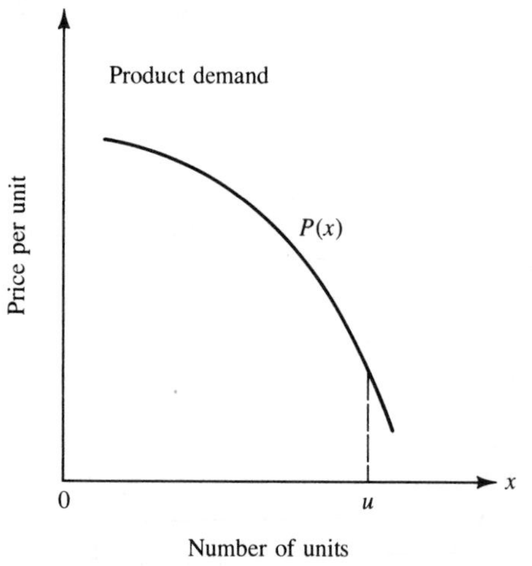

(c) $\int_0^u P(x)\,dx$ represents the total number of dollars spent to purchase u units.

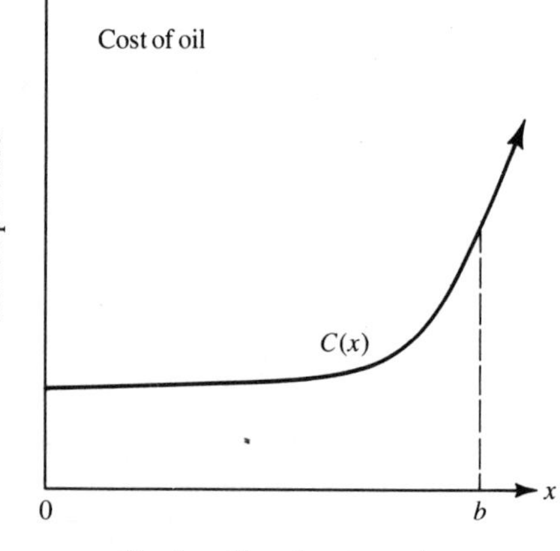

(d) $\int_0^b C(x)\,dx$ represents the total cost of b barrels of oil.

FIGURE 7.20

Depletion of Petroleum and Other Natural Resources Recent history has shown that the world's use or depletion of natural resources such as coal, oil, gas, etc. is proceeding at a rate that can be approximated by an exponential function. Thus, if C_0 represents the current annual consumption of a natural resource and $C(t)$ represents the annual consumption at some time t in the future, then $C(t)$ is given by the formula

$$C(t) = C_0 e^{rt} \tag{7.21}$$

where $r =$ the growth rate.

Eq. (7.21) is a rate function similar to the one discussed in Example 7.29. To determine the total amount of the resource consumed from time $t=0$ to a time $t=T$, we evaluate the definite integral $\int_0^T C(t)\,dt$ as follows

$$\begin{aligned}\int_0^T C(t)\,dt &= \int_0^T C_0 e^{rt}\,dt \\ &= \frac{C_0}{r} e^{rt}\Big|_0^T \\ &= \frac{C_0}{r} e^{rT} - \frac{C_0}{r} e^0\end{aligned}$$

Therefore,

$$\int_0^T C_0 e^{rt}\,dt = \frac{C_0}{r}\left[e^{rT} - 1\right] \tag{7.22}$$

For example, in 1977 the annual consumption of petroleum was 20,000 million barrels and the annual growth rate was 1%. The consumption in t years (from 1977) is given by

$$C(t) = 20{,}000 e^{.01t} \text{ million barrels/year}$$

and the total amount of oil consumed in the next 10 years is found by evaluating the definite integral $\int_0^{10} 20{,}000 e^{.01t}\,dt$. Substituting $T=10$, $r=.01$, and $C_0 = 20{,}000$ into Eq. (7.22), we find

$$\begin{aligned}\int_0^{10} 20{,}000 e^{.01t}\,dt &= \frac{20{,}000}{.01}\left[e^{(.01)(10)} - 1\right] \\ &= 2{,}000{,}000\left[e^{.1} - 1\right]\end{aligned}$$

Approximating $e^{.1}$ by using Table A.3, we obtain

$$\begin{aligned}\int_0^{10} 20{,}000 e^{.01t}\,dt &\doteq 2{,}000{,}000\left[1.1052 - 1\right] \\ &\doteq 210{,}400\end{aligned}$$

Hence, from 1977 to 1987 approximately 210,400 million barrels of oil will be consumed.

Example 7.30 Depletion of Petroleum Reserves In 1977 the total petroleum reserve was estimated to be 666,000 million barrels. Based on the assumption that the annual growth rate of petroleum consumption continues at 1%, determine the calendar year in which all the petroleum reserves will be depleted.

Solution If $t=0$ in 1977, then $C_0=20{,}000$, $r=.01$, and the total amount of petroleum consumed in T years is given by

$$\int_0^T 20{,}000e^{.01t}\,dt$$

To determine the number of years it would take to deplete all the reserves, we equate this value to the total reserves; that is,

$$\int_0^T 20{,}000e^{.01t}\,dt = 666{,}000 \tag{7.23}$$

Evaluating the left side of Eq. (7.23) by using Eq. (7.22), we have

$$\left(\frac{20{,}000}{.01}\right)[e^{.01T}-1] = 666{,}000$$

$$e^{.01T}-1 = .333$$

$$e^{.01T} = 1.333$$

Using the definition of the natural logarithm, we can write this equation as

$$.01T = \ln 1.333$$

Evaluating $\ln 1.333$ by using Table A.2, we have

$$T = \frac{.2874}{.01} = 28.7 \text{ years}$$

Hence, the total petroleum reserves will be depleted in 1977+28.7 years or sometime late in the year 2005.

We should add that in the above calculation we have assumed a continued increase in the annual rate of consumption of petroleum, and that there will be no further oil discoveries. It is obvious that the law of supply and demand will eventually decrease the annual consumption as world reserves dwindle and the cost of oil increases. Although we may not run out of petroleum in 2005, the seriousness of the problem is not diminished.

The Definite Integral in Marginal Analysis The definite integral can also be used in certain types of marginal analysis problems. For example, if we are given the marginal cost function $c(x)$ and wish to determine the total cost of increasing x from $x=a$ to $x=a+k$ units, we can determine the indefinite integral of the marginal cost function and then determine the difference between the cost for $x=a$ units and for $a+k$ units. However, this involves the same steps as determining the definite integral

$$\int_a^{a+k} c(x)\,dx = C(a+k) - C(a)$$

This concept can also be extended to total profit or revenue functions.

Example 7.31 The marginal cost in dollars of manufacturing x units of a particular product is given by

$$c(x) = 3x^2 - 42x + 360$$

Determine the increase in cost of manufacturing 13 units as opposed to manufacturing 10 units.

Solution The increase in cost is given by

$$\begin{aligned}\int_{10}^{13}(3x^2-42x+360)\,dx &= x^3-21x^2+360x\Big|_{10}^{13}\\ &= 3{,}328 - 2{,}500\\ &= \$828\end{aligned}$$

In Example 7.31 we should note that \$3,328 and \$2,500 are not total costs but total variable costs. In order to determine total cost, we would need to know the fixed costs also.

Consumers' and Producers' Surplus

In Sec. 2.2 we introduced the concept of supply and demand functions and discussed the meaning of the equilibrium point. In the area of economics it is conventional to consider the quantity of the commodity demanded by the consumer or supplied by the producer to be the independent variable x, and the price per unit that the consumer is willing to pay or the producer charges to be the dependent variable. Typical demand functions are decreasing functions denoting that as the unit price decreases the number of items demanded increases, as depicted in Fig. 7.21a. Analogously, typical supply functions are increasing functions denoting that as the unit price increases, the supply increases, as depicted in Fig. 7.21b.

Sketching the graphs of the supply and demand functions depicted in Fig. 7.21a and b on the same coordinate system, we can graphically determine the equilibrium point $(x_E, f(x_E))$ or $(x_E, g(x_E))$ as shown in Fig. 7.22. In an open competition market where there is no price fixing, etc., eventually the commodity is sold at the equilibrium price. From Fig. 7.22 we conclude

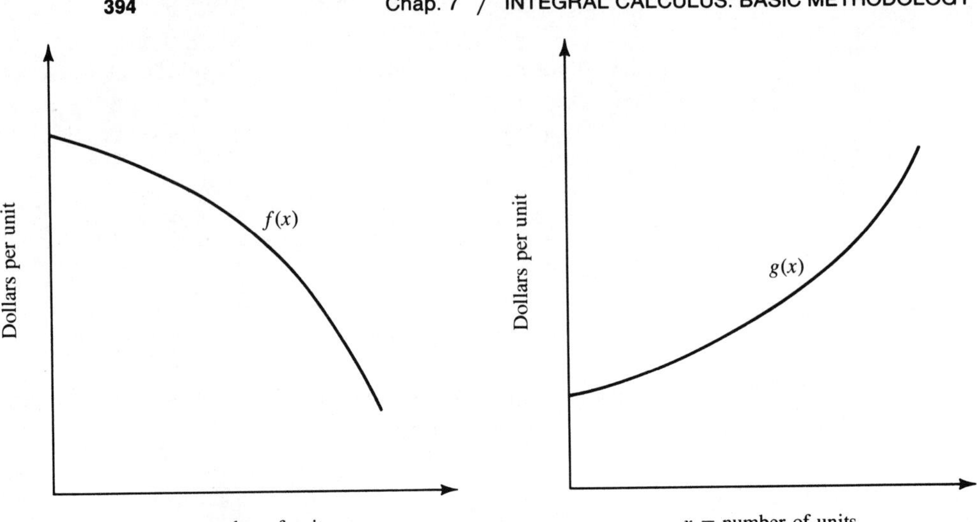

FIGURE 7.21

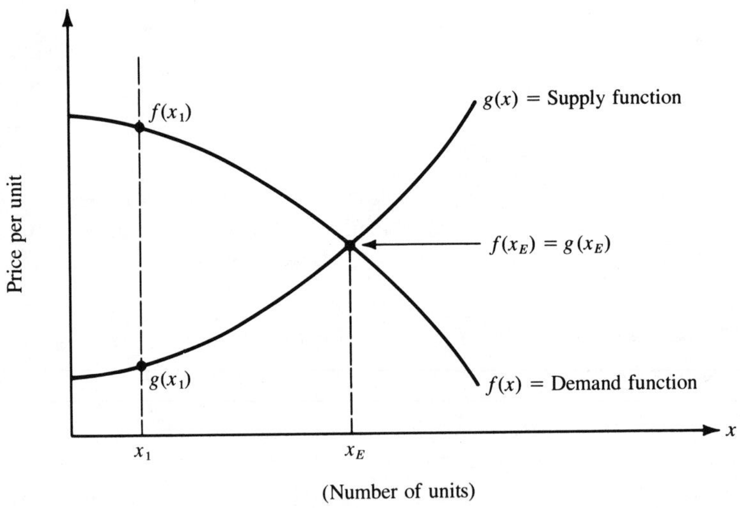

FIGURE 7.22

that the consumer is willing to pay $f(x_1)$ dollars/unit for x_1 units, and the supplier is able to supply x_1 units at $g(x_1)$ dollars/unit.

These definitions of the supply function $g(x)$ and the demand function $f(x)$ are such that they satisfy our descriptions of rate functions (see, for example, Fig. 7.20c). Thus, the definite integral $\int_0^{x_E} f(x)\,dx$ can be interpreted as the total number of dollars spent by the consumer to purchase x_E units prior to equilibrium being established. Geometrically, the definite integral $\int_0^{x_E} f(x)\,dx$ is the area of the shaded region shown in Fig. 7.23.

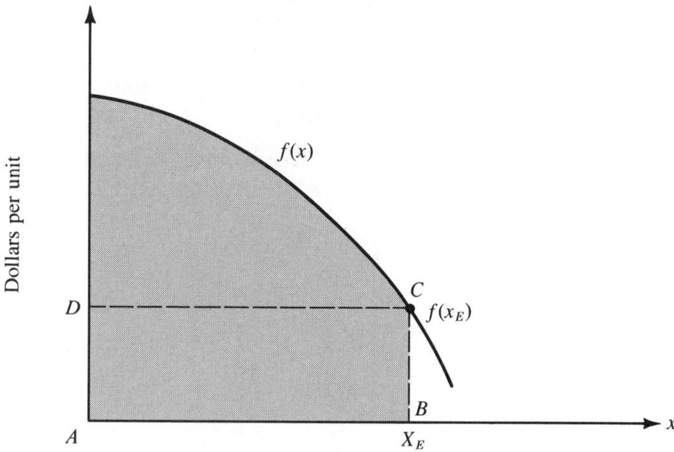

FIGURE 7.23

Once equilibrium has been established the consumer can purchase x_E units at a price of $f(x_E)$ dollars/unit or his total expense is $f(x_E)x_E$. (*Note*: $f(x_E)x_E$ is the area of the rectangular region $ABCD$ shown in Fig. 7.23.) Hence, the dollars saved by the consumer in a competitive system where equilibrium is allowed to be reached is $\int_0^{x_E} f(x)\,dx - f(x_E)x_E$. Geometrically this is the area of the shaded region shown in Fig. 7.24. This value is called the *consumers' surplus*.

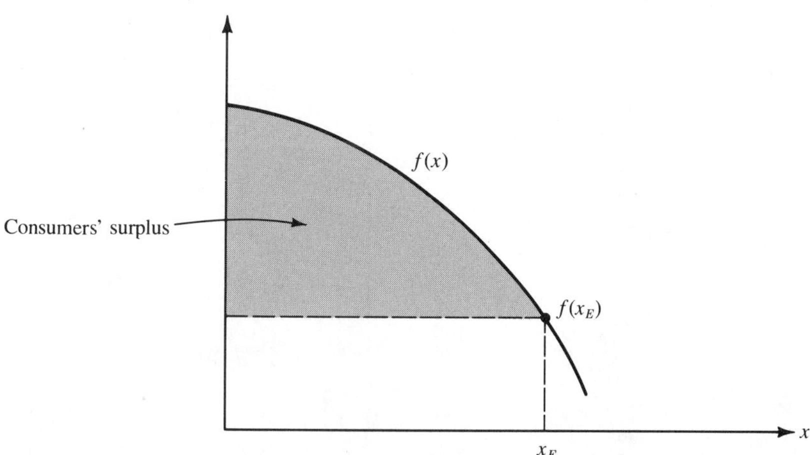

FIGURE 7.24

Analogously, $\int_0^{x_E} g(x)\,dx$ is the dollar income to the producer from the sale of x_E units prior to equilibrium being established. Geometrically the definite integral $\int_0^{x_E} g(x)\,dx$ is the shaded area shown in Fig. 7.25. Since the producer can sell all x_E units at $g(x_E)$ dollars/unit at the equilibrium point, his income would be $g(x_E)x_E$ dollars. [*Note*: $g(x_E)x_E$ is the area of the rectangular region $ABCD$ in Fig. 7.25.] Hence, the producers' income is increased by $[g(x_E)x_E - \int_0^{x_E} g(x)\,dx]$ dollars if equilibrium is allowed to be reached. Geometrically this is the area of the shaded region shown in Fig. 7.26 and its value is called the *producers' surplus*.

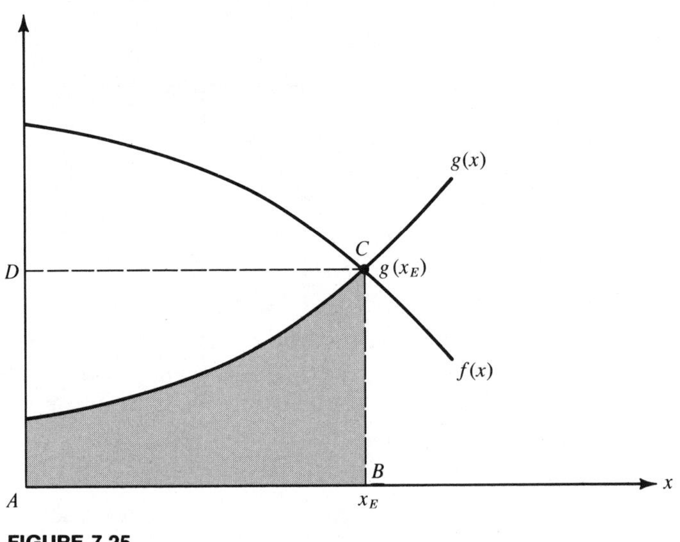

FIGURE 7.25

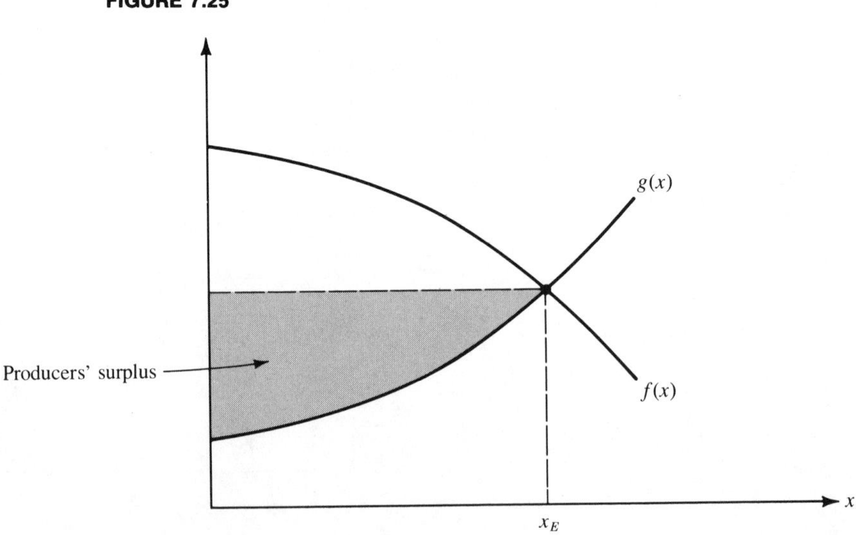

FIGURE 7.26

Example 7.32 Consumers' and Producers' Surplus The supply and demand functions for a certain commodity are given by

$$g(x) = 50 + x$$

and

$$f(x) = 200 - \tfrac{1}{5}x^2$$

where $g(x)$ is the price/unit in dollars at which the producer can supply x units, and $f(x)$ is the price/unit in dollars at which the consumer will buy x units.

(a) Sketch the graphs of the supply and demand functions on the same coordinate system.
(b) Determine the equilibrium price as well as the number of units produced and purchased at this price.
(c) Determine the consumers' surplus.
(d) Determine the producers' surplus.

Solution

(a) The graphs of the linear supply function $g(x) = 50 + x$ and the quadratic demand function $f(x) = 200 - \frac{1}{5}x^2$ are shown in Fig. 7.27.
(b) Since $g(x)$ and $f(x)$ are equal at the equilibrium point, we determine the equilibrium value x_E by solving

$$50 + x = 200 - \tfrac{1}{5}x^2$$

Rewriting this quadratic equation, we obtain

$$x^2 + 5x - 750 = 0$$

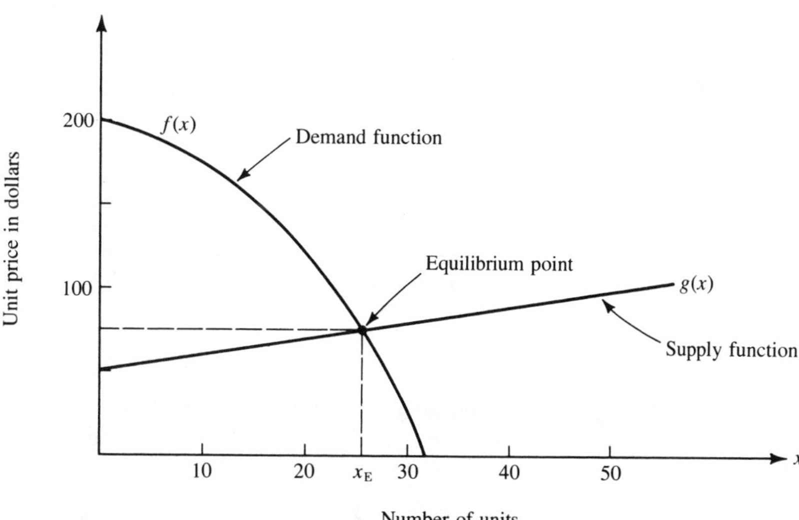

FIGURE 7.27

Factoring we have $(x+30)(x-25)=0$. Therefore, $x_E=25$ or $x_E=-30$ (not a feasible solution). Hence, equilibrium is reached at an output of 25 units and the unit price at this output is

$$g(25) = f(25) = 75 \text{ dollars}$$

(c) The consumers' surplus is given by

$$\begin{aligned}
\int_0^{x_E} f(x)\,dx - x_E f(x_E) &= \int_0^{25}\left(200-\frac{1}{5}x^2\right)dx - (25)(75)\\
&= 200x - \frac{1}{15}x^3\Big|_0^{25} - 1{,}875\\
&= \left(200(25) - \frac{(25)^3}{15}\right) - 1{,}875\\
&= 5{,}000 - 1{,}041.67 - 1{,}875\\
&= 2{,}083.33
\end{aligned}$$

Hence, the consumers' surplus is \$2,083.33.

(d) The producers' is surplus given by

$$\begin{aligned}
x_E g(x_E) - \int_0^{x_E} g(x_E)\,dx &= 1{,}875 - \int_0^{25}(50+x)\,dx\\
&= 1{,}875 - \left(50x + \frac{1}{2}x^2\right)\Big|_0^{25}\\
&= 1{,}875 - \left[50(25) + \frac{(25)^2}{2}\right]\\
&= 1{,}875 - 1{,}250 - 312.50\\
&= 312.50
\end{aligned}$$

Hence, the producers' surplus is \$312.50.

EXERCISES

1. A leasing company has determined that the monthly maintenance expense on each of their copy machines can be approximated by the continuous function

$$M(t) = 15 + .01t^2$$

where $M(t)$ is in dollars and t represents the time in months that the machine has been rented. Determine the total number of dollars that will be spent on maintenance of a machine that is leased for 5 years.

2. **Expenditure for Education in Connecticut** The annual per-pupil expenditure for education in the State of Connecticut since 1960 is closely approximated by the formula

$$E(t) = 350e^{.09t}$$

where $E(t)$ is the dollar cost/year after t years assuming $t=0$ in 1960. Using this formula, determine the total cost to educate one student for 12 years assuming he or she started in (a) 1960, (b) 1970.

3. **Divorce Rate** The divorce rate in a certain community over the next t years is projected to be given by

$$D(t) = 200e^{.04t}$$

where $D(t)$ is the number of divorces/year t years from now. Determine the total number of divorces that will occur in the next 8 years in this community.

4. **Depreciated Value of Typewriters** The World-Wide Insurance Company depreciates new typewriters over a period of 10 years. When a typewriter is t years old the rate of depreciation is given by

$$D(t) = 15(10-t)$$

where $D(t)$ is in dollars/year.

(a) Determine the total amount of depreciation on 100 new typewriters that will be held for 10 years.
(b) If the cost of each typewriter is $900, determine the book value of the 100 typewriters at the end of the 10th year.

5. **Total Interest Income from an Investment** Determine the total interest income from $20,000 invested for a period of 10 years at 9% compounded continuously.

6. Determine the total interest income from $20,000 invested at 9% simple interest for a period of 10 years and compare this to the result obtained in Exercise 5.

7. **Tackle Shop Profits** As a result of the great interest in freshwater bass fishing, Doug Cushing, owner of a small tackle shop in Orlando, Florida, has found that his net profit has been increasing since he built the shop 5 years ago. Examining his accounts, he concluded that the net profit rate is closely approximated by the continuous function

$$P(t) = 10{,}000 + 5{,}000t^{1/3}$$

$P(t)$ is the net profit rate in dollars/year at time t where t is in years and $t=0$ five years ago.

(a) Determine his net profit rate 5 years ago.
(b) Determine his current net profit rate.
(c) Determine the total net profit that he has made during the last 5 years.

8. **National Health Care Expenses** Since 1965 the total national health care expense has increased exponentially at a rate of 12%/year compounded continuously. At the beginning of 1965 the total national health care expenditure rate was 40 billion dollars/year.

(a) Express $E(t)$, the health care expenditure rate, as a function of t measured in years from 1965.
(b) Using this formula, estimate the health care expenditure rate in 1980.
(c) Determine the total number of dollars spent on national health care from 1965 to 1980.

9. **National Hospital Expenditures** Hospital services constitute not only the largest items of health care expenditures, but one of the fastest growing in terms of increasing costs (see Exercise 8). In 1965 national hospital costs were 13 billion dollars/year and have increased exponentially from that time at the rate of 13%/year compounded continuously.

(a) Express the national hospital expenditure rate, $H(t)$, as a function of the time t measured in years. Assume $t=0$ in 1965.
(b) Using this formula, estimate the national hospital expenditure rate in 1980.
(c) Determine the total number of dollars spent on hospital care from 1965 to 1980.

10. **Petroleum Consumption and Depletion of Reserves** Suppose the current annual consumption of petroleum is 25,000 million barrels and that consumption is increasing exponentially at the rate of 2%/year compounded continuously.

(a) Determine the total amount of petroleum that will be consumed in the next 10 years.
(b) If the petroleum reserves are 700,000 million barrels, determine the calendar year in which these reserves will be depleted assuming that the above rate of consumption continues.

11. **Bauxite Consumption and Depletion of Reserves** In 1977 the world's annual consumption of bauxite, or aluminum ore, was 100 million tons and the exponential growth rate was 10%/year compounded continuously.

(a) Assuming this growth continues, determine the number of millions of tons of ore that will be used between 1977 and 1982.
(b) If the known world's reserves were 15,000 million tons in 1977, when will these reserves be depleted assuming the above growth of consumption continues?

12. **Copper Consumption and Depletion of Reserves** Suppose that the current world use of copper is 8 million tons/year and that the demand is growing exponentially at the rate of 5%/year compounded continuously.

(a) If this growth in demand continues, determine the number of tons of copper that will be used in the next 10 years.
(b) If the known world's reserves of copper are 400 million tons, determine when these reserves will be depleted assuming the above growth in demand continues.

13. In Exercise 2, Sec. 7.4, determine the increase in profit obtained from increasing the advertising expenditure from \$9,000 to \$12,000.

14. In Exercise 3, Sec. 7.4, determine the increase in cost that results from increasing the production level from 1,000 to 1,100 units.

15. In Exercise 4, Sec. 7.4, determine the decrease in cost that results from decreasing the production level from 210 to 190 units.

16. In Exercise 5, Sec. 7.4, determine the increase in cost that results from increasing the production level of radios from 6 to 8.

17. In Exercise 6, Sec. 7.4, determine the increase in profit that results from increasing the construction of the number of apartment houses from 3 to 4.

18. In Exercise 7, Sec. 7.4, determine the decrease in profit that results from decreasing the number of sales representatives from 225 to 210.

19. In Exercise 8, Sec. 7.4, determine the increase in profit that results from increasing the number of widgets from 15 to 25.

20. Assume that the demand function for a certain product is given by

$$f(x) = 150 - .01x^2$$

where $f(x)$ is the unit price in dollars at which the consumer will purchase x units. If the supply and demand are in equilibrium when 100 units are produced and sold, find the consumer surplus.

21. Suppose the supply function for a certain product is given by

$$g(x) = e^{.01x}$$

where $g(x)$ is the unit price in dollars at which the producer can supply x units. If the supply and demand are in equilibrium when 200 units are produced and sold, find the producer's surplus.

In Exercises 22–27, $g(x)$ is the price/unit in dollars at which a producer can supply x units of a product and $f(x)$ is the price/unit at which the consumer will purchase x units of the product. For each exercise:

(a) *Sketch the graphs of the supply and demand functions on the same coordinate system.*
(b) *Determine the equilibrium price and the number of units produced and sold at this price.*
(c) *Determine the consumers' surplus.*
(d) *Determine the producers' surplus.*

22. $g(x) = 25 + 2x$
$f(x) = 200 - \frac{1}{2}x$

23. $g(x) = 10 + \frac{1}{5}x^2$
$f(x) = 290 - 10x$

24. $g(x) = 60 + x$
$f(x) = 300 - x$

25. $g(x) = \sqrt{800 + x}$
$f(x) = \sqrt{1{,}000 - x}$

26. $g(x) = 20 + \frac{1}{10}x^2$
$f(x) = 120 - \frac{3}{20}x^2$

27. $g(x) = 4 + .02x$
$f(x) = 40 - .01x$

IMPORTANT TERMS AND CONCEPTS

Antiderivative
Antidifferentiation
Area of a region
Constant of integration
Consumers' surplus
Definite integral
Depletion of natural resources
Depletion of petroleum
Differential
Fixed cost
Fundamental theorem of integral calculus
Indefinite integral
Integral sign
Integrand
Integration
Limits of integration
  lower limit
  upper limit
Marginal analysis
Partitioning
Producers' surplus
Rate function
Variable cost

SUMMARY OF RULES AND FORMULAS

Integration Rules

Rule 7.1. $\int x^n\,dx = \frac{x^{n+1}}{n+1} + C, \qquad n \neq -1$

Rule 7.2. $\int kf(x)\,dx = k\int f(x)\,dx$

Rule 7.3. $\int [f(x)+g(x)]\,dx = \int f(x)\,dx + \int g(x)\,dx$

Rule 7.4. $\int [u(x)]^n u'(x)\,dx = \frac{[u(x)]^{n+1}}{n+1} + C, \qquad n \neq -1$

Rule 7.5. $\int e^{u(x)}u'(x)\,dx = e^{u(x)} + C$

Rule 7.6. $\int \frac{u'(x)}{u(x)}\,dx = \ln|u(x)| + C$

Rule 7.7. The Fundamental Theorem of Integral Calculus

$$\int_a^b f(x)\,dx = F(b) - F(a) \quad \text{where} \quad F'(x) = f(x)$$

Rule 7.8. $\int_a^b f(x)\,dx = \int_a^c f(x)\,dx + \int_c^b f(x)\,dx$, where $a \leqslant c \leqslant b$

Rule 7.9. $\int_a^b kf(x)\,dx = k\int_a^b f(x)\,dx$

Rule 7.10. $\int_a^b [f(x)+g(x)]\,dx = \int_a^b f(x)\,dx + \int_a^b g(x)\,dx$

Rule 7.11. The area of the region between the continuous nonnegative function $f(x)$ and the x axis bounded by $x=a$ and $x=b$ is given by

$$A = \int_a^b f(x)\,dx$$

8 INTEGRAL CALCULUS: ADVANCED METHODOLOGY

8.1 INTRODUCTION

In Chapter 7 we developed several rules of integration. We found that after inspecting each function and applying the appropriate rule we could rather easily integrate these functions. However, there are many useful functions that we cannot integrate with the rules developed so far. For example, as noted previously, the "simple" indefinite integral $\int xe^x\,dx$ cannot be determined by the direct use of Rules 7.1 through 7.6 (see Summary of Rules and Formulas at the end of Chapter 7).

In this chapter we shall consider additional methods of integration that can be used to express many integrands of this type in such a way that we can apply our rules of integration. We will also find that there are many functions where no simple integral can be determined and that the best way to evaluate the definite integral of such functions is to use a numerical procedure.

8.2 CHANGE OF VARIABLE

One of the easiest ways of solving any problem is to express it in terms of a simple problem with which we are familiar. This is precisely the approach used when integrating a function by the *change-of-variable* method. Let us begin by restating Rule 7.4 in a less complicated form; that is, in formula (7.5)

$$\int [u(x)]^n u'(x)\,dx = \frac{[u(x)]^{n+1}}{n+1} + C \qquad \text{for} \quad n \neq -1 \tag{8.1}$$

we are integrating the derivative of a composite function. Thus, if we let $u = u(x)$ and $du = u'(x)\,dx$, the integrand in (8.1) becomes u^n, and the integral is

$$\int u^n\,du$$

We immediately recognize this form and integrating obtain

$$\int u^n\,du = \frac{u^{n+1}}{n+1} + C \qquad \text{for} \quad n \neq -1$$

Since we have simply changed the variable of integration from x to u it should be easy to see how this method received its name.

Example 8.1 Determine the indefinite integral

$$\int 2x(x^2+2)^3\,dx$$

Solution Let $u=x^2+2$; then du, which equals $u'(x)dx$, is $2x\,dx$. Rearranging the terms and substituting, we have

$$\int (x^2+2)^3 \cdot 2x\,dx = \int u^3\,du$$

Integrating with respect to u, we obtain

$$\int u^3\,du = \frac{1}{4}u^4 + C$$

Since the original integrand is a function of x, it is necessary to determine the indefinite integral as a function of x. Therefore, substituting (x^2+2) for u, we obtain

$$\int (x^2+2)^3 \cdot 2x\,dx = \frac{1}{4}(x^2+2)^4 + C$$

The previous example demonstrates the concepts associated with this method. For the general situation where we wish to determine the integral of a function that is of the chain rule type, we can simplify the integration process by changing the variable of integration; that is, if we wish to determine the integral

$$\int f(u(x)) \cdot u'(x)\,dx \tag{8.2}$$

we can let the variable u represent the function $u(x)$ and let the term du represent $u'(x)dx$. Thus, the integral (8.2) becomes

$$\int f(u)\,du \tag{8.3}$$

which is less complicated to determine. After determining the antiderivative of $f(u)$, we simply substitute $u(x)$ for u in order to express the result as a function of x.

Example 8.2 Determine the indefinite integral

$$\int (2x-4)e^{(x^2-4x)}\,dx$$

Solution Let $u=(x^2-4x)$; then $du=u'(x)\,dx=(2x-4)\,dx$. With this substitution, we obtain

$$\int (2x-4)e^{(x^2-4x)}\,dx = \int e^u\,du$$

Integrating with respect to u we obtain

$$\int e^u\,du = e^u + C$$

Substituting (x^2-4x) for u in the right side of this equation, we obtain

$$\int (2x-4)e^{(x^2-4x)}\,dx = e^{(x^2-4x)} + C$$

This method is especially advantageous when the integrand is "off" by a constant factor as we discussed in Sec. 7.3. It eliminates the need to fix or preadjust the integral as we did in Sec. 7.3. Consider the following example.

Example 8.3 Determine the indefinite integral

$$\int (1-x)^{3/2}\,dx$$

Solution Let $u=(1-x)$; then $du=u'(x)\,dx=(-1)\,dx$. Solving for dx this becomes $dx=(-1)\,du$. Substituting u for $(1-x)$ and $(-1)\,du$ for dx and integrating with respect to u, we have

$$\begin{aligned}\int (1-x)^{3/2}\,dx &= \int u^{3/2}(-1)\,du\\ &= -1\int u^{3/2}\,du\\ &= (-1)\frac{2}{5}u^{5/2}+C\end{aligned}$$

Substituting $(1-x)$ in place of u in the resulting indefinite integral, we obtain

$$\int (1-x)^{3/2}\,dx = -\frac{2}{5}(1-x)^{5/2}+C$$

This method is easy to apply if we can select $u=u(x)$ such that $u'(x)$ or a constant times $u'(x)$ is a factor in the integrand. Note the choice of $u(x)$ in the following example.

Example 8.4 Determine the indefinite integral

$$\int (x^2-2x)^3(x-1)\,dx$$

Solution If we let $u=(x^2-2x)$, then $u'(x)=(2x-2)$ or $2(x-1)$, which is twice one of the factors in the integrand. With this choice for u, $du=2(x-1)\,dx$ or $(x-1)\,dx=\frac{1}{2}du$. Making these substitutions, we have

$$\begin{aligned}\int (x^2-2x)^3(x-1)\,dx &= \int u^3\cdot\frac{1}{2}\,du\\ &= \frac{1}{2}\int u^3\,du\\ &= \frac{1}{2}\cdot\frac{1}{4}u^4+C\end{aligned}$$

Resubstituting (x^2-2x) for u, we obtain

$$\int (x^2-2x)^3(x-1)\,dx = \frac{1}{8}(x^2-2x)^4+C$$

The change-of-variable technique can also be used to find indefinite integrals of functions that are not of the chain rule type. Frequently we encounter integrands which are products of functions that are neither of the form $[u(x)]^n u'(x)$ nor of the form $e^{u(x)}u'(x)$. In such cases a change of variable may make it possible to write the integrand in a form that is easy to integrate.

Example 8.5 Determine the indefinite integral

$$\int (4-x)^{1/2}x\,dx$$

Solution Although the integrand may at first appear to be of the form $[u(x)]^n u'(x)$ it is soon evident that this is not the case. To write the integrand in a different form we let $u=(4-x)$. With this change of variable $u'(x)=-1$, $du=-1\,dx$, and $x=(4-u)$. Making the appropriate substitutions, we have

$$\begin{aligned}\int (4-x)^{1/2}x\,dx &= \int u^{1/2}(4-u)(-1\,du)\\ &= \int u^{1/2}(u-4)\,du\\ &= \int (u^{3/2}-4u^{1/2})\,du\\ &= \frac{2}{5}u^{5/2}-\frac{8}{3}u^{3/2}+C\end{aligned}$$

Resubstituting $(4-x)$ for u, we obtain

$$\int (4-x)^{1/2} x\,dx = \frac{2}{5}(4-x)^{5/2} - \frac{8}{3}(4-x)^{3/2} + C$$

Definite Integrals To evaluate a definite integral it would appear at first that the indefinite integral must be written in terms of the original variable of integration, since the limits of integration are values of this variable. This is, in fact, a perfectly legitimate approach. For example, if we wish to evaluate the definite integral

$$\int_1^2 2x(x^2+2)^3\,dx$$

we can evaluate the result of Example 8.1 at a and b and take the appropriate difference; that is,

$$\int_1^2 2x(x^2+2)^3\,dx = \left.\frac{(x^2+2)^4}{4}\right|_1^2 = 324 - \frac{81}{4} = 303\frac{3}{4} \tag{8.4}$$

However, in many situations, it is easier to evaluate the definite integral for the new variable of integration than for the original variable. To employ the change-of-variable method to evaluate

$$\int_a^b f[u(x)]u'(x)\,dx$$

we make the substitution $u = u(x)$. If we let u_1 and u_2 be the values of u such that $u_1 = u(a)$ and $u_2 = u(b)$, then it can be proven that

$$\int_a^b f[u(x)]u'(x)\,dx = \int_{u_1}^{u_2} f(u)\,du \tag{8.5}$$

provided that as x changes continuously from a to b, $u(x)$ increases or decreases continuously from u_1 to u_2.

Example 8.6 Evaluate $\int_1^2 2x(x^2+2)^3\,dx$ by changing the variable in the integrand and making the appropriate changes in the limits of integration.

Solution Let $u = x^2+2$; then $du = 2x\,dx$. Since x^2+2 increases continuously as x increases from 1 to 2, we can change the limits of integration. Evaluating $u(x)$ at $x=1$ and $x=2$ we have $u(1)=3$ and $u(2)=6$; hence, by (8.5) we obtain

$$\int_1^2 2x(x^2+2)^3\,dx = \int_3^6 u^3\,du$$

$$= \left.\frac{u^4}{4}\right|_3^6$$

$$= 324 - \frac{81}{4}$$

$$= 303\tfrac{3}{4}$$

which is exactly the same value obtained by not changing the limits of integration as shown in Eq. (8.4).

Example 8.7 Evaluate $\int_0^1 (1-x)^{3/2}\,dx$ by changing the variable in the integrand and making the appropriate change of limits.

Solution Let $u=1-x$; then $du=(-1)dx$ or $dx=(-1)du$. Since $(1-x)$ decreases continuously as x increases from 0 to 1, we can change the limits of integration. Evaluating $u(0)=1$ and $u(1)=0$, we obtain

$$\int_0^1 (1-x)^{3/2}\,dx = \int_1^0 (-1)u^{3/2}\,du$$

$$= -\left.\frac{2}{5}u^{5/2}\right|_1^0$$

$$= 0 + \frac{2}{5}$$

$$= \frac{2}{5}$$

As we have shown, integration by change of variable is particularly useful when the integrand involves a composite function. There are also many situations for which this method is not particularly helpful. In the next section we will examine some of these and develop another method of integration to solve these problems.

EXERCISES

Determine each of the indefinite integrals in Exercises 1–12.

1. $\int (3x+6)^4\,dx$

2. $\int \sqrt{x+4}\;dx$

3. $\int (x+2)^6\,dx$

4. $\int (x^2+3)^2 2x\,dx$

5. $\int 3x^2(x^3+2)^4\,dx$

6. $\int \frac{x+1}{\sqrt{2x^2+4x}}\,dx$

7. $\int \frac{5t}{\sqrt{2-3t^2}}\,dt$

8. $\int \frac{t}{(4-t^2)^3}\,dt$

9. $\int \frac{y+2}{(y^2+4y+5)}\,dy$

10. $\int 6y\sqrt{2y^2+2}\,dy$

11. $\int \frac{z^3}{\sqrt{z^4+2}}\,dz$

12. $\int \frac{(2+\sqrt{z})^5}{2\sqrt{z}}\,dz$

13. $\int xe^{x^2}\,dx$

14. $\int \frac{x+1}{\sqrt{x+2}}\,dx$

15. $\int \frac{t}{\sqrt{9-t}}\,dt$

16. $\int \frac{e^{\sqrt{x}}}{\sqrt{x}}\,dx$

Evaluate each of the definite integrals in Exercises 17–24 *by change of variable and change of limits.*

17. $\int_0^2 (x-2)^3\,dx$

18. $\int_0^5 \sqrt{x+4}\,dx$

19. $\int_0^3 x(x^2+16)^{1/2}\,dx$

20. $\int_2^{10} \frac{4}{\sqrt{t-1}}\,dt$

21. $\int_0^5 \frac{t+2}{\sqrt{9-t}}\,dt$

22. $\int_1^3 \frac{t+2}{(t^2+4t)^2}\,dt$

23. $\int_0^2 e^{x^2-1}x\,dx$

24. $\int_0^1 ye^{y^2}\,dy$

25. The marginal cost function in dollars for manufacturing a particular model airplane is given by

$$c(x) = x(x^2+16)^{1/2}$$

where x is the number of units made and the fixed cost is \$250. Determine the total cost function and the total cost for three units.

26. The marginal profit function in thousands of dollars for a brand of aftershave lotion is given by

$$p(x) = \frac{x+2}{\sqrt{x+4}}$$

where x represents the thousands of dollars spent on promotional advertising. Furthermore, historically, when \$5,000 has been spent on advertising, the total profit has been about \$40,000. Determine the total profit function and the total profit when \$12,000 is spent on advertising. Also, determine the increase in profit for increasing advertising expenditures from \$12,000 to \$21,000.

27. The marginal profit for manufacturing and selling x thousand widgets is given by the function $f(x)=x^3(\sqrt{x^4+3{,}600})$. Fixed cost is \$25,500. Determine the total profit function and the total profit if 10,000 widgets are manufactured and sold. Also, determine the increase in profit for increasing sales volume from 10,000 to 12,000 units.

28. Determine the area of the region bounded by the graph of $f(x)=x/\sqrt{x+4}$, the x axis, the line $x=0$, and the line $x=5$.

8.3 INTEGRATION BY PARTS

In many situations we are confronted with the problem of integrating the product of two functions, neither of which appears to be the derivative of the other. For example, consider

$$\int xe^x\,dx$$

At first glance the integrand appears to be of the form $u'(x)e^{u(x)}$; however, if we let $u(x)=x$, we find that $u'(x)=1$, which is not a constant times x. Hence, we cannot use Rule 7.5 directly as the integrand xe^x and $u'(x)e^{u(x)}$ differ by more than a constant factor. If we attempt to apply the change of variable method of Sec. 8.2, we find that by letting $u=x$ we obtain

$$\int xe^x\,dx=\int ue^u\,du$$

which is exactly the same integral with just a different variable of integration. Hence, this method of attack is not fruitful. How then can we integrate problems of this form?

When considering such integrands, we often use the method of *integration by parts*. This method is based on the formula for the differentiation of a product. From Rule 4.11 we have the equation for the derivative of a product; that is

$$\begin{aligned} D_x[f(x)\cdot g(x)] &= f(x)\cdot D_xg(x)+g(x)\cdot D_xf(x)\\ &= f(x)\cdot g'(x)+g(x)\cdot f'(x)\end{aligned}$$

If we integrate both sides of this equation with respect to x, we obtain

$$\int D_x[f(x)\cdot g(x)]\,dx=\int f(x)\cdot g'(x)\,dx+\int g(x)\cdot f'(x)\,dx$$

We note that the integral to the left of the equality has an integrand for which an antiderivative is $[f(x)\cdot g(x)]$ since the derivative of $[f(x)\cdot g(x)]$ is simply

$D_x[f(x)\cdot g(x)]$. Therefore,

$$f(x)\cdot g(x) = \int f(x)\cdot g'(x)\,dx + \int g(x)\cdot f'(x)\,dx \tag{8.6}$$

In order to simplify the notation, let $u=f(x)$ and $v=g(x)$; then, as in Sec. 8.2, $du=f'(x)dx$ and $dv=g'(x)dx$. Substituting these changes into Eq. (8.6), we have

$$u\cdot v = \int v\,du + \int u\,dv$$

Solving for one of the integrals, say, $\int u\,dv$, we obtain

$$\int u\,dv = u\cdot v - \int v\,du \tag{8.7}$$

which is the *integration by parts formula.*

We can use this formula (8.7) to our advantage if, by choosing u and dv to correspond to the two factors of the integrand under consideration, we can integrate dv easily and the integral $\int v\,du$ is easier to determine than the original integral.

Example 8.8 Determine the indefinite integral

$$\int xe^x\,dx$$

Solution Let $dv=e^x\,dx$ and $u=x$; then,

$$v = \int dv = \int e^x\,dx = e^x$$

and $$du = u'(x)\,dx = (1)\,dx$$

Substituting into formula (8.7) and integrating the second term on the right, we obtain

$$\begin{aligned}\int xe^x\,dx &= xe^x - \int e^x\,dx \\ &= xe^x - e^x + C\end{aligned}$$

The choice of u and dv is extremely important. If we were to choose $dv=x\,dx$ and $u=e^x$ in Example 8.8, then

$$du = u'(x)\,dx = e^x\,dx$$

and
$$v=\int dv=\int x\,dx=\frac{1}{2}x^2$$

which when substituted into formula (8.7) yields

$$\int xe^x\,dx=\frac{x^2}{2}e^x-\int x^2e^x\,dx$$

In this case, the term $\int x^2e^x\,dx$ is more difficult to integrate than the original integral $\int xe^x\,dx$ and we have not improved the situation. If this happens, we make a new choice for u and dv and often obtain more favorable results. There are some functions that allow a variety of choices, and unfortunately there is no single way of choosing u and dv that works every time. In all choices, however, dv must be selected so that it may be integrated easily and in general u and dv are selected so that the integral $\int v\,du$ is simpler than the original. The ability to make good choices improves with practice.

Example 8.9 Determine the indefinite integral

$$\int x\sqrt{x+3}\;dx$$

Solution The two obvious choices for u and dv are $u=x$ and $dv=(x+3)^{1/2}dx$ or $u=(x+3)^{1/2}$ and $dv=x\,dx$. (There is a third choice; that is, $u=x(x+3)^{1/2}$ and $dv=dx$, which is not quite as obvious. The third choice appears to be a poor one as du will be very complicated since it involves a product and we will not pursue it.) Suppose we continue with the second choice. If we let $u=(x+3)^{1/2}$ and $dv=x\,dx$, then $du=\frac{1}{2}(x+3)^{-1/2}dx$ and $v=\frac{1}{2}x^2$. With this choice, the integral $\int v\,du=\int\frac{1}{4}x^2(x+3)^{-1/2}dx$, which is more complicated than our original integral. Since this did not work out, let us pursue the first choice. If we let $u=x$ and $dv=(x+3)^{-1/2}dx$, then

$$du=u'(x)\,dx=(1)\,dx$$

and
$$v=\int(x+3)^{1/2}dx=\frac{2}{3}(x+3)^{3/2}$$

by Rule 7.4. Substituting in formula (8.7) and integrating the second term on the right by Rule 7.4, we obtain

$$\begin{aligned}\int x\sqrt{x+3}\;dx&=\frac{2}{3}x(x+3)^{3/2}-\int\frac{2}{3}(x+3)^{3/2}dx\\&=\frac{2}{3}x(x+3)^{3/2}-\frac{2}{3}\int(x+3)^{3/2}dx\\&=\frac{2}{3}x(x+3)^{3/2}-\frac{2}{3}\cdot\frac{2}{5}(x+3)^{5/2}+C\\&=\frac{2}{3}x(x+3)^{3/2}-\frac{4}{15}(x+3)^{5/2}+C\end{aligned}$$

In the next example we have only one choice for u and dv.

Example 8.10 Determine the indefinite integral

$$\int \ln x \, dx$$

Solution Let $u = \ln x$ and $dv = dx$; then, $du = (1/x)dx$ and $v = \int dx = x$. Substituting into formula (8.7) we obtain

$$\begin{aligned}\int \ln x \, dx &= x \ln x - \int \frac{1}{x} x \, dx \\ &= x \ln x - \int dx \\ &= x \ln x - x + C\end{aligned}$$

Example 8.11 Determine the indefinite integral

$$\int x(\ln x) \, dx$$

Solution Let $u = \ln x$ and $dv = x\,dx$; then $du = (1/x)dx$ and $v = x^2/2$. Substituting, we obtain

$$\begin{aligned}\int x \ln x \, dx &= \frac{x^2}{2} \ln x - \int \frac{x^2}{2} \cdot \frac{1}{x} dx \\ &= \frac{x^2}{2} \ln x - \frac{x^2}{4} + C\end{aligned}$$

Occasionally, it is necessary to apply the process of integration by parts two or more times. When this is the case, we should be very careful to make certain that the new integral $\int v\,du$ is improving in a direction that will eventually lead to a function instead of an integral.

Example 8.12 Determine the indefinite integral

$$\int x^2 e^x \, dx$$

Solution Let $u = x^2$ and $dv = e^x dx$; then $du = 2x\,dx$ and $v = e^x$. Substituting, we obtain

$$\begin{aligned}\int x^2 e^x \, dx &= x^2 e^x - \int e^x 2x \, dx \\ &= x^2 e^x - 2\int x e^x \, dx\end{aligned}$$

Next, we need to integrate $\int xe^x\,dx$. Since we did this in Example 8.8, we substitute the result and obtain

$$\begin{aligned}\int x^2e^x\,dx &= x^2e^x - 2(xe^x - e^x + C)\\ &= x^2e^x - 2xe^x + 2e^x + C'\end{aligned}$$

To determine the definite integral of a function of x from $x=a$ to $x=b$, where the indefinite integral must be found by integration by parts, we follow the same procedure we used when we first considered definite integrals. The method of integration by parts does not change the variable of integration and hence the limits remain values of x. The definite integral form of (8.7) is

$$\int_{x=a}^{x=b} u\,dv = u\cdot v\Big|_a^b - \int_a^b v\,du$$

where u and v are functions of x.

Example 8.13 Determine the definite integral

$$\int_0^2 xe^{-2x}\,dx$$

Solution Let $u=x$ and $dv=e^{-2x}\,dx$; then $du=dx$ and $v=-\frac{1}{2}e^{-2x}$. Substituting, we obtain

$$\begin{aligned}\int_0^2 xe^{-2x}\,dx &= -\frac{1}{2}xe^{-2x}\Big|_0^2 - \int_0^2 -\frac{1}{2}e^{-2x}\,dx\\ &= \left(-\frac{1}{2}xe^{-2x}\right)\Big|_0^2 - \left(\frac{1}{4}e^{-2x}\right)\Big|_0^2\\ &= (-e^{-4}-0) - \left(\frac{1}{4}e^{-4} - \frac{1}{4}e^0\right)\\ &= -\frac{5}{4}e^{-4} + \frac{1}{4}\end{aligned}$$

Approximating e^{-4} by Table A.3 we obtain

$$\begin{aligned}\int_0^2 xe^{-2x}\,dx &\doteq -\frac{5}{4}(0.01832) + \frac{1}{4}\\ &\doteq 0.22710\end{aligned}$$

EXERCISES

In Exercises 1–16 determine the indefinite integral.

1. $\int x(x+2)^6\,dx$
2. $\int (x+1)(x+3)^3\,dx$
3. $\int xe^{3x}\,dx$
4. $\int x\ln x^2\,dx$
5. $\int x^2(x+1)^4\,dx$
6. $\int (\ln x)^2\,dx$
7. $\int x^2\ln x\,dx$
8. $\int x^3\ln x\,dx$
9. $\int (x+2)^2(x+5)^8\,dx$
10. $\int x^2e^{3x}\,dx$
11. $\int x^3e^{2x^2}\,dx$
12. $\int x^n\ln x\,dx,\quad n\neq -1$
13. $\int x(x+5)^{1/3}\,dx$
14. $\int x\sqrt{4-x}\,dx$
15. $\int \frac{x}{\sqrt{x+4}}\,dx$
16. $\int \frac{x}{e^x}\,dx$

In Exercises 17–20 determine the value of the definite integral.

17. $\int_0^1 x(x+1)^4\,dx$
18. $\int_1^2 (x-1)^2(x-2)^4\,dx$
19. $\int_0^2 3x^2e^{2x}\,dx$
20. $\int_{-1}^1 (x+1)^2(x-1)^5\,dx$

21. The marginal profit function in dollars for growing and selling corn is given by

$$p(x) = (x-1)(x-5)^2$$

where x is the number of bushels grown and sold in tens. Fixed cost for the farmer is \$25. Determine the total profit function and the total profit for selling 100 bushels of corn. Determine the amount of increase in profits by increasing sales from 130 bushels to 160 bushels.

22. A sales department has a monthly marginal cost function in dollars that is given by

$$c(x) = (x-3)^2(x+2)^2$$

where x is the number of salespeople employed, and fixed cost is \$75. Determine the total cost function and the cost of having 10 salespeople employed. Determine the amount that could be saved by decreasing the sales force from 10 to 8.

8.4 INTEGRAL TABLES

There are many more methods of integration than those discussed in Secs. 8.2 and 8.3; however, we shall not develop any other methods. Rather we will present a *table of integrals* that lists integration formulas in a systematic order according to the type of functions in the integrand. No table of integrals, however large, contains formulas for all integrals. This is true for two reasons. First, the table is designed to reflect the level of skill of the user, and secondly, there are many integrals for which no antiderivative exists.

Table 8.1 on pages 418–419 is a short table of integrals (many contain hundreds of formulas)* that will satisfy our needs and will not require any new skills. Note that several of the formulas involve logarithms of absolute values of an expression. This removes the restriction that such expressions must be nonnegative for the logarithm to be defined. To use the table we first identify the integrand of concern as one of the forms in the table, make the proper substitutions, and write the answer. Following are several examples to exhibit the techniques for using a table of integrals.

Example 8.14 Determine the indefinite integral

$$\int \frac{x}{2+3x}\,dx$$

Solution The integral is of the form of Formula 6 in Table 8.1, where $a=2$ and $b=3$. Substituting, we have

$$\int \frac{x}{2+3x}\,dx = \frac{1}{9}(2+3x-2\ln|2+3x|) + C$$

Example 8.15 Determine the indefinite integral

$$\int (3+4x)^6\,dx$$

Solution The integral is of the form of Formula 7 in Table 8.1, where $a=3$, $b=4$, and $n=6$. Substituting, we obtain

$$\int (3+4x)^6\,dx = \frac{1}{28}(3+4x)^7 + C$$

Example 8.16 Determine the indefinite integral

$$\int \frac{1}{x^2-25}\,dx$$

*There exist a number of books which contain extensive collections of integrals. Two of the better-known volumes are the following:

H. B. Dwight, *Tables of Integrals and Other Mathematical Data* (New York: Macmillan, 1961).

B. O. Pierce, *A Short Table of Integrals* (Boston: Ginn and Company, 1957).

Solution The integral is of the form of Formula 10 in Table 8.1, where $a=5$. Thus,

$$\int \frac{1}{x^2-25}\,dx = \frac{1}{10}\ln\left|\frac{x-5}{x+5}\right| + C$$

Example 8.17 Determine the indefinite integral

$$\int 5^{3x}\,dx$$

Solution Formula 18 in Table 8.1 is appropriate, where $a=5$ and $b=3$. Substituting, we have

$$\int 5^{3x}\,dx = \frac{5^{3x}}{3\ln 5} + C$$
$$= \frac{5^{3x}}{4.8283} + C$$

EXERCISES

In each of Exercises 1–18 use the appropriate integral formula given in Table 8.1 to determine the integral.

1. $\int \frac{x}{5+x}\,dx$

2. $\int (10+2x)^3\,dx$

3. $\int x(3+8x)^3\,dx$

4. $\int x(2+5x)^{-4}\,dx$

5. $\int \frac{1}{\sqrt{x^2-16}}\,dx$

6. $\int \frac{x}{\sqrt{9-x^2}}\,dx$

7. $\int e^{-6x}\,dx$

8. $\int xe^{2x}\,dx$

9. $\int 3^{2x}\,dx$

10. $\int \ln(4x)\,dx$

11. $\int 3\ln(2x)\,dx$

12. $\int x(3-x^2)^{-1/2}\,dx$

13. $\int x(x^2+4)^{1/2}\,dx$

14. $\int \frac{x}{x+2}\,dx$

15. $\int_1^2 x\ln x\,dx$

16. $\int_4^6 \frac{1}{x(\ln x)}\,dx$

17. $\int (\ln 3x)^2\,dx$

18. $\int x\ln(4x)\,dx$

Table 8.1 Table of Integrals

Elementary Forms

1. $\int kf(x)\,dx = k\int f(x)\,dx$

2. $\int [f(x)+g(x)]\,dx = \int f(x)\,dx + \int g(x)\,dx$

3. $\int [f(x)-g(x)]\,dx = \int f(x)\,dx - \int g(x)\,dx$

4. $\int x^n\,dx = \frac{x^{n+1}}{n+1} + C$ (for $n \neq -1$)

$= \ln|x| + C$ (for $n = -1$)

5. $\int [u(x)]^n \cdot u'(x)\,dx = \frac{[u(x)]^{n+1}}{n+1} + C$ (for $n \neq -1$)

$= \ln|u(x)| + C$ (for $n = -1$)

Forms Containing $a+bx$

6. $\int \frac{x}{a+bx}\,dx = \frac{1}{b^2}(a+bx-a\ln|a+bx|) + C$

7. $\int (a+bx)^n\,dx = \frac{1}{b(n+1)}(a+bx)^{n+1} + C$ (for $n \neq -1$)

$= \frac{1}{b}\ln|a+bx| + C$ for $n = -1$)

8. $\int \frac{x}{(a+bx)^n}\,dx = \frac{1}{b^2}\left[\frac{-1}{(n-2)(a+bx)^{n-2}} + \frac{a}{(n-1)(a+bx)^{n-1}}\right]$

$+ C$ (for $n \neq 1$ and $n \neq 2$)

$= \frac{1}{b^2}(a+bx-a\ln|a+bx|) + C$ (for $n = 1$)

$= \frac{1}{b^2}\left(\ln|a+bx| + \frac{a}{a+bx}\right) + C$ (for $n = 2$)

9. $\int x(a+bx)^n\,dx = \frac{1}{b^2(n+2)}(a+bx)^{n+2} - \frac{a}{b^2(n+1)}(a+bx)^{n+1}$

$+ C$ (for $n \neq -1, -2$)

$= \frac{1}{b^2}[a+bx-a\ln|a+bx|] + C$ (for $n = -1$)

$= \frac{1}{b^2}\left[\ln|a+bx| + \frac{a}{a+bx}\right] + C$ (for $n = -2$)

Table 8.1 (continued)

Forms Containing $x^2 \pm a^2$, or $a^2 \pm x^2$

10. $\int \frac{1}{x^2 - a^2}\, dx = \frac{1}{2a} \ln\left|\frac{x-a}{x+a}\right| + C$

11. $\int \frac{1}{a^2 - x^2}\, dx = \frac{1}{2a} \ln\left|\frac{a+x}{a-x}\right| + C$

12. $\int \sqrt{x^2 \pm a^2}\, dx = \frac{1}{2}[x\sqrt{x^2 \pm a^2} \pm a^2 \ln|x + \sqrt{x^2 \pm a^2}\,|] + C$

13. $\int \frac{1}{\sqrt{x^2 \pm a^2}}\, dx = \ln|x + \sqrt{x^2 \pm a^2}\,| + C$

14. $\int x\sqrt{x^2 \pm a^2}\, dx = \frac{1}{3}\sqrt{(x^2 \pm a^2)^3} + C$

15. $\int \frac{x}{\sqrt{x^2 \pm a^2}}\, dx = \sqrt{x^2 \pm a^2} + C$

16. $\int x\sqrt{a^2 - x^2}\, dx = -\frac{1}{3}\sqrt{(a^2 - x^2)^3} + C$

17. $\int \frac{x}{\sqrt{a^2 - x^2}}\, dx = -\sqrt{a^2 - x^2} + C$

Exponential Forms

18. $\int a^{bx}\, dx = \frac{a^{bx}}{b(\ln a)} + C$

19. $\int e^{bx}\, dx = \frac{e^{bx}}{b} + C$

20. $\int xe^{bx}\, dx = \frac{e^{bx}}{b^2}(bx - 1) + C$

Logarithmic Forms

21. $\int x(\ln x)\, dx = \frac{x^2}{2}(\ln x) - \frac{x^2}{4} + C$

22. $\int \frac{1}{x(\ln x)}\, dx = \ln(\ln x) + C$

23. $\int (\ln x)^2\, dx = x(\ln x)^2 - 2x(\ln x) + 2x + C$

24. $\int \ln(ax)\, dx = x(\ln ax) - x + C$

8.5 NUMERICAL INTEGRATION

As we mentioned in Sec. 8.4, there are functions for which no integral can be determined by any method of integration or by searching in the most extensive set of tables. An example of such a function is the normal distribution function, $f(x)=(1/\sqrt{2\pi}\,\sigma)e^{-\frac{1}{2}[(x-\mu)/\sigma]^2}$, discussed in Sec. 8.7. To compute the value of the definite integral of these functions we are forced to use numerical approximation methods. Many such *numerical integration* methods have come into use with the availability of hand-held calculators and computers. In this section we will consider two of these methods; the *trapezoidal rule* and *Simpson's rule*. Each is based on the fact that the definite integral $\int_a^b f(x)\,dx$ can be interpreted as the area of the region between $f(x)$ and the x axis bounded by the lines $x=a$ and $x=b$. Hence, any approximation of this area is an approximation of the definite integral $\int_a^b f(x)\,dx$.

Trapezoidal Rule

To develop this method let us consider approximating the value of the definite integral

$$\int_1^6 x^3\,dx$$

We have selected this integral because we can easily evaluate it in the usual manner and hence compare our approximation to the actual value. Figure 8.1*a* shows the area of the region between the graph of $f(x)=x^3$ and the x axis bounded by the lines $x=1$ and $x=6$. If the function $f(x)$ is approximated by the straight line segments shown in Fig. 8.1*a*, the area of the shaded region and hence the definite integral $\int_1^6 x^3\,dx$ can be approximated by the five trapezoids shown in this figure. From elementary geometry it is known that the area of a trapezoid is equal to half the product of its altitude and the sum of the lengths of the parallel bases; that is, $A=\frac{1}{2}h(b_1+b_2)$ (see Fig. 8.1*b*).

Since each of the trapezoids in Fig. 8.1*a* has an altitude of 1, we have

$$\int_1^6 x^3\,dx \doteq \left[\left(\frac{f(1)+f(2)}{2}\right)\cdot 1+\left(\frac{f(2)+f(3)}{2}\right)\cdot 1 + \left(\frac{f(3)+f(4)}{2}\right)\cdot 1+\left(\frac{f(4)+f(5)}{2}\right)\cdot 1+\left(\frac{f(5)+f(6)}{2}\right)\cdot 1\right]$$

Regrouping the right side, we obtain

$$\int_1^6 x^3\,dx \doteq \left[\frac{f(1)}{2}+f(2)+f(3)+f(4)+f(5)+\frac{f(6)}{2}\right]$$

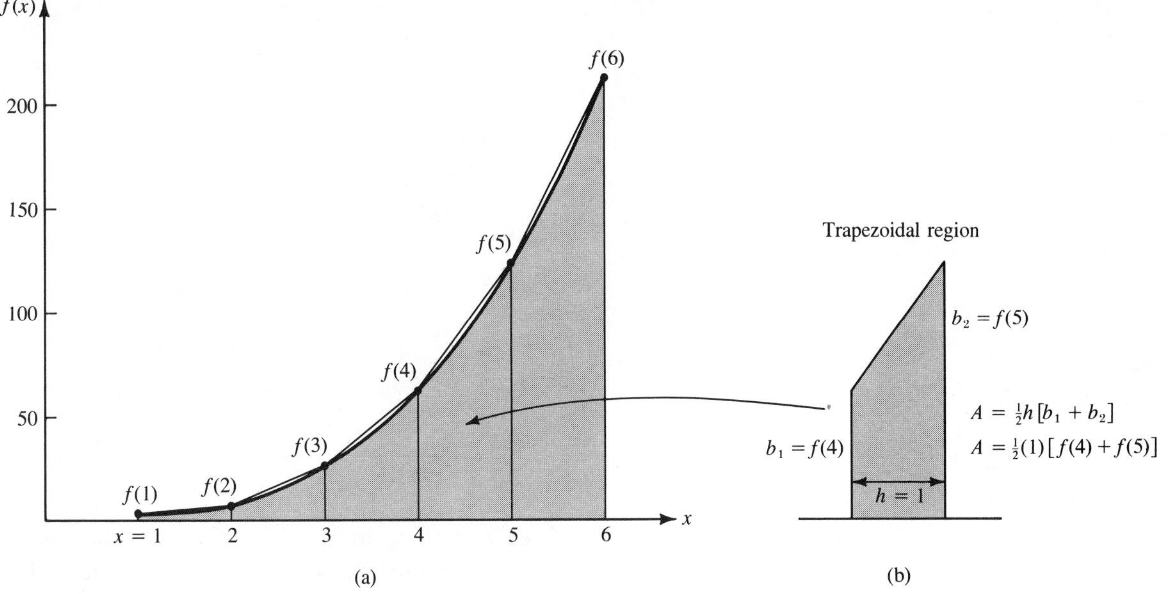

FIGURE 8.1

Evaluating $f(1), f(2)$, etc., by substituting into the function $f(x)=x^3$, we have

$$\int_1^6 x^3\,dx \doteq \left[\frac{1}{2}+8+27+64+125+\frac{216}{2}\right]$$
$$\doteq 332.5$$

We can compare this approximation to the actual value of $\int_1^6 x^3\,dx$ since

$$\int_1^6 x^3\,dx = \frac{x^4}{4}\bigg|_1^6 = \frac{1296}{4}-\frac{1}{4}=\frac{1295}{4}=323.75$$

Our approximate value is greater as we might have expected by noting that each trapezoid is an overestimate of the desired area, since the graph is concave upward.

To improve our approximation, we should increase the number of trapezoids by dividing the interval $[1,6]$ into more subintervals. The larger the number of such subintervals, the better will be the approximation.

We can generalize the previous example to obtain a formula for the approximate value of the definite integral $\int_a^b f(x)\,dx$. If an interval $[a,b]$ is divided into n subintervals of equal length by the points $x_0, x_1, x_2, \ldots, x_n$ in order from $x_0=a$ to $x_n=b$, the length of each subinterval will be $(b-a)/n$, as

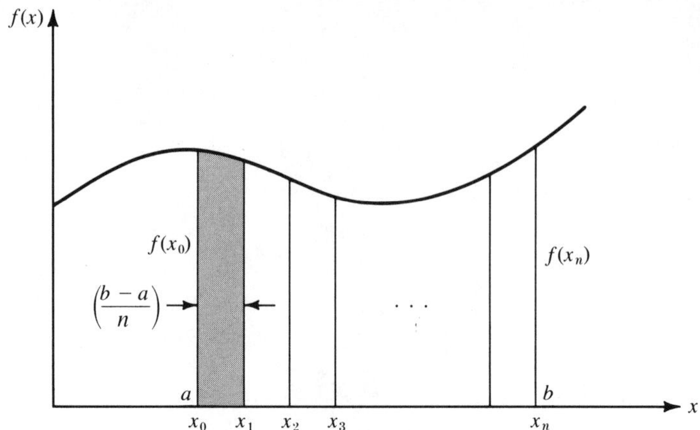

FIGURE 8.2

is illustrated in Fig. 8.2. An approximate value of the definite integral $\int_a^b f(x)\,dx$ found by summing the area of the n trapezoids indicated in Fig. 8.2 is given by the following rule.

Rule 8.1 *The trapezoidal rule:* An approximation to the definite integral of the continuous function $f(x)$ from $x=a$ to $x=b$ is given by

$$\int_a^b f(x)\,dx \doteq \left(\frac{b-a}{n}\right)\left[\frac{f(a)}{2}+f(x_1)+f(x_2)+\cdots +f(x_{n-1})+\frac{f(b)}{2}\right] \tag{8.8}$$

where the interval $[a,b]$ has been divided into n equal subintervals by the points $x_0, x_1, x_2, \ldots, x_n$ in order from $x_0=a$ to $x_n=b$.

To illustrate the application of formula (8.8), we will determine an improved approximation of $\int_1^6 x^3\,dx$ in the next example by increasing n to 10.

Example 8.18 Determine an approximate value of $\int_1^6 x^3\,dx$ by using formula (8.8) by choosing $n=10$.

Solution If $n=10$, the length of each subinterval given by $(b-a)/n$ is $(6-1)/10$ or 0.5. Table 8.2 summarizes the important data needed for substitution into formula (8.8). Each x_i in Table 8.2 is found by adding the subinterval length of 0.5 to the previous x. Substituting the appropriate

values from this table into formula (8.8), we obtain

$$\int_1^6 x^3\,dx \doteq (0.5)\left[\frac{1}{2}+3.375+8.0+15.625+27.0\right.$$

$$\left.+42.875+64+91.125+125+166.325+\frac{216}{2}\right]$$

$$\doteq (0.5)(651.875)$$

$$\doteq 325.94$$

This approximation is considerably closer to the actual value of 323.75 than our first approximation of 332.5 where n was 5.

Table 8.2

x	x_0	x_1	x_2	x_3	x_4	x_5	x_6	x_7	x_8	x_9	x_{10}
	1	1.5	2.0	2.5	3.0	3.5	4.0	4.5	5.0	5.5	6
$f(x)=x^3$	1	3.375	8.0	15.625	27.0	42.875	64	91.125	125	166.375	216

To improve our approximation, we could increase n or use the following method which is, in general, more accurate than the trapezoidal rule.

Simpson's Rule

This numerical integration method is also based on dividing the interval $[a,b]$ into n equal subintervals of length $(b-a)/n$. However, there is an added restriction that n must be an even whole number. The reason for this will become evident as we develop this method.

Rather than approximating the curve of $f(x)$ by straight line segments as in the case of the trapezoidal rule, consecutive portions of the curve are approximated with parabolas as illustrated in Fig. 8.3. The graph of $f(x)$ defined on the interval $[x_0,x_2]$ is approximated by the parabola, P, that passes through points P_0, P_1, and P_2. The graph of $f(x)$ defined on $[x_2,x_4]$ is approximated by the parabola, Q, that passes through the points P_2, P_3, and P_4, etc. An approximation to the total area and hence to $\int_a^b f(x)\,dx$ is found by summing the areas under each parabola in each of the intervals. The formula to find this area is given by the following rule.

Rule 8.2 *Simpson's rule*: An approximation to the definite integral of the continuous function $f(x)$ from $x=a$ to $x=b$ is given by

$$\int_a^b f(x)\,dx \doteq \left(\frac{b-a}{3n}\right)[f(x_0)+4f(x_1)+2f(x_2)+4f(x_3)+\cdots$$

$$+4f(x_{n-1})+f(x_n)] \qquad (8.9)$$

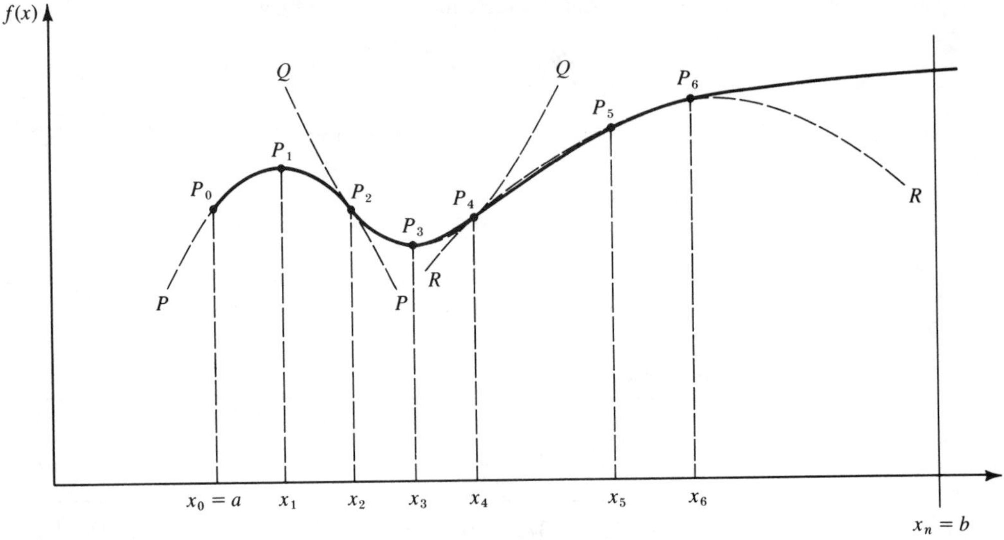

FIGURE 8.3

where the interval $[a,b]$ has been divided into n (even) equal subintervals by the points $x_0, x_1, x_2, \ldots, x_n$ in order from $x_0 = a$ to $x_n = b$.

Note the arrangement of the terms in the parentheses: $f(x_0)$ and $f(x_n)$ occur with a coefficient of 1; the remaining $f(x_i)$ occur with a coefficient of 2 when i is even, and with 4 when i is odd.

As mentioned, Simpson's rule usually is more accurate than the trapezoidal rule. As an illustration of this, consider the following example.

Example 8.19 Using Simpson's rule, approximate $\int_1^6 x^3\,dx$ by choosing $n = 10$.

Solution As in Example 8.18, the length of each subinterval given by $(b-a)/n$ is $(6-1)/10$ or 0.5. The important data required for formula (8.9) is summarized in Table 8.2. Substituting the appropriate values from Table 8.2 into formula (8.9) we obtain

$$\begin{aligned}\int_1^2 x^3\,dx &\doteq \left(\frac{6-1}{3(10)}\right)[1+4(3.375)+2(8.0)+4(15.625)\\ &\quad +2(27.0)+4(42.875)+2(64)+4(91.125)\\ &\quad +2(125)+4(166.375)+216]\\ &\doteq \tfrac{1}{6}[1+13.5+16+62.5+54+171.5+128\\ &\quad +364.5+250+665.5+216]\\ &\doteq \tfrac{1}{6}(1{,}942.5)\\ &\doteq 323.75\end{aligned}$$

This approximation is the exact value 323.75 and is, of course, much better than the value of 325.96 found by the use of the trapezoidal rule.

Since the work involved is relatively the same in Simpson's rule and the trapezoidal rule, the former is usually preferred.

EXERCISES

In Exercises 1–4 approximate the definite integral by the trapezoidal rule using the given value of n. Carry the work to five decimal places and round off your answer to four places. Compare the results with the exact value found by integrating.

1. $\int_0^2 x^2\,dx, \quad n=4$

2. $\int_1^3 x^3\,dx, \quad n=4$

3. $\int_1^5 \frac{1}{x}\,dx, \quad n=8$

4. $\int_2^6 (x^2-1)\,dx, \quad n=10$

5. Repeat Exercise 1 using Simpson's rule and compare the results.

6. Repeat Exercise 2 using Simpson's rule and compare the results.

7. Repeat Exercise 3 using Simpson's rule and compare the results.

8. Repeat Exercise 4 using Simpson's rule and compare the results.

9. Approximate the value of $\int_0^2 \frac{1}{(1+x^2)}\,dx$ with $n=4$

(a) by the trapezoidal rule
(b) by Simpson's rule.

Carry the work to five decimal places and round off the answers to four decimal places.

10. Approximate the value of $\int_0^2 \frac{1}{(1+x^2)}\,dx$ with $n=10$

(a) by the trapezoidal rule
(b) by Simpson's rule.

Carry the work to five decimal places and round off the answers to four decimal places.

11. Oil Tanker Spill To determine the size of an oil spill from a grounded tanker, the Coast Guard has provided the dimensions, in meters, of the spill as shown in the following figure. Approximate the surface area (square meters) of the spill using Simpson's rule. (See sketch on page 426.)

12. Assume the following graph reflects the increase in the cost of a barrel of oil as a function of x, the number of thousands of millions of barrels consumed. Approximate the total value of 600,000 million barrels of oil by evaluating $\int_0^{600} c(x)\,dx$ by using the trapezoidal rule. (See sketch on page 426.)

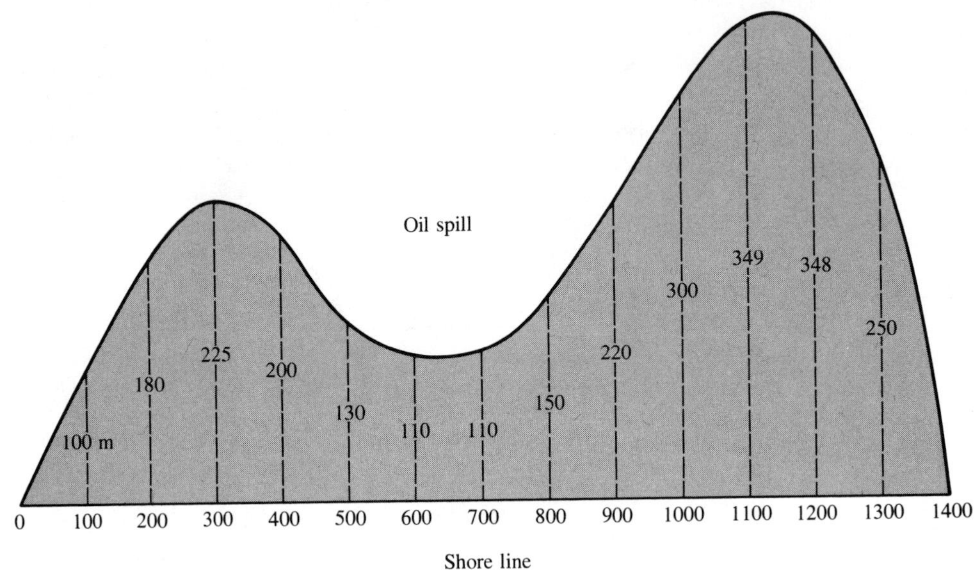

EXERCISE 11

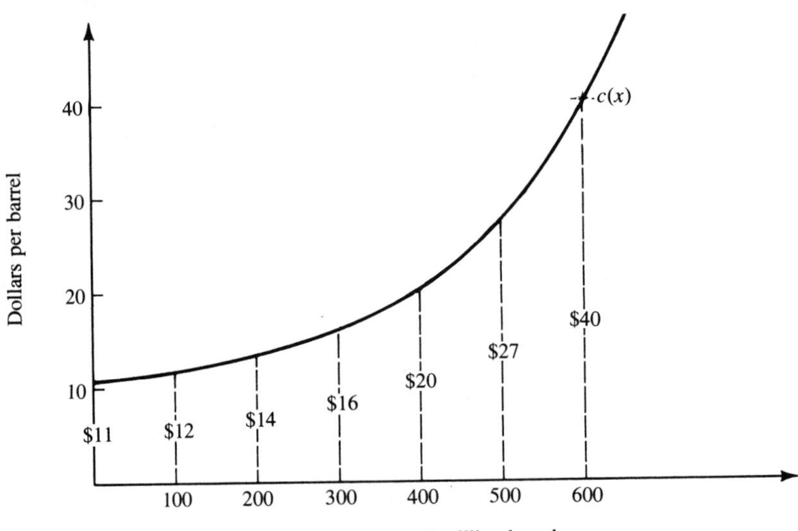

EXERCISE 12

8.6 IMPROPER INTEGRALS

Let us consider the area of the region between the graph of the function $f(x)=1/x^2$ and the x axis bounded on the left by the line $x=1$ and unbounded on the right as shown in Fig. 8.4. It would seem appropriate to denote this area by the integral $\int_1^{\infty}(1/x^2)\,dx$. Note, however, that this integral is not a definite integral since the upper limit is not a constant. An integral of this form is called an *improper integral*. However, we will define it in terms of a definite integral.

To determine the area of the region shown in Fig. 8.4 let us find the area between the same curve and the x axis bounded by the lines $x=1$ and $x=b$ as shown in Fig. 8.5, and investigate what happens as b increases without bound. The area, $A(b)$, of this region is given by

$$A(b) = \int_1^b \frac{1}{x^2}\,dx = \int_1^b x^{-2}\,dx = -\frac{1}{x}\bigg|_1^b = -\frac{1}{b} - \left(\frac{-1}{1}\right)$$
$$= \left(1 - \frac{1}{b}\right)$$

Now as $b\to\infty$, $(1/b)\to 0$; hence, the area $A(b)\to 1$, or in terms of limits the area of the region shown in Fig. 8.4 is

$$\lim_{b\to\infty} A(b) = \lim_{b\to\infty} \int_1^b \frac{1}{x^2}\,dx$$
$$= \lim_{b\to\infty} [1-(1/b)] = 1$$

Hence, although the region is unbounded, its area is finite and equal to 1.

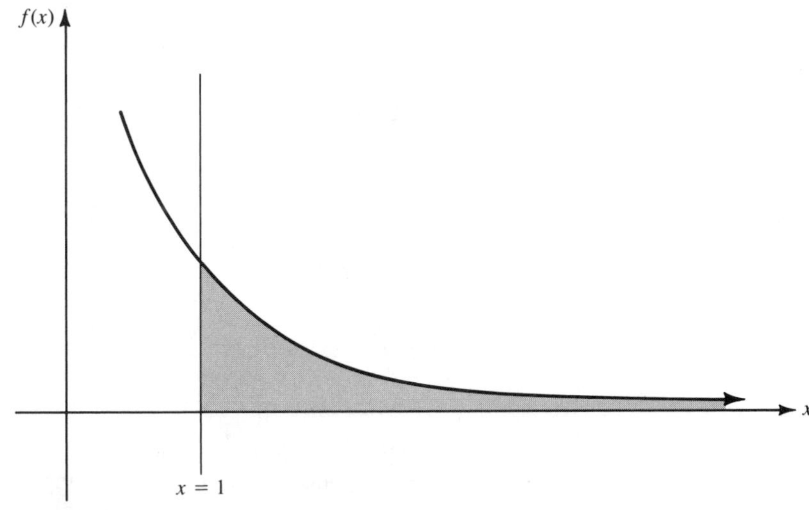

FIGURE 8.4

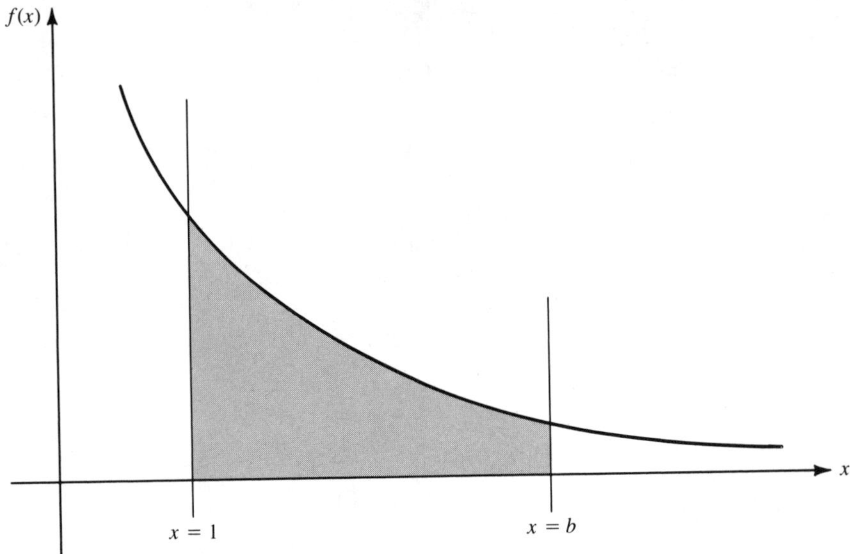

FIGURE 8.5

This discussion suggests we define the improper integral $\int_1^\infty (1/x^2)\,dx$ in the following manner

$$\int_1^\infty \frac{1}{x^2}\,dx = \lim_{b\to\infty} \int_1^b \frac{1}{x^2}\,dx$$
$$= 1$$

Generalizing, we make the following definition.

Definition 8.1 If $f(x)$ is continuous on $[a, \infty)$, then

$$\int_a^\infty f(x)\,dx = \lim_{b\to\infty} \int_a^b f(x)\,dx$$

provided the limit exists.

If the $\lim_{b\to\infty} \int_a^b f(x)dx$ exists, i.e., is a number, then we say the improper integral $\int_a^\infty f(x)dx$ *converges*; however, if it does not exist, then the improper integral *diverges*.

For example, consider the improper integral $\int_1^\infty (1/x)dx$. We can interpret this integral as the area, if the limit exists, of the region between the

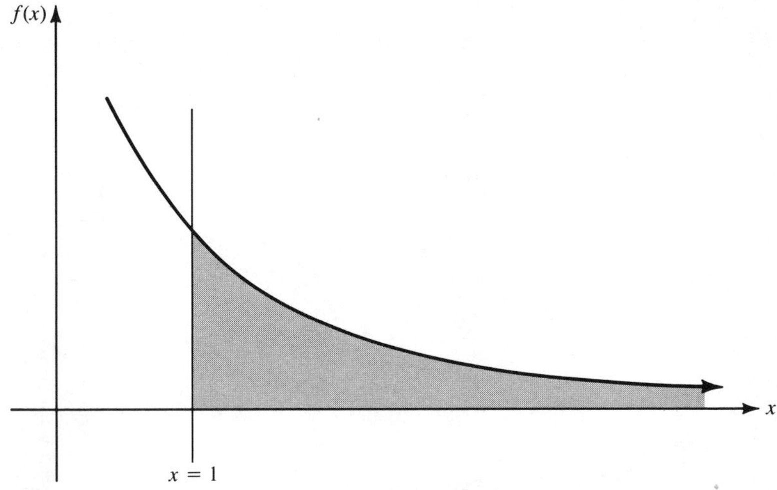

FIGURE 8.6

graph of the function $f(x)=1/x$ and the x axis bounded on the left by the line $x=1$ and unbounded on the right as shown in Fig. 8.6.

By Definition 8.1, we have

$$\begin{aligned}\int_1^\infty \frac{1}{x}\,dx &= \lim_{b\to\infty}\int_1^b \frac{1}{x}\,dx \\ &= \lim_{b\to\infty}\left(\ln x\Big|_1^b\right) \\ &= \lim_{b\to\infty}\left[\ln b - \ln 1\right]\end{aligned}$$

Since $\ln 1 = 0$, we have

$$\int_1^\infty \frac{1}{x}\,dx = \lim_{b\to\infty}(\ln b)$$

Now as $b\to\infty$, we know by properties of the logarithmic function that ln b increases without bound; therefore, this limit does not exist, i.e. it is not a number, and the improper integral, $\int_1^\infty (1/x)\,dx$ diverges.

Since ln b increases without bound as $b\to\infty$, we also conclude that the area of the region shown in Fig. 8.6 is infinite. It is interesting to note that although the regions defined by Figs. 8.4 and 8.6 look similar, the area of one is finite and the other is not. This can be attributed to the fact that the graph of $f(x)=1/x^2$ approaches the x axis "faster" than the graph of $f(x)=1/x$.

There are two other similar forms of improper integrals. Their definitions follow.

Definition 8.2 If $f(x)$ is continuous on $(-\infty, b]$, then

$$\int_{-\infty}^{b} f(x)\,dx = \lim_{a\to-\infty} \int_{a}^{b} f(x)\,dx$$

provided the limit exists.

Definition 8.3 If $f(x)$ is continuous for all real values of x, and c is any real number, then

$$\int_{-\infty}^{\infty} f(x)\,dx = \lim_{a\to-\infty} \int_{a}^{c} f(x)\,dx + \lim_{b\to\infty} \int_{c}^{b} f(x)\,dx$$

provided each limit exists.

Example 8.20 Determine whether each of the following improper integrals is convergent or divergent and if convergent, calculate its value.

(a) $\int_{0}^{\infty} e^{-x}\,dx$ (b) $\int_{-\infty}^{0} 2e^{x}\,dx$

Solution

(a) By Definition 8.1

$$\begin{aligned}
\int_{0}^{\infty} e^{-x}\,dx &= \lim_{b\to\infty} \int_{0}^{b} e^{-x}\,dx \\
&= \lim_{b\to\infty} \left(-e^{-x}\Big|_{0}^{b}\right) \\
&= \lim_{b\to\infty} [-e^{-b}-(-e^{-0})] \\
&= \lim_{b\to\infty} [1-e^{-b}] \\
&= \lim_{b\to\infty} \left[1-\frac{1}{e^{b}}\right]
\end{aligned}$$

As $b\to\infty$, $e^{b}\to\infty$; therefore, $1/e^{b}\to 0$. Hence,

$$\int_{0}^{\infty} e^{-x}\,dx = 1$$

Therefore, the integral converges and its value is 1.

(b) By Definition 8.2

$$\begin{aligned}
\int_{-\infty}^{0} 2e^{x}\,dx &= \lim_{a\to-\infty} \int_{a}^{0} 2e^{x}\,dx \\
&= \lim_{a\to-\infty} \left(2e^{x}\Big|_{a}^{0}\right) \\
&= \lim_{a\to-\infty} (2e^{0}-2e^{a}) \\
&= \lim_{a\to-\infty} (2-2e^{a})
\end{aligned}$$

Now as $a \to -\infty$, $2e^a \to 0$; hence,

$$\int_{-\infty}^{0} 2e^x\, dx = 2$$

Therefore, the integral converges and its value is 2.

EXERCISES

In Exercises 1–10 *determine whether each of the improper integrals is convergent or divergent and if convergent, calculate its value.*

1. $\int_1^\infty \frac{1}{\sqrt{x}}\, dx$

2. $\int_4^\infty \frac{1}{\sqrt{x^3}}\, dx$

3. $\int_0^\infty \frac{x}{(x^2+1)^2}\, dx$

4. $\int_{-\infty}^{0} \frac{x}{\sqrt{x^2+1}}\, dx$

5. $\int_{-\infty}^{-4} \frac{dx}{(x+1)^2}$

6. $\int_0^\infty e^{-2x}\, dx$

7. $\int_{-\infty}^{\infty} \frac{x}{(x^2+1)^2}\, dx$

8. $\int_0^\infty \frac{1}{1+x}\, dx$

9. $\int_1^\infty xe^{-x^2}\, dx$

10. $\int_{-\infty}^{-2} \frac{x^2}{(x^3+1)^2}\, dx$

11. Determine the area, if possible, of the region between the graph of $f(x) = 1/(x+1)^2$ and the x axis bounded on the left by the line $x=0$.

12. Determine the area, if possible, of the region between the graph of the function $f(x) = e^{-x}$ and the x axis bounded on the left by the line $x = -1$.

13. Determine the area, if possible, of the region between the graph of the function $f(x) = x/(x^2+1)^2$ and the x axis.

14. Determine the area, if possible, of the region between the graph of $f(x) = 1/x^{3/2}$ and the x axis bounded on the left by the line $x=1$.

15. For what values of p will $\int_1^\infty [1/(x^{1+p})]dx$ be convergent?

8.7 PROBABILITY DENSITY FUNCTIONS

In previous sections we discussed applications of the definite integral that involved finding the area between a function and the x axis in a given interval. We shall now consider another such application, where the function is called a *probability density function* and the area corresponds to the *probability* that a *continuous random variable* X takes a value within the interval.

Before continuing, let us examine some of the fundamental notions underlying this application of the definite integral.

A *continuous random variable* is a variable which may, at any given time, assume *any* value within some subset of the real numbers. For instance, let X represent the height of a first grader at South Shore Elementary School. As we measure the heights of these first graders, we are liable to encounter *any* measurement from perhaps 100 cm to about 160 cm, and these measurements occur randomly, according to no pattern (unless, of course, the children are lined up by height to await measurement).

The *probability* of an occurrence, or "event," is a number, between 0 and 1 inclusive, that indicates how likely it is that the event will actually occur. (An event that cannot occur has probability 0, and an event that always occurs has probability 1.) Thus the probability that a random variable, X, falls in a given interval is a numerical indication of the likelihood that X takes a value between the limits of the interval.

For example, returning to our first graders, suppose that there are 120 children in the group and that 42 of them have heights between 130 and 140 cm. We then say that the probability that a first grader at South Shore Elementary School will be between 130 cm and 140 cm tall is 42/120, or .35.

Now let us consider the properties that a function must possess in order for it to be a probability density function for a continuous random variable, X. The function, $f(x)$, is a probability density function for the continuous random variable, X, if, and only if, the following properties are satisfied:

1. The function is nowhere negative; that is, $f(x) \geqslant 0$ for all possible values of X.
2. The probability $P(a \leqslant X \leqslant b)$, where a and b are any two possible values of X such that $a \leqslant b$, is determined by the area under the graph of $f(x)$ between the values a and b.
3. The total area under the graph of $f(x)$ is equal to unity (1).

Property 1 is obviously necessary, since negative probabilities do not exist. (An event cannot occur less often than never!) Property 2 has been discussed above, and Property 3 is required to meet the condition that an event that will surely occur has probability equal to unity (1). The probability that the random variable, X, will take a value falling somewhere in the interval between its lowest and highest possible values, inclusive, must be equal to one. Looking again at our first graders, all 120 of them have heights between 100 cm and 160 cm; therefore, the probability that a first grader at South Shore will be between 100 and 160 cm tall is $\frac{120}{120}$, or 1.

Example 8.21 Throwing Darts Suppose we are throwing darts at a target, that we are expert dart tossers, and that the random variable, X, which is the distance from the center of the bullseye to the point of impact of the dart (measured in centimeters), has a probability density function given by

$$f(x) = \begin{cases} -\frac{1}{8}x+\frac{1}{2}, & 0 \leqslant x \leqslant 4 \\ 0 & \text{otherwise} \end{cases}$$

Find $P(1 \leqslant X \leqslant 2)$, the probability that the dart lands at least 1 cm from the center, but no more than 2 cm away.

Solution It can easily be seen that Properties 1 and 3 hold by considering the graph of $f(x)$, shown in Fig. 8.7.

The shaded area represents the probability in which we are interested. This area can be obtained through geometric means, but let us find it by applying our knowledge of the definite integral. We then have

$$\begin{aligned} P(1 \leqslant X \leqslant 2) = A_{\text{shaded}} &= \int_1^2 \left(-\frac{1}{8}x+\frac{1}{2}\right) dx \\ &= -\frac{1}{16}x^2+\frac{1}{2}x\Big|_1^2 \\ &= \left[-\frac{1}{16}(4)+\frac{1}{2}(2)\right]-\left[-\frac{1}{16}(1)+\frac{1}{2}(1)\right] \\ &= \frac{3}{4}-\frac{7}{16}=\frac{5}{16} \end{aligned}$$

That is, our darts hit between 1 cm and 2 cm away from the center of the target about $\frac{5}{16}$ of the time, or about 5 of every 16 darts hit in that interval.

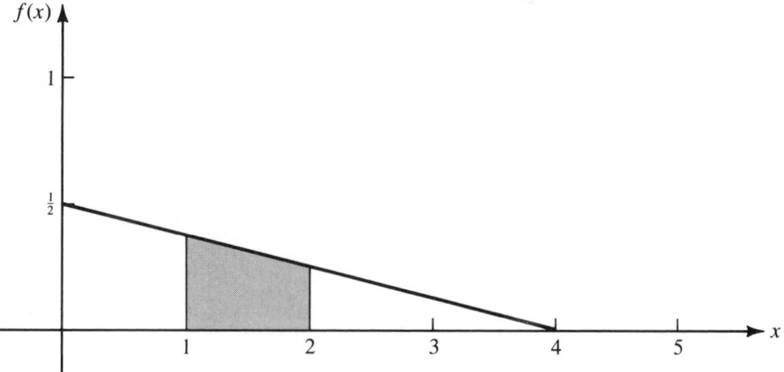

FIGURE 8.7

Notice that a probability density function is usually stated for a specific domain, often a proper subset of the real numbers, but that we can broaden the domain to include all real numbers by letting $f(x)=0$ for those values of x that are not of interest. Let us now state the properties of a probability density function using integral calculus, and then we shall consider how they are used for specific probability density functions.

1. For all possible values of X,

$$f(x) \geqslant 0 \tag{8.10}$$

2. For $a \leqslant b$,

$$P(a \leqslant X \leqslant b) = \int_a^b f(x)\,dx \tag{8.11}$$

3.

$$\int_{-\infty}^{\infty} f(x)\,dx = 1 \tag{8.12}$$

Thus, if $f(x)$ satisfies (8.10) and (8.12), we can use (8.11) to compute associated probabilities for X.

Example 8.22 For the probability density function

$$f(x) = \begin{cases} \dfrac{x}{9} + \dfrac{1}{6}, & 0 \leqslant x \leqslant 3 \\ 0 & \text{otherwise} \end{cases}$$

determine $P(0 \leqslant X \leqslant 2)$. The graph of this function is shown in Fig. 8.8.

Solution We note that $f(x) \geqslant 0$ for all possible values of x and that

$$\int_{-\infty}^{\infty} \left(\frac{x}{9} + \frac{1}{6}\right) dx = \int_{-\infty}^{0} 0\,dx + \int_0^3 \left(\frac{x}{9} + \frac{1}{6}\right) dx + \int_3^{\infty} 0\,dx$$

$$= 0 + \left(\frac{x^2}{18} + \frac{x}{6}\bigg|_0^3\right) + 0 = \frac{9}{18} + \frac{3}{6} = 1$$

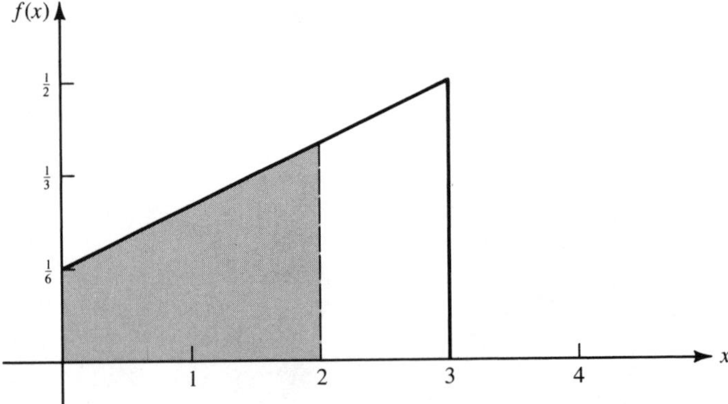

FIGURE 8.8

Thus,

$$P(0 \leqslant X \leqslant 2) = \int_0^2 \left(\frac{x}{9} + \frac{1}{6}\right) dx = \frac{x^2}{18} + \frac{x}{6}\bigg|_0^2 = \frac{5}{9}$$

Generally, we do not write the function or the definite integrals for those values of x for which $f(x)=0$; that is, for the previous example we would usually write

$$f(x) = \frac{x}{9} + \frac{1}{6} \text{ for } 0 \leqslant x \leqslant 3 \text{ and } \int_{-\infty}^{\infty} f(x)\,dx = \int_0^3 \left(\frac{x}{9} + \frac{1}{6}\right) dx.$$

Example 8.23 Show that $f(x)=x^2/8+x/6+1/6$ for $0 \leqslant x \leqslant 2$ satisfies (8.10) and (8.12) and determine $P(0 \leqslant X \leqslant 1)$.

Solution To show that (8.10) is satisfied, we determine the discriminant of the quadratic function to be

$$b^2 - 4ac = \left(\tfrac{1}{6}\right)^2 - 4\left(\tfrac{1}{8}\right)\left(\tfrac{1}{6}\right) = -\tfrac{1}{18}$$

Thus, recalling from Chapter 2 that a quadratic function with a negative discriminant has no real roots, $f(x)$ is either always positive or always negative; furthermore, $f(0)=\frac{1}{3}$, which is positive. Hence the function is always positive, thus satisfying (8.10) (see Fig. 8.9). To show that (8.12) is satisfied, we determine

$$\int_0^2 \left(\frac{x^2}{8} + \frac{x}{6} + \frac{1}{6}\right) dx = \frac{x^3}{24} + \frac{x^2}{12} + \frac{x}{6}\bigg|_0^2 = 1$$

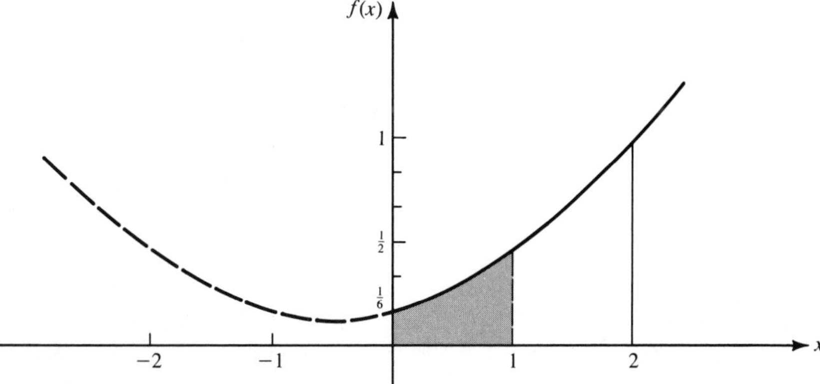

FIGURE 8.9

Furthermore,

$$\begin{aligned} P(0 \leqslant X \leqslant 1) &= \int_0^1 \left(\frac{x^2}{8} + \frac{x}{6} + \frac{1}{6} \right) dx \\ &= \left. \frac{x^3}{24} + \frac{x^2}{12} + \frac{x}{6} \right|_0^1 \\ &= \frac{7}{24} \end{aligned}$$

Example 8.24 Determine the value of k such that $f(x)=k(x+2)$ for $0 \leqslant x \leqslant 3$ satisfies (8.10) and (8.12). Determine $P(1 \leqslant X \leqslant 2)$.

Solution By inspection, we see that $f(x) \geqslant 0$ for all x values in the interval $0 \leqslant x \leqslant 3$ provided $k \geqslant 0$; thus (8.10) is satisfied if $k \geqslant 0$. To determine k such that (8.12) is satisfied requires the following integration:

$$\begin{aligned} \int_0^3 k(x+2)\,dx &= k \int_0^3 (x+2)\,dx \\ &= k\left(\frac{x^2}{2} + 2x \right)\Bigg|_0^3 \\ &= k\frac{21}{2} \end{aligned}$$

However, in order for the integral to equal unity, we must have

$$\frac{21}{2}k = 1$$

which is true if and only if $k = \frac{2}{21}$. Thus, we obtain a probability density function if $k = \frac{2}{21}$, and it is given by

$$f(x) = \frac{2x}{21} + \frac{4}{21} \qquad \text{for} \quad 0 \leqslant x \leqslant 3$$

We determine the probability to be

$$\begin{aligned} P(1 \leqslant X \leqslant 2) &= \int_1^2 \left(\frac{2x}{21} + \frac{4}{21} \right) dx \\ &= \left. \frac{x^2}{21} + \frac{4x}{21} \right|_1^2 \\ &= \frac{12}{21} - \frac{5}{21} \\ &= \frac{7}{21} \end{aligned}$$

Expected Value of a Continuous Random Variable

The *average*, *mean*, or *expected value* of a continuous random variable, X, for which a probability density function, $f(x)$, is known, is given by

$$E(X) = \int_{-\infty}^{\infty} xf(x)\,dx \tag{8.13}$$

Thus if the domain of $f(x)$ is given by $a \leqslant x \leqslant b$, then the mean, or expected value, of the random variable X is determined by

$$E(X) = \int_{a}^{b} xf(x)\,dx \tag{8.14}$$

Consider the function in Example 8.21.

$$\begin{aligned} E(X) &= \int_0^4 x\left(-\frac{1}{8}x+\frac{1}{2}\right)dx = \int_0^4\left(-\frac{x^2}{8}+\frac{x}{2}\right)dx \\ &= -\frac{x^3}{24}+\frac{x^2}{4}\bigg|_0^4 = -\frac{64}{24}+\frac{16}{4} = -\frac{8}{3}+4 \\ &= \frac{4}{3} \end{aligned}$$

That is, the average distance from our dart to the center of the bullseye is $\frac{4}{3}$ cm.

Example 8.25 Determine the expected value of the random variable X when the probability density function is given by $f(x)=3x^2$ for $0 \leqslant x \leqslant 1$.

Solution Using Eq. (8.14), we obtain

$$\begin{aligned} E(X) &= \int_0^1 x(3x^2)\,dx \\ &= \int_0^1 3x^3\,dx \\ &= \frac{3x^4}{4}\bigg|_0^1 \\ &= \frac{3}{4} \end{aligned}$$

Example 8.26 Arrival Pattern of Bank Customers The Third National Bank has determined that the arrival pattern of its customers during the business day is given by $f(x)=kx(1-x)$ for $0 \leqslant x \leqslant 1$, where x represents the time of the customer's arrival as a fraction of the business day (arrival time of 10:00 A.M. gives $x=0$; arrival time of 1:00 P.M., $x=.5$; $x=1$ means the customer arrived at 4:00 P.M. or closing time). Determine the value of k that makes $f(x)$ a probability density function, determine the average arrival time for the customers, and determine $P(0 \leqslant X \leqslant \frac{1}{4})$.

Solution To show that $f(x)$ is a probability density function, we note that $f(x) \geqslant 0$ for all values of x in the interval $0 \leqslant x \leqslant 1$ if $k \geqslant 0$. For the total area under the curve of $f(x)$ to equal unity, we have

$$\begin{aligned}\int_0^1 kx(1-x)\,dx &= k\int_0^1 (x-x^2)\,dx\\ &= k\left(\frac{x^2}{2}-\frac{x^3}{3}\right)\Bigg|_0^1\\ &= k\frac{1}{6}\end{aligned}$$

which must equal unity; thus, $f(x)$ is a probability density function if and only if $k=6$.

The average arrival time for the customers is given by

$$\begin{aligned}E(X) &= \int_0^1 x6x(1-x)\,dx\\ &= 6\int_0^1 (x^2-x^3)\,dx\\ &= 6\left(\frac{x^3}{3}-\frac{x^4}{4}\right)\Bigg|_0^1\\ &= \frac{1}{2}\end{aligned}$$

Thus, the *average* arrival time for the customers would be in the middle of the working day.

To determine $P(0 \leqslant X \leqslant \frac{1}{4})$, we calculate

$$\begin{aligned}P\left(0 \leqslant X \leqslant \frac{1}{4}\right) &= \int_0^{1/4} 6x(1-x)\,dx\\ &= 6\left(\frac{x^2}{2}-\frac{x^3}{3}\right)\Bigg|_0^{1/4}\\ &= 6\cdot\frac{5}{192}\\ &= \frac{5}{32}\end{aligned}$$

For continuous random variables, often it is either difficult or impossible to evaluate the integral of a probability density function, in which case tables for selected values are determined. Nevertheless, even when using tables, it is most convenient to consider the probability for continuous variables as the area under the curve and remember that the total area must equal 1.

A probability distribution that is very widely used in statistical analysis is the *normal distribution*. The *normal probability density function* is given by

$$f(x) = \frac{1}{\sqrt{2\pi}\,\sigma} e^{-\frac{1}{2}[(x-\mu)/\sigma]^2}$$

for $-\infty < x < \infty$, where μ and σ are parameters that represent the mean and standard deviation of the distribution. The graph of this distribution has the familiar bell shape, given in Fig. 8.10. The probability that the random variable X will fall in the interval $a \leqslant X \leqslant b$ is given by

$$P(a \leqslant X \leqslant b) = \int_a^b \frac{1}{\sqrt{2\pi}\,\sigma} e^{-\frac{1}{2}[(x-\mu)/\sigma]^2}\,dx$$

and the shaded area in Fig. 8.10 represents this probability. The value of this integral can be determined by the numerical methods discussed in Section 8.5. However, the widespread use of the normal distribution requires the availability of tables from which we can obtain values of the definite integral for a wide variety of values of a and b. An in-depth familiarity with such tables will be gained by those who study statistics. (The study of the use of such tables and of other special probability density functions is beyond the scope of this textbook.)

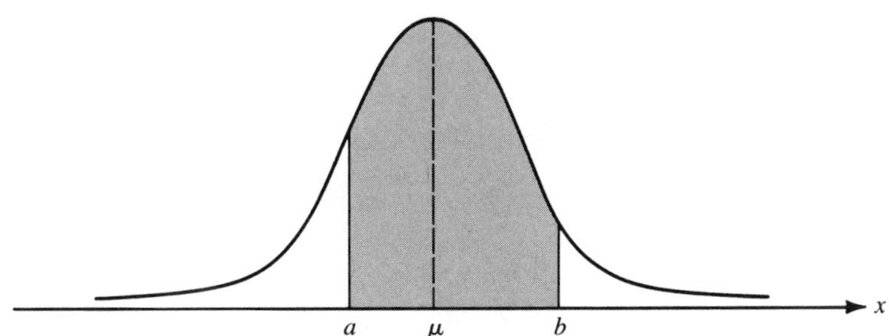

FIGURE 8.10

EXERCISES

In each of Exercises 1–12, *determine whether the function could be a probability density function. If not, state why.*

1. $f(x) = \frac{1}{8}$ for $0 \leqslant x \leqslant 8$

2. $f(x) = 6$ for $0 \leqslant x \leqslant 6$

3. $f(x) = \frac{1}{10}$ for $0 \leqslant x \leqslant 5$

4. $f(x)=2$ for $0 \leqslant x \leqslant \frac{1}{2}$

5. $f(x)=1/n$ for $0 \leqslant x \leqslant n$ and $n>0$

6. $f(x)=-x+2$ for $0 \leqslant x \leqslant 2$

7. $f(x)=-2x/3+1/3$ for $0 \leqslant x \leqslant 1$

8. $f(x)=2x^2$ for $0 \leqslant x \leqslant 1$

9. $f(x)=3x^2/16$ for $0 \leqslant x \leqslant 2$

10. $f(x)=6x^2-12x+\frac{9}{2}$ for $0 \leqslant x \leqslant 2$

11. $f(x)=12x^2-12x^3$ for $0 \leqslant x \leqslant 1$

12. $f(x)=30(x^2-2x^4+x^6)$ for $0 \leqslant x \leqslant 1$

In each of Exercises 13–22, *determine the value of* k *such that* f(x) *is a probability density function. Also, determine the expected value of* X *for each probability density function in Exercises* 13–21.

13. $f(x)=k$ for $0 \leqslant x \leqslant 2$

14. $f(x)=1/k$ for $0 \leqslant x \leqslant 5$

15. $f(x)=\frac{1}{3}$ for $0 \leqslant x \leqslant k$

16. $f(x)=k(-x+3)$ for $0 \leqslant x \leqslant 2$

17. $f(x)=k(-x+4)$ for $0 \leqslant x \leqslant 4$

18. $f(x)=kx^2$ for $0 \leqslant x \leqslant 2$

19. $f(x)=k(x^2-x+4)$ for $0 \leqslant x \leqslant 3$

20. $f(x)=k(x-x^2)$ for $2 \leqslant x \leqslant 5$

21. $f(x)=k(2x-x^2)$ for $0 \leqslant x \leqslant 2$

22. $f(x)=ke^{-2x}$ for $0 \leqslant x \leqslant \infty$

In each of Exercises 23–30, *for the given probability density function, determine the respective probabilities.*

23. $f(x)=\frac{1}{4}$ for $0 \leqslant x \leqslant 4$

(a) $P(0 \leqslant X \leqslant 1)$
(b) $P(1 \leqslant X \leqslant 2)$
(c) $P(2 \leqslant X \leqslant 4)$
(d) $P(\frac{1}{2} \leqslant X \leqslant \frac{5}{2})$

24. $f(x)=-x/4+3/4$ for $0 \leqslant x \leqslant 2$

(a) $P(0 \leqslant X \leqslant 1)$
(b) $P(\frac{1}{2} \leqslant X \leqslant \frac{3}{2})$
(c) $P(X \geqslant \frac{3}{2})$
(d) $P(X \leqslant \frac{1}{2})$

25. $f(x)=-x/8+1/2$ for $0 \leqslant x \leqslant 4$

(a) $P(1 \leqslant X \leqslant 3)$
(b) $P(X \geqslant 2)$
(c) $P(X \leqslant 4)$
(d) $P(2 \leqslant X \leqslant 6)$

26. $f(x) = -2x/21 + 8/21$ for $0 \leqslant x \leqslant 3$

(a) $P(1 \leqslant X \leqslant \frac{3}{2})$ (b) $P(X \geqslant 1)$
(c) $P(X \geqslant 4)$ (d) $P(2 \leqslant X \leqslant 3)$

27. $f(x) = \frac{3}{65}x^2$ for $-1 \leqslant x \leqslant 4$

(a) $P(1 \leqslant X \leqslant 2)$ (b) $P(X \geqslant 2)$
(c) $P(X \leqslant 3)$ (d) $P(0 \leqslant X \leqslant 2)$

28. $f(x) = 20x^3 - 20x^4$ for $0 \leqslant x \leqslant 1$

(a) $P(0 \leqslant X \leqslant \frac{1}{2})$ (b) $P(X \geqslant \frac{1}{2})$
(c) $P(X \leqslant \frac{1}{3})$ (d) $P(X = \frac{1}{2})$

29. $f(x) = 12x - 24x^2 + 12x^3$ for $0 \leqslant x \leqslant 1$

(a) $P(X \leqslant \frac{1}{2})$ (b) $P(0 \leqslant X \leqslant 2)$
(c $P(X \geqslant \frac{1}{4})$ (d) $P(X \leqslant 1)$

30. $f(x) = 3e^{-3x}$ for $0 \leqslant x \leqslant \infty$

(a) $P(X \leqslant 1)$ (b) $P(1 \leqslant X \leqslant 2)$
(c) $P(X \geqslant 3)$ (d) $P(0.2 \leqslant X \leqslant 0.4)$

31. **Transistor Radio Battery Failure** The time in hours to failure of a battery in a transistor radio is described by the probability density function $f(t) = t^3/2500$ for $0 \leqslant t \leqslant 10$. Determine the probability that a randomly selected battery will fail within the first 4 hr of operation; also, determine the probability that it will not fail before 8 hr of operation. Finally, determine the average time to failure.

32. **Aircraft Arrival Pattern** The time in minutes between arrivals of aircraft at Los Angeles International Airport is described by the probability density function $f(t) = ke^{-30t}$ for $0 \leqslant t \leqslant \infty$. Determine the value of k. Moreover, determine the probability that an aircraft will land within 2 min behind the aircraft preceding it. Finally, determine the probability that an aircraft will land within 30 sec behind the aircraft preceding it.

33. The time in minutes between customer arrivals at the checkout counter of the Super Saver Discount Store is described by the probability density function $f(t) = (30t - 6t^2)/125$ for $0 \leqslant t \leqslant 5$. Determine the probability that two customers arrive within 1 min of each other, and determine the probability that they arrive within 3 min of each other. Finally, determine the average time between arrivals.

34. **Target Practice** The distance in feet by which a shot from an antiaircraft gun misses its target is described by the probability density $f(x) = \frac{1}{50}$ for $0 \leqslant x \leqslant 50$. Determine the probability of hitting within 20 ft of the target, and determine the probability of hitting within 30 ft, but not closer than 10 ft. Finally, determine the average distance from the target.

35. **Expected Life of a Light Bulb** The time in weeks to failure of a light bulb is described by the probability density function $f(t) = (30 - 2t)/225$ for $0 \leqslant t \leqslant 15$. Determine the probability that a randomly selected light bulb will fail within the first 2 weeks of operation; also, determine the probability that it will fail in the fifth or sixth week of operation. Finally, determine the average time to failure.

36. The time in hours between fire alarms in a small town is described by the probability density function $f(t) = k(10t - .5t^2)$ for $0 \leqslant t \leqslant 16$. Determine the value of k. Also, determine the probability that there will be at least 14 hr between 2 alarms; also, determine the probability that the time between 2 alarms is between 8 and 12 hr. Finally, determine the average time between alarms.

37. The distance in yards between defects in weaving material is described by the probability density function $f(x) = k/x^2$ for $40 \leqslant x \leqslant 120$. Determine the value of k. Also, determine the probability that 2 defects are within 50 yards of each other, and determine the probability that the distance between 2 defects is between 100 and 115 yards.

38. Tire Lifetime The Rubber Tire Company has determined that the distance in miles that their tires last is described by the probability density function $f(x) = k(x^2 + x)$ for $1 \leqslant x \leqslant 6$ and x is in 10,000's. A trucking company will not purchase tires unless the probability of the tires lasting 40,000 miles is at least 0.75. Will the trucking company purchase tires from the Rubber Tire Company?

IMPORTANT TERMS AND CONCEPTS

- Average
- Change of variable
- Continuous random variable
- Converges
- Diverges
- Expected value
- Improper integral
- Integration by parts
- Integration by parts formula
- Mean
- Normal distribution
- Normal probability density function
- Numerical integration
- Probability
- Probability density function
- Simpson's rule
- Table of integrals
- Trapezoidal rule

SUMMARY OF RULES AND FORMULAS

Eq. 8.7. Integration by parts formula

$$\int u\,dv = u \cdot v - \int v\,du$$

Rule 8.1. Trapezoidal rule

$$\int_a^b f(x)\,dx \doteq \frac{b-a}{n}\left[\frac{f(a)}{2} + f(x_1) + F(x_2) + \cdots + f(x_{n-1}) + \frac{f(b)}{2}\right]$$

Rule 8.2. Simpson's rule

$$\int_a^b f(x)\,dx \doteq \frac{b-a}{3n}[f(x_0) + 4f(x_1) + 2f(x_2) + 4f(x_3) + \cdots + 4f(x_{n-1}) + f(x_n)]$$

Definition 8.1. $\int_a^{\infty} f(x)\,dx = \lim_{b\to\infty} \int_a^b f(x)\,dx$

Definition 8.2. $\int_{-\infty}^{b} f(x)\,dx = \lim_{a\to-\infty} \int_a^b f(x)\,dx$

Definition 8.3. $\int_{-\infty}^{\infty} f(x)\,dx = \lim_{a\to-\infty} \int_a^c f(x)\,dx + \lim_{b\to\infty} \int_c^b f(x)\,dx$

9 FUNCTIONS OF SEVERAL VARIABLES

9.1 INTRODUCTION

So far in this textbook, all the functions we have considered have involved only one independent variable. There are many practical situations where the value of one quantity depends on the values of several independent variables. For example, a large service station may sell four grades of fuel, diesel, regular, no-lead, and premium, at prices of 62, 68, 73, and 76¢/gallon, respectively. If we wish to express the total revenue from the sale of fuel, we can let x_1, x_2, x_3, and x_4 be the number of gallons sold of diesel, regular, no-lead, and premium, respectively. Then, the total revenue in cents can be expressed by the relationship:

$$\text{Total revenue} = 62x_1 + 68x_2 + 73x_3 + 76x_4$$

Using the functional notation that we have been using for functions with a single variable, we could let f denote the total revenue and write

$$f(x_1, x_2, x_3, x_4) = 62x_1 + 68x_2 + 73x_3 + 76x_4$$

In this case the function f is a function of four independent variables x_1, x_2, x_3, and x_4. Thus, we can use similar notation to represent functions of several variables. Such functions are called *multivariate functions*.

9.2 FUNCTIONS AND GRAPHS

A function $f(x, y)$ of two variables is a correspondence that associates with each pair of possible values of the independent variables (x, y) one and only one value of the dependent variable $f(x, y)$. This concept can be extended to include as many independent variables as desired, but, regardless of the number of independent variables, there is only one *dependent* variable.

When working with functions of several variables, we can manipulate them in the same way as we do functions of one variable. However, an increase in the number of variables tends to increase the complexity of the problem, which makes performing the operations more difficult and increases the possibilities for confusion. Nevertheless, the algebraic operations are the same as for the single-variable function, and the procedures for differentiating multivariate functions are practically the same as for one-variable (univariate) functions. Here is an example of performing algebraic operations on two-variable functions.

Example 9.1 If $f(x, y)=x^2+3xy+y^2$ and $g(x, y)=2x^2+4xy$, determine the function $h(x, y)$ obtained by each of the following algebraic operations:

(a) $f+g$ (b) $f-g$

(c) fg (d) $\frac{f}{g}$

Solution

$$\begin{aligned}
\text{(a) } h(x,y) &= f(x,y)+g(x,y)\\
&= x^2+3xy+y^2+2x^2+4xy\\
&= 3x^2+7xy+y^2\\
\text{(b) } h(x,y) &= f(x,y)-g(x,y)\\
&= x^2+3xy+y^2-(2x^2+4xy)\\
&= -x^2-xy+y^2\\
\text{(c) } h(x,y) &= f(x,y)g(x,y)\\
&= (x^2+3xy+y^2)(2x^2+4xy)\\
&= 2x^4+10x^3y+14x^2y^2+4xy^3\\
\text{(d) } h(x,y) &= \frac{f(x,y)}{g(x,y)}\\
&= \frac{x^2+3xy+y^2}{2x^2+4xy}
\end{aligned}$$

Next, let us consider two examples in which we consider the domain of a function and evaluate the function for some values of the independent variables.

Example 9.2 If $f(x, y)=x^2+3xy+y^2$, determine the domain of f and compute $f(2, 1)$ and $f(3, 6)$.

Solution The domain of f consists of all ordered pairs (x, y) of real numbers. To compute the values of the function we simply substitute $x=2$ and $y=1$ into the rule for f to determine

$$f(2,1) = (2)^2+3(2)(1)+(1)^2 = 11$$

and substitute $x=3$ and $y=6$ to determine

$$f(3,6) = (3)^2 + 3(3)(6) + (6)^2 = 99$$

Example 9.3 If $f(x, y, z) = (2z + y^2)/(x - y)$, determine the domain of f and compute $f(2,1,0)$ and $f(1,0,2)$.

Solution The function is defined for all ordered triples (x, y, z) except for those where the denominator equals zero; that is, where $x = y$. For example, the triple $(3,3,1)$ is not in the domain of f since $3-3=0$. Thus, the triples $(2,1,0)$ and $(1,0,2)$ are in the domain of f, and we determine that

$$f(2,1,0) = \frac{2(0)+(1)^2}{2-1} = 1$$

and

$$f(1,0,2) = \frac{2(2)+(0)^2}{1-0} = 4$$

Occasionally, a certain value is given as a function of two variables, each of which can be expressed in terms of a third variable. In such situations, we can express the initial value as a function of the third variable by using a *composite function*, as illustrated in the next example.

Example 9.4 Wages as a Function of Two Variables The Centex Trucking Company employs 25 drivers and 30 dock workers. The union contract specifies that t months from now the drivers will receive $7+0.2t$ dollars/hour and the dock workers will receive $5+0.01t^2$ dollars/hour. Express the total hourly wages to be paid by Centex to the drivers and dock workers as a function of time t. Determine functions $f(x, y)$, $g(t)$, and $h(t)$ so that the total hourly wages can be expressed as the composite function $f(g(t), h(t))$. Express the total hourly wages paid as a function of time.

Solution Let $x = 7+0.2t$ and $y = 5+0.01t^2$ represent the hourly wages for the drivers and dock workers, respectively. The function $f(x, y)$ that yields the total hourly wages to be paid is

$$f(x, y) = 25x + 30y$$

where $x = g(t) = 7+0.2t$ and $y = h(t) = 5+0.01t^2$. Hence,

$$\begin{aligned} f(g(t), h(t)) &= 25(7+0.2t) + 30(5+0.01t^2) \\ &= 175 + 5t + 150 + 0.3t^2 \\ &= 325 + 5t + 0.3t^2 \end{aligned}$$

Thus the function of time which yields total hourly wages is

$$c(t) = 325 + 5t + 0.3t^2$$

In Example 9.4 we had two variables x and y that were functions of a third variable t. We formed a composite function that allowed us to express the original function of two variables as a function of a third variable. Conversely, we occasionally start with a function of one variable and then express the variable as a function of two other variables, thus yielding a composite function of two variables. Let us consider an example of this type of change.

Example 9.5 Let $g(x)=(x+2)^2$ and $h(y,z)=3y+2z+4$. Determine the composite function $f(y,z)=g(h(y,z))$ and its domain.

Solution In the function $g(x)$ we replace x by $h(y,z)$ and obtain

$$\begin{aligned} f(y,z) &= g(h(y,z)) = (h(y,z)+2)^2 = [(3y+2z+4)+2]^2 \\ &= [(3y+2z+4)+2]^2 \\ &= (3y+2z+6)^2 \end{aligned}$$

The domain of f consists of all ordered pairs (y,z) since the function is defined for all real values of y and z.

When studying functions of one variable, we made extensive use of the geometrical representation of the functions. For functions of two variables, we can use a three-dimensional coordinate system and graphically represent the functions as surfaces.

For functions of one variable we used the variable y to represent the function and wrote $y=f(x)$. For functions of two variables it is customary to use z to represent the function and write $z=f(x, y)$. To graph a function of two variables, we use the three-dimensional coordinate system to represent the ordered triples (x, y, z). In this system we let the xy plane be horizontal and attach a third axis (the z axis) that is perpendicular to the xy plane. The positive direction of the z axis is up. The coordinate system is shown in Fig. 9.1. Part a shows the way the coordinate system is usually shown, and part b shows the negative portions of the three axes as dotted lines, indicating that they are out of view.

In order to plot a point (x, y, z) in the three-dimensional system we simply take three steps. For example, the point $(2,3,4)$ is plotted by going 2 units along the x axis in a positive direction from the origin, then going 3 units to the right (the positive direction) parallel to the y axis, and finally we go up parallel to the z axis four units. Fig. 9.2 shows the points $(2,3,4)$, $(-1,2,1)$, $(2,-1,2)$, and $(-2,-3,1)$.

To graph a particular function we can specify numerous (x, y) pairs and evaluate $z=f(x, y)$ to obtain the z coordinates. Proceeding in this manner we can determine the general shape of the graph if we plot a sufficient number of points. In general, the number of points that is sufficient will vary from equation to equation. Most three-variable equations are rather difficult to

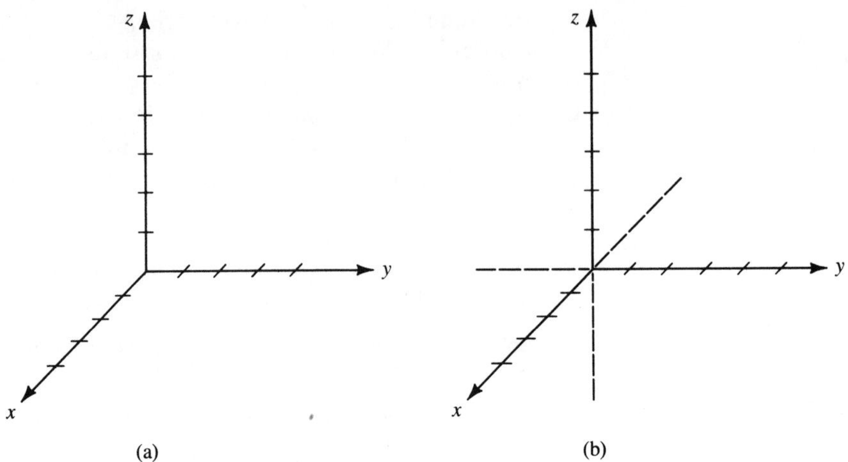

FIGURE 9.1

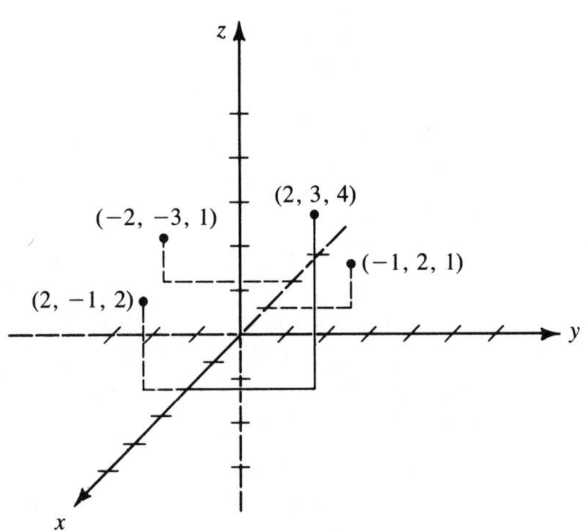

FIGURE 9.2

graph. However, if it is a first-degree equation of the form $ax + by + cz = d$, we need determine only three points to establish its graph. The three-dimensional graph of a first-degree equation is a *plane*, for which the two-dimensional counterpart is a straight line. The three points that are usually easiest to determine are those for which the plane intersects the three axes; that is, $(x, 0, 0)$, $(0, y, 0)$ and $(0, 0, z)$. Often we need only sketch the graph in the first octant (i.e., where all values are positive).

Example 9.6 Graph the following equations.

(a) $z = -3x - y + 4$
(b) $2x + 3y + z = 6$

Solution (a) We can write the equation as $3x + y + z = 4$ and find some representative ordered triples to plot. The ordered triples include $(0,0,4)$, $(\frac{4}{3},0,0)$, and $(0,4,0)$. If we plot these triples, we get the graph in Fig. 9.3.

(b) For the equation $2x+3y+z=6$ we determine the triples $(3,0,0)$, $(0,2,0)$, and $(0,0,6)$ and obtain the graph in Fig. 9.4.

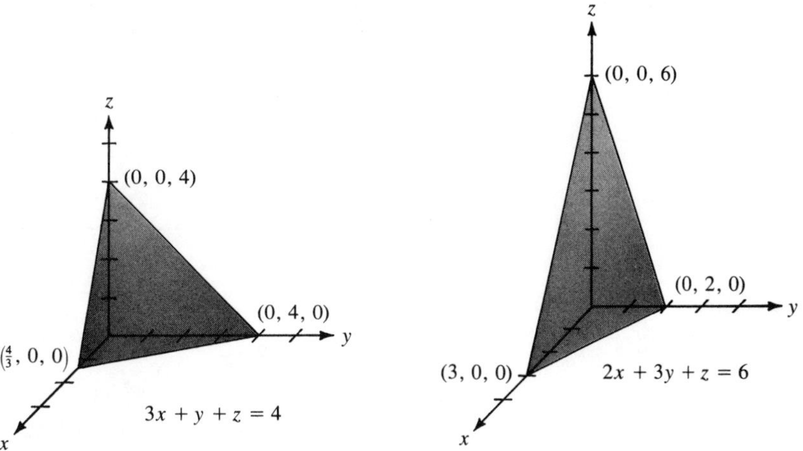

FIGURE 9.3 **FIGURE 9.4**

Example 9.7 Graph the following equations.

(a) $x+y=4$
(b) $x=4$

Solution (a) For $x+y=4$ any value of z is acceptable, thus we find those triples for which $z=0$. This results in a straight line in the xy plane passing through the points $(0,4,0)$ and $(4,0,0)$. Next we can let z vary, and we obtain the graph in Fig. 9.5.

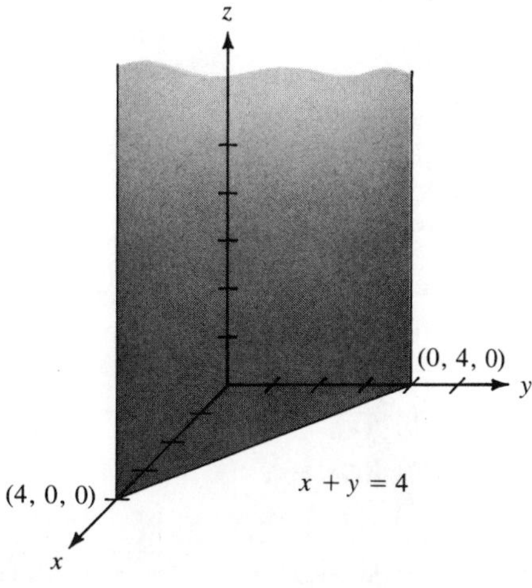

FIGURE 9.5

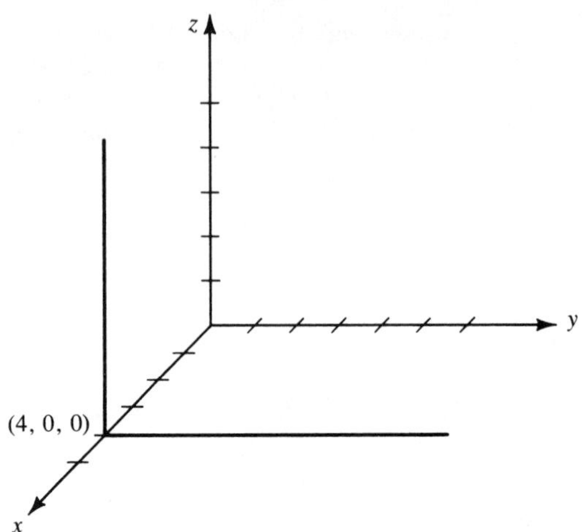

FIGURE 9.6

(b) For $x=4$ any values of y and z will satisfy the equation. Thus we must maintain $x=4$, but letting the other variables assume many other values, we obtain the graph in Fig. 9.6.

We generally are interested in determining the maxima and minima, so it is helpful to know that such functions can be represented by surfaces such as those given in Fig. 9.7.

EXERCISES

1. Given $f(x, y)=2x^2+4xy+2x+y^2$ and $g(x, y)=x^2-xy+3y-x+2y^2$, determine

(a) $f+g$ (b) $f-g$

(c) fg (d) $\dfrac{f}{g}$

2. Given $f(x,y)=e^{x^2+2xy+y^2}$ and $g(x,y)=e^{x^2+y^2}$, determine

(a) $f+g$ (b) $f-g$

(c) fg (d) $\dfrac{f}{g}$

3. For each of the following functions, specify the domain and determine the function values for the points indicated.

(a) $f(x, y)=3x^2-2xy+y^2$; $f(2,-1)$, $f(-1,2)$

(b) $f(x_1,x_2,x_3)=2(x_1-2)^2-4x_1^2(x_2-2)+(x_3-1)^2$; $f(2,2,1)$, $f(1,1,2)$

(c) $f(r,s)=\sqrt{r(s-2)^3}$; $f(2,3)$, $f(3,2)$

(d) $g(u,v)=\dfrac{u+5v}{3u-2v}$; $g(1,1)$, $g(1,4)$

(e) $h(x,y,z)=z+\sqrt{y^2-x^2}$; $h(1,2,3)$, $h(4,5,6)$

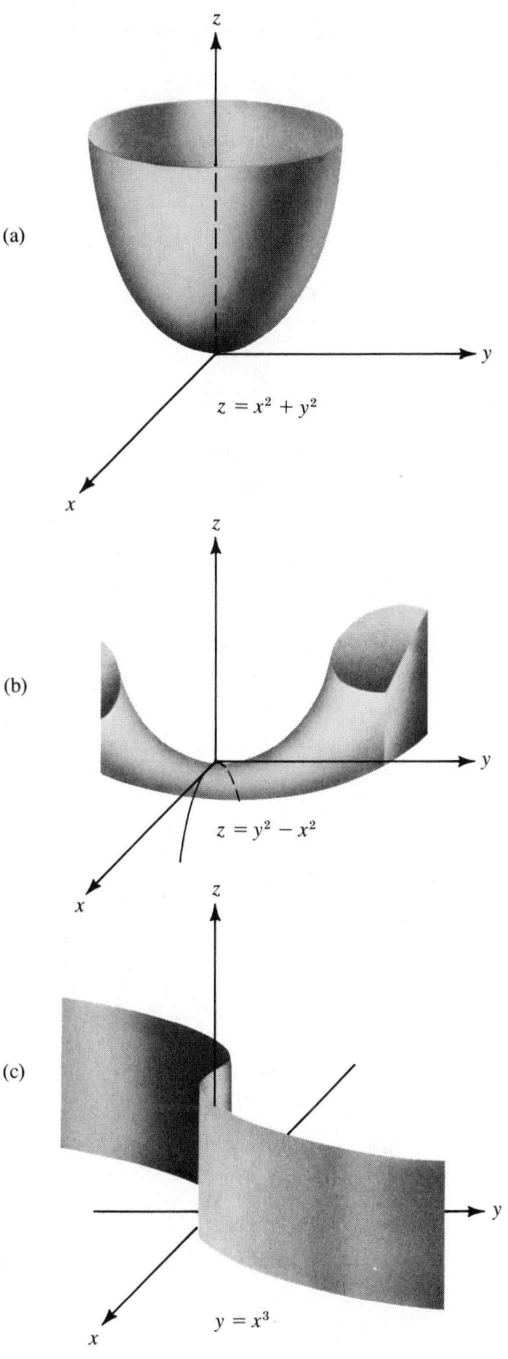

FIGURE 9.7

4. Determine the composite function and specify its domain for each of the following.

(a) $f(g(x), h(x))$, where $f(u,v) = u^2 + 3uv$, $g(x) = x^2$, $h(x) = x + 4$
(b) $f(g(x, y))$, where $f(u) = u^2$, $g(x, y) = \sqrt{xy^2}$
(c) $g(f(x, y), h(x, y))$, where $g(u,v) = u^2 - 3v^2$, $f(x, y) = x + y$, $h(x, y) = x - y$
(d) $g(h(x, y))$, where $g(w) = w^2 - 5$, $h(x, y) = \sqrt{x + y - 5}$

5. Graph each of the following planes.

(a) $x + y + z = 8$ (b) $x + y + z = 10$
(c) $2x + 4y + 3z = 12$ (d) $4x + 8y + z = 16$
(e) $x + y = 8$ (f) $x + z = 4$
(g) $y + z = 10$ (h) $x = 2$
(i) $y = 10$ (j) $y = 5 - z$

6. The Hansen Electronics Company makes electronic pocket calculators. The scientific model costs \$15 to make, and the financial model costs \$12.

(a) Determine Hansen's total weekly production cost as a function of the number of scientific models and the number of financial models.
(b) Determine the total weekly production cost if 1,000 scientific models and 1,500 financial models are made in one week.
(c) Determine the change Hansen should make in the weekly production of financial models if the production of scientific models is increased to 2,000 units and the total production cost is to remain the same as in part (b).

7. Suppose the Hansen Electronic Company in Exercise 6 wishes to maintain a fixed weekly production cost of \$33,000.

(a) Determine the equation involving the production levels of scientific and financial models that must be satisfied to do this.
(b) Draw the graph of the equation obtained in part (a).
(c) From the graph in part (b), determine the maximum number of scientific models that Hansen can make and still maintain the fixed cost. Determine the maximum number of financial models.

8. The Hansen Electronics Company operates a calculator assembly line where 15 skilled workers and 22 unskilled workers are assigned. The company has determined that by using x skilled and y unskilled workers they can assemble $x^2 + 10y$ calculators per day.

(a) Determine the current daily level of production.
(b) How much would the addition of one skilled worker increase daily production?
(c) If two skilled workers are ill and cannot work, what would be the daily level of production if the company were to hire two skilled and three unskilled workers?

9.3 PARTIAL DERIVATIVES

In a function of two variables, say $u = f(x, y)$, each of the independent variables can be varied *independently* of the other. In changing x for a fixed y, we are determining changes in u that are caused by the corresponding

First-Order Partial Derivatives

changes in x. Essentially we have reduced the two-variable function $u=f(x, y)$ to a one-variable function, say $z=g(x)$, because y is fixed and consequently is being treated as a constant. Thus we may refer to the derivative of $f(x, y)$ with respect to x and mean essentially the same as the derivative of $g(x)$ with respect to x; the only difference is that the actual function for $g(x)$ can change for different values of y. Nevertheless, the derivative of $f(x, y)$ with respect to x refers to the instantaneous rate of change of the function for a constant value of y.

Derivatives of multivariate functions are called *partial derivatives* because the derivatives are taken with respect to a single variable while the remaining variables are temporarily held constant. If we represent the function by $u=f(x,y)$, the derivative of the function with respect to x is generally denoted by one of the following symbols:

$$\frac{\partial f}{\partial x} \qquad \frac{\partial f(x,y)}{\partial x} \qquad f_x \qquad f_x(x,y) \qquad \frac{\partial u}{\partial x} \qquad u_x$$

You may have already noticed that we have replaced D_x by the symbol $\partial/\partial x$ in order to denote that we are determining *partial* derivatives instead of derivatives of one-variable functions. Since we are temporarily holding constant the variables other than the one with respect to which we are differentiating, the definition of partial derivatives should not be completely unfamiliar.

Definition 9.1 The partial derivative of $f(x, y)$ with respect to x is

$$\frac{\partial f}{\partial x} = \lim_{h\to 0} \frac{f(x+h,y)-f(x,y)}{h}$$

and the partial derivative of $f(x, y)$ with respect to y is

$$\frac{\partial f}{\partial y} = \lim_{h\to 0} \frac{f(x,y+h)-f(x,y)}{h}$$

provided the limits exist.

We could employ Definition 9.1 to determine the partial derivatives of any particular two-variable function; however, this is not necessary. Since we are temporarily holding one variable constant, we can differentiate with respect to the remaining variable as though the function were a one-variable function. Thus, to determine the partial derivatives, we can simply apply the rules of differentiation that were developed for the one-variable case; however, the result may still be a function of the variable that was being held constant. For example, if $f(x, y)=6xy+5y+3x$, the partial derivative with respect to x is obtained by considering y to be constant and differentiating with respect to x; thus $\partial f/\partial x = 6y + 0 + 3 = 6y + 3$. Similarly, the partial derivative with respect to y is obtained by considering x to be constant and differentiating with respect to y; differentiating, we obtain $\partial f/\partial y = 6x + 5 + 0 = 6x + 5$.

We note that a two-variable function has two first-order partial derivatives $\partial f/\partial x$ and $\partial f/\partial y$, whereas a one-variable function has only one first-order derivative. Similarly, a function of n variables will have n first-order partial derivatives.

Example 9.8 Determine the two first-order partial derivatives for $f(x, y) = 6x^2y + 3xy + y^2$.

Solution

$$\frac{\partial f}{\partial x} = 12xy + 3y + 0 = 12xy + 3y$$

$$\frac{\partial f}{\partial y} = 6x^2 + 3x + 2y$$

Example 9.9 Determine the two first-order partial derivatives for $f(x, y) = (x^2y + xy^2)^2$.

Solution

$$\frac{\partial f}{\partial x} = 2(x^2y + xy^2)(2xy + y^2)$$

$$\frac{\partial f}{\partial y} = 2(x^2y + xy^2)(x^2 + 2xy)$$

Example 9.10 Determine all first-order partial derivatives for $f(x, y) = e^{x^3+y^2}$.

Solution

$$\frac{\partial f}{\partial x} = 3x^2e^{x^3+y^2}$$

$$\frac{\partial f}{\partial y} = 2ye^{x^3+y^2}$$

For multivariate functions of three or more variables, we determine the partial derivatives with respect to each variable by temporarily holding *all* other variables constant and differentiating in the usual manner.

Example 9.11 Determine all first-order partial derivatives for $f(x, y, z) = 6x^2y + 4y^2z + 3xyz$.

Solution

$$\frac{\partial f}{\partial x} = 12xy + 0 + 3yz = 12xy + 3yz$$

$$\frac{\partial f}{\partial y} = 6x^2 + 8yz + 3xz$$

$$\frac{\partial f}{\partial z} = 0 + 4y^2 + 3xy = 4y^2 + 3xy$$

Example 9.12 Determine all first-order partial derivatives for $f(x, y, z, w) = e^{x^2y} + 2xzw + 4w^2z^2$.

Solution

$$\frac{\partial f}{\partial x} = 2xye^{x^2y} + 2zw + 0 = 2xye^{x^2y} + 2zw$$

$$\frac{\partial f}{\partial y} = x^2e^{x^2y} + 0 + 0 = x^2e^{x^2y}$$

$$\frac{\partial f}{\partial z} = 0 + 2xw + 8w^2z = 2xw + 8w^2z$$

$$\frac{\partial f}{\partial w} = 0 + 2xz + 8wz^2 = 2xz + 8wz^2$$

Higher-Order Partial Derivatives

When considering higher-order derivatives for multivariate functions, we encounter a considerable increase in the number of partial derivatives of a particular order. For example, a two-variable function has two first-order partial derivatives. Consequently, to determine all second-order derivatives, we must differentiate *each* first-order partial derivative with respect to each variable, which yields four second-order partial derivatives, as shown in Fig. 9.8. Furthermore, if we were to determine the third-order partial derivatives, we would differentiate each of the second-order partial derivatives with respect to each of the two variables, which would yield eight third-order partial derivatives. Thus, for a two-variable function, there are 2^n nth-order partial derivatives. In general, for a k-variable function, there are k^n nth-order partial derivatives. (Fortunately, we shall not consider any derivatives beyond second-order partial derivatives.)

The process of determining higher-order partial derivatives is essentially the same as determining higher-order derivatives for one-variable functions. That is, we simply determine the partial derivative of each function that represents a partial derivative of the preceding order; also, we must do this

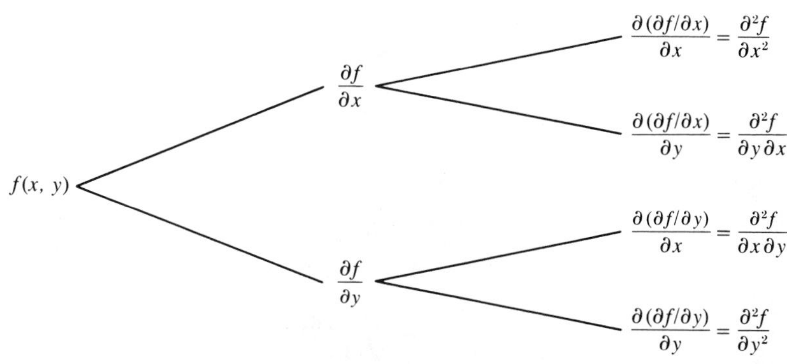

FIGURE 9.8

for *each* variable. Notationally this is demonstrated in Fig. 9.8. For higher-order partial derivatives the notation used is such that the superscript in the notation $\partial^2 f/\partial x \partial y$ represents the *order* of the partial derivative. The lower portion of the symbol (that appears to be the denominator but is not because the term is not a fraction) represents the order in which the derivatives were taken. The variable on the *right* represents the *first* variable with respect to which the partial derivative was taken. The variable on the *left* represents the *last* variable of differentiation.

Thus $\partial^2 f/\partial x \partial y$ represents the function that is determined by first differentiating $f(x, y)$ with respect to y and then differentiating the resulting first-order partial derivative $\partial f/\partial y$ with respect to x. Conversely, $\partial^2 f/\partial y \partial x$ represents the second-order partial derivative obtained by first differentiating $f(x, y)$ with respect to x and then differentiating $\partial f/\partial x$ with respect to y. Furthermore, for all functions that we shall consider, these two second-order partial derivatives will be equal, i.e., $\partial^2 f/\partial x \partial y = \partial^2 f/\partial y \partial x$. This provides a useful check on the accuracy of our differentiation. The second-order partial derivatives denoted by $\partial^2 f/\partial x^2$ and $\partial^2 f/\partial y^2$ are obtained by differentiating both the function and the resulting first-order partial derivative with respect to x and y, respectively.

Example 9.13 Determine all first- and second-order partial derivatives for $f(x, y) = x^3y^2 + 3x^2y + 2y^3$.

Solution The first-order partial derivatives are

$$\frac{\partial f}{\partial x} = 3x^2y^2 + 6xy$$

and

$$\frac{\partial f}{\partial y} = 2x^3y + 3x^2 + 6y^2$$

The second-order partial derivatives are

$$\frac{\partial^2 f}{\partial x^2} = \frac{\partial}{\partial x}\left(\frac{\partial f}{\partial x}\right) = 6xy^2 + 6y$$

$$\frac{\partial^2 f}{\partial y \partial x} = \frac{\partial}{\partial y}\left(\frac{\partial f}{\partial x}\right) = 6x^2y + 6x$$

$$\frac{\partial^2 f}{\partial x \partial y} = \frac{\partial}{\partial x}\left(\frac{\partial f}{\partial y}\right) = 6x^2y + 6x + 0 = 6x^2y + 6x$$

$$\frac{\partial^2 f}{\partial y^2} = \frac{\partial}{\partial y}\left(\frac{\partial f}{\partial y}\right) = 2x^3 + 0 + 12y = 2x^3 + 12y$$

Example 9.14 Determine all first- and second-order partial derivatives for $f(x, y) = (2xy + x^2y)^2$.

Solution The first-order partial derivatives are

$$\frac{\partial f}{\partial x} = 2(2xy + x^2y)(2y + 2xy)$$
$$= 4xy^2(2+x)(1+x)$$
$$= 4y^2(2x + 3x^2 + x^3)$$

and

$$\frac{\partial f}{\partial y} = 2(2xy + x^2y)(2x + x^2)$$
$$= 2y(2x + x^2)(2x + x^2)$$
$$= 2y(2x + x^2)^2$$

The second-order partial derivatives are

$$\frac{\partial^2 f}{\partial x^2} = 4y^2(2 + 6x + 3x^2)$$

$$\frac{\partial^2 f}{\partial y \partial x} = 8y(2x + 3x^2 + x^3)$$

$$\frac{\partial^2 f}{\partial x \partial y} = 2y(2)(2x + x^2)(2 + 2x) = 8y(2x + 3x^2 + x^3)$$

$$\frac{\partial^2 f}{\partial y^2} = 2(2x + x^2)^2$$

Example 9.15 Determine all first- and second-order partial derivatives for $f(x, y) = e^{xy}$.

Solution The first-order partial derivatives are

$$\frac{\partial f}{\partial x} = ye^{xy}$$

and

$$\frac{\partial f}{\partial y} = xe^{xy}$$

The second-order partial derivatives are

$$\frac{\partial^2 f}{\partial x^2} = y^2 e^{xy}$$

$$\frac{\partial^2 f}{\partial y \partial x} = xye^{xy} + e^{xy} = (xy + 1)e^{xy}$$

$$\frac{\partial^2 f}{\partial x \partial y} = xye^{xy} + e^{xy} = (xy + 1)e^{xy}$$

$$\frac{\partial^2 f}{\partial y^2} = x^2 e^{xy}$$

Example 9.16 Determine all first- and second-order partial derivatives for $f(x, y, z) = x^2 + y^2 + z^2 + 2xyz$.

Solution The three first-order partial derivatives are

$$\frac{\partial f}{\partial x} = 2x + 2yz$$

$$\frac{\partial f}{\partial y} = 2y + 2xz$$

$$\frac{\partial f}{\partial z} = 2z + 2xy$$

The nine second-order partial derivatives are

$$\frac{\partial^2 f}{\partial x^2} = 2 \qquad \frac{\partial^2 f}{\partial x \partial y} = 2z \qquad \frac{\partial^2 f}{\partial x \partial z} = 2y$$

$$\frac{\partial^2 f}{\partial y^2} = 2 \qquad \frac{\partial^2 f}{\partial y \partial x} = 2z \qquad \frac{\partial^2 f}{\partial y \partial z} = 2x$$

$$\frac{\partial^2 f}{\partial z^2} = 2 \qquad \frac{\partial^2 f}{\partial z \partial x} = 2y \qquad \frac{\partial^2 f}{\partial z \partial y} = 2x$$

The partial derivatives $\partial^2 f/\partial x \partial y$ and $\partial^2 f/\partial y \partial x$ are called *cross partial derivatives* and, as mentioned earlier, are equal if their respective functions are continuous in some region, which means simply that the order of differentiation is immaterial if the continuity condition is satisfied. (This continuity condition is satisfied for nearly all functions that we shall consider in this book.) In each of Examples 9.13 through 9.16 the reader should verify that the corresponding cross partial derivatives are, in fact, equal.

EXERCISES

In Exercises 1–16, *determine all first-order partial derivatives.*

1. $f(x, y) = 3y + 2$

2. $f(x, y) = 2x$

3. $f(x, y) = 2x + 5$

4. $f(x, y) = 6x - 3y$

5. $f(x, y) = x^2 + 2xy$

6. $f(x,y) = y^2 - 3xy + 4x^2$

7. $f(x, y) = y^2 + 6xy + x^2 - 4x + 3y$

8. $f(x, y) = 4x^2 + 5x^{-6} + 2y^3$

9. $f(x, y) = e^{x^2 y}$

10. $f(x, y) = e^{4xy + x^2}$

11. $f(x, y) = (3x^2 + 2xy)^3$

12. $f(x, y) = 11^{6xy}$

13. $f(x, y) = 4^{x^2 - y^2}$

14. $f(x, y) = \ln(xy + 2)$

15. $f(x, y, z) = xy/z$

16. $f(x, y, z) = x^2 + \ln(y^2z^2)$

17. Using f and g as given in Exercise 1, Section 9.2, verify that $\partial(f+g)/\partial x = \partial f/\partial x + \partial g/\partial x$ and $\partial(f+g)/\partial y = \partial f/\partial y + \partial g/\partial y$.

18. Using f and g as given in Exercise 2, Section 9.2, verify that $\partial(fg)/\partial x = f(\partial g/\partial x) + g(\partial f/\partial x)$, and $\partial(fg)/\partial y = f(\partial g/\partial y) + g(\partial f/\partial y)$.

In each of the functions in Exercises 19–28, determine all first- and second-order partial derivatives and verify that the respective cross partial derivatives are equal.

19. $f(x, y) = x^2 + 3y^2$

20. $f(x, y) = x^2 + 2y^2 + 3x + 4y + 1$

21. $f(x, y) = xy + 2x^2 - 3y^2$

22. $f(x, y) = e^{3x-2y}$

23. $f(x, y) = 2x - 4y + 3x/y$

24. $f(x, y) = (x+1)\ln(y+1)$

25. $f(x, y) = 2x^2 - y + ye^x$

26. $f(x, y) = \ln(xy)$

27. $f(x, y) = \ln x^y + 4y^3x$

28. $f(x, y, z) = 4xyz + 2xz$

29. For $f(x, y) = 2x^2 + 3y^2$ at $(2, 2)$, determine the instantaneous rate of change for (a) x and (b) y.

30. For $f(x, y) = 4xy$ at $(2, 1)$, determine the instantaneous rate of change for (a) x and (b) y.

31. For $f(x, y) = 2x^2 + 3xy - y^2$ at $(2, 3)$, determine the instantaneous rate of change for (a) x and (b) y.

32. Product A has a contribution margin of \$5/unit, and product B has a contribution margin of \$6/unit. Fixed costs are \$250/month.

(a) Determine the function that represents monthly profits. (Profit is equal to the sum of the contribution margin of each product minus fixed costs.)
(b) Determine the function that represents the instantaneous rate of change for the contribution margin of product A.
(c) Determine the function that represents the instantaneous rate of change for the contribution margin of product B.
(d) Determine the instantaneous rate of change for product A and product B at the point $(50, 40)$.

33. The number of minutes required in department A to produce products X and Y each day is equal to the number of units of product X produced squared, plus the number of units of product Y produced cubed, minus 4 times the number of units of product X produced, multiplied by the number of units of product Y produced.

(a) Determine the function that represents the number of minutes required to process products X and Y in department A.
(b) Determine the function that represents the instantaneous rate of change for product X.
(c) Determine the function that represents the instantaneous rate of change for product Y.

(d) Determine the instantaneous rate of change for product X and product Y at the point $(6,5)$.
(e) Determine all of the second-order partial derivatives for the function in part (a).

34. The increase in the demand for widgets is equal to the increase in the amount, in thousands of dollars, of advertising, times the natural logarithm of the number of salespeople.

(a) Determine the function that represents the increase in the demand for widgets.
(b) Determine the function that represents the instantaneous rate of change for the amount, in thousands of dollars, spent on advertising.
(c) Determine the function that represents the instantaneous rate of change for the number of salespeople.
(d) Determine all of the second-order partial derivatives for the function in part (a).

9.4 EXTREMA FOR BIVARIATE FUNCTIONS

In Sec. 5.3 we considered the application of differential calculus to determine the relative maxima and/or relative minima for a one-variable function. We shall now consider the corresponding problem for a two-variable (bivariate) function. We shall restrict our attention to those functions that are continuous since this is the type we will most commonly encounter.

If the function $f(x, y)$ is a continuous function with continuous partial derivatives $\partial f/\partial x$ and $\partial f/\partial y$, the function can be represented by a smooth unbroken surface whose equation is $u=f(x, y)$. The function $f(x, y)$ has a *relative maximum* at a point (a,b) if the number $f(a,b)$ is greater than the values of $f(x, y)$ at all points in the neighborhood of (a,b). The *relative minimum* can be defined in the same manner by replacing "greater" with "less." Both maxima and minima are included under the term *extreme values*. The geometric illustration of a relative maximum is given in Fig. 9.9, with the maximum occurring at the point (a,b,c), where $c=f(a,b)$.

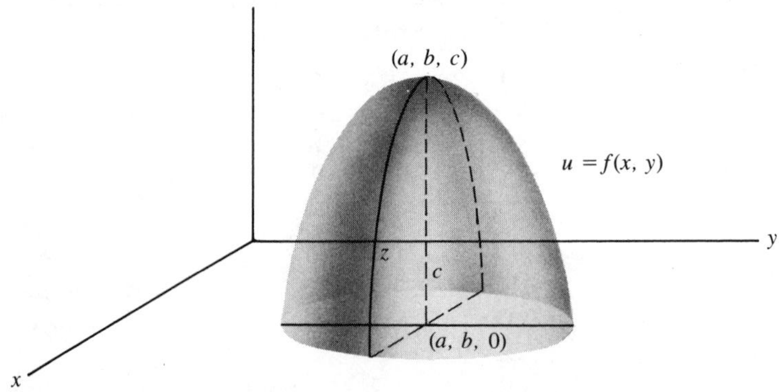

FIGURE 9.9

If $f(x, y)$ and its first derivatives are continuous in a region that includes the point (a,b), a *necessary condition* for $f(a,b)$ to be a relative extremum (maximum or minimum) of the function $f(x, y)$ is that

$$\frac{\partial f}{\partial x} = 0 \qquad \text{and} \qquad \frac{\partial f}{\partial y} = 0$$

where the partial derivatives are evaluated at (a,b). If we denote $\partial f/\partial x$ by $f_x(x, y)$ and $\partial f/\partial y$ by $f_y(x, y)$, the necessary condition is that

$$f_x(a,b) = 0 \qquad \text{and} \qquad f_y(a,b) = 0$$

It must be understood that this necessary condition does not guarantee that we have a relative extremum at (a,b). This is analogous to the one-variable function where it is necessary that $f'(x^*)=0$ or does not exist for $f(x^*)$ to be a relative extremum.) We need a test for the bivariate case to enable us to distinguish among a relative maximum, a relative minimum, and the situation where we have neither. Such a test is now given, and it provides *sufficient conditions* for the existence of a relative extremum.

Extending the above notation for partial derivatives to the second-order partial derivatives, we denote $\partial^2 f/\partial x^2$ by $f_{xx}(x,y)$ and $\partial^2 f/\partial y^2$by $f_{yy}(x,y)$; for the cross partial derivatives we denote $\partial^2 f/\partial x\partial y$ by $f_{xy}(x,y)$ and $\partial^2 f/\partial y\partial x$ by $f_{yx}(x,y)$. However, for the types of problems that are of concern to us, we shall have $\partial^2 f/\partial x\partial y = \partial^2 f/\partial y\partial x$ because they will satisfy the required condition given in Sec. 9.3. The extreme value test for bivariate functions is given below:

If at the point (a,b)

$$f_x(a,b) = 0 \qquad \text{and} \qquad f_y(a,b) = 0 \tag{9.1}$$

and if

$$D(a,b) = f_{xx}(a,b)f_{yy}(a,b) - [f_{xy}(a,b)]^2 > 0$$

then $f(x, y)$ has a relative *maximum* value at (a,b) if

$$f_{xx}(a,b) < 0$$

or a relative *minimum* value if

$$f_{xx}(a,b) > 0$$

If condition (9.1) holds and

$$D(a,b) < 0$$

then $f(a,b)$ is neither a relative maximum nor a relative minimum. If

$$D(a,b) = 0$$

the test fails to give any information.

The previous test can be stated in the following form.

Rule 9.1 *Extreme value test for bivariate functions*: Given a point (a,b) that satisfies Eq. (9.1)

1. If $D(a,b)>0$ and $f_{xx}(a,b)<0$, then $f(a,b)$ is a relative maximum.
2. If $D(a,b)>0$ and $f_{xx}(a,b)>0$, then $f(a,b)$ is a relative minimum.
3. If $D(a,b)<0$, then $f(a,b)$ is not an extremum.
4. If $D(a,b)=0$, the test fails.

When a point satisfies Eq. (9.1) and extreme value test condition 3, it is called a *saddle point*. This name evolves from the analogy between the graph of the function at this point and the characteristics of a saddle; that is, from the point in question, the function increases in both directions along one axis and decreases in both directions along the other axis. (A graphic illustration of such a point is given in Fig. 9.10.)

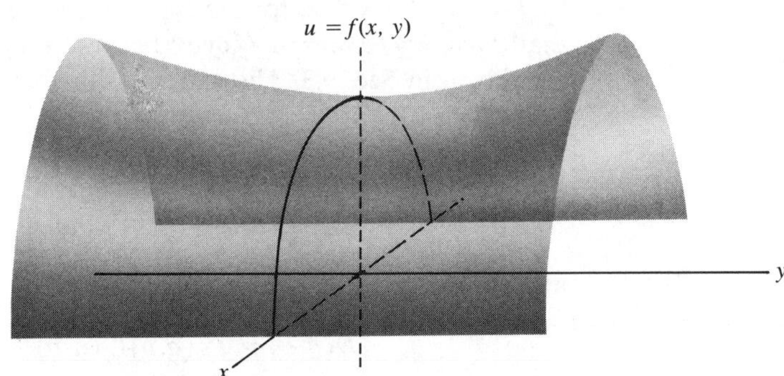

FIGURE 9.10

Example 9.17 Examine the function $f(x, y)=x^2+y^2+xy$ for relative extrema.

Solution To determine the points for which condition (9.1) is satisfied, we determine the first derivatives and set them equal to zero:

$$\frac{\partial f}{\partial x} = 2x + y$$

$$\frac{\partial f}{\partial y} = 2y + x$$

Rearranging the two equations and setting them equal to zero, we have

$$2x + y = 0$$
$$x + 2y = 0$$

Simultaneously solving this system of equations, we determine all points that are candidates for extreme values. The system has only one point that satisfies it, namely (0,0). Thus (0,0) is the only point that satisfies condition (9.1).

Determining the second-order partial derivatives, we obtain $\partial^2 f/\partial x^2 = 2$, $\partial^2 f/\partial y^2 = 2$, and $\partial^2 f/\partial x \partial y = 1$; therefore,

$$\begin{aligned} D(0,0) &= f_{xx}(0,0) f_{yy}(0,0) - [f_{xy}(0,0)]^2 \\ &= 2 \cdot 2 - (1)^2 \\ &= 3 \end{aligned}$$

Furthermore, $f_{xx}(0,0) = 2 > 0$, which means that extreme value test condition 2 is satisfied; hence $f(x, y) = x^2 + y^2 + xy$ is a relative minimum at the point (0,0) and the minimum value is $f(0,0) = 0$.

Example 9.18 Examine the function $f(x, y) = x^3 - 3x - y^2$ for relative extrema.

Solution First we determine the points that satisfy condition (9.1); thus we obtain the first-order partial derivatives

$$\frac{\partial f}{\partial x} = 3x^2 - 3$$
$$\frac{\partial f}{\partial y} = -2y$$

Setting the two partial derivatives equal to zero and solving simultaneously, we have

$$3x^2 - 3 = 0$$
$$-2y = 0$$

The first equation yields $x = -1$ and $x = 1$, while the second equation yields $y = 0$; thus the two points $(-1,0)$ and $(1,0)$ are candidates for yielding any extreme values for $f(x, y)$.

Next we determine the status of $(-1,0)$. The second-order partial derivatives are $\partial^2 f/\partial x^2 = 6x$, $\partial^2 f/\partial y^2 = -2$, and $\partial^2 f/\partial x \partial y = 0$; thus

$$D(-1,0) = [6(-1)](-2) - (0)^2 = 12 > 0 \quad \text{and} \quad f_{xx}(-1,0) = -6 < 0$$

Therefore, extreme value test condition 1 is satisfied, and $f(-1,0)$ is a relative maximum. Checking the status of (1,0), we determine

$$D(1,0) = [6(1)](-2) - (0)^2 = -12 < 0$$

Thus extreme value test condition 3 is satisfied, and $f(1,0)$ is *not* an extreme value. Hence the point $(1,0,-3)$ is a saddle point of the graph of $f(x, y)$.

In our discussion of functions of one variable we found that a continuous function $f(x)$, defined on a closed interval $[a,b]$, has an absolute extremum either at an endpoint or at a critical value corresponding to a relative extremum. Similarly, an absolute extremum of a continuous bivariate function $f(x,y)$, defined on a region consisting of a set of points (x,y) in the xy plane, will occur either at a boundary point of the region in the xy plane or at a point (a,b) corresponding to a relative extremum of the function. To determine if the absolute extremum occurs at a boundary point or is a relative extremum requires careful analysis of the function being considered. Fortunately, in many real-world problems the absolute extremum is a relative extremum. In this text we will limit our discussion to problems of this nature. Hence the absolute maximum will be the greatest of the relative maxima and the absolute minimum will be the least of the relative minima.

EXERCISES

In Exercises 1–12, examine the given function for relative extrema.

1. $f(x, y)=4x^2+y^2-4y$
2. $f(x, y)=4x^2+y^2+8x-2y-1$
3. $f(x, y)=xy$
4. $f(x, y)=x^2+y^3-3y$
5. $f(x, y)=x^2+xy+y^2-6x-2y+5$
6. $f(x, y)=x^3+x^2+y^2+xy+10$
7. $f(x, y)=x^3+y^3-3xy$
8. $f(x, y)=x^2y-4x^2+y^3$
9. $f(x, y)=y^2-x^2$
10. $f(x, y)=y^3-3y-x^2$
11. $f(x, y)=3x^3-5x^2+x-5y^3+6y^2+3y-10$
12. $f(x, y)=\ln x^6+x^2+y^2+4xy+10$
13. The labor cost of a certain construction project depends upon the number of skilled workers x and the number of unskilled workers y. The labor cost function has been determined to be

$$C(x,y) = 90{,}000 + 60x^2 - 240xy + 10y^3$$

(a) Examine $C(x, y)$ for extreme values.
(b) Determine the numbers of skilled and unskilled workers that yield the minimum labor cost.
(c) Determine the minimum labor cost.

14. The Resthaven Funeral Home determines that the profit per month in hundreds of dollars is approximated by

$$P(x, y) = 2{,}800 + 8x + 10y - x^2 + 6xy - 10y^2$$

where x is salary expense and y is advertising expense, both in thousands of dollars.

(a) Examine $P(x, y)$ for extreme values.
(b) Determine the salary and advertising expenses that yield maximum profit.
(c) Determine the maximum profit.

15. The Peyton Manufacturing Company can approximate their monthly sales S in thousands of dollars by the function

$$S(x, y) = 100x + 10y + 6xy$$

and they can approximate their monthly cost C in thousands of dollars by the function

$$C(x, y) = 16x + 2y + 9x^2 + 2xy + y^2$$

where x is inventory value in thousands of dollars and y is floor space in thousands of square feet.

(a) Determine the profit function $P(x, y)$.
(b) Examine the profit function for extreme values.
(c) Determine the inventory value and floor space that yield maximum profit.
(d) Determine the maximum profit.

16. The Green Acres Cattle Ranch fattens steers to sell to meat-processing plants. The total profit per head in dollars can be approximated by

$$P(x, y) = 2.4x + 3y + 0.5xy - 0.5x^2 - 0.3y^2$$

where x represents the amount of especially mixed feed in hundreds of pounds and y represents the number of bales of hay required to fatten one steer.

(a) Examine the profit function for extreme values.
(b) Determine the amounts of feed and hay that yield maximum profit.
(c) Determine the maximum profit.

17. The cost of making 1 unit of product X is given by the function

$$C(x, y) = 3 + 4x^3 - 10xy + 6y^2$$

where x is the dollar cost of materials and y is the dollar cost of labor.

(a) Examine $C(x, y)$ for extreme values.
(b) Determine the dollar cost of materials and labor to use in each unit of product X to minimize costs.
(c) Determine the minimum cost for a unit of product X.

18. The daily cost of Gizmo Manufacturing Company is given by

$$C(x, y) = x^3 + 13x^2 - 20x + y^3 + 15y^2 - 25y + 85$$

where x is the number of widgets produced in thousands and y is the number of gadgets produced in thousands.

(a) Determine the number of widgets and gadgets to produce in order to minimize costs by examining $C(x, y)$ for extreme values.
(b) Determine the minimum cost.

19. The profit for making and selling 5,000 boxes of screws is given by

$$P(x, y) = 0.5x^2 + 0.9xy - y^2 + 10x + 15y$$

where x is the time in hours required to make the screws and y is the number of boxes of screws made per batch.

(a) Examine $P(x, y)$ for extreme values.
(b) Determine the number of boxes per batch and the time required to make the screws that will maximize the total profit.
(c) Determine the maximum profit.
(d) Determine the profit per box of screws.

20. The XYZ Corporation has weekly revenue R in hundreds of dollars that can be represented by the function

$$R(x, y) = 50x + 60y + 10xy$$

and weekly cost C in hundreds of dollars that can be represented by the function

$$C(x, y) = 4x^2 + 5y^2 + 5x + 8y + 2xy$$

where x is the number of vending machines, in hundreds, filled per week and y is the number of people employed by XYZ Corporation.

(a) Determine the profit function $P(x, y)$.
(b) Examine the profit function for extreme values.
(c) Determine the numbers of vending machines and employees that yield maximum profit.
(d) Determine the maximum profit.

9.5 CONSTRAINED OPTIMA

There are many situations in which the problem of minimizing production cost is constrained by certain demand requirements. Similarly, the problem of maximizing profit is quite often constrained by limitations on working capital or staff, and the maximization of sales is generally constrained by production capacity and advertising budget. Such problems of determining the optimum values of functions that are subject to constraints can be solved by differential calculus.

The problem of determining extreme values of a function $f(x,y)$ that is subject to a constraint can be stated as that of finding the extreme values of

$f(x,y)$ subject to the constraint $g(x,y)=0$. One method of solving for the extreme values of a constrained optima problem is to solve the constraining equation for one variable in terms of the remaining variable, and substitute the result into $f(x,y)$. This reduces the function $f(x,y)$ to a function of one variable, which can be solved by the methods given in Sec. 5.4. This procedure is called the *method of substitution*.

Example 9.19 Determine the relative extrema of $f(x,y)=x^2+y^2+xy$ subject to $x+y=10$.

Solution Solving $x+y=10$ for y in terms of x and substituting into $f(x,y)=x^2+y^2+xy$, we obtain $y=10-x$ and

$$\begin{aligned} f(x,y) &= f(x,10-x) \\ &= x^2+(10-x)^2+x(10-x) \\ &= x^2+100-20x+x^2+10x-x^2 \\ &= x^2-10x-100 \end{aligned}$$

Thus, with the constraint substituted into $f(x,y)$, the function is dependent only on x, and hence we need simply to determine the extrema of $h(x)=x^2-10x+100$.

To find the critical values, we determine the first derivative

$$h'(x) = 2x - 10$$

Setting it equal to zero and solving for x, we obtain

$$2x - 10 = 0$$

or

$$x = 5$$

The second derivative of $h(x)$ is $h''(x)=2$; thus $h''(5)=2>0$, which means that $h(x)$ is a relative minimum at $x=5$.

Returning to the original two-variable problem, we determine $y=10-x=10-5=5$. Hence, when the function $f(x,y)=x^2+y^2+xy$ is constrained by $x+y=10$, the minimum occurs at the point $(5,5)$ and is computed in the usual manner to be

$$f(5,5) = (5)^2+(5)^2+(5)(5) = 75$$

In Example 9.17 the minimum of the function $f(x,y)=x^2+y^2+xy$ was determined to occur at the point $(0,0)$ and was found to be $f(0,0)=0$. Thus we see that when the function is constrained by $x+y=10$, the minimum value is considerably increased, which is the penalty paid to satisfy the constraint.

Unfortunately the method of substitution is not widely applicable because in nonlinear equations it is generally difficult to solve for one variable in terms of the remaining variable. For example, the method of substitution quickly encounters difficulty in a constraint such as $3x^2-4xy+y^3+4x=0$. Fortunately an alternative method is available; it is called the *method of lagrangian multipliers*, named after the famous mathematician Joseph Louis Lagrange, who discovered it during the eighteenth century.

The method of lagrangian multipliers is simply a technique for expressing the original function $f(x,y)$ and the constraint $g(x,y)=0$ together as a single function. Since $g(x,y)=0$, we can let λ (lambda) be a third variable, called the *lagrangian multiplier*, and form the new three-variable function

$$F(x,y,\lambda) = f(x,y) + \lambda g(x,y)$$

We see that the value of $F(x,y,\lambda)$ is the same as the value of $f(x,y)$ for any set of values of x,y and λ because $g(x,y)=0$; thus we have not altered the function value. Furthermore, taking the first-order partial derivatives with respect to each of the three variables, we obtain

$$\frac{\partial F}{\partial x} = \frac{\partial f}{\partial x} + \frac{\lambda \partial g}{\partial x}, \qquad \frac{\partial F}{\partial y} = \frac{\partial f}{\partial y} + \frac{\lambda \partial g}{\partial y}, \qquad \text{and} \qquad \frac{\partial F}{\partial \lambda} = g(x,y)$$

If we extend to our three-variable function the *necessary* conditions for a function to attain an extreme value, we have

$$\frac{\partial F}{\partial x} = 0, \qquad \frac{\partial F}{\partial y} = 0, \qquad \text{and} \qquad \frac{\partial F}{\partial \lambda} = 0$$

However, expressing these conditions in terms of the original objective function and the constraint, we have

$$\frac{\partial f}{\partial x} + \frac{\lambda \partial g}{\partial x} = 0, \qquad \frac{\partial f}{\partial y} + \frac{\lambda \partial g}{\partial y} = 0, \qquad \text{and} \qquad g(x,y) = 0$$

Thus, in order for the function $F(x,y,\lambda)$ to have an extreme value at the point (a,b,α), it is necessary that the following conditions be satisfied:

$$f_x(a,b) + \alpha g_x(a,b) = 0, \qquad f_y(a,b) + \alpha g_y(a,b) = 0, \qquad \text{and} \qquad g(a,b) = 0 \tag{9.2}$$

In order to determine the critical values of x, y, and λ, we would solve simultaneously the three equations of (9.2) to determine the values a, b, and α.

Necessary conditions for $F(x,y,\lambda)$ to assume an extreme value are given in (9.2), and these conditions are very similar to those for bivariate functions. Note, however, when $F(x,y,\lambda)$ assumes an extreme value, our original function $f(x,y)$ will assume an extreme value since $g(x,y)=0$. Furthermore, the extreme value of $F(x,y,\lambda)$ is not *directly* dependent upon the value of λ because $g(x,y)=0$, which leads us to the consideration of sufficient conditions for $F(x,y,\lambda)$ to assume an extreme value.

We note that $\partial F/\partial\lambda=0$ is the constraint $g(x,y)=0$, and so it is unnecessary to differentiate F with respect to λ. However, it is necessary that we use this constraint in solving the system of equations given in (9.2), which means that the critical values are influenced by the value of λ because it is one of the variables in the system of equations that represents the necessary conditions. Furthermore, the sufficient conditions that are given below are directly influenced by the value of λ.

Rule 9.2 *Extreme value test for constrained bivariate functions*: Given $F(x,y,\lambda)=f(x,y)+\lambda g(x,y)$, where $g(x,y)=0$, and $D(x,y,\lambda)=F_{xx}(x,y,\lambda)F_{yy}(x,y,\lambda)-[F_{xy}(x,y,\lambda)]^2$, then the critical point (a,b,α), which satisfies the equations of (9.2), is such that

1. If $D(a,b,\alpha)>0$ and $F_{xx}(a,b,\alpha)<0$, then $f(a,b)$ is a relative maximum.
2. If $D(a,b,\alpha)>0$ and $F_{xx}(a,b,\alpha)>0$, then $f(a,b)$ is a relative minimum.
3. If $D(a,b,\alpha)\leqslant 0$, the test fails.

Example 9.20 Determine the relative extrema of $f(x, y)=x^2+y^2+xy$ subject to $x+y=10$.

Solution From the constraint, we determine $g(x, y)=x+y-10=0$. Thus, using a lagrangian multiplier, we have

$$F(x,y,\lambda) = x^2+y^2+xy+\lambda(x+y-10).$$

Determining the first-order partial derivatives of $F(x, y,\lambda)$ and setting them equal to zero, we obtain

$$\frac{\partial F}{\partial x} = 2x+y+\lambda = 0$$

$$\frac{\partial F}{\partial y} = 2y+x+\lambda = 0$$

$$\frac{\partial F}{\partial \lambda} = x+y-10 = 0$$

Subtracting the second equation from the first, we eliminate λ and combine the resulting equation with the third equation to have two equations in two unknowns; that is,

$$x-y = 0$$

$$x+y = 10$$

Solving this reduced system for x and y, we have $x=5$ and $y=5$; substituting these values into the equation for $\partial F/\partial x=0$ (or $\partial F/\partial y)=0$), we obtain $\lambda=-15$. Thus the only critical point (x, y,λ) is $(5,5,-15)$.

Having determined a critical point, next we examine the value of $D(x, y, \lambda)$. Obtaining all second-order partial derivatives of $F(x, y, \lambda)$ with respect to x and y, we have

$$\frac{\partial^2 F}{\partial x^2} = 2, \qquad \frac{\partial^2 F}{\partial y^2} = 2, \qquad \text{and} \quad \frac{\partial^2 F}{\partial x \partial y} = 1$$

Thus regardless of the specific values of the critical point, we have

$$D(x, y, \lambda) = (2)(2) - (1)^2 = 3 > 0$$

which means that $D(5, 5, -15) = 3 > 0$. Furthermore, $\partial^2 F/\partial x^2 = 2 > 0$, which means that $F_{xx}(5, 5, -15) = 2 > 0$ and condition 2 of the extreme value test is satisfied. Hence $f(x, y) = x^2 + y^2 + xy$ assumes a relative minimum value of

$$f(5,5) = (5)^2 + (5)^2 + (5)(5) = 75$$

We should note that Example 9.20 presents the same problem as Example 9.19; however, the method of solution utilizes lagrangian multipliers.

Example 9.21 Determine the relative extrema of $f(x, y) = x + y$ subject to $x^2 + y^2 = 2$.

Solution Using a lagrangian multiplier, we have

$$F(x, y, \lambda) = x + y + \lambda(x^2 + y^2 - 2)$$

Determining the critical values, we have

$$\frac{\partial F}{\partial x} = 1 + 2\lambda x = 0$$

$$\frac{\partial F}{\partial y} = 1 + 2\lambda y = 0$$

$$\frac{\partial F}{\partial \lambda} = x^2 + y^2 - 2 = 0$$

Solving the system of equations, we obtain from the first two equations $\lambda = -1/2x$ and $\lambda = -1/2y$, which yields $x = y$. Substituting into the third equation, we have $x^2 + x^2 - 2 = 0$, or $x^2 = 1$, which yields $x = -1$ and $x = 1$ as solutions. Thus the critical points are $(-1, -1, \frac{1}{2})$ and $(1, 1, -\frac{1}{2})$. Determining the values of $D(x, y, \lambda)$ for the two critical points, we determine the required second-order partial derivatives and substitute the values for each point:

$$\frac{\partial^2 F}{\partial x^2} = 2\lambda \quad \frac{\partial^2 F}{\partial y^2} = 2\lambda \quad \frac{\partial^2 F}{\partial x \partial y} = 0$$

Thus $D(x, y, \lambda) = (2\lambda)(2\lambda) - (0)^2 = 4\lambda^2$, which is positive for any nonzero real value of λ. For the critical point $(-1, -1, \frac{1}{2})$, $D(-1, -1, \frac{1}{2}) = 1$, and $F_{xx}(-1, -1, \frac{1}{2}) = 1 > 0$; therefore, $f(-1, -1) = -2$ is a relative minimum. Similarly, $D(1, 1, -\frac{1}{2}) = 1$, and $F_{xx}(1, 1, -\frac{1}{2}) = -1 < 0$; hence $f(1, 1) = 2$ is a relative maximum.

In the event a two-constraint problem is encountered, we may employ a process that uses two lagrangian multipliers, the mechanics of which are quite similar to those for the one-constraint problem.

EXERCISES

Determine the relative extrema of each of the functions in Exercises 1–6 *subject to the given constraints.*

1. $f(x, y) = 4x^2 + 3y^2 - xy$ (if $x + 2y = 21$)
2. $f(x, y) = x^2 + y^2 - xy$ (if $x - y = 8$)
3. $f(x, y) = 10x^2 + y^2$ (if $x - y = 22$)
4. $f(x, y) = 3x + 2y$ (if $x^2 + y^2 = 125$)
5. $f(x, y) = 2x^2 + y^2 - xy$ (if $x + y = 8$)
6. $f(x, y) = 6x^2 + 5y^2 - xy$ (if $2x + y = 24$)
7. The Hayden Manufacturing Company has determined that the number of units of production, as a function of the number of tons of raw material x and the number of person-hours y in hundreds of hours, is given by

$$P(x, y) = -4x^2 + 5xy + y^2$$

   Determine the amounts of raw material and labor that maximize production if $3x + 2y = 74$. Also, determine the maximum number of units of production.

8. The Fritter-Lee Potato Chip Company uses an automatic packaging machine that has two critical parts. The number of breakdowns of the machine, as a function of the numbers of replacements x and y of the parts A and B, respectively, is given by

$$f(x, y) = x^2 + 3y^2 + 2xy - 11x + 15$$

   Determine the number of replacements that should be made for each part in order to minimize the number of breakdowns if for every replacement of part B we must make two replacements of part A. Also determine the minimum number of breakdowns.

9. The cost of repairs for the machine in Exercise 8 is a function of the numbers x and y of inspections per week of the two parts A and B, respectively. The cost

function is given by

$$f(x,y) = 2x^2 + 5xy + 4y^2 - 40y + 100$$

If the total number of inspections is 20, determine the number of inspections per week that should be made for each part in order to minimize repair costs.

10. For the Resthaven Funeral Home of Exercise 14, Sec. 9.4, determine the salary and advertising expenses that yield maximum profit if the total amount spent on salary and advertising must equal \$55,000. Also determine the maximum profit.

11. For the XYZ Corporation of Exercise 20, Sec. 9.4, determine the number of vending machines and employees that will yield maximum profit if an employee can only service 100 vending machines per week. Also determine the maximum profit.

12. The number of person-hours required per day in a department is given by

$$T(x,y) = 5x^2 + 4y^2 - 2x - y + 2xy$$

where x is the number of tons of raw material A used and y is the number of tons of raw material B used. Determine the amounts of raw materials A and B that minimizes time if $x+3y=50$. Also determine the minimum amount of time required per day.

13. The weekly demand for widgets in units, as a function of the material input x in dollars and the selling price y in dollars, is given by

$$f(x,y) = -y^2 + 25y - 2x^2 + 5x + xy$$

Determine the dollar value of material input and the selling price that maximizes demand if the price is three times the dollar value of material. Also determine the maximum demand for widgets.

9.6 DOUBLE INTEGRALS

In Chapter 7 we considered the problem of finding the area of a region bounded by one or more single-variable functions and some additional constraints. For example, we saw that the definite integral of a function $y=f(x)>0$ from $x=a$ to $x=b$ could be used to determine the area bounded by the curve for $y=f(x)$, the x axis, and the lines for $x=a$ and $x=b$. It would appear that finding the *volume* under a surface would be a natural generalization of finding the area under a curve. This is precisely the case. To find the volume under a surface, we must integrate twice; thus, the process is called *double integration*.

A rigorous definition of the double integral is beyond the scope of this text; however, we shall consider a geometrical argument to explain why we must integrate twice.

Suppose we have an area A in the xy plane that is bounded by the functions $y=f_1(x)$ and $y=f_2(x)$ and the lines $x=a$ and $x=b$. Furthermore, suppose the surface for the two-variable function $z=f(x, y)$ lies above the area A as shown in Fig. 9.11, and we wish to determine the volume between $z=f(x\,y)$ and the xy plane, which is bounded by $f_1(x), f_2(x)$, $x=a$, and $x=b$.

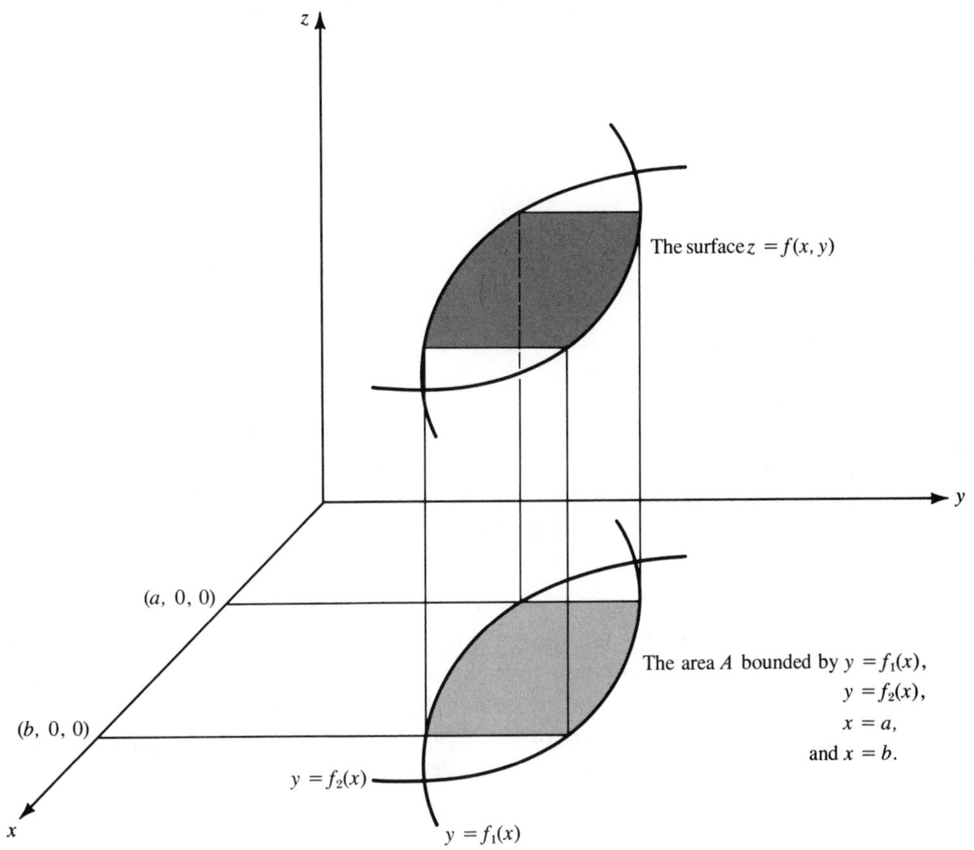

FIGURE 9.11

To determine the volume under $z=f(x\,y)$, we use a plane $x=c$ to section the surface and obtain a plane as shown in Fig. 9.12. This is analogous to slicing a piece of bread. The area in the plane that is bounded by $z=f(c, y)$ (the two-dimensional function that describes the intersection of the surface $z=f(x, y)$ and the plane $x=c$), the xy plane, and the two values $f_1(c)$ and $f_2(c)$ is determined by the definite integral.

$$\int_{f_1(c)}^{f_2(c)} f(c, y)\, dy$$

We can do this for every possible value of c (e.g., every possible thinly sliced piece of bread) obtaining an area for each value. However, if $x=a$ and $x=b$ are the values that bound the possible values of c (i.e., the end slices of

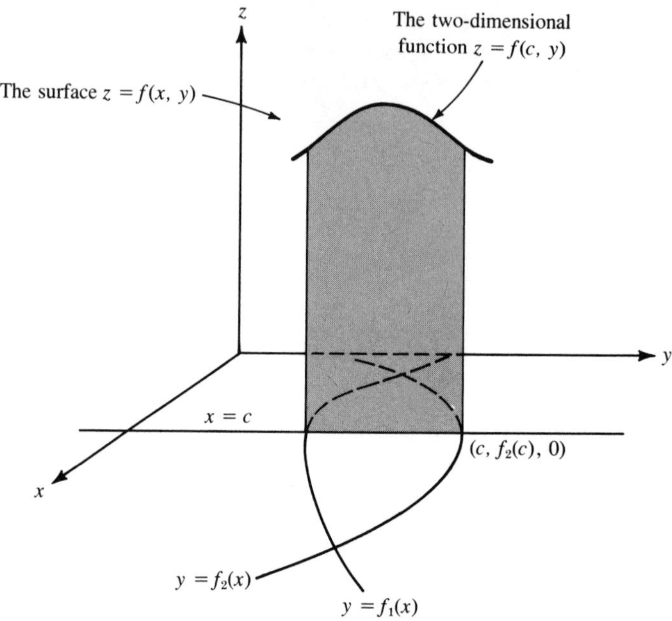

FIGURE 9.12

the loaf of bread), we can simply sum the individual areas to obtain the volume of the solid under $z=f(x, y)$ (i.e., the volume of the loaf of bread). As we saw when discussing definite integrals for single-variable functions, the summation process is done through integration. Thus, the cross-sectional area we obtain for each plane $x=c$ passing through the solid is given by $A(x)$ where x can range over any value from a to b.

$$A(x) = \int_{f_1(x)}^{f_2(x)} f(x, y)\, dy$$

Furthermore, "summing" all of these cross-sectional areas between $x=a$ and $x=b$ we obtain the definite integral

$$\int_a^b A(x)\, dx$$

Substituting for $A(x)$ we get the following *double integral* of $z=f(x, y)$ over the area A bounded by $f_1(x)$ and $f_2(x)$:

$$\int_a^b A(x)\, dx = \int_a^b \left[\int_{f_1(x)}^{f_2(x)} f(x, y)\, dy \right] dx$$

Hence the volume V contained in the solid bounded by $z=f(x\,y), f_1(x), f_2(x)$, $x=a$, $x=b$, and the xy plane is given by the following equation.

$$V=\int_a^b\left[\int_{f_1(x)}^{f_2(x)} f(x,y)\,dy\right]dx \tag{9.3}$$

There are a few points that we should note about the double integral in Eq. (9.3). The inner integral is defined for x constant and should be treated accordingly. The brackets around the inner integral can be removed if the integration is done first with respect to y with x considered constant. This antiderivative is evaluated in the normal manner by replacing y in the antiderivative by the upper limit $f_2(x)$ and the lower limit $f_1(x)$. The result is a single-variable function of x that is then integrated with respect to x in the usual manner. In general, in a double integral, the antiderivative of the inner integral is found with respect to the variable in the inner differential with the other variable considered constant. Let us consider some examples.

Example 9.22 Evaluate the double integral

$$\int_2^3\int_1^x (3x-2y)\,dy\,dx.$$

Solution

$$\begin{aligned}\int_2^3\int_1^x (3x-2y)\,dy\,dx &= \int_2^3 (3xy-y^2)\Big|_1^x dx\\ &= \int_2^3 [(3x^2-x^2)-(3x-1)]dx\\ &= \int_2^3 (2x^2-3x+1)dx\\ &= \left(\frac{2}{3}x^3-\frac{3}{2}x^2+x\right)\Big|_2^3\\ &= 6\tfrac{1}{6}\end{aligned}$$

Example 9.23 Evaluate

$$\int_0^2\int_0^4 (x+2y)\,dy\,dx$$

Solution

$$\begin{aligned}\int_0^2\int_0^4 (x+2y)\,dy\,dx &= \int_0^2 (xy+y^2)\Big|_0^4 dx\\ &= \int_0^2 (4x+16)dx\\ &= (2x^2+16x)\Big|_0^2\\ &= 40\end{aligned}$$

Example 9.24 Evaluate

$$\int_0^3 \int_y^{2y} xy^2\,dx\,dy$$

Solution

$$\int_0^3 \int_y^{2y} xy^2\,dx\,dy = \int_0^3 \frac{1}{2}x^2y^2\Big|_y^{2y} dy$$
$$= \int_0^3 \left(2y^4 - \frac{1}{2}y^4\right) dy$$
$$= \int_0^3 \frac{3}{2}y^4\,dy$$
$$= \frac{3}{10}y^5\Big|_0^3$$
$$= 72\tfrac{9}{10}$$

Example 9.25 Evaluate

$$\int_0^2 \int_0^x e^{x^2}dy\,dx$$

Solution

$$\int_0^2 \int_0^x e^{x^2}dy\,dx = \int_0^2 (ye^{x^2})\Big|_0^x dx$$
$$= \int_0^2 xe^{x^2}dx$$
$$= \frac{1}{2}e^{x^2}\Big|_0^2$$
$$= \frac{1}{2}(e^4 - 1)$$
$$= 26.799$$

Example 9.26 Evaluate

$$\int_1^3 \int_x^{x+3} y\,dy\,dx$$

Solution

$$\int_1^3 \int_x^{x+3} y\,dy\,dx = \int_1^3 \frac{y^2}{2}\Big|_x^{x+3} dx$$
$$= \int_1^3 \left[\frac{(x+3)^2}{2} - \frac{x^2}{2}\right] dx$$
$$= \frac{1}{2}\int_1^3 (6x+9)\,dx$$
$$= \frac{1}{2}(3x^2+9x)\Big|_1^3$$
$$= 21$$

Example 9.27 The Volume of a Solid Determine the volume of the solid in the first octant under the plane $2x+3y+z=12$.

Solution First we determine the function $z=f(x\ y)$ by solving the equation for z in terms of x and y to get $z=12-2x-3y$. The volume we wish to determine, the tetrahedron shown in Fig. 9.13, is formed by the intersection of the plane, and the xy, yz and xz planes. The functions $f_1(x)$ and $f_2(x)$ are the functions that form the area on the xy plane. Thus, $f_1(x)=0$ and $f_2(x)=\frac{1}{3}(12-2x)$. The limits on x are 0 and 6, which are the a and b values, respectively. Thus, the volume is given by the double integral

$$\int_0^6 \int_0^{\frac{1}{3}(12-2x)} (12-2x-3y)\,dy\,dx = \int_0^6 \left(12y-2xy-\frac{3}{2}y^2\right)\Big|_0^{\frac{1}{3}(12-2x)} dx$$

$$= \int_0^6 \left[12\frac{(12-2x)}{3} - 2x\frac{(12-2x)}{3} - \frac{3}{2}\frac{(12-2x)^2}{9} \right] dx$$

$$= \int_0^6 \frac{2}{3}(6-x)^2 dx$$

$$= -\frac{2}{9}(6-x)^3 \Big|_0^6$$

$$= 48$$

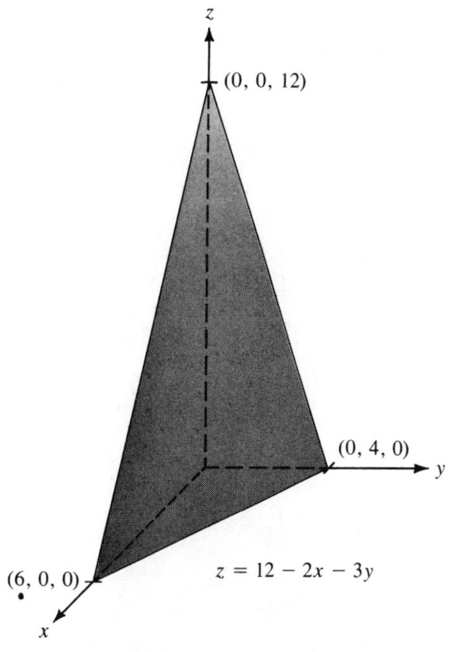

FIGURE 9.13

Example 9.28 Determine the volume of the solid in the first octant under the plane $z=16-4y-12x$ and bounded by the functions $y=x^2$ and $x=0$.

Solution First we sketch the graph in Fig. 9.14 and note that the range on y is from $y=x^2$ to $y=4-3x$. Also the range on x is from 0 to 1. The volume is then determined by the double integral

$$\begin{aligned}\int_0^1\int_{x^2}^{4-3x}(16-4y-12x)\,dy\,dx &= \int_0^1(16y-2y^2-12xy)\Big|_{x^2}^{4-3x}dx\\ &= \int_0^1\big[16(4-3x)-2(4-3x)^2-12x(4-3x)\\ &\quad -16x^2+2x^4+12x^3\big]dx\\ &= \int_0^1(32-48x+2x^2+12x^3+2x^4)\,dx\\ &= \left(32x-24x^2+\frac{2x^3}{3}+3x^4+\frac{2x^5}{5}\right)\Big|_0^1\\ &= 12\tfrac{1}{15}\end{aligned}$$

In our previous discussions concerning integrals and probability, we noted that the probability of an event could be determined by evaluating a definite integral for a single-variable function. Double integrals (or multiple integrals for functions involving two or more variables) can be used to determine probabilities for joint events involving two variables. The applications of double integrals are found in many areas of business, economics, science, and engineering.

EXERCISES

Evaluate the following integrals.

1. $\int_0^3\int_1^2(3x^2y+x+2y)\,dy\,dx$
2. $\int_1^3\int_0^2(xy+2y)\,dy\,dx$
3. $\int_1^4\int_2^3(\sqrt{x}+5xy^2+xy)\,dy\,dx$
4. $\int_1^2\int_0^1(xy+x^2y+y^2+x^2)+2)\,dy\,dx$
5. $\int_1^5\int_0^3(x+2y-x^2y)\,dy\,dx$
6. $\int_2^4\int_1^4(\sqrt{x}+x\sqrt{y}+\sqrt{y})\,dy\,dx$
7. $\int_0^2\int_0^{x+3}(3y^2+x^2+xy-2)\,dy\,dx$
8. $\int_0^1\int_{x^2}^{\sqrt{4}}2\,dy\,dx$
9. $\int_0^1\int_{y^2}^{\sqrt{y}}x\,dx\,dy$
10. $\int_0^1\int_y^{2y}x\,dx\,dy$
11. $\int_1^2\int_{x^2}^{x+2}dy\,dx$
12. $\int_1^2\int_x^{2x}xy\,dy\,dx$
13. $\int_0^2\int_y^2(x^2+y^2)\,dx\,dy$
14. $\int_0^1\int_0^x x^2y\,dy\,dx$

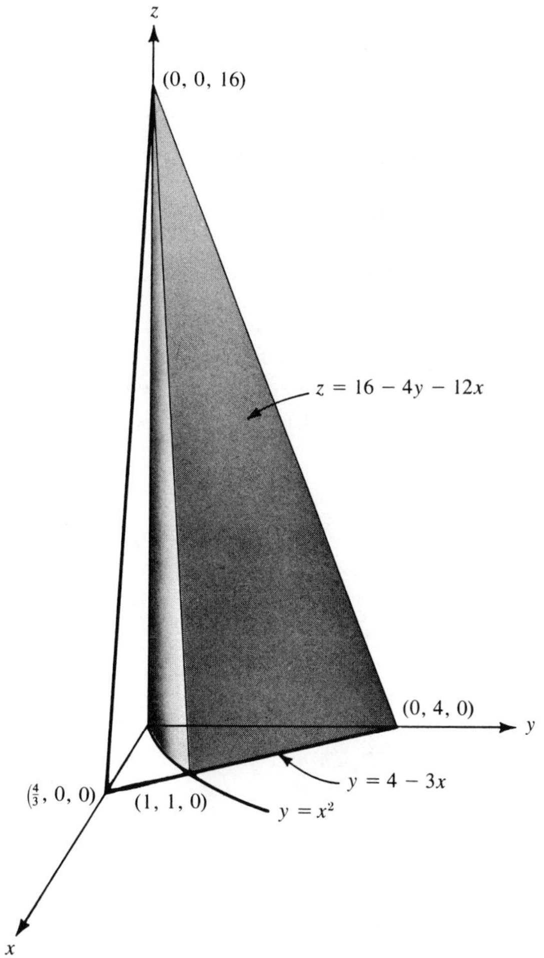

FIGURE 9.14

15. $\int_2^4 \int_1^{x^2} \frac{x}{y^2}\, dy\, dx$

16. $\int_0^3 \int_{9-x}^{\sqrt{9-x^2}} y\, dy\, dx$

Sketch the figures and determine the volume in the first octant for each of the following.

17. The solid bounded above by the surface defined by $z=10-2x-y$ and below by the rectangular region on the xy plane bounded by the lines for $x=1$, $x=3$, $y=2$, and $y=5$.

18. The solid bounded above by the surface defined by $z=12-x-y$ and below by the rectangular region on the xy plane bounded by the lines for $x=3$, $x=6$, $y=5$, and $y=9$.

19. The solid bounded above by the surface defined by $z=7-2x-y$ and below by the triangular region on the xy plane defined by the lines for $x=0$, and $y=x$.

20. The solid bounded above by the surface defined by $z=10-5x-y$ and below by the region on the xy plane defined by the lines for $x=1$, $x=0$, and $y=x^2$.

IMPORTANT TERMS AND CONCEPTS

Constrained optima
Cross partial derivatives
Derivative of
  multivariate functions
  partial derivatives
Double integral
Extreme value tests
  bivariate functions
  constrained bivariate functions
Higher-order partial derivatives
Lagrangian multiplier
Multiple integral
Multivariate function
Necessary conditions for extreme values
Optimum
Partial derivative
Plane
Saddle point
Volume

SUMMARY OF RULES AND FORMULAS

Rule 9.1. Extreme value test for bivariate functions

Given a point (a,b) that satisfies $f_x(a,b)=0$ and $f_y(a,b)=0$; and where $D(a,b)=f_{xx}(a,b)f_{yy}(a,b)-[f_{xy}(a,b)]^2$

1. If $D(a,b)>0$ and $f_{xx}(a,b)<0$, then $f(a,b)$ is a relative maximum.
2. If $D(a,b)>0$ and $f_{xx}(a,b)>0$, then $f(a,b)$ is a relative minimum.
3. If $D(a,b)<0$, then $f(a,b)$ is not an extremum.
4. If $D(a,b)=0$, the test fails.

Rule 9.2. Extreme value test for constrained bivariate functions

Given $F(x, y,\lambda)=f(x, y)+\lambda g(x, y)$, where $g(x, y)=0$, and $D(x, y,\lambda)=F_{xx}(x, y,\lambda)F_{yy}(x, y,\lambda)-[F_{xy}(x, y,\lambda)]^2$, then the critical point (a,b,α), which satisfies each of the following

$$f_x(a,b) + \alpha g_x(a,b) = 0$$

$$f_y(a,b) + \alpha g_y(a,b) = 0$$

and

$$g(a,b) = 0$$

is such that

1. If $D(a,b,\alpha)>0$ and $F_{xx}(a,b,\alpha)<0$, then $f(a,b)$ is a relative maximum.
2. If $D(a,b,\alpha)>0$ and $F_{xx}(a,b,\alpha)>0$, then $f(a,b)$ is a relative minimum.
3. If $D(a,b,\alpha)\leqslant 0$, the test fails.

Eq. (9.3). *The volume V*, contained in the solid bounded by the surfaces $z=f(x, y)$, $y=f_2(x)$, $y=f_1(x)$, $x=a$ and $x=b$ and the xy plane is given by

$$V = \int_a^b \left[\int_{f_1(x)}^{f_2(x)} f(x, y)\,dy \right] dx$$

ANSWERS TO SELECTED EXERCISES

CHAPTER 1

Section 1.2

1. (a) $\{(-2,3), (-2,4), (-2,5), (-1,3), (-1,4), (-1,5), (0,3), (0,4), (0,5)\}$
(b) $\{(3,-2), (3,-1), (3,0), (4,-2), (4,-1), (4,0), (5,-2), (5,-1), (5,0)\}$
(c) $\{(-2,2), (-2,4), (-1,2), (-1,4), (0,2), (0,4)\}$
(d) $\{(2,-2), (2,-1), (2,0), (4,-2), (4,-1), (4,0)\}$
(e) $\{(3,2), (3,4), (4,2), (4,4), (5,2), (5,4)\}$
(f) $\{(2,3), (2,4), (2,5), (4,3), (4,4), (4,5)\}$

3. $g(-10)=93$; $g(-3)=9$; $g(0)=3$; $g(3)=15$; $g(4)=23$

5. $R=\{8,10,12\}$; $\{(1,8), (2,10), (3,12)\}$; a function

7. $R=\{4,5,6\}$; $\{(16,4), (25,5), (36,6)\}$; a function

9. $R=\{1,\frac{1}{2},\frac{1}{3},\frac{1}{4}\}$; $\{(1,1), (2,\frac{1}{2}), (3,\frac{1}{3}), (4,\frac{1}{4})\}$; a function

11. $R=\{-2,1,2\}$; $\{(-2,-2), (-1,1), (0,2), (1,1), (2,-2)\}$; a function

13. $R=\{11,9,5,15\}$; $\{(-2,11), (-1,9), (0,5), (1,5), (2,15)\}$; a function

15. $R=\{\sqrt{5}, -\sqrt{5}, 3, -3, 5, -5\}$; $\{(0,\sqrt{5}), (0,-\sqrt{5}), (1,3), (1,-3), (5,5), (5,-5)\}$; *not* a function

17. (a) R_1, R_2
(b) R_1 is one-to-one; R_2 is many-to-one; R_3 is one-to-many; R_4 is many-to-many

19. $\mathbb{R}$. 21. $\mathbb{R}$, $x \neq -2$ 23. $\mathbb{R}$ 25. (a) $A(x)=x^2$ (b) relation and a function

27. (a) $S(e)=6e^2$ (b) relation and a function

29. (a) $A=3{,}600\left(1+\dfrac{0.10n}{360}\right)$ (b) Yes (c) \$3,780

31. (a) $S(x)=4x$ (b) $p(x)=3.2x$ (c) $p(40)=128$

33. (a) relation and a function (b) \$1,105 (c) \$12.01

35. (a) $T(x)=0.04+0.02x^2-0.01x$ (b) 0.19 sec; 0.70 sec

37. $I(x)=50x^2+200-100x$, $I(x)$ in dollars

Section 1.3

1. (a) $(3,3)$ (b) $(5,2)$ (c) $(-6,5)$ (d) $(-4,0)$ (e) $(-4,-6)$ (f) $(-6,-3)$
(g) $(4,-2)$ (h) $(2,-4)$

3.

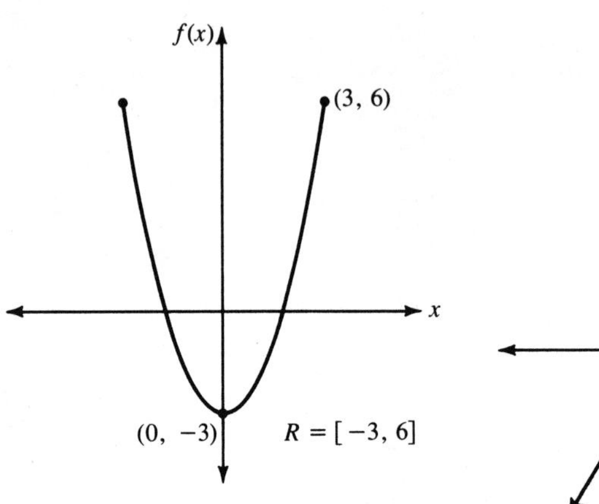

5.

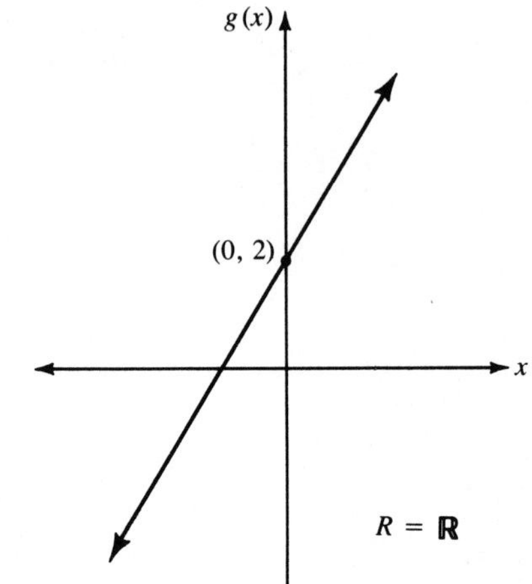

7. (a), (c), and (d)
9. (a)

x	-5	-4	-3	-2	-1	0	1	2	3	4	5
$f(x)$	-16	-14	-12	-10	-8	-6	-4	-2	0	2	4

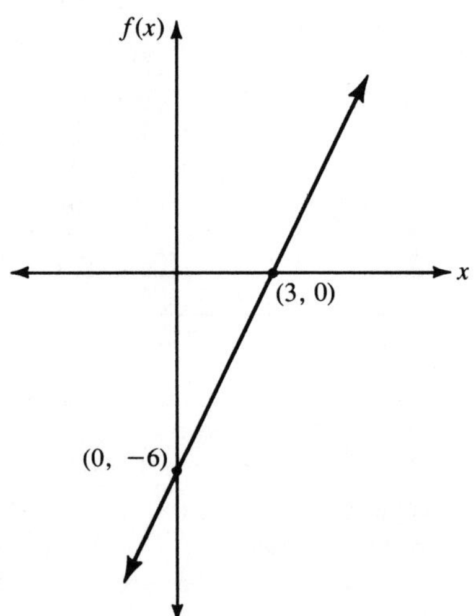

(b)

x	-5	-4	-3	-2	-1	0	1	2	3	4	5
$f(x)$	43	26	13	4	-1	-2	1	8	19	34	53

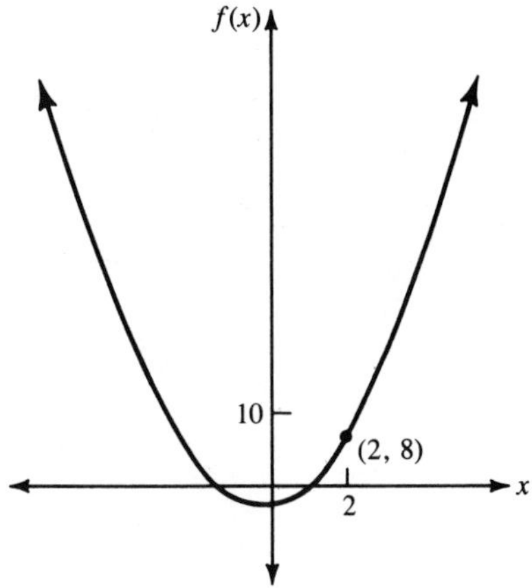

(c)

x	-5	-4	-3	-2	-1	0	1	2	3	4	5
$f(x)$	-125	-64	-27	-8	-1	0	1	4	27	64	125

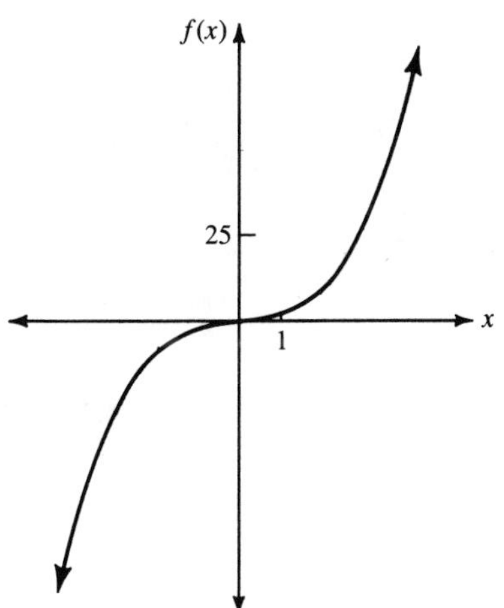

(d)

x	-5	-4	-3	-2	-1	0	1	2	3	4	5
$f(x)$	-19	-10	-3	2	5	6	5	2	-3	-10	-19

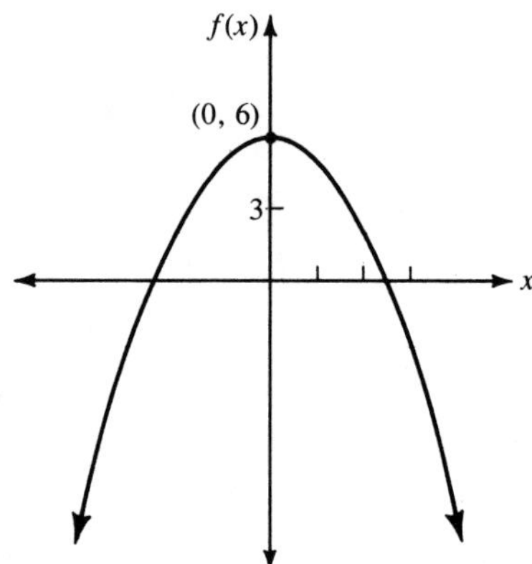

(e)

x	-5	-4	-3	-2	-1	0	1	2	3	4	5
$f(x)$	-2	$-\frac{5}{2}$	$-\frac{10}{3}$	-5	-10	—	10	5	$\frac{10}{3}$	$\frac{5}{2}$	2

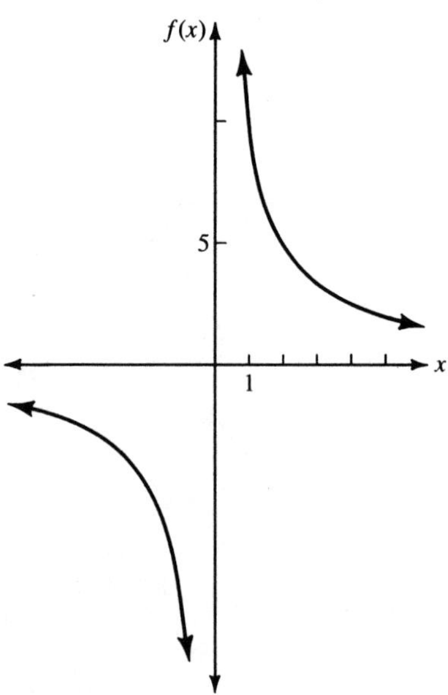

(f)

x	−5	−4	−3	−2	−1	0	1	2	3	4	5
$f(x)$	9	8	7	6	5	4	3	2	1	0	−1

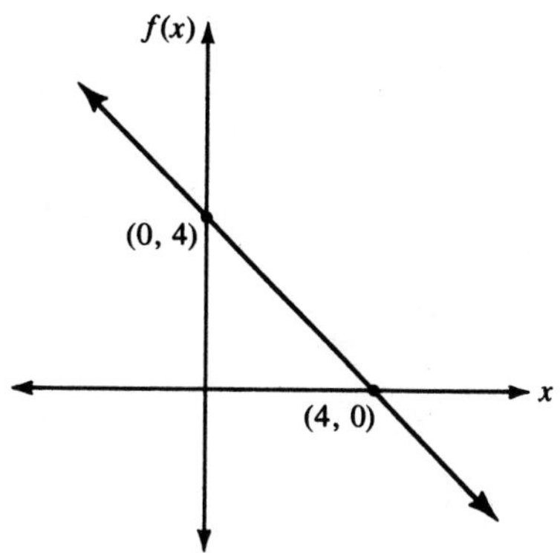

(g)

x	−5	−4	−3	−2	−1	0	1	2	3	4	5
$f(x)$	0	1	$\sqrt{2}$	$\sqrt{3}$	2	$\sqrt{5}$	$\sqrt{6}$	$\sqrt{7}$	$\sqrt{8}$	3	$\sqrt{10}$

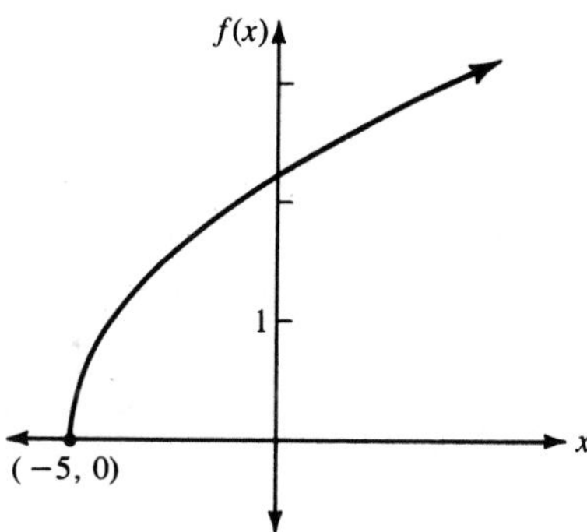

(h)

x	-5	-4	-3	-2	-1	0	1	2	3	4	5
$f(x)$	-6	-5	-4	-3	-2	-1	—	1	2	3	4

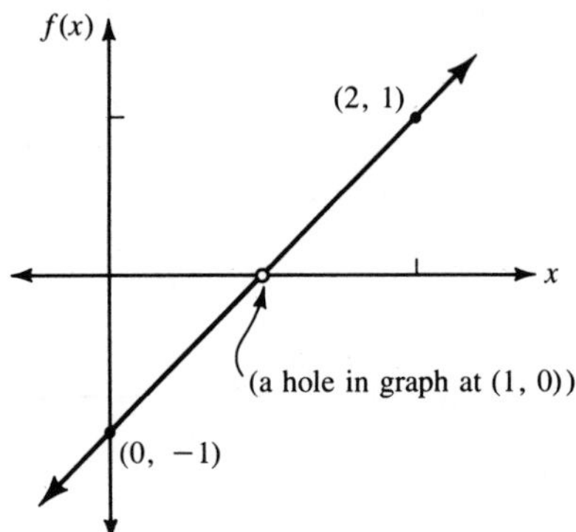

11. $A = 1800\left(1 + \frac{0.12n}{360}\right)$
$= 1800\left(1 + \frac{n}{3000}\right)$

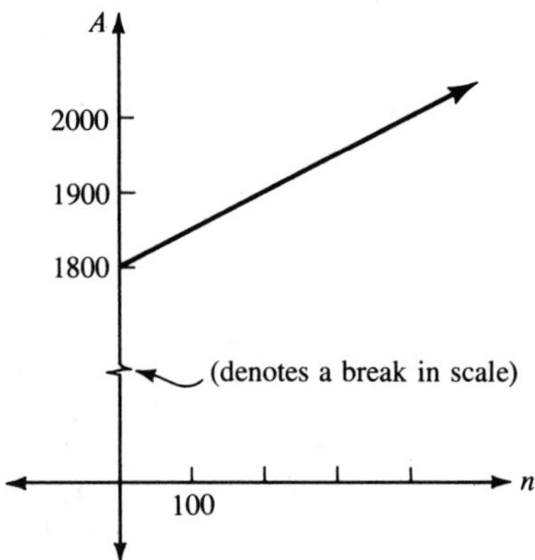

13. 74.074 cu in. (when $x = 1.67$)

15. $S(h)=(10-2h)^2+4h(10-2h)=100-4h^2$; (a) 84 sq in.
(b) (c) no

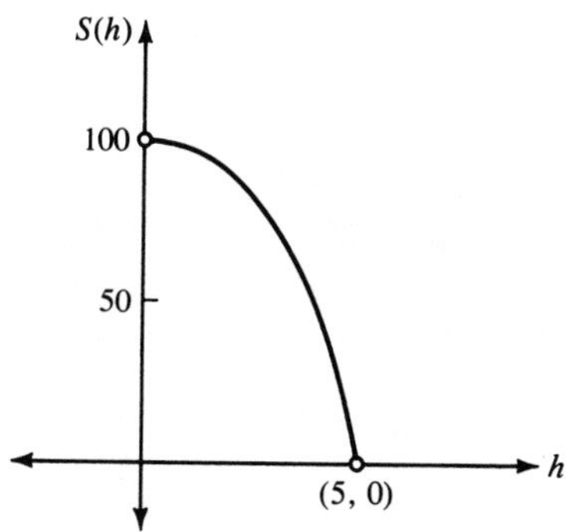

Section 1.4

1. 66 3. 115 5. -1 7. 4 9. 63 11. 0
13. $(f+g)(x)=x^2+1+\sqrt{x-1}$; $D=[1,\infty)$ 15. $(f\cdot g)(x)=(x^2+1)\sqrt{x-1}$; $D=[1,\infty)$
17. $\left(\frac{f}{g}\right)(x)=\frac{x^2+1}{\sqrt{x-1}}$; $D=(1,\infty)$ 19. $g[f(x)]=\sqrt{x^2}=|x|$; $D=\mathbb{R}$
21. $g[u(x)]=\sqrt{\frac{1}{x}-1}$; $D=(0,1]$ 23. $f(g[u(x)])=\frac{1}{x}$; $D=(0,1]$
25. $f(x)=\sqrt{x}$, $u(x)=x^2+5$ 27. $f(x)=\frac{1}{\sqrt{x}}$, $u(x)=x+3$
29. $f(x)=x^3$, $u(x)=x^2+2x+5$
31. (a) $P(x)=150x-100$ (b) $P(10)=\$1{,}400$
33. (a) $f[x(t)]=1.4t+15$ (b) $f[x(5)]=22$

Section 1.5

1. $y=\frac{x+5}{2}$

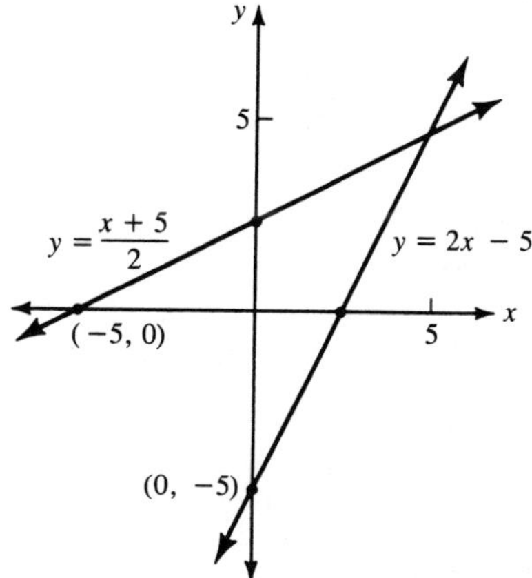

3. $f^{-1}(x) = 2x - 12$

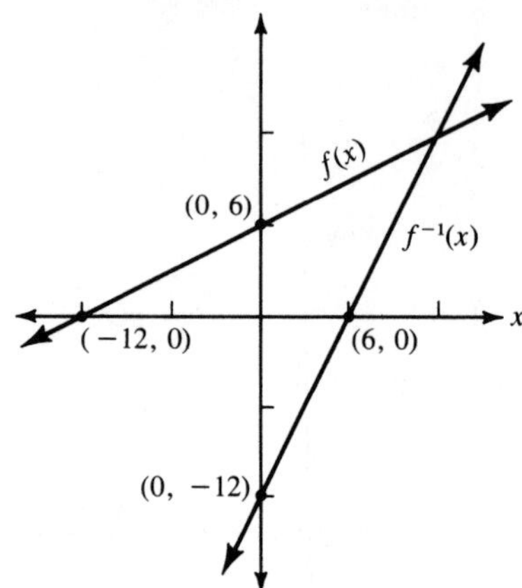

5. $g^{-1}(x) = x + 4$

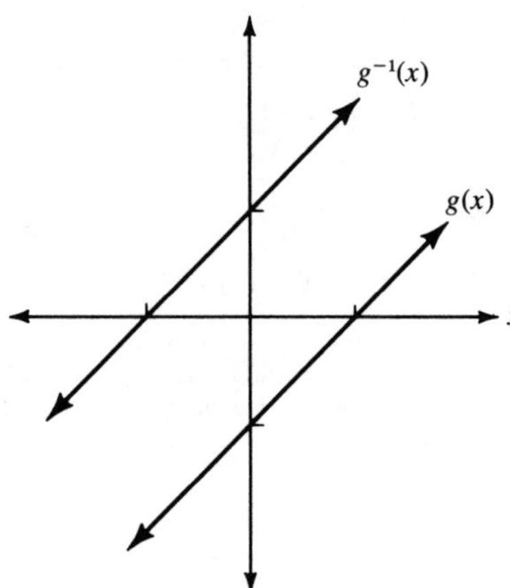

7. $f^{-1}(x) = \sqrt[3]{x}$

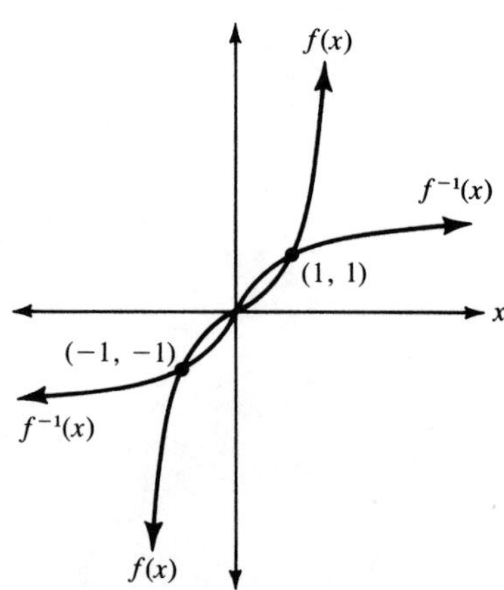

9. $g^{-1}(x) = \frac{1}{x}$

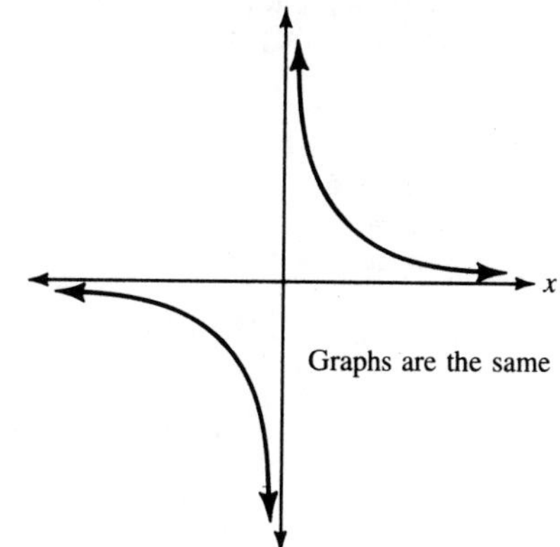

11.

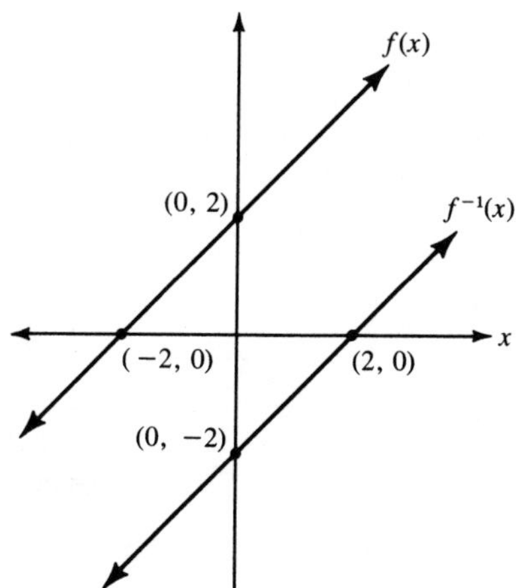

13.

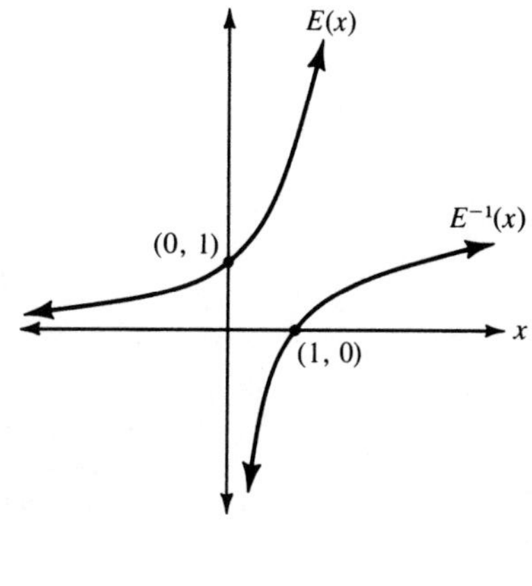

15.

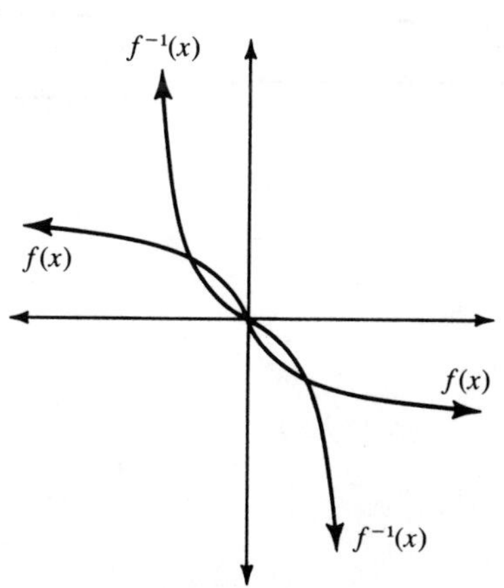

17. (a), (c), and (d)

19. (a) $g^{-1}(x) = -\frac{1}{2}x + \frac{3}{2}$ (b) $g[g^{-1}(6)] = 6$ (c) $g^{-1}[g(6)] = 6$
(d) $g[g^{-1}(x)] = g^{-1}[g(x)] = x$ for any function $g(x)$ and its inverse function $g^{-1}(x)$.

21. $y = \frac{1}{2}(x - 15)$, where x = IQ score, y = ATMM score; when $x = 119$, $y = 52$; when converting IQ scores to ATMM scores

CHAPTER 2

Section 2.2

1. $y = 6x + 16$
slope = 6
y-intercept = 16

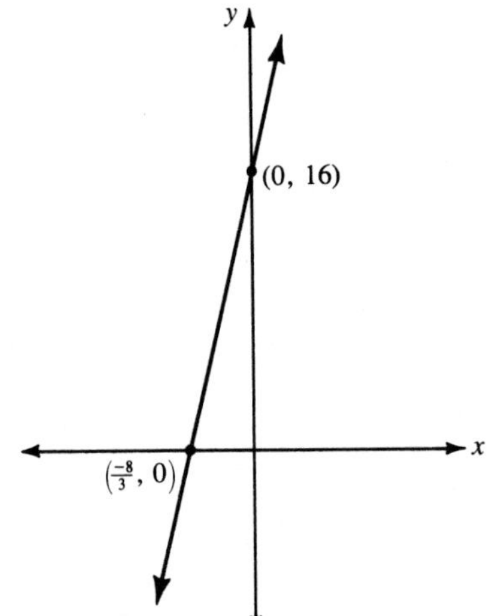

3. $y = 3x + 5$
slope = 3
y-intercept = 5

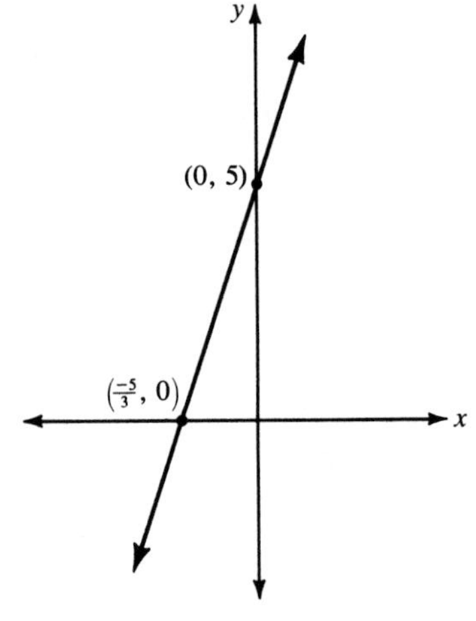

5. $y = \frac{1}{10}x + 5$
slope = $\frac{1}{10}$
y-intercept = 5

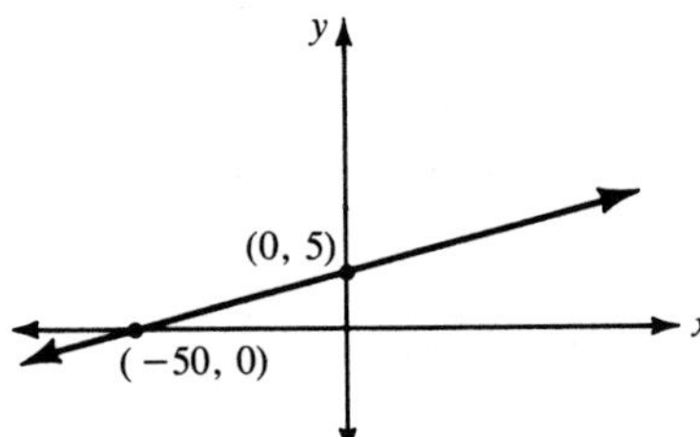

7. $y = 5$
slope = 0
y-intercept = 5

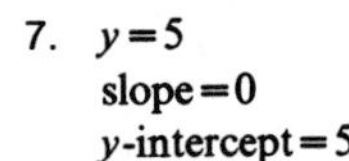
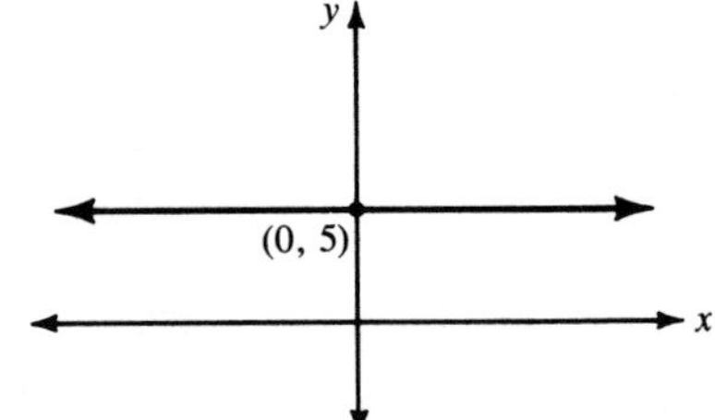

9. $y=\frac{3}{4}x-4$
slope $=\frac{3}{4}$
y-intercept $=-4$

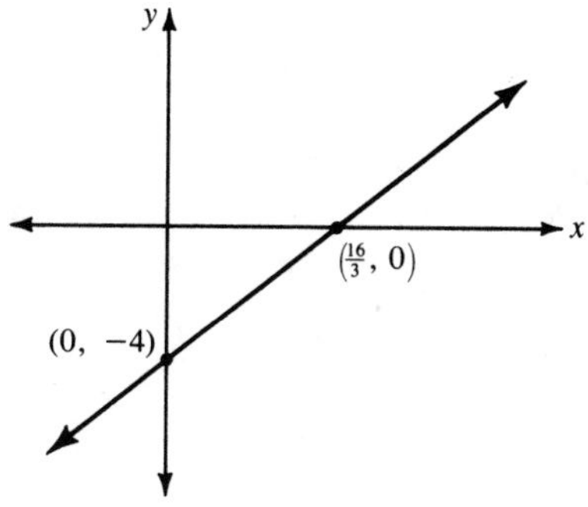

11. x-intercept $=\frac{7}{2}$
y-intercept $=7$

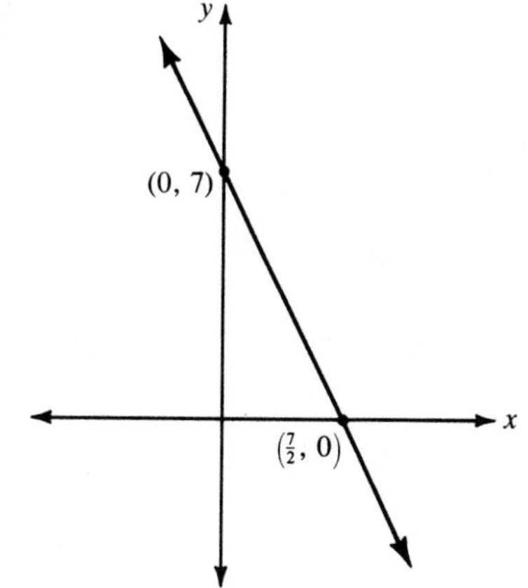

13. x-intercept $=-15$
y-intercept $=12$

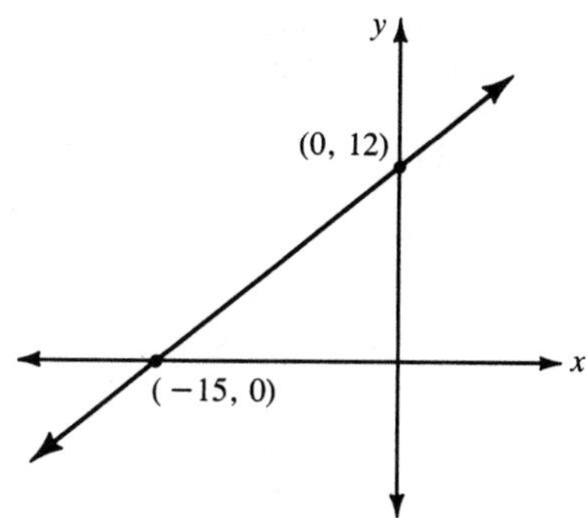

15. x-intercept $=\frac{9}{2}$
y-intercept $=9$

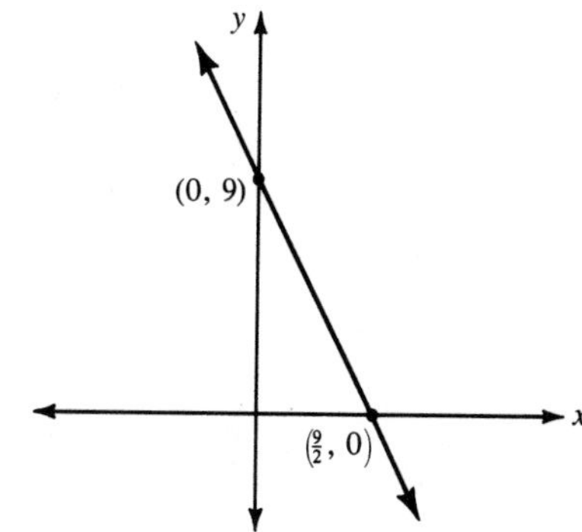

17. x-intercept $=-10$
y-intercept $=5$

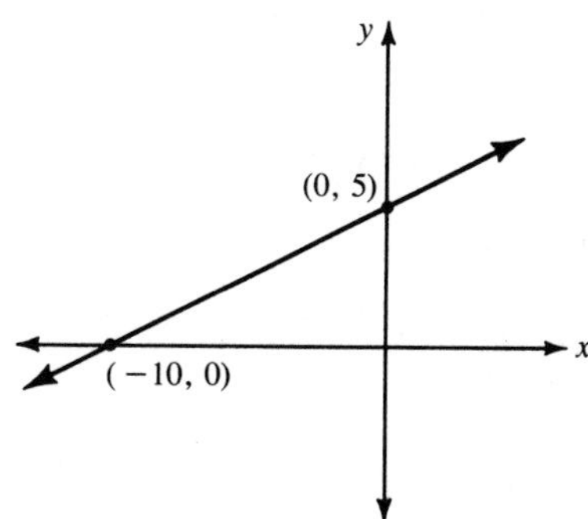

19. x-intercept $= -6$
y-intercept $= -12$

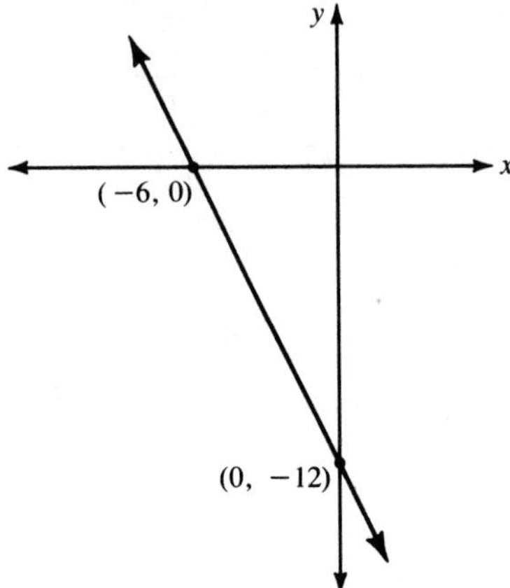

21. $m=\frac{4}{3}$; $y=\frac{4}{3}x$ 23. $m=-2$; $y=-2x$ 25. $m=\frac{1}{8}$; $y=\frac{1}{8}x+5$
27. $y=4x-11$ 29. $y=3x+6$ 31. $y=-4x$
33. $y=-\frac{3}{2}x+9$ 35. (a) $x=2$; $y=5$ (b) 0; there is none (c) no (d) yes
37. $y=-3x-2$; yes; yes 39. $x=-\frac{4}{3}$ 41. $x=\frac{7}{2}$ 43. $x=-\frac{14}{5}$
45. $y=\frac{1}{2}x-200$, where $x=$ the number of miles traveled
47. $P(x)=4x-1200$; loss $=\$400$; profit $=\$1200$
49. (a) $F(x)=240-20x$ (b) $F(2)=\$200$; $F(5)=\$140$

(c)

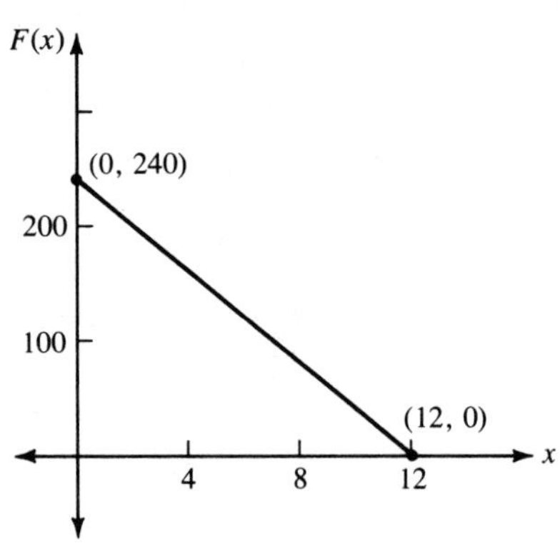

51. (a) 28 units (b) 70 units (c) $x=\$60$

53. (a) (b) \$60 (c) 40 thousand

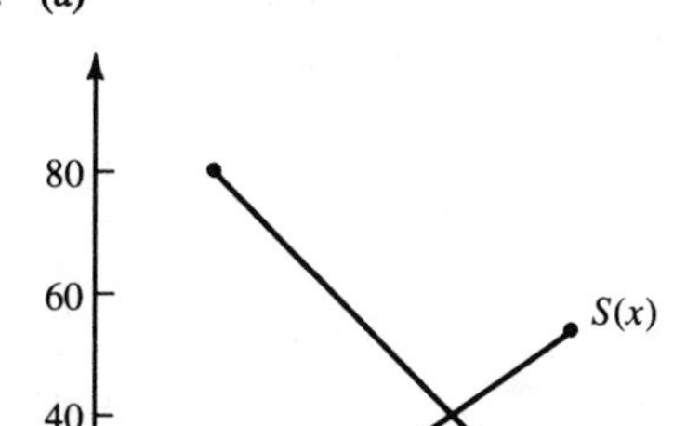

55. $D^{-1}(x)=\dfrac{400-5x}{3}$ (a) $x=$ the number of units, $D^{-1}(x)=$ the unit price

(b)

(c) $D=[20,74]$
$R=[10,100]$

Section 2.3

1. $y=9x^2-6x+3$

3. $y=8x^2-3x-7$

5. (a) Upward (b) $(-3,-25)$ (c) $x=2$ and $x=-8$ (d) $[-25,\infty)$ (e) $x=-3$

7. (a) Upward (b) $(-1,-16)$ (c) $t=3$ and $t=-5$ (d) $[-16,\infty)$ (e) $t=-1$

9. (a) Upward (b) $(\frac{1}{12},-\frac{49}{12})$ (c) $u=\frac{2}{3}$ and $u=-\frac{1}{2}$ (d) $[-\frac{49}{12},\infty)$ (e) $u=\frac{1}{12}$

11. (a) Downward (b) $(0,-32)$ (c) No zeros (d) $(-\infty,-32]$ (e) $x=0$

13. (a) Upward (b) Intersects
(c) (3,0) (d) $x=3$ (e) and (f)

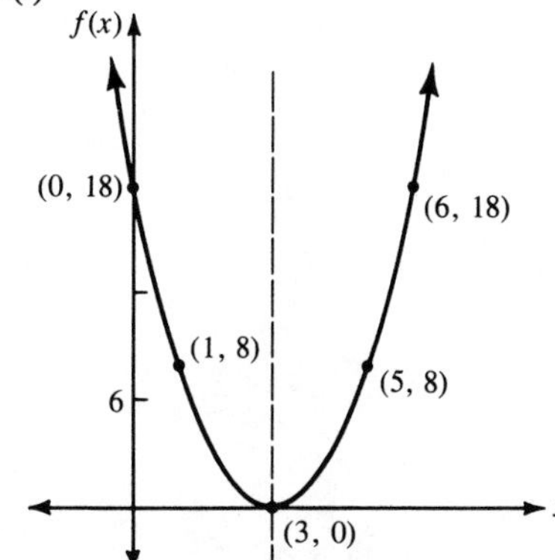

15. (a) Upward (b) Does not intersect (c) (3,8)
(d) No zeros (e) and (f)

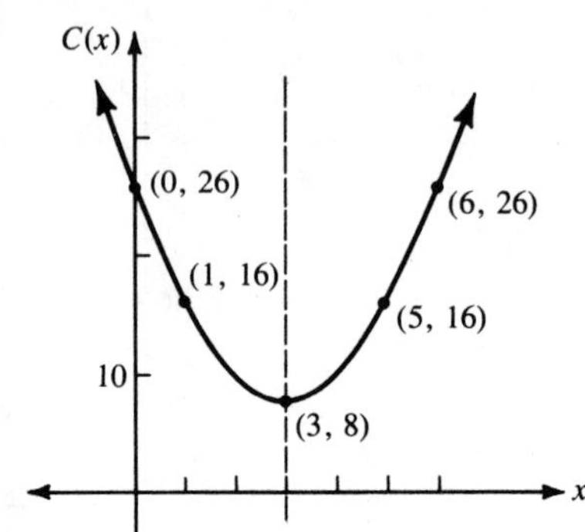

17. (a) Downward (b) Intersects (c) (6,3)
(d) $x=7$ and $x=5$ (e) and (f)

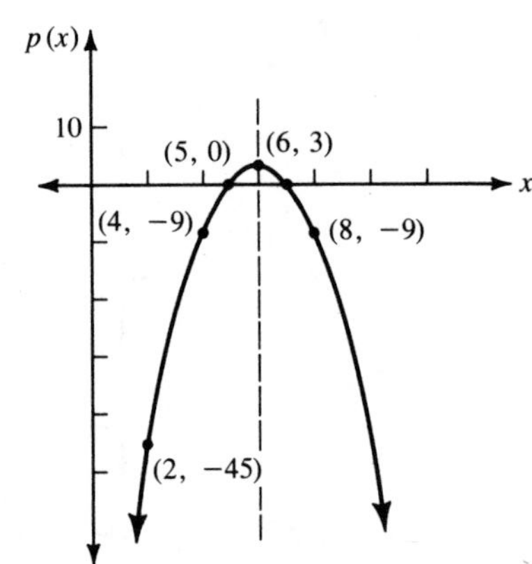

19. In 2 weeks; $1,800

21.
$$C = B$$
$$Q^2 - 600Q + 50{,}000 = 0$$
$$Q = 100 \quad \text{or} \quad Q = 500$$
so
$$Q = 500 \quad B = \$475{,}000$$

23. (a) $I(x) = -50x^2 + 600x$

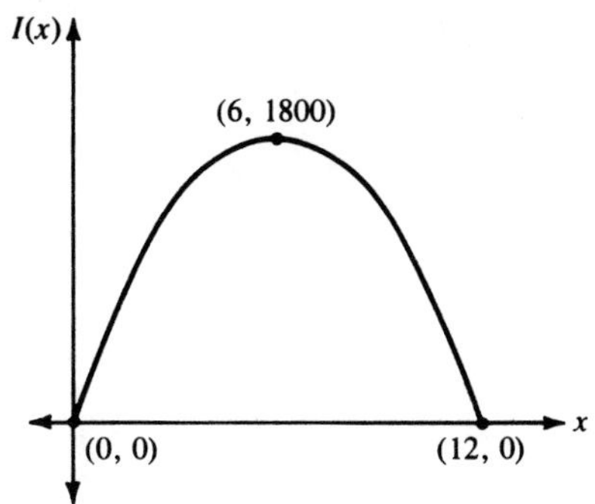

(b) $x = \$6$, and $I(x) = \$1{,}800$ (c) 300 cakes

(d) $P(x) = -50x^2 + 750x - 1{,}800$

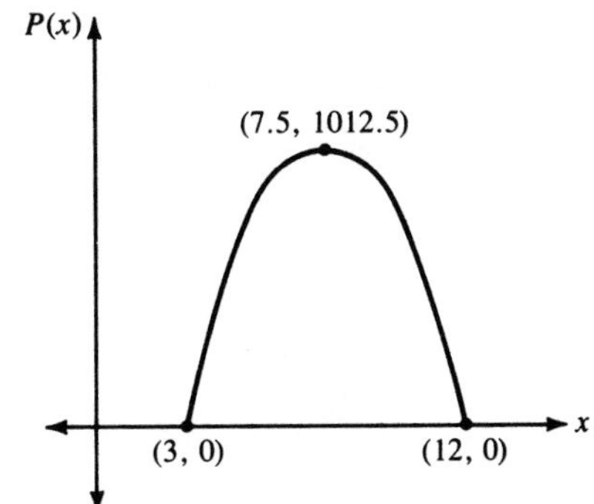

(e) $P = \$1{,}012.50$, and $x = \$7.50$ (f) 225 cakes (g) \$112.50

25. (a) 2 (b) 40 (c) 4

Section 2.4

1. $\{\pm 8, \pm 4, \pm 2, \pm 1\}$
3. $\{\pm \frac{3}{2}, \pm 3, \pm 6, \pm \frac{3}{4}, \pm \frac{1}{2}, \pm 1, \pm 2, \pm \frac{1}{4}\}$
5. $\{\pm 8, \pm 4, \pm 2, \pm 1, \pm \frac{1}{2}\}$
7. (a) $+\infty$ (b) $+\infty$ (c) $x = 5$ and $x = -7$

(d) 

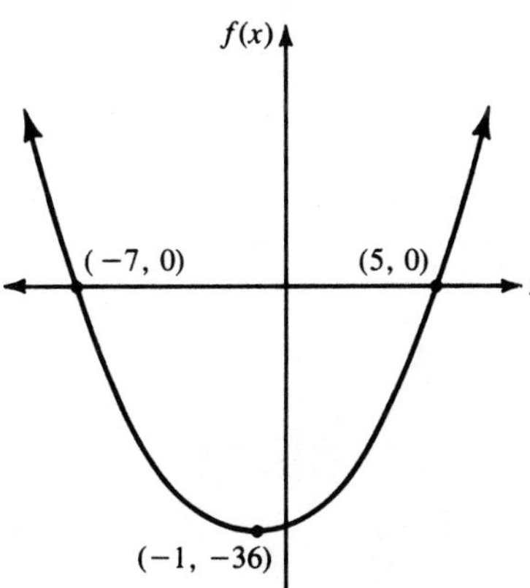

9. (a) $+\infty$ (b) $-\infty$ (c) $x=4$, $x=-4$, and $x=1$
(d)

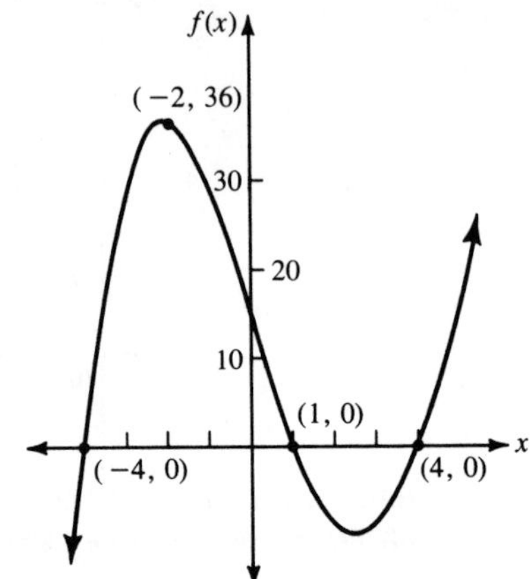

11. (a) $+\infty$ (b) $-\infty$ (c) $x=-\frac{1}{2}$, $x=-5$, and $x=3$
(d)

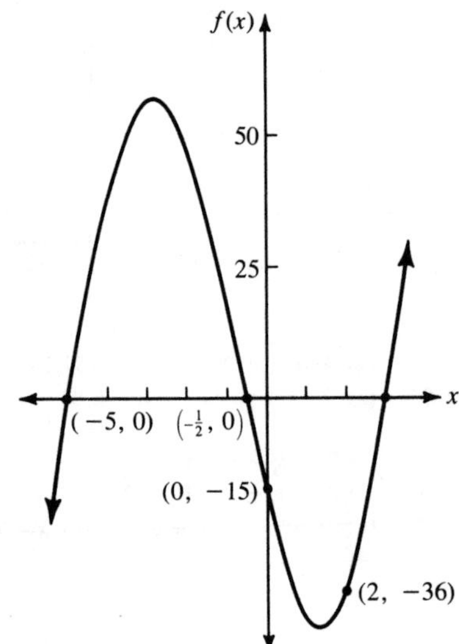

13. (a) $+\infty$ (b) $+\infty$ (c) No zeros
(d)

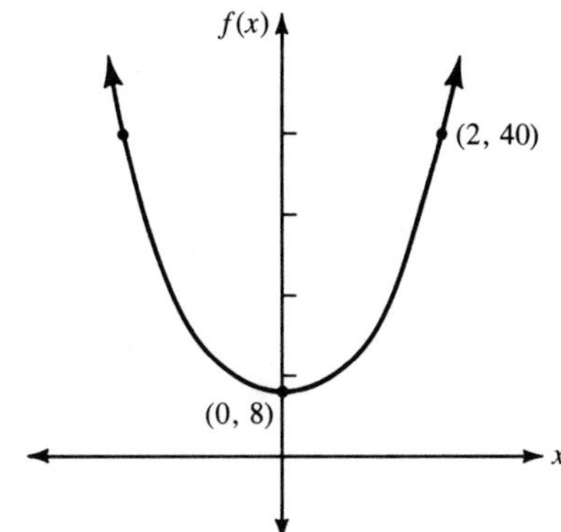

15. (a) $+\infty$ (b) $+\infty$ (c) $x=3$, $x=-3$, $x=1$, and $x=-1$
(d)

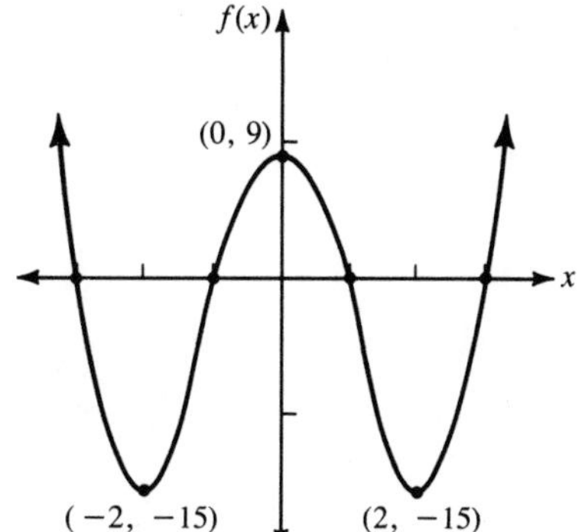

17. $[-\frac{1}{2}, \frac{4}{3}]$
19. $(-\infty, -2) \cup (1, \infty)$
21. $[-2, 3]$
23. $(-\infty, -5) \cup (0, 5)$
25. $[-4, 1] \cup [4, \infty)$

Section 2.5

1.	(a) 1	(b) 1	(c) $y=1$	(d) $x=1$
3.	(a) 0	(b) 0	(c) $y=0$	(d) $x=\frac{5}{2}$
5.	(a) $+\infty$	(b) $-\infty$	(c) none	(d) $x=0$
7.	(a) $\frac{3}{5}$	(b) $\frac{3}{5}$	(c) $y=\frac{3}{5}$	(d) $x=-\frac{4}{5}$
9.	(a) -1	(b) -1	(c) $y=-1$	(d) $x=-2$

11. (f) $D=(-\infty,1)\cup(1,\infty)$ or $\mathbb{R}, x\neq 1$
$R=(-\infty,1)\cup(1,\infty)$ or $\mathbb{R}, y\neq 1$

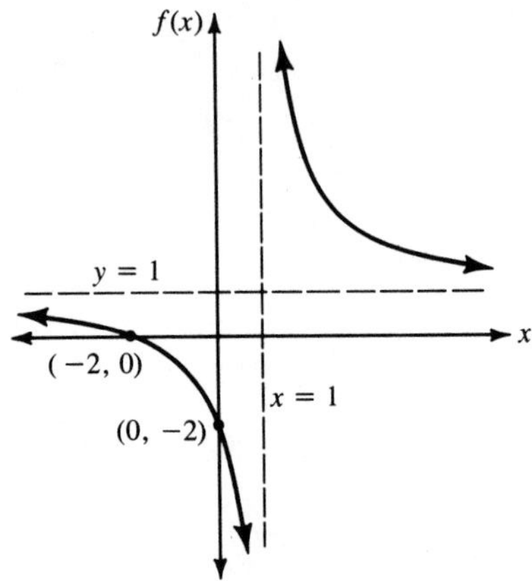

13. (f) $D=(-\infty,\frac{5}{2})\cup(\frac{5}{2},\infty)$ or $\mathbb{R}, x\neq \frac{5}{2}$
$R=(-\infty,0)\cup(0,\infty)$ or $\mathbb{R}, y\neq 0$

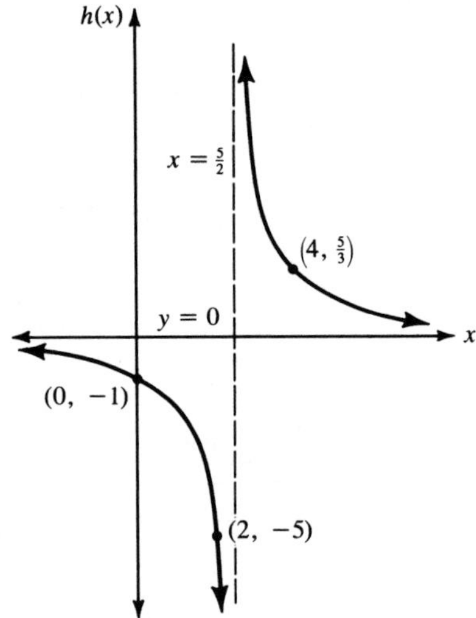

15.

(f) $D=(-\infty,4)\cup(4,\infty)$ or $\mathbb{R}, x\neq 4$
$R=(-\infty,-3)\cup(-3,\infty)$ or $\mathbb{R}, y\neq -3$

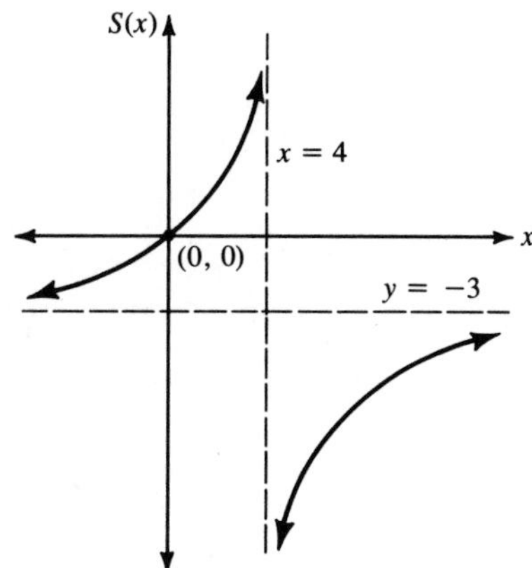

17.

(f) $D=(-\infty,3)\cup(3,\infty)$ or $\mathbb{R}, x\neq 3$
$R=(-\infty,-4)\cup(-4,\infty)$ or $\mathbb{R}, y\neq -4$

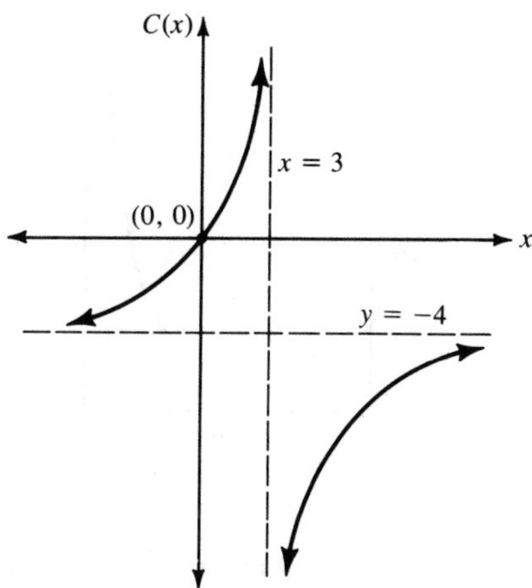

19. (f) $D=(-\infty,0)\cup(0,\infty)$ or $\mathbb{R}, x\neq 0$
$R=\mathbb{R}$

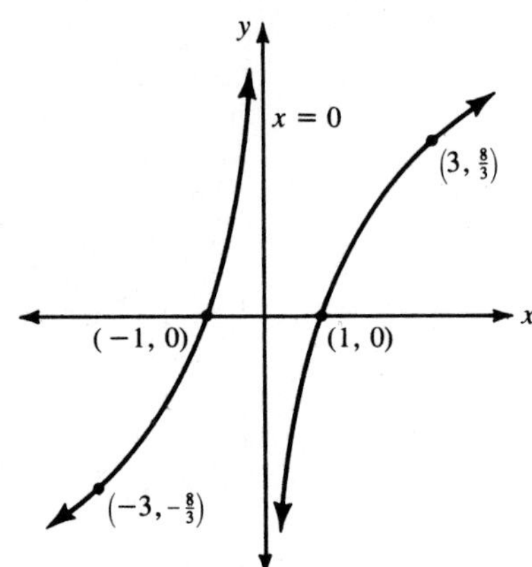

21. (a) \$50, \$90, \$110, \$143.33, \$210, \$410
(b)
(c) 89.5% (rounded to nearest tenth)
(d) no

23. (a) $C(x)=0.10x^2+\dfrac{100}{x}$, where $C(x)$ is in dollars
(c) Approximately 8 inches
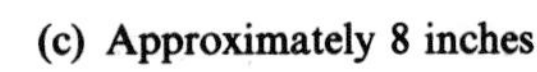
(b)

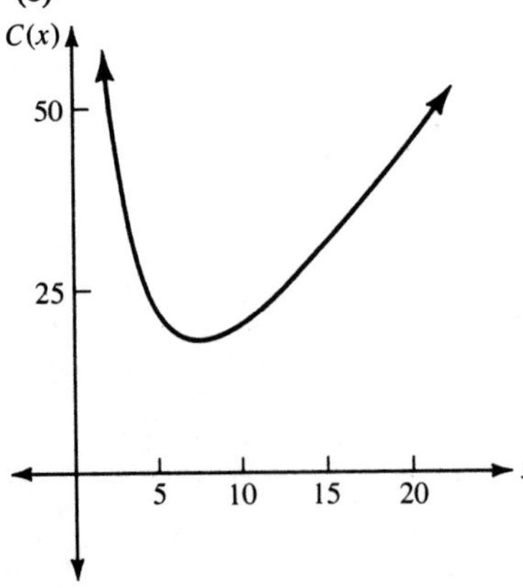

Section 2.6

1. $D = \mathbb{R}$, $R = \mathbb{R}$

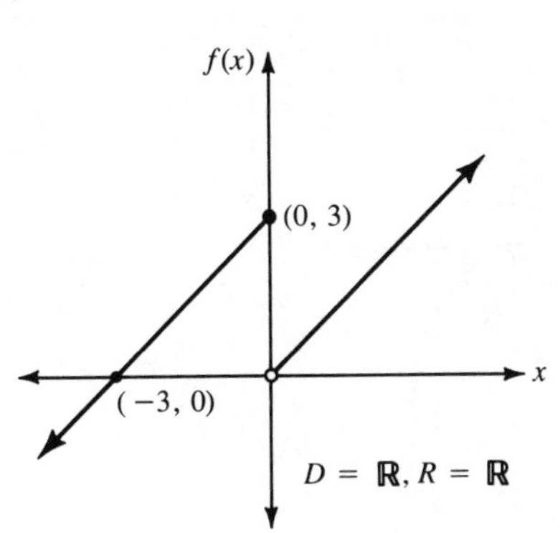

3. $D = \mathbb{R}$ $R = (-4, \infty)$

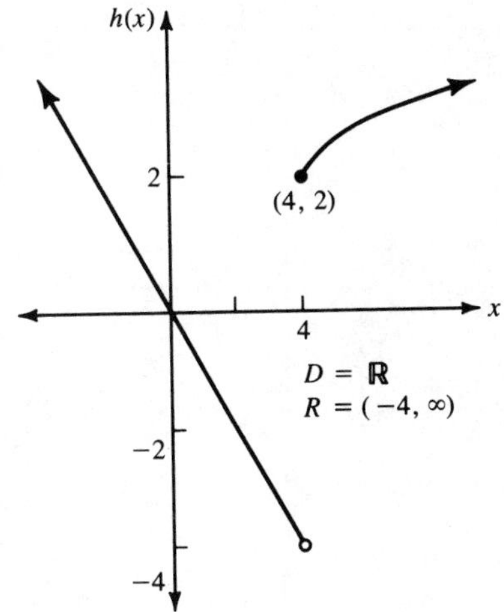

5. $D = \mathbb{R}$ $R = (-\infty, 0]$

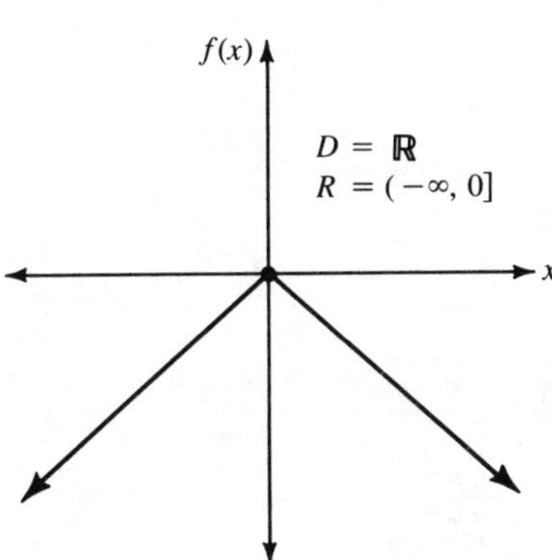

7. $D = (0, 4]$ $R = \{0, 2, 4, 6\}$

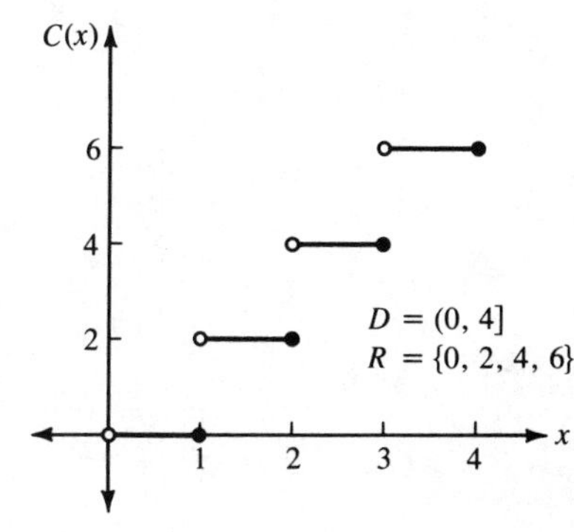

9. $D=\mathbb{R}\ R=\mathbb{R}$

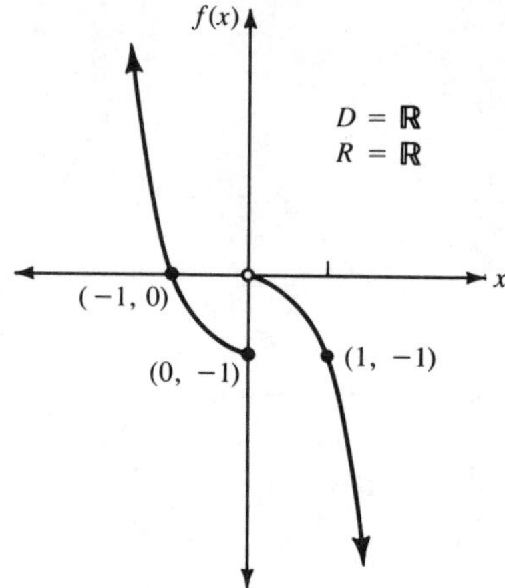

11. $D=\mathbb{R}\ R=(-\infty,8]$

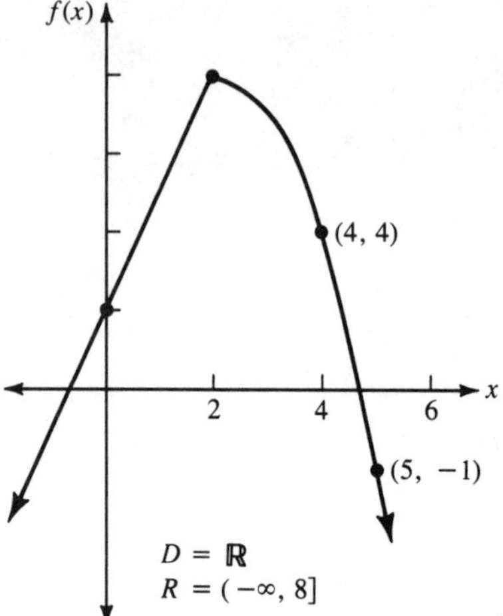

13. $W(t)=\begin{cases} 6t & \text{if } 0\leqslant t\leqslant 8 \\ 9(t-8)+48 & \text{if } 8<t\leqslant 10 \\ 12(t-10)+66 & \text{if } 10<t\leqslant 12 \end{cases}$

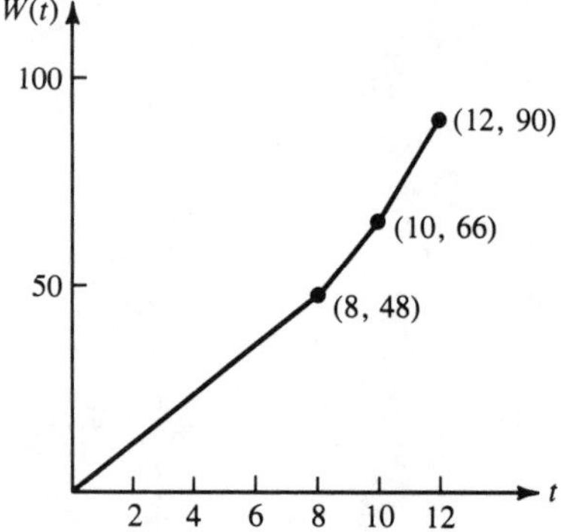

15. $I(n)=\begin{cases} 5n & \text{if } 2\leqslant n\leqslant 5 \\ 4n & \text{if } 6\leqslant n\leqslant 9 \\ 3.5n & \text{if } 10\leqslant n\leqslant 14 \\ \text{where } n \text{ is an integer} \end{cases}$

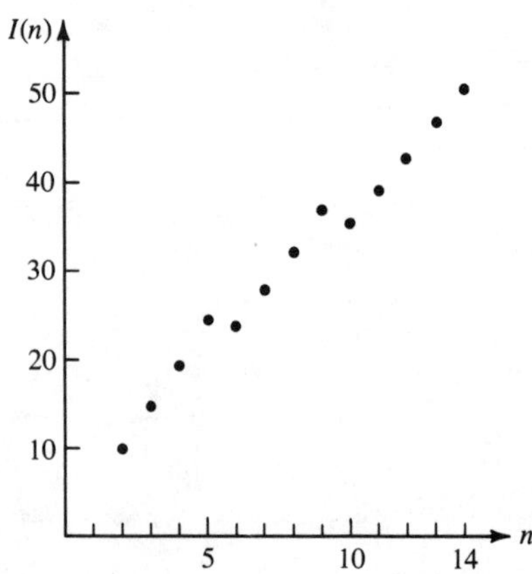

17. $w(x)=\begin{cases} 6.00 & \text{if } 0 \leqslant x \leqslant 30 \\ 6.00+0.50(x-30) & \text{if } x>30 \end{cases}$

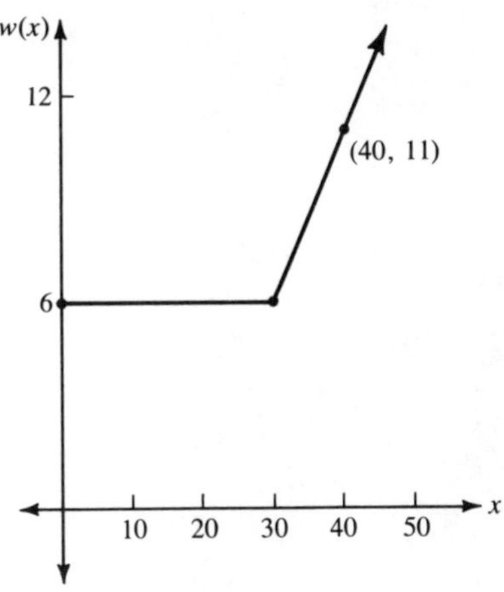

19. $R(x)=400x$ where $15 \leqslant x \leqslant 40$

CHAPTER 3

Section 3.2

1.	(a) $\frac{8}{3}$	(b) 6	(c) $\frac{27}{2}$	(d) 8
3.	(a) $\dfrac{3y^2}{xz}$	(b) $\dfrac{x^4}{y}$	(c) $(2x+1)(y+3)^6$	(d) $x^3y^5w^{10}$
5.	(a) $y^{-2/3}$	(b) $x^{-3/2}y^{-1/2}$	(c) $x^{-1/2}y^{13/12}$	(d) $x^{3/4}y^{3/4}$
7.	(a) $x^{3/2}$	(b) $(x^2+5x+3)^{1/3}$	(c) $(x+2)^{-3/2}$	(d) $(x+5)^{-2/3}$
9.	(a) 9^x	(b) 6^x	(c) $(\frac{3}{2})^x$	(d) $3^{(2^x)}$
	(e) $\frac{1}{9}$	(f) 1	(g) 81	(h) 512

Section 3.3

1. 200% increase

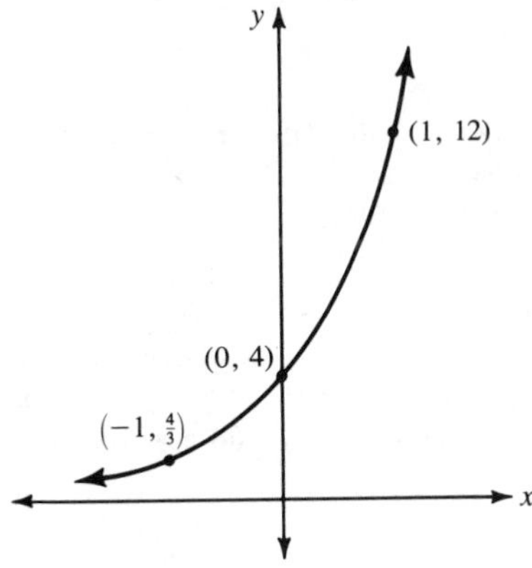

3. 400% decrease

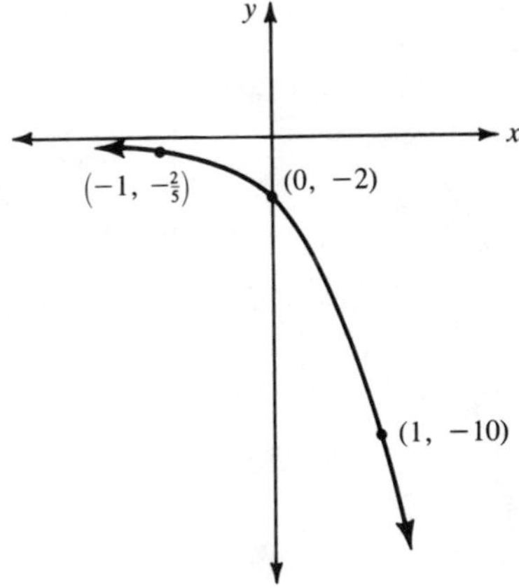

5. 582% increase

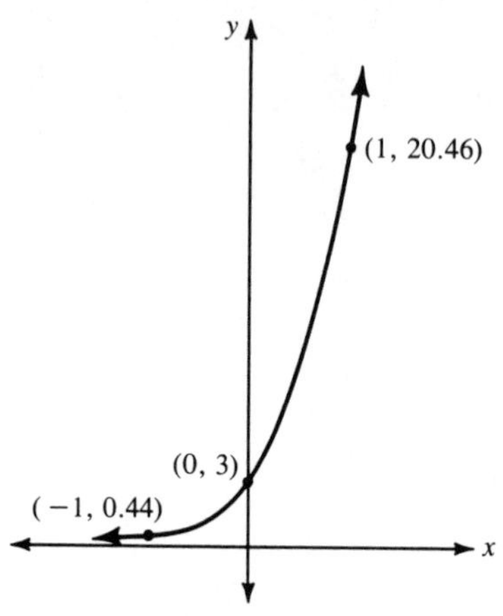

7. 1,100% increase

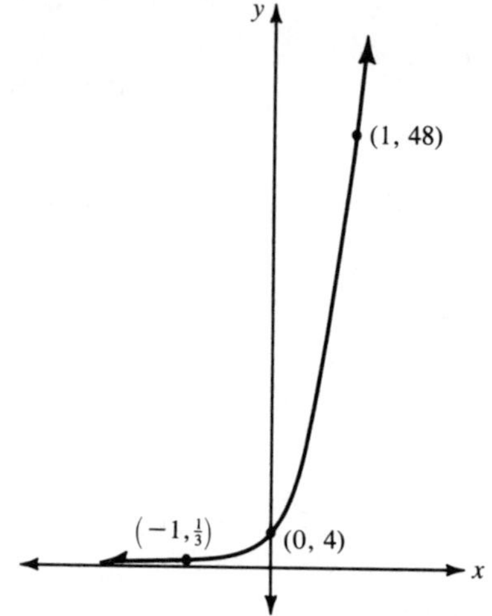

9. 36% decrease

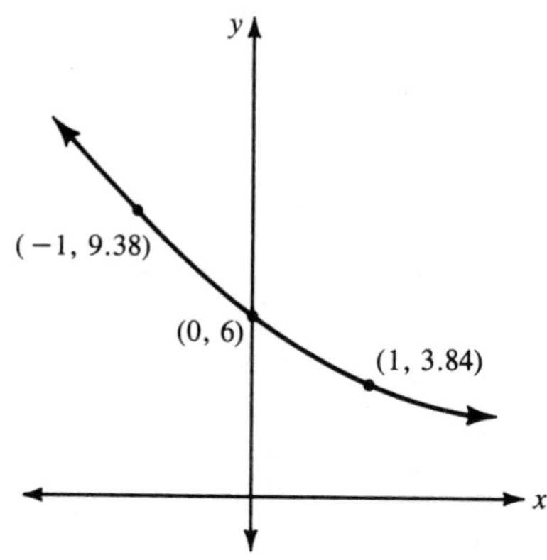

11. $y=6(1.4)^x$; $y=3(1.4)^x$; $y=10(1.4)^x$ where x is the number of units of stimuli and y is the response time in seconds
13. 80 (rounded in the units place)
15. 20 ft

Section 3.4

1. \$6,658.50; \$5,158.50
3. \$1,528.30
5. 7.18%
7. (a) 10.25% (b) 10.52%
9. $N(4)=48\times10^6$; $r=69.3\%$
11. approximately 12 years
13. 60%
15. 28.9%
17. (a) $1.4918B_0$ (b) 49.18% (c) approximately 17 years

19.

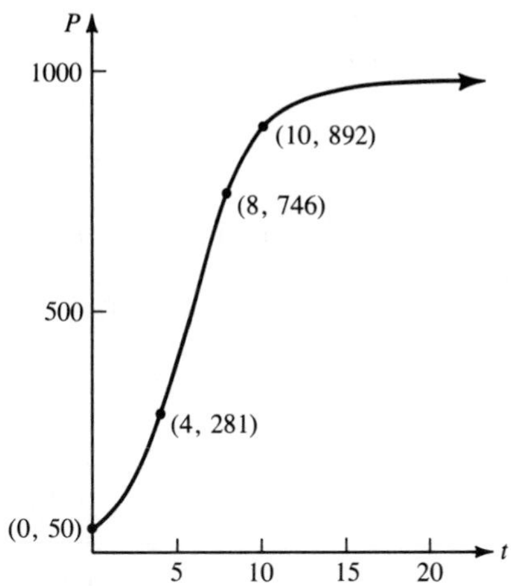

Section 3.5

1. (a) $\log_3 9=2$ (b) $\log_{10}1000=3$ (c) $\log_{10}0.1=-1$ (d) $\log_8(\frac{1}{2})=-\frac{1}{3}$
3. (a) 2 (b) 3 (c) $\frac{1}{2}$ (d) 4 (e) 10
5. (a) 0.6021 (b) 0.9031 (c) 0.4771 (d) 0.9542
7. (a) 2.30259 (b) 1.60944 (c) 1.94591 (d) 0.69315
   (e) 0.18232 (f) 1.46326
9. $y=5^x$: $D=\mathbb{R}$, $R=(0,\infty)$
   $y=\log_5 x$: $D=(0,\infty)$, $R=\mathbb{R}$

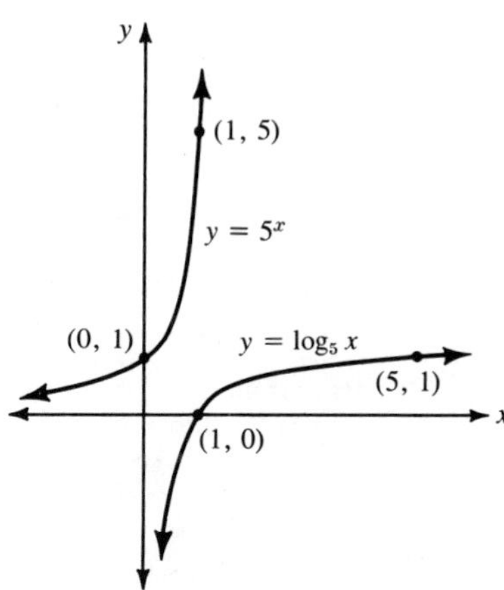

11. $F(x)=x$, domain of $F(x)=\mathbb{R}$; $G(x)=x$, domain of $G(x)=(0,\infty)$
13. (a) $I=10I_0$ (b) $I=100I_0$; no, it is 10 times the answer in part (a)

CHAPTER 4

Section 4.2

1. (a) 2 (b) None exists (c) None exists (d) 1 (e) 2 (f) 2
3. (a) 11 (b) 4 (c) 3 (d) 2
5. Results should be the same
7. (a) 1 (b) 4 (c) 0 (d) $\frac{1}{3}$ (e) 18 (f) 8 (g) $-\frac{1}{3}$ (h) $\frac{5}{4}$

9. (a) $x=-1$ (b) $x=0$ (c) $x=2, x=-2$ (d) $x=-4, x=1$
(e) $x=1, x=-1$ (f) $x=0$ (g) $x=0$ (h) $x=-2$

11. (a) $\dfrac{10h+h^2}{h}$ (b) 10

13. (a) $\dfrac{6xh+3h^2+2h}{h}$ (b) $6x+2$

15. (a) $\dfrac{\dfrac{1}{(x+h)^2}-\dfrac{1}{x^2}}{h}=\dfrac{-2hx-h^2}{x^2(x+h)^2h}$ (b) $\dfrac{-2}{x^3}$

17. (a) $f(x)$ continuous in the domain $[2,12]$; greatest value of $f(x)$ is 10, least value of $f(x)$ is -5

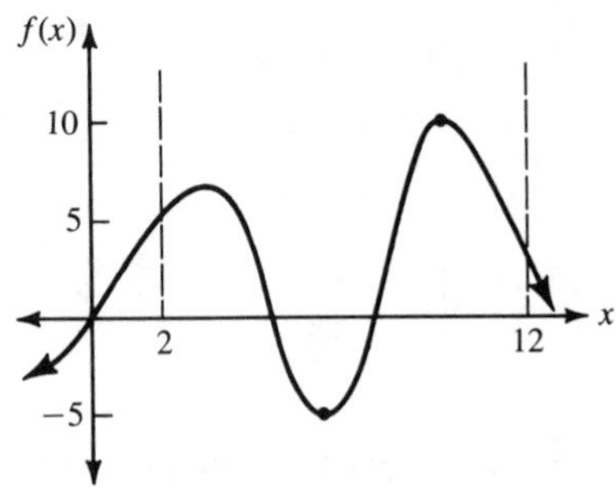

(b) $f(x)$ continuous in the interval $[0,8]$; 2 and 6 belong to the range of $f(x)$ in this interval. Note there exist x's in the interval $[0,8]$ such that $f(x)$ assumes all values between 2 and 6

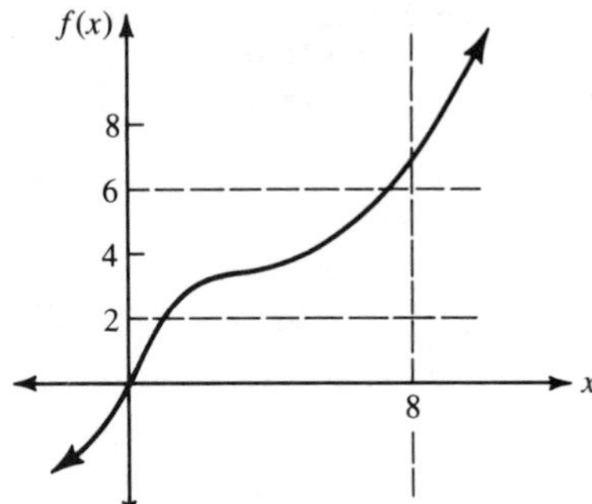

(c) $f(x)$ continuous in the interval $[0,10]$; $f(7)$ is positive and $f(8)$ is negative. There exists an x in $[0,10]$ such that $f(x)=0$. Note $f(7.5)=0$

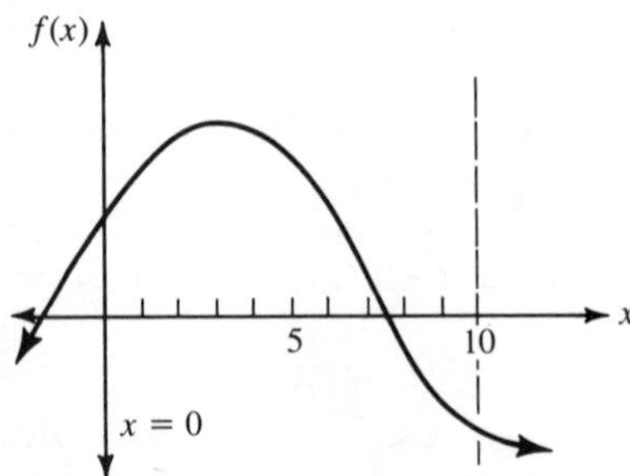

Section 4.3

1. $f'(x)=0$ 3. $f'(x)=2x$ 5. $f'(x)=0$

7. $f'(x)=\dfrac{-1}{x^2}$ 9. $f'(x)=\dfrac{1}{2\sqrt{x}}$ 11. $f'(x)=6x$

13. $f'(x)=-6x$ 15. $f'(x)=\dfrac{-1}{x^2}$ 17. $f'(x)=\dfrac{1}{2\sqrt{x}}$

19. $f'(x)=0; f'(4)=0$ 21. $f'(x)=2x; f'(1)=2$ 23. $f'(x)=\dfrac{-2}{x^3}; f'(-1)=2$

25. $f'(x)=\frac{1}{2}x^{-1/2}; f'(4)=\frac{1}{4}$ 27. $\dfrac{dy}{dx}=\dfrac{-4}{x^5}$ 29. $\dfrac{dr}{ds}=0.2S^{-0.8}=\dfrac{0.2}{S^{0.8}}$

31. $g'(z)=3z^2$

33. (a)

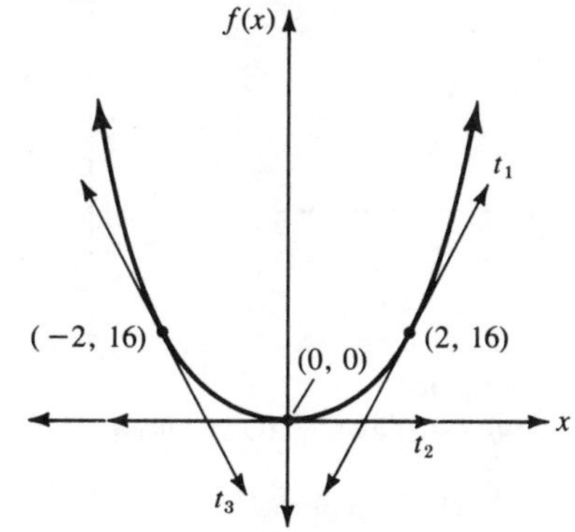

(b) $f'(-2)=-32; f'(0)=0; f'(2)=32$

(c) $y=-32x-48; y=0; y=32x-48$

35. (a) \$16,000 (b) \$22,000 in 1980 (c) \$600/year; \$600/year
(d) not particularly

37. (a) 106 hundred thousand dollars or \$10,600,000
(b) 46 hundred thousand dollars/hundred thousand dollars.
(c) Yes; yes; no; no

Section 4.4

1. $f'(x)=12x^3$ 3. $f'(x)=3+6x^2$ 5. $g'(x)=2x^{-1/2}$ 7. $h'(x)=20x-2x^{-3}$

9. $g'(t)=8t+5-\frac{2}{3}t^{-4/3}$ 11. $\dfrac{du}{dx}=4x^3+9x^2-2x+\frac{1}{2}$ 13. $f'(1)=4$

15. $\left.\dfrac{du}{dz}\right|_{\text{at } z=0}=4$ 17. $g'(-2)=\frac{5}{4}$ 19. $s'(8)=\frac{19}{192}$ 21. $y=5x-3$

23. $y=4x+5$ 25. $y=7x-7$ 27. $f'(x)=2x-2$

29. $g'(t)=\frac{5}{2}t^{3/2}-\frac{9}{2}t^{1/2}$ 31. $\dfrac{du}{dz}=\dfrac{16}{3}z^{1/3}-\dfrac{2}{3}z^{-2/3}$

33. $f'(u)=-6u^{-1/2}$ 35. $g'(x)=10x^4-24x^3-24x^2$

37. $f(x)=\dfrac{-2}{(x-1)^2}$ 39. $g'(x)=\dfrac{-8}{x^3}$

41. $s'(t)=\dfrac{3t^2+6t+1}{(t+1)^2}$ 43. $\dfrac{dy}{dx}=\dfrac{2x^2+2x-7}{(2x+1)^2}$

45. $h'(t)=\dfrac{-1\left(\sqrt{t}+\dfrac{3}{\sqrt{t}}\right)}{2(t-3)^2}$ or $\dfrac{-1(t+3)}{2\sqrt{t}\,(t-3)^2}$ 47. $f'(2)=-1$

49. $\left.\dfrac{dy}{dx}\right|_{\text{at } x=2}=\dfrac{2}{27}$ 51. $\left.\dfrac{ds}{dz}\right|_{\text{at } z=3}=6$ 53. $x=-1$ 55. $u=-\frac{1}{3}, u=-1$

57. $x=-5, x=1$

59. (a) $T(2)=4°C$ (b) $T'(2)=\frac{-4}{3}$; two hundred meters downstream the temperature is decreasing at the rate of $1\frac{1}{3}$°C per hundred meters.

61. (a) \$1,650 (b) $C'(50)=\$18$ (c) $C'(60)=\$18.40$

63. (a) $D_vP=10-2v$ (b) 2;0 (c) -2; -4

(d)

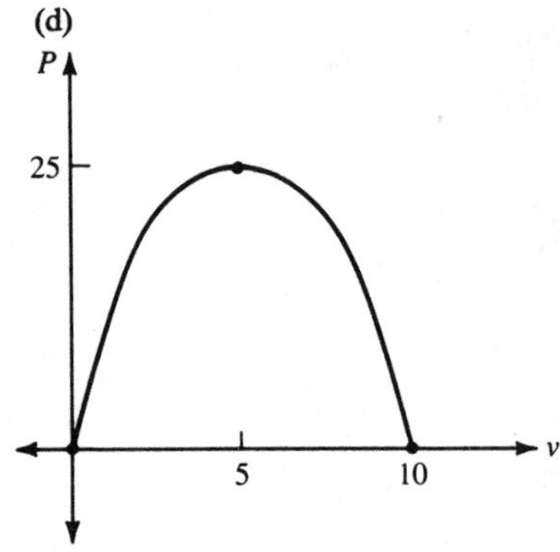

(e)

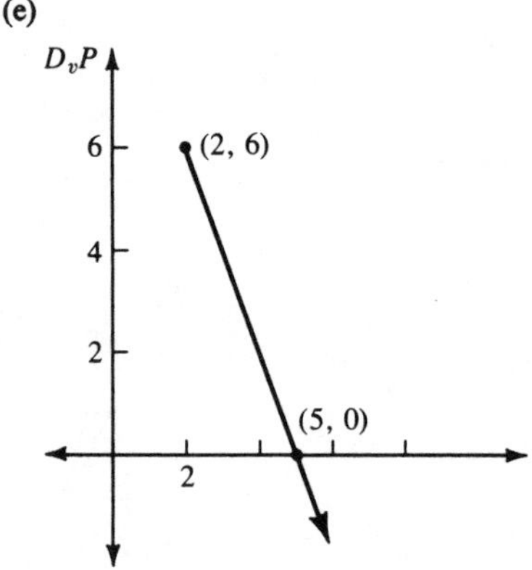

(f) maximum profit when $v=5$

65. (a) 10 mg/liter (b) 2 mg/liter (c) -1.6 mg/liter per hundred meters

Section 4.5

1. $f(u)=u^3$; $u(x)=3x+1$; $D_x[f(u(x))]=9(3x+1)^2$

3. $f(u)=u^8$; $u(x)=2x+6$; $D_x[f(u(x))]=16(2x+6)^7$

5. $f(u)=u^{3/2}$; $u(x)=10x^2-3x+2$; $D_x[f(u(x))]=\frac{3}{2}(10x^2-3x+2)^{1/2}(20x-3)$

7. $f(u)=2u^{-1/2}$; $u(x)=(7x^2+3)$; $D_x[f(u(x))]=-14x(7x^2+3)^{-3/2}$

17. $f(g)=g^{-2}$; $g(u)=u^3+1$; $u(x)=x+2$; $D_x[f(g(u(x)))]=-6(x+2)^2[(x+2)^3+1]^{-3}$

19. $f(g)=g^4$; $g(u)=u^{2/3}+5$; $u(x)=x^2-4$; $D_x[f(g(u(x)))]=\frac{16}{3}x[(x^2-4)^{2/3}+5]^3(x^2-4)^{-1/3}$

25. $f'(x)=8(4x+3)$ 27. $g'(x)=-15(5x+7)^{-4}$

29. $h'(x)=\frac{3}{2}x(3x+5)^{-1/2}+(3x+5)^{1/2}$

31. $g'(x)=\dfrac{(x^2+1)^{1/2}-x^2(x^2+1)^{-1/2}}{(x^2+1)}$ or $\dfrac{1}{(x^2+1)^{3/2}}$

33. $\dfrac{dy}{dx}=\dfrac{1}{2}\left(\dfrac{x}{x+1}\right)^{-1/2}\cdot\dfrac{1}{(x+1)^2}$ 35. $f'(x)=\frac{2}{3}(3x^2+4x-2)^{-1/3}(6x+4)$

37. $s'(x)=3x^2(x-100)^2+(x-100)^3(2x)$
$=x(x-100)^2(5x-200)$

39. $f'(x)=(3x-5)^{1/3}(2x)+x^2(3x-5)^{-2/3}$

41. $g'(x)=\dfrac{x}{2}(x^2+4)^{-3/4}$ 43. $f'(x)=\dfrac{6x(2x+3)^2(x^2-4)^2-4(x^2-4)^3(2x+3)}{(2x+3)^4}$
$=\dfrac{2(x^2-4)^2(4x^2+9x+8)}{(2x+3)^3}$

45. (a) 0.00174 milligrams/gram
(b) -0.000579 milligrams/gram per hr

47. (a) 4 (b) $-\frac{4}{9}$ (decreasing at a rate of approximately $\frac{1}{2}$ egg per female)
49. (a) 100 hundred (b) -0.5 hundred (c) 99.5 hundred units or 9,950 units

Section 4.6

1. $f'(x)=2x+3;\ f'(-2)=-1$
$f''(x)=2;\ f''(-2)=2$
$f'''(x)=0;\ f'''(-2)=0$
3. $f'(x)=4x^3+15x^2-14x+3;\ f'(1)=8$
$f''(x)=12x^2+30x-14;\ f''(1)=28$
$f'''(x)=24x+30;\ f'''(1)=54$
$f^{iv}(x)=24;\ f^{iv}(1)=24$
$f^{v}(x)=0;\ f^{v}(1)=0$
5. $f'(x)=(x-1)^3+3x(x-1)^2;\ f'(1)=0$
$f''(x)=6(x-1)^2+6x^2-6x;\ f''(1)=0$
$f'''(x)=24x-18;\ f'''(1)=6$
$f^{iv}(x)=24;\ f^{iv}(1)=24$
$f^{v}(x)=0;\ f^{v}(0)=0$
7. $z'(t)=3t^2+6t+2;\ z'(-1)=-1$
$z''(t)=6t+6;\ z''(-1)=0$
$z'''(t)=6;\ z'''(-1)=6$
$z^{iv}(t)=0;\ z^{iv}(-1)=0$
9. $f'(x)=-\frac{1}{2}x^{-3/2};\ f''(x)=\frac{3}{4}x^{-5/2}$
11. $g'(x)=(x^2-4)/x^2;\ g''(x)=8/x^3$
13. $h'(x)=\frac{3}{2}(3x-x^3)^{-1/2}(1-x^2);\ h''(x)=-\frac{9}{4}(3x-x^3)^{-3/2}(1-x^2)^2-3x(3x-x^3)^{-1/2}$
15. $f'(x)=-2(x^2+3x+2)^{-3}(2x+3);\ f''(x)=-4(x^2+3x+2)^{-3}+6(2x+3)^2(x^2+3x+2)^{-4}$
17. (a)

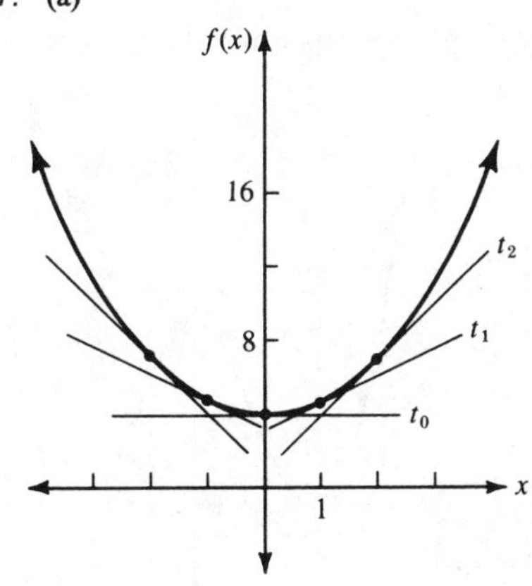

(b)

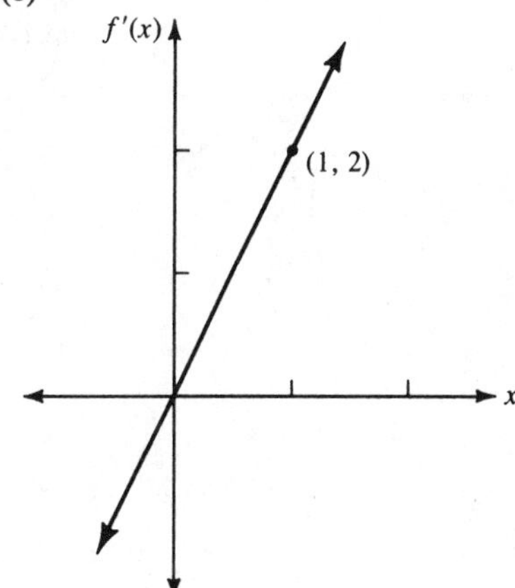

(c) $f''(x)=2>0$

CHAPTER 5

Section 5.2

1. Increasing on $(3,\infty)$; decreasing on $(-\infty,3)$
3. Increasing on $\mathbb{R}$
5. Increasing on $(-\infty,-2)\cup(1,\infty)$; decreasing on $(-2,1)$
7. Increasing on $(3,\infty)$; decreasing on $(-\infty,3)$
9. Increasing on $(-2,2)$; decreasing on $(-\infty,-2)\cup(2,\infty)$
11. Increasing on $(-\infty,-1)\cup(1,\infty)$; decreasing on $(-1,0)\cup(0,1)$
13. Increasing on $(0,\infty)$; decreasing on $(-\infty,0)$
15. (a) $(0,16)$ (b) \$270/unit
17. (a) $E(x)=25x-4x^2$ (b) $(0,\frac{25}{8})$

(c) \$9; when the wholesale price of coffee is \$2/lb the per capita expenditure is increasing at the rate of \$9/dollar increase in a lb of coffee.

19. (a) 2°C (b) $(0,2)$ (c) $(2,\infty)$

Section 5.3

1. $(-\frac{3}{2}, -12.5)$; $f(-\frac{3}{2}) = -12.5$ (rel. min.)

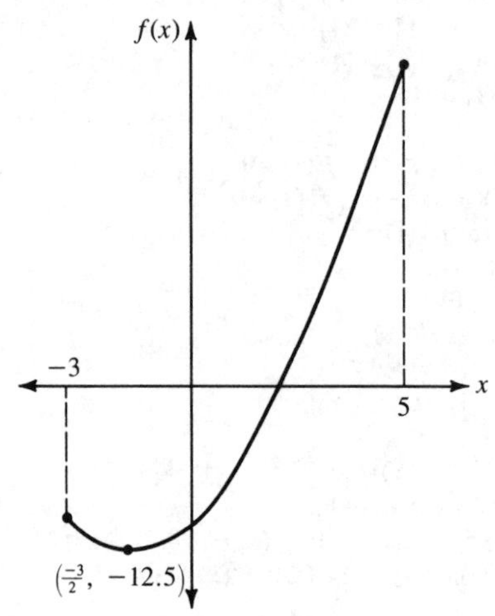

3. $(5, -96)$, $(-1, 12)$; $g(5) = -96$ (rel. min.), $g(-1) = 12$ (rel. max.)

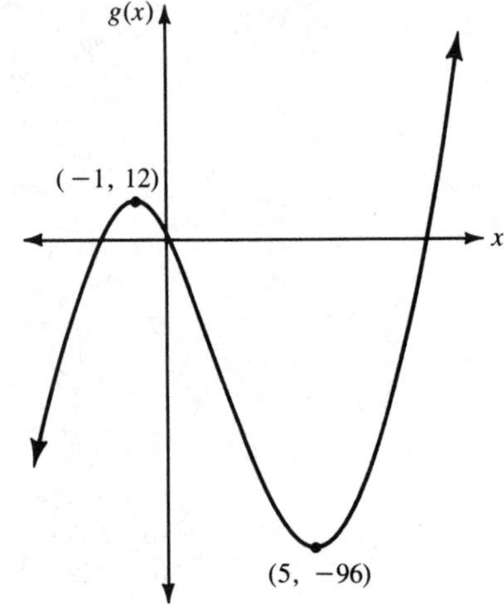

5. $(0, 16)$, $(-2, 0)$, $(2, 0)$; $s(0) = 16$ (rel. max.), $s(-2) = 0$ (rel. min.), $s(2) = 0$ (rel. min.)

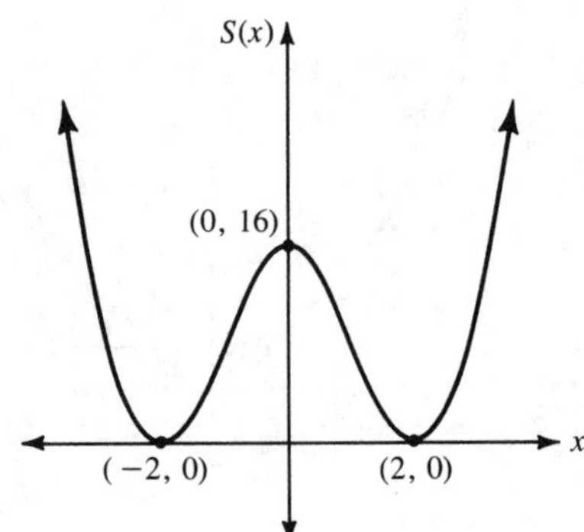

7. $(-2,-9)$; $f(-2)=-9$ (rel. min.)

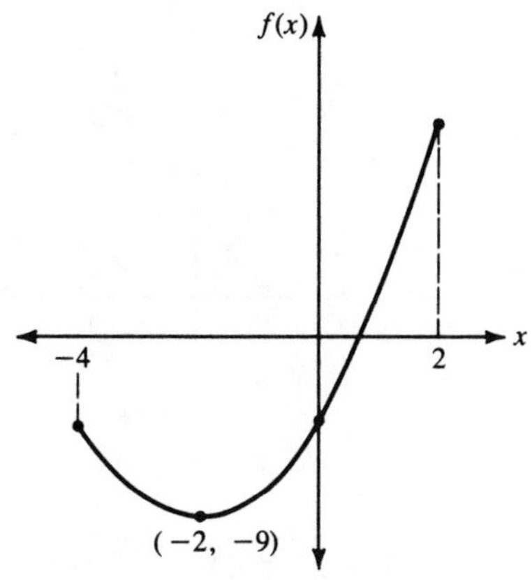

9. $(3,-\frac{27}{2})$, $(0,0)$; $g(3)=-\frac{27}{2}$ (rel. min.), $g(0)=0$ (neither)

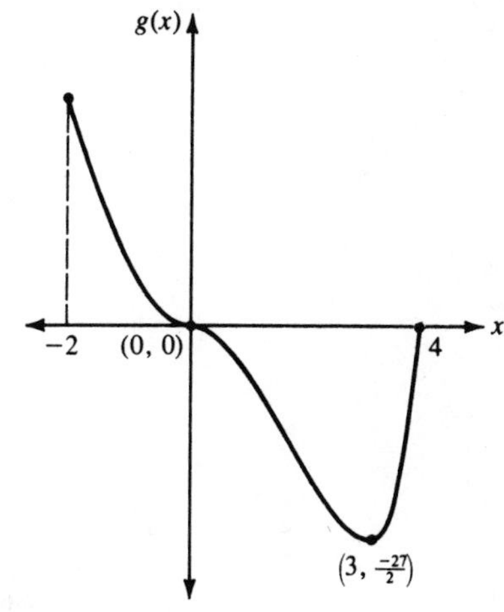

11. $(2,-16)$, $(-2,-16)$, $(0,0)$; $s(2)=-16$ (rel. min.), $s(-2)=-16$ (rel. min.), $s(0)=0$ (rel. max.)

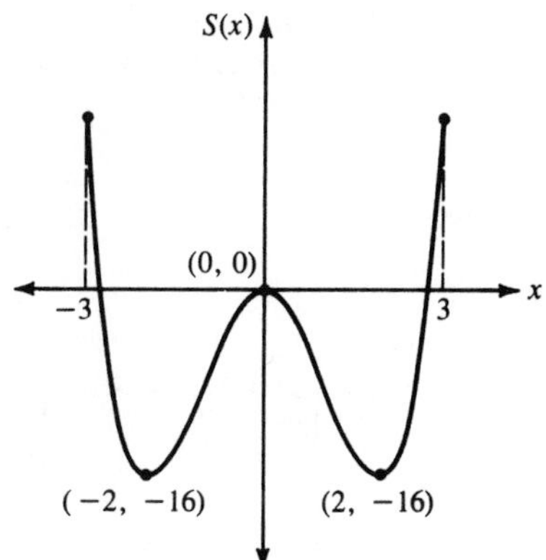

13. $(1,4)$, $(-1,0)$; $h(1)=4$ (rel. max.), $h(-1)=0$ (rel. min.)

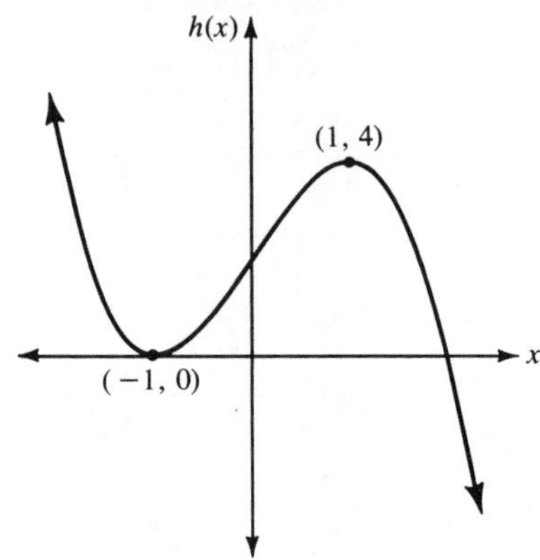

15. $(0,2)$, $(-2,6)$; $f(0)=2$ (rel. min.), $f(-2)=6$ (rel. max.)
17. $(2,\frac{-13}{3})$, $(-2,\frac{19}{3})$; $h(2)=\frac{-13}{3}$ (rel. min.), $f(-2)=\frac{19}{3}$ (rel. max.)
19. $(1,2)$; $f(1)=2$ (rel. min.). Note $x=-1$ is not in the domain of $f(x)$
21. $(0,0)$, $(3,\frac{27}{2})$; $h(0)=0$ (neither), $h(3)=\frac{27}{2}$ (rel. max.)
23. $f(-\frac{3}{2})=\frac{-29}{4}$ (abs. min.); $f(6)=49$ (abs. max.)
25. $h(0)=0$ (abs. min.); $h(5)=\frac{25}{9}$ (abs. max.)
27. $f(0)=1$ (abs. max.); $f(10)=\frac{1}{121}$ (abs. min.)
29. $P(0)=0$ (abs. min.); $P(1600)=(1600)^2(2400)^3$ (abs. max.); $P(4000)=0$ (abs. min.)
31. 12,500 units; \$312,500
33. (a) $P(x)=-0.04x^2+8x-80$
(b) 100 units
(c) \$320
35. 44
37. $T(2)=10°C$

Section 5.4

1. $x=50, y=50$
3. $s(4)=48$
5. (a)

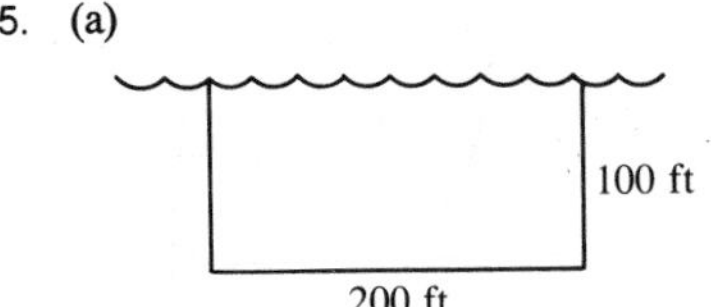

(b)

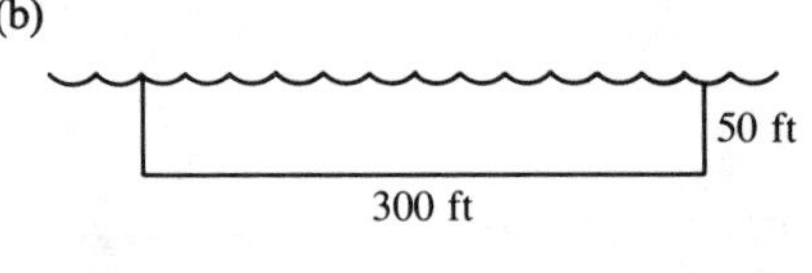

7. $P(x)=-2x^2+900x-20{,}000$; 225; $P(225)=81{,}250$
9. $t=4.85$ weeks
11. \$280; \$15,680
13. 5 weeks; \$4,900
15. 4 ft×4 ft×2 ft
17. 5 days; \$62.50
19. 44
21. 44
23. 200 ft
25. 3,000 yards; \$38,400
27. $Q=1200$; $B=\$720{,}000$; Deficit $=\$770{,}000$
29. 4,000; 25/year

Section 5.5

1. (a) \$50 (b) \$50,000
3. (a) $C'(x)=2{,}000-30x+3x^2$ (b) \$2,000 (c) \$69,500
(d) \$1,925 (e) \$59,750
5. (a) \$157,500 (b) \$1,150
7. (a) $C'(x)=12-0.12x+0.0003x^2$ (b) \$300
(c) \$80,000 (d) 0 (e) \$80,000
9. (a) $P'(x)=8-0.34x-0.006x^2$ (b) \$4

Section 5.6

1.

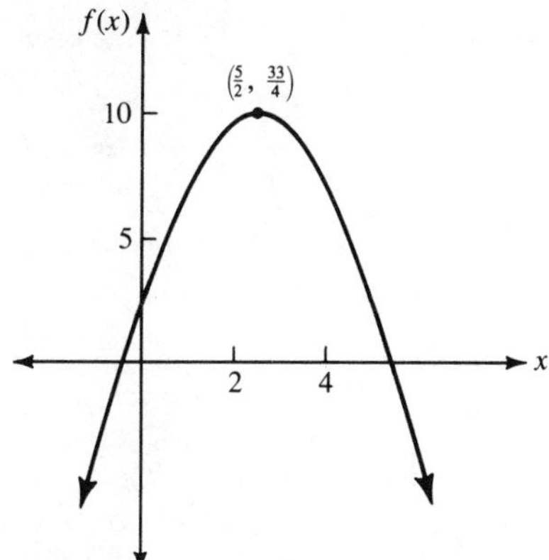

Increasing on $(-\infty, \frac{5}{2})$
Decreasing on $(\frac{5}{2}, \infty)$
Concave downward on $\mathbb{R}$
No inflection points.

3

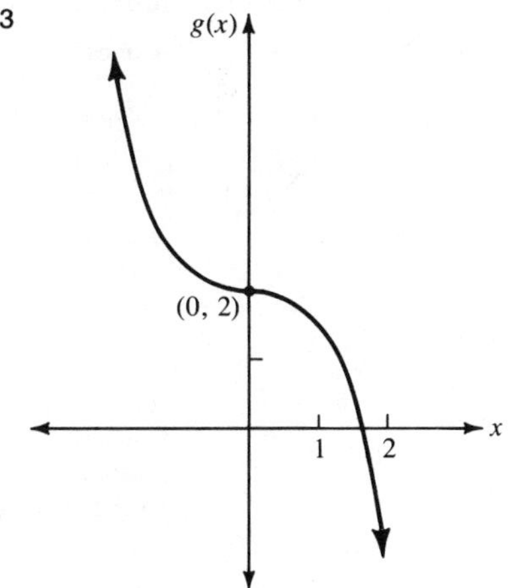

Decreasing on $\mathbb{R}$
Concave upward on $(-\infty, 0)$
Concave downward on $(0, \infty)$
Inflection point $(0, 2)$

5.

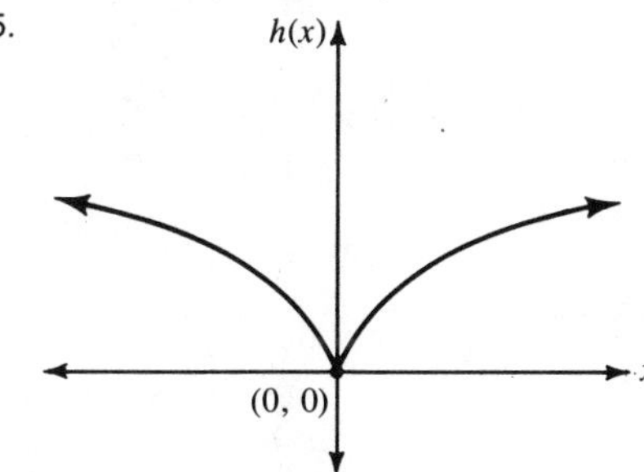

Decreasing on $(-\infty, 0)$
Increasing on $(0, \infty)$
Concave downward on $(-\infty, 0) \cup (0, \infty)$
No inflection points

7.

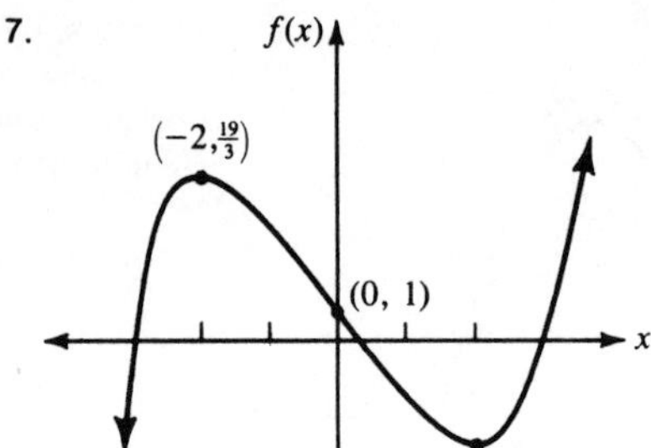

Increasing on $(-\infty, -2) \cup (2, \infty)$
Decreasing on $(-2, 2)$
Concave downward on $(-\infty, 0)$
Concave upward on $(0, \infty)$
Inflection point $(0, 1)$

9.

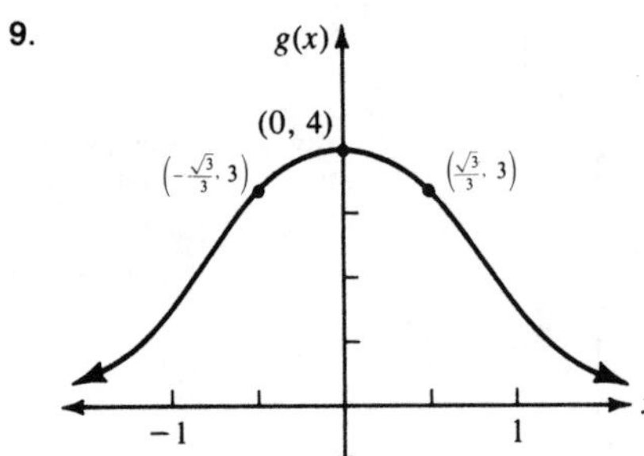

Increasing on $(-\infty, 0)$;
Decreasing on $(0, \infty)$
Concave upward on $(-\infty, -\frac{\sqrt{3}}{3}) \cup (\frac{\sqrt{3}}{3}, \infty)$
Concave downward on $(-\frac{\sqrt{3}}{3}, \frac{\sqrt{3}}{3})$
Inflection points: $\left(-\frac{\sqrt{3}}{3}, 3\right), \left(\frac{\sqrt{3}}{3}, 3\right)$

11.

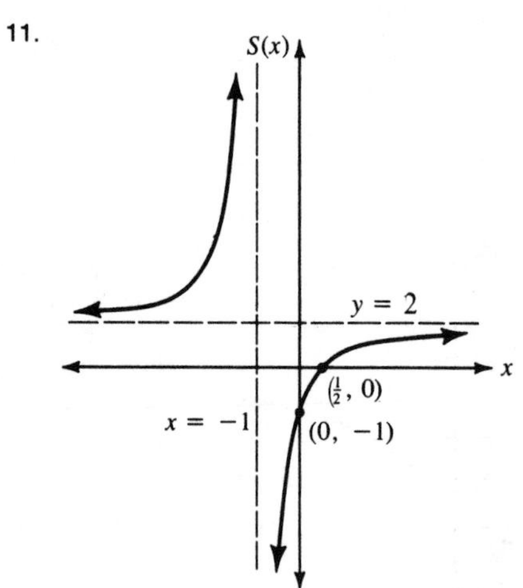

Increasing on $(-\infty, -1) \cup (-1, \infty)$
Concave upward on $(-\infty, -1)$
Concave downward on $(-1, \infty)$
No inflection points

13.

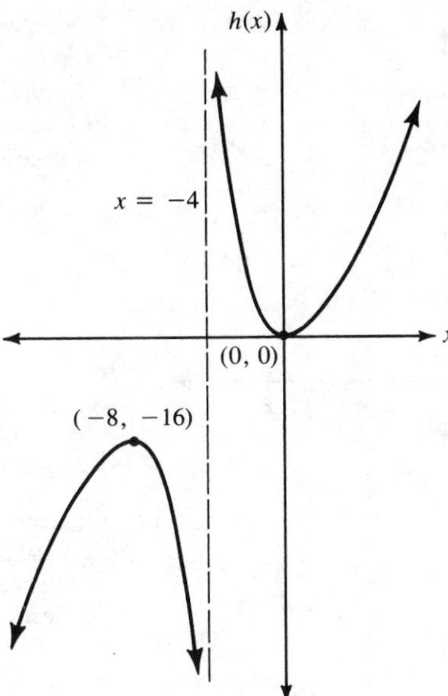

Increasing on $(-\infty, -8) \cup (0, \infty)$
Decreasing on $(-8, -4) \cup (-4, 0)$
Concave downward on $(-\infty, -4)$
Concave upward on $(-4, \infty)$
No inflection points

15.

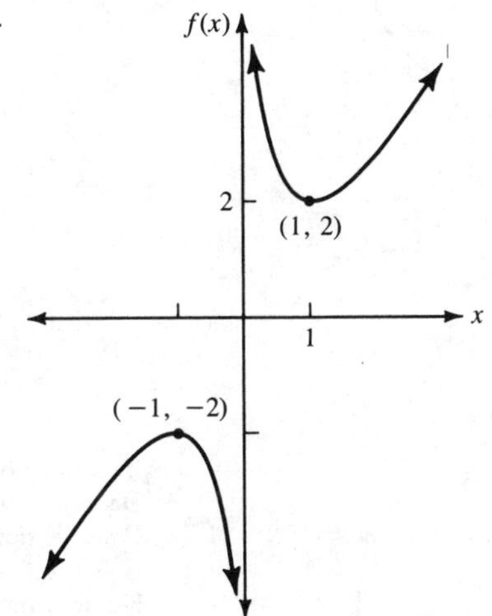

Increasing on $(-\infty, -1) \cup (1, \infty)$
Decreasing on $(-1, 0) \cup (0, 1)$
Concave downward on $(-\infty, 0)$
Concave upward on $(0, \infty)$
No inflection points

17.

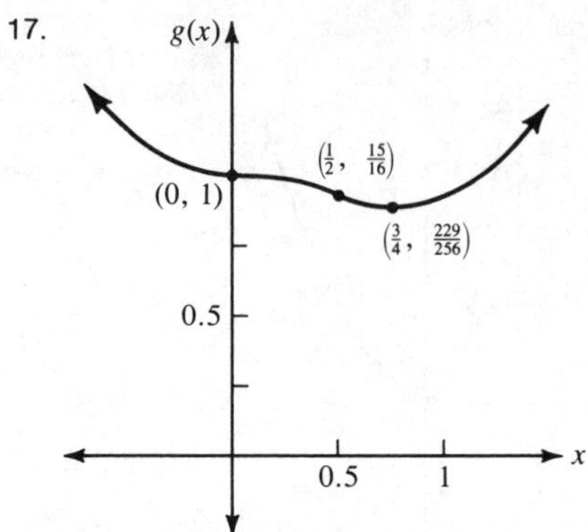

Decreasing on $(-\infty, \frac{3}{4})$
Increasing on $(\frac{3}{4}, \infty)$
Concave upward on $(-\infty, 0) \cup (\frac{1}{2}, \infty)$
Concave downward on $(0, \frac{1}{2})$
Inflection points: $(0, 1)$, $(\frac{1}{2}, \frac{15}{16})$

19.

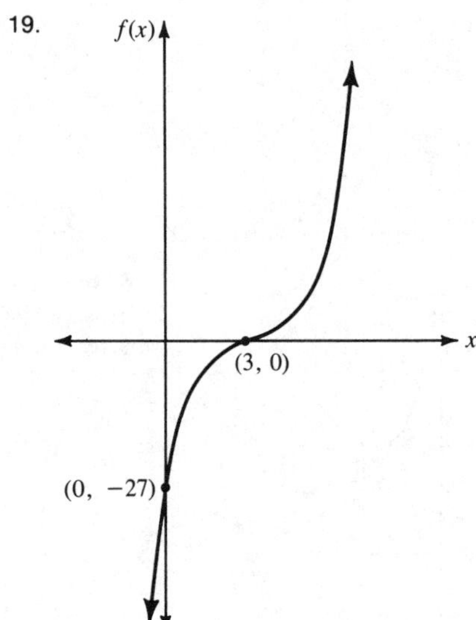

Increasing on $\mathbb{R}$
Concave downward on $(-\infty, 3)$
Concave upward on $(3, \infty)$
Inflection point $(3, 0)$

21.

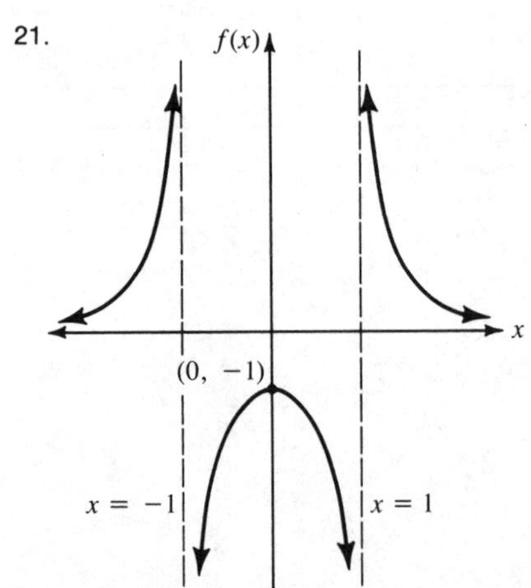

Increasing on $(-\infty, -1) \cup (-1, 0)$
Decreasing on $(0, 1) \cup (1, \infty)$
Concave upward on $(-\infty, -1) \cup (1, \infty)$
Concave downward on $(-1, 1)$
No inflection points

23. (a) Increasing on $(-\infty,-3)\cup(4,\infty)$; decreasing on $(-3,4)$
(b) Concave upward on $(\frac{1}{2},\infty)$; concave downward on $(-\infty,\frac{1}{2})$
(c) $x=\frac{1}{2}$; since $f(x)$ is not given, $f(\frac{1}{2})$ cannot be determined
25. (a) Increasing on $(1,\infty)$; decreasing on $(-\infty,0)\cup(0,1)$
(b) Concave downward on $(-\infty,0)\cup(2,\infty)$; concave upward on $(0,2)$
(c) $x=2$ and $x=0$; since $h(x)$ is not given, $h(2)$ and $h(0)$ cannot be determined

Section 5.7

1. 1.861 3. 2.640 5. 1.659 7. 1.639
9. Rel. maximum $\doteq 6.324$ 11. Rel. minimum $\doteq -19.06$

CHAPTER 6

Section 6.2

1. $f'(x)=3e^{3x}$ 3. $g'(x)=2xe^{x^2}$ 5. $\frac{dy}{dx}=(3x^2-x)e^{x^3-\frac{1}{2}x^2}$
7. $S'(t)=te^t+e^t$
$=e^t(t+1)$
9. $f'(x)=(x^2-1)e^{-x}(-1)+e^{-x}(2x)$
$=e^{-x}(1+2x-x^2)$
11. $f'(x)=\dfrac{(1+x^2)e^x-e^x(2x)}{(1+x^2)^2}$
$=\dfrac{(x-1)^2e^x}{(1+x^2)^2}$
13. $g'(x)=\frac{1}{2}(2x+e^x)^{-1/2}(2+e^x)$
15. $f'(x)=\dfrac{-2}{x^2}e^{\frac{1}{x}}$
17. $\frac{dy}{dx}=10^{3x}(\ln 10)(3)$
$\doteq(6.90777)\cdot 10^{3x}$
19. $f'(x)=8^{x^2+x}(2x+1)\ln 8$
$\doteq(2.07944)(2x+1)8^{x^2+x}$
21. $f'(t)=40e^{-0.5t}$
23. $f'(x)=\dfrac{50e^{-.01x}}{(10+5e^{-.01x})^2}$
25. $f'(x)=2xe^{x^2}$; $f''(x)=4x^2e^{x^2}+2e^{x^2}$
$=2e^{x^2}(2x^2+1)$
27. $g'(x)=x^2e^x+e^x(2x)$; $g''(x)=x^2e^x+e^x(2x)+e^x(2)+2xe^x$
$=xe^x(x+2)$ $\quad=e^x(x^2+4x+2)$
29. Increasing on $\mathbb{R}$; concave upward on $\mathbb{R}$
31. Increasing on $(-1,\infty)$; decreasing on $(-\infty,-1)$; concave upward on $(-2,\infty)$; concave downward on $(-\infty,-2)$
33. Increasing on $\mathbb{R}$; concave downward on $(-\infty,0)$; concave upward on $(0,\infty)$
35. Increasing on $\mathbb{R}$; concave downward on $\mathbb{R}$
37. $\frac{1}{e}$; 1; e
39. $-2e$; 0; $2e$

Section 6.3

1. $f'(x)=\dfrac{3}{x}$ 3. $g'(x)=\dfrac{2}{x}$ 5. $s'(x)=\dfrac{6x^2}{x^3-1}$ 7. $s'(t)=1+\ln t$
9. $f'(x)=\dfrac{2}{2x+5}\cdot\dfrac{1}{\ln 10}$
$\doteq\dfrac{0.8686}{2x+5}$
11. $h'(x)=e^x\left(\dfrac{1}{x}+\ln x\right)$
13. $s'(t)=t\left[\dfrac{t}{t-1}+2\ln(t-1)\right]$
15. $\dfrac{dy}{dx}=\dfrac{6x}{(x^2+3)}[\ln(x^2+3)]^2$
17. $\dfrac{dy}{dx}=\dfrac{1}{(x+1)}\cdot\dfrac{1}{\ln 5}$
$\doteq\dfrac{1}{(1.60944)(x+1)}\doteq\dfrac{0.6213}{(x+1)}$
19. $g'(x)=\dfrac{3x}{(3x^2+4)}$
21. $f'(x)=\dfrac{2}{x}$; $f''(x)=\dfrac{-2}{x^2}$

23. $g'(x) = \dfrac{x^2-1}{x(x^2+2)}$; $g''(x) = \dfrac{-x^4+8x^2+4}{x^2(x^2+2)^2}$

25. $f'(x) = \dfrac{2\ln(x+1)}{x+1}$; $f''(x) = \dfrac{2-2\ln(x+1)}{(x+1)^2}$

27. $D=(0,\infty)$; increasing on $(0,\infty)$; concave downward on $(0,\infty)$

29. $D=(0,\infty)$; decreasing on $\left(0,\dfrac{1}{e}\right)$; increasing on $\left(\dfrac{1}{e},\infty\right)$; concave upward on $(0,\infty)$

31. $f'(\frac{1}{2})=2$; $f'(1)=1, f'(2)=\frac{1}{2}$

33. $f'(\frac{1}{2})=-2$; $f'(1)=2$; $f'(-2)=-1, f'(2)=1$

35. $\left.\dfrac{dy}{dx}\right|_{\text{at } x=0}=2$; $\left.\dfrac{dy}{dx}\right|_{\text{at } x=1}=1$; $\left.\dfrac{dy}{dx}\right|_{\text{at } x=2}=\dfrac{2}{3}$

Section 6.4

1.

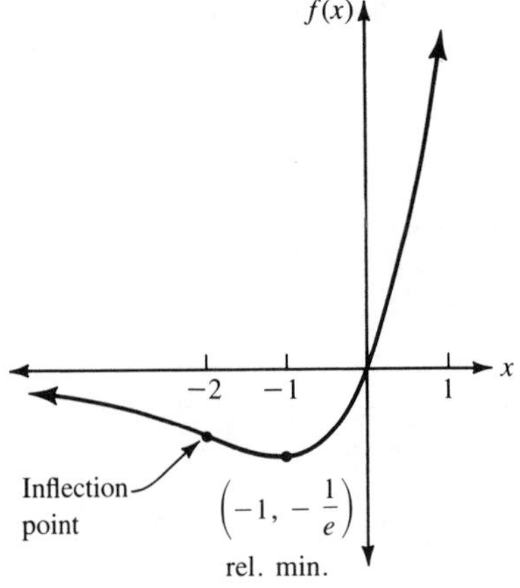

Decreasing on $(-\infty,-1)$
Increasing on $(-1,\infty)$
Concave downward on $(-\infty,-2)$
Concave upward on $(-2,\infty)$

3.

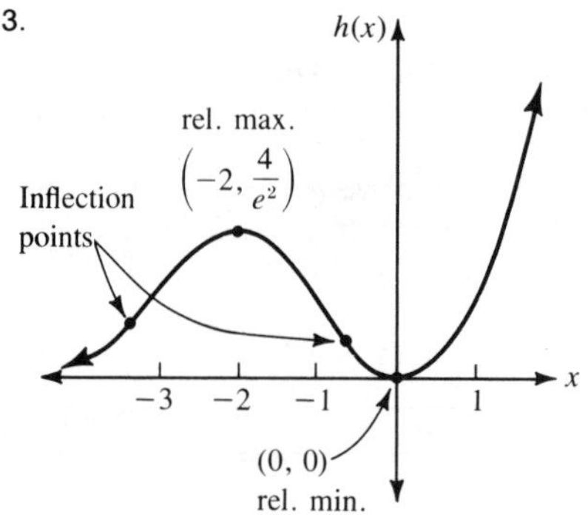

Increasing on $(-\infty,-2)\cup(0,\infty)$
Decreasing on $(-2,0)$
Concave upward on
$(-\infty,-2-\sqrt{2})\cup(-2+\sqrt{2},\infty)$
Concave downward on $(-2-\sqrt{2},-2+\sqrt{2})$

5.

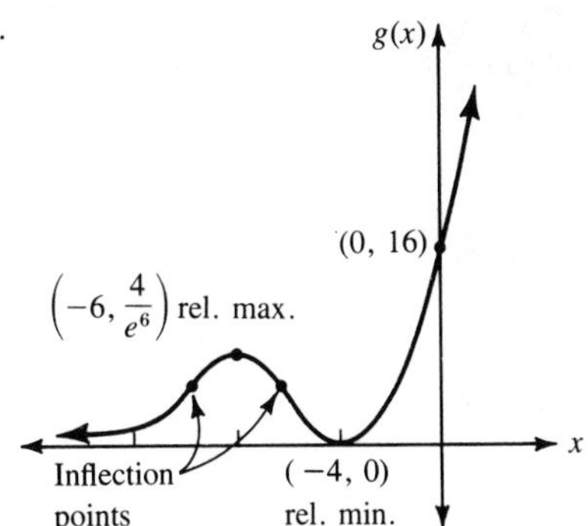

Increasing on $(-\infty, -6) \cup (-4, \infty)$
Decreasing on $(-6, -4)$
Concave upward on
$(-\infty, -6-\sqrt{2}) \cup (-6+\sqrt{2}, \infty)$
Concave downward on $(-6-\sqrt{2}, -6+\sqrt{2})$

7.

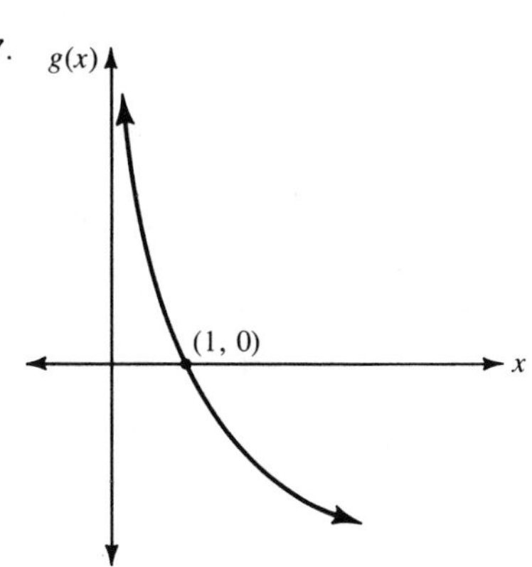

Decreasing on $(0, \infty)$
Concave upward on $(0, \infty)$
No inflection points

9. $f(0) = 2$ absolute minimum

11. (a) 100

(b)

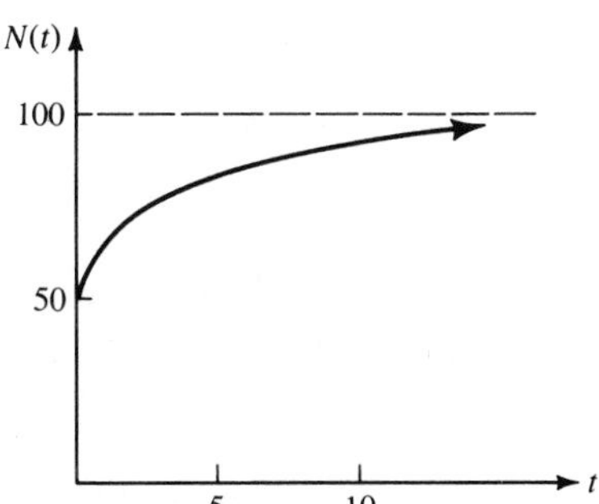

(c) $N'(5) \doteq 3.7$; after 5 weeks of training a worker's performance is increasing at the rate of 3.7 parts/hr/week

13. \$1,649/year; \$2,718/year; \$7,389/year (to the nearest dollar)

15. (a) 45¢ (b) Decreasing at the rate of 4¢/year

17. (a) 2.5 g (b) −.14 g/hr (to the nearest hundredth)

19. (a) 739 cells
(b) $N'(100) \doteq 15$; after 100 minutes the culture is growing at the rate of 15 cells/min

21. (a) $E(t) = 350e^{.09t}$ (b) \$549; \$861; \$1,350 (c) \$122/year

23. (a) $C_0 = 101$; $r = 0.069315$
(c) \$17.22; in 1980 the minimum weekly income is increasing at the rate of \$17.22/year

25. (a)

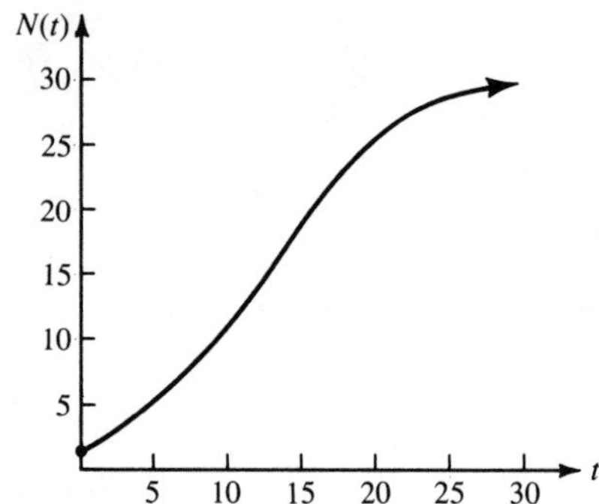

(b) 13.2 years (to the nearest tenth)

Section 6.5

1. $D_xy = \dfrac{1-2y}{2x}$ 3. $D_xy = \dfrac{2-3y}{3x}$ 5. $D_xy = \dfrac{4xy-y}{(x-2y-2x^2)}$

7. $D_xy = \dfrac{2yx-2y^2x-y^2}{2x^2y+2xy-x^2}$ 9. $D_xy = -\dfrac{y^2+3yx^2}{x^3+2xy}$ 11. -1

13. $-\frac{4}{3}$ 15. $D_xy = x^{(x+1)}\left[\dfrac{x+1}{x} + \ln x\right]$

17. $D_xy = x^{(x^2+4x+1)}\left[\dfrac{x^2+4x+1}{x} + (2x+4)\ln x\right]$

19. $D_xy = (5x+4)^{5x+4}[5 + 5\ln(5x+4)]$ 21. $D_xy = 10^x x^{10}\left[ln\,10 + \dfrac{10}{x}\right]$

23. $D_xy = x^{(4x+5)}\left[\dfrac{4x+5}{x} + 4\ln x\right]$ 25. $D_xy = 7x^6$

27. $D_xy = -\frac{1}{2}(2x+3)(x^2+3x+2)^{-3/2}$ 29. $D_xy = b^x \ln b$

31. $D_xy = 1$ 33. $D_xy = 24(6x+5)^3(x+3)^{-2} - 2(6x+5)^4(x+3)^{-3}$

35. $D_xy = (x+1)(x^2+2x-3)^{-1/2}$

CHAPTER 7

Section 7.2

1. $x^8 + C$ 3. $\frac{1}{3}x^3 + C$ 5. $e^{2x} + C$ 7. $x + C$
9. $x^{-1} + C$ 11. $\frac{1}{3}x^6 + C$ 13. $\frac{2}{3}x^{3/2} + C$ 15. $\frac{3}{2}x^{2/3} + C$
17. $\frac{1}{4}e^{4x} + C$ 19. $2\ln x + C$

Section 7.3

1. $\frac{3}{5}x^5 + C$ 3. $-\frac{1}{2}x^{-4} + C$ 5. $\frac{12}{5}u^{5/2} + C$ 7. $\frac{10}{3}z^{3/2} + C$
9. $\frac{3}{2}x^{4/3} + C$ 11. $\frac{2}{3}y^3 - 2y^2 + 3y + C$ 13. $-y^{-3} + y^{-4} + C$
15. $\frac{4}{3}y^{3/4} + 5y^{-1} + C$
17. $-\frac{1}{3}x^{-3} - \frac{1}{2}x^{-2} + C$ 19. $y - \frac{1}{2}y^4 + \frac{1}{7}y^7 + C$ 21. $e^x - \frac{15}{4}x^{4/3} + 2x + C$
23. $\frac{1}{6}(x^2-2x+1)^6 + C$ 25. $\frac{2}{3}(x^2+1)^{3/2} + C$ 27. $2(3x^2+4)^{1/2} + C$
29. $\frac{1}{11}(3t^2-1)^{11} + C$ 31. $\frac{1}{6}(x+1)^6 + C$ 33. $(x^2+1)^{1/2} + C$
35. $-\frac{1}{2}(x^2+2x+1)^{-1} + C$ 37. $e^{2x} + C$ 39. $e^{x^2} + C$
41. $-2e^{-\frac{1}{2}x} + C$ 43. $\ln(x^2+1) + C$ 45. $\frac{1}{8}\ln(4t^2+1) + C$
47. $\frac{1}{3}\ln|3x+1| + C$ 49. $\ln(e^x+1) + C$

Section 7.4

1. $C(x) = x^3 - 21x^2 + 360x + 3{,}025$; $C(10) = \$5{,}525$
3. $C(x) = .0002x^3 - .3x^2 + 50x + 100{,}000$; $C(1{,}000) = \$50{,}000$
5. $C(x) = x^3 - 15x^2 + 2{,}000x + 4{,}000$; $C(10) = \$23{,}500$
7. $P(x) = 60x^{3/2} - 200x$; $P(250) = 187{,}170.82$ hundred dollars or \$18,717,082
9. $C(x) = .01x^3 + .1x^2 + .05x + 150$; $C(50) = \$1{,}652.50$

11. $C(x) = \frac{10}{3}x^3 + 25x^2 - 180x$; $C(18) = \$24{,}300$
13. $R(x) = \frac{3}{2}x^{2/3} + 0.1x + 10$; $R(27) = 26.2$ thousand dollars or \$26,200

Section 7.5

1. 63   3. $4\frac{1}{2}$ 5. $20\frac{1}{4}$ 7. 4   9. $2\frac{2}{3}$
11. $61\frac{4}{5}$ 13. 0   15. 1   17. 4   19. 2.773
21. $\frac{52}{3} = \frac{5}{3} + \frac{47}{3}$; true   23. $\frac{3}{2} - \frac{1}{e} = \frac{1}{2} + -\frac{1}{e} + 1$; true   25. $\sqrt{2} - 1 = \sqrt{2} - 1$; true

Section 7.6

1. $A = 28$ 3. 

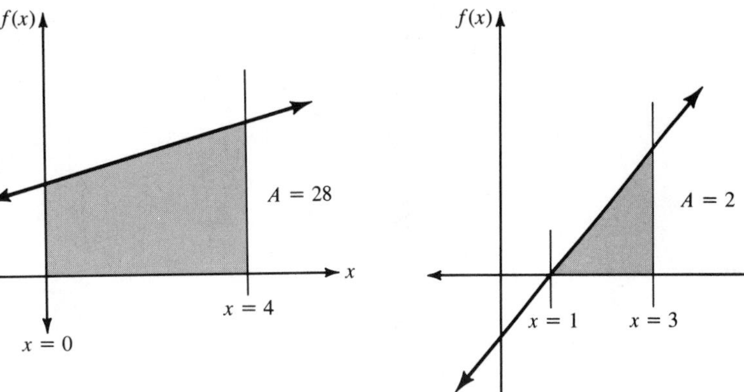

   7. $A = 2\frac{2}{3}$

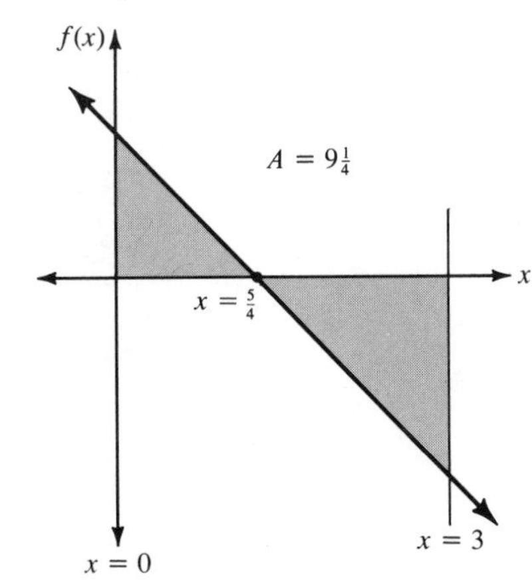

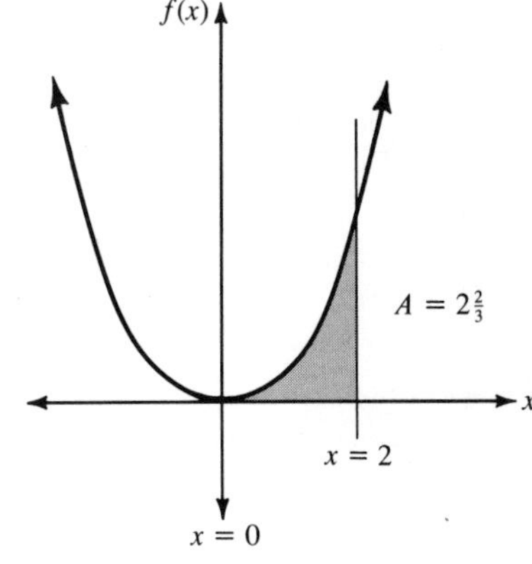

9. $A = 6$

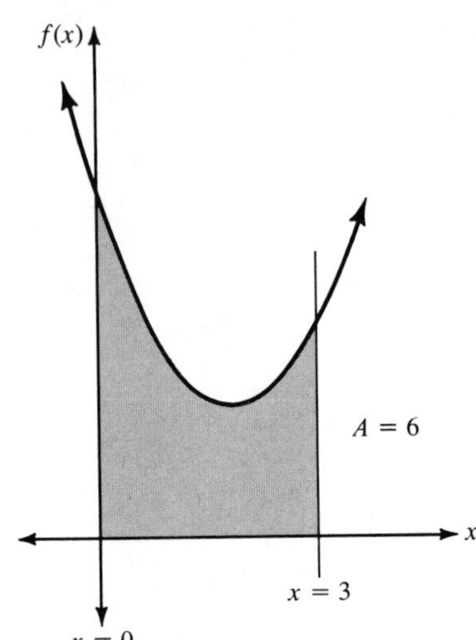

11. $A = 13$

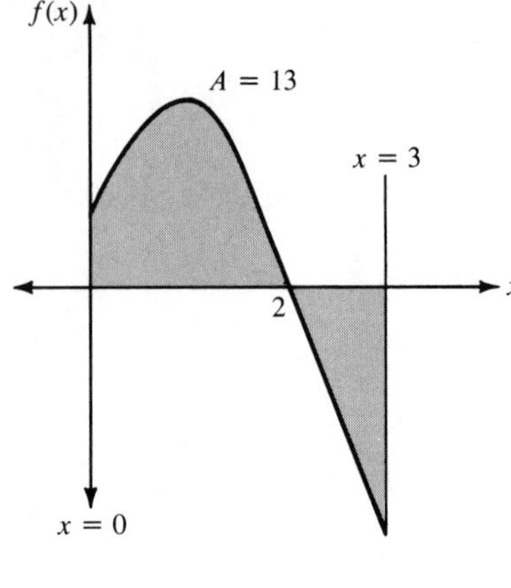

13. $A = \frac{1}{4}$

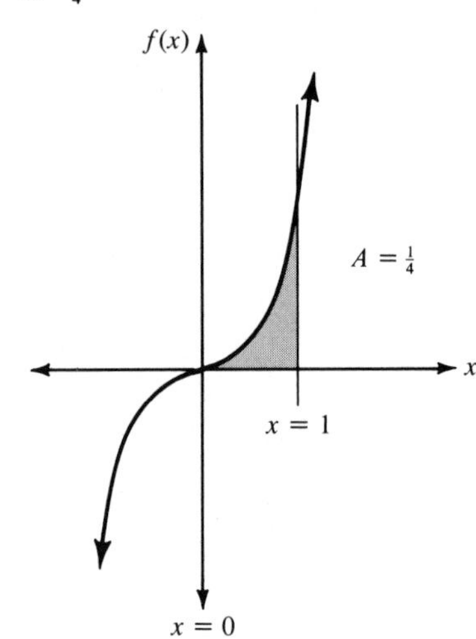

15. $A = \frac{11}{12}$

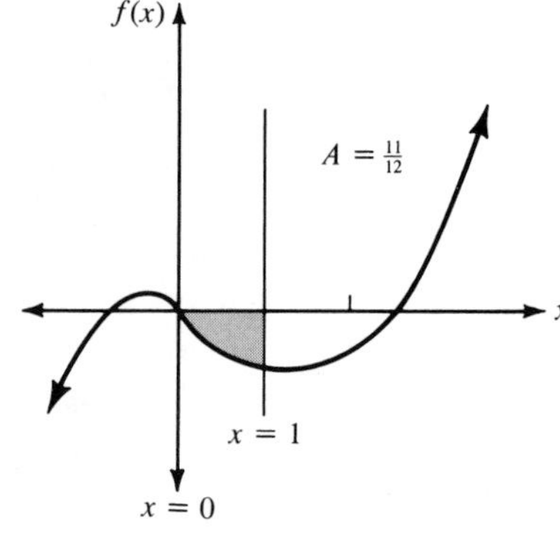

17. $A = \dfrac{e^6 - 1}{2}$
$\doteq 201.2$

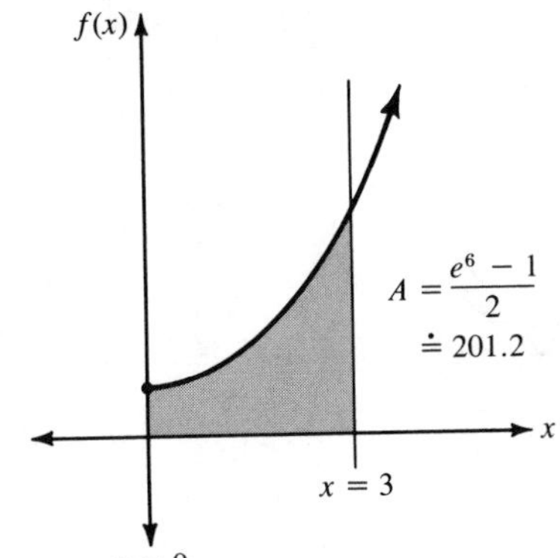

19. $A = 2 \ln 2$
$\doteq 1.3863$

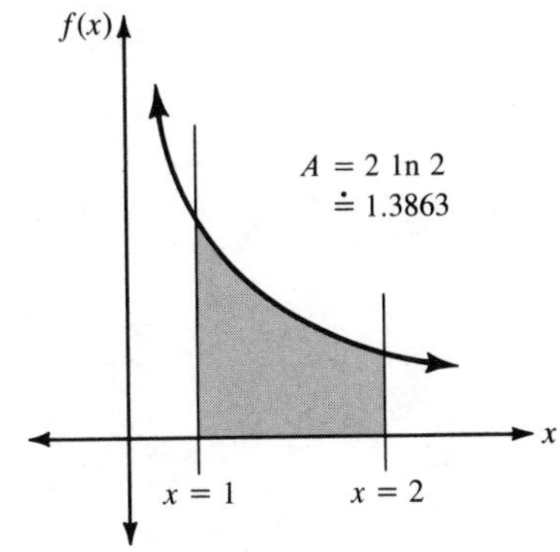

21. $A = 10\frac{2}{3}$

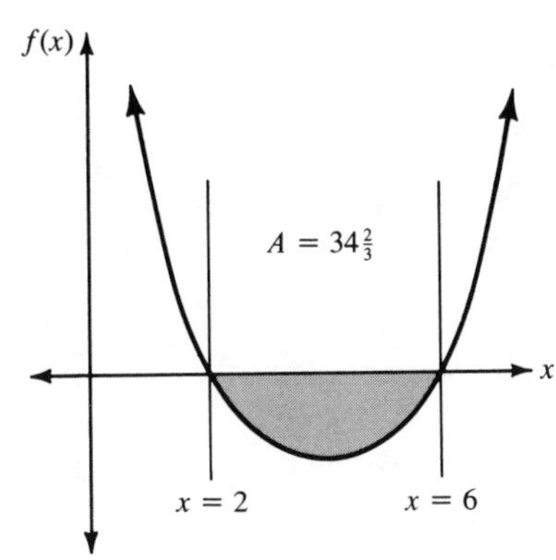

23. $A = 1$

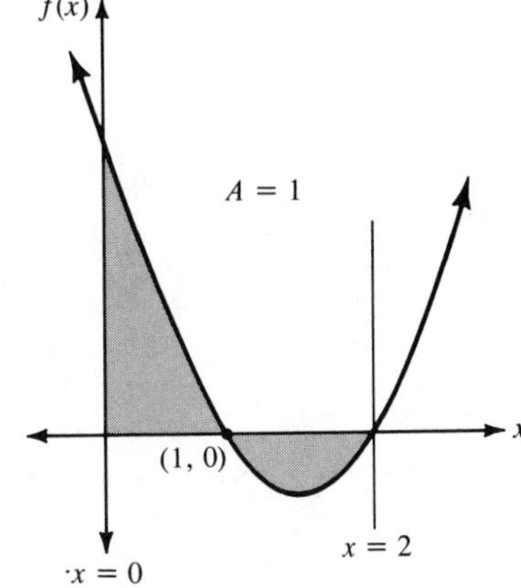

25. $A = \frac{1}{12}$

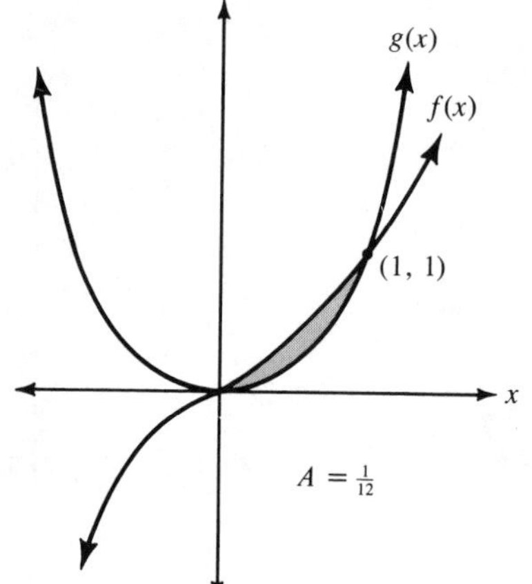

27. $A = \frac{1}{3}$

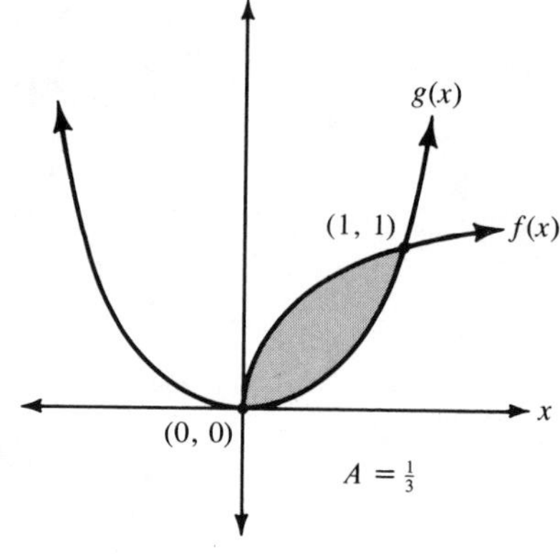

29. $A = 7\frac{1}{3}$

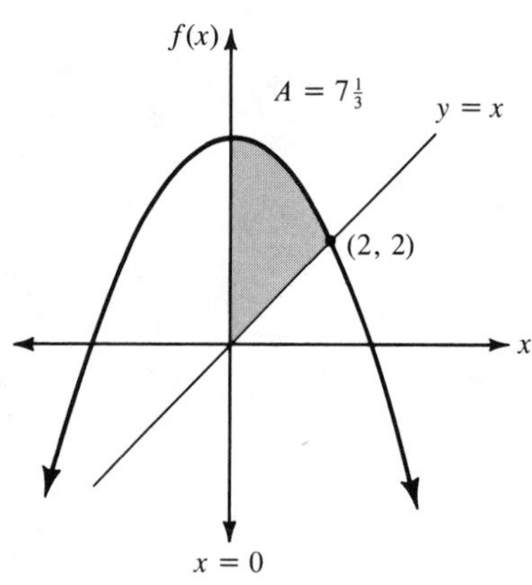

31. $A = 9$

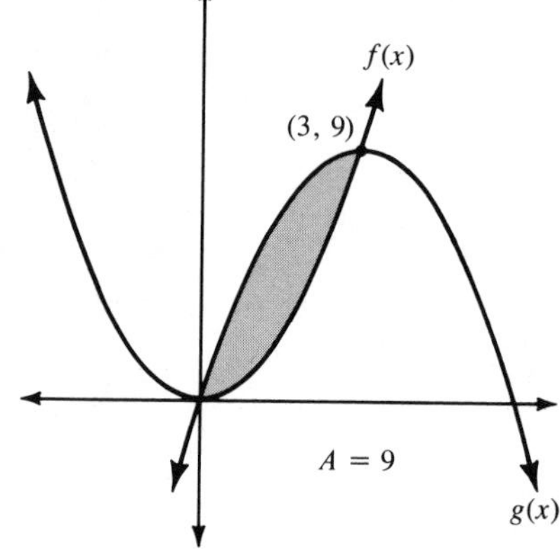

33. $A=\frac{5}{12}$

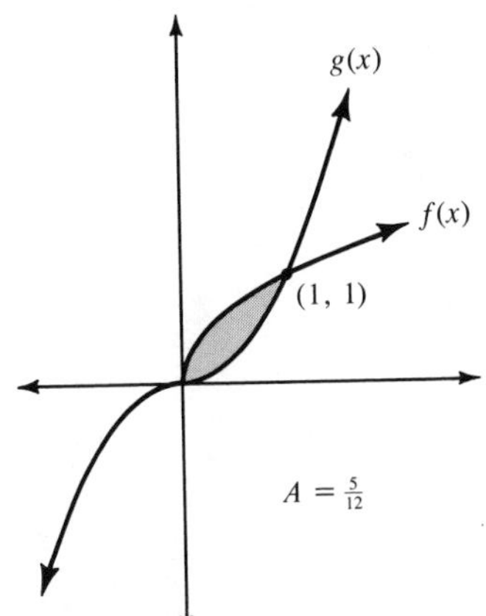

35. $A=72$

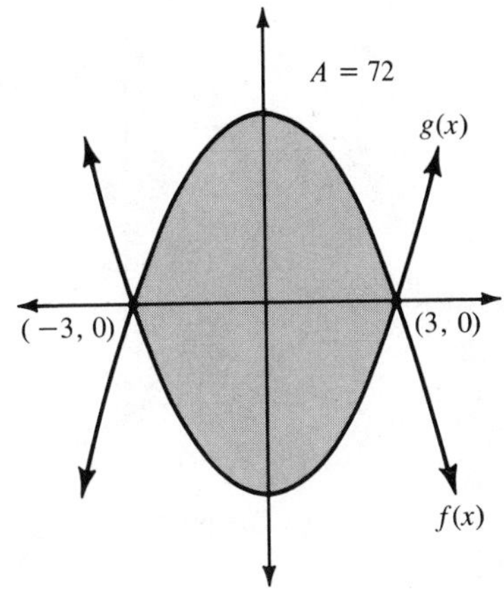

Section 7.7

1. \$1,620 3. 1,886 5. \$29,192
7. (a) $P(0)=\$10{,}000/\text{year}$ (b) $P(5)=\$18{,}550/\text{year}$ (c) \$82,062
9. (a) $H(t)=13e^{.13t}$ (billion dollars/year) (b) 91.4 billion dollars
(c) 602.9 billion dollars
11. (a) 648.7 million tons (b) in the year 2004 13. \$2,691 15. \$1.80
17. \$11,843 19. \$100 21. \$838.92
23. (a)

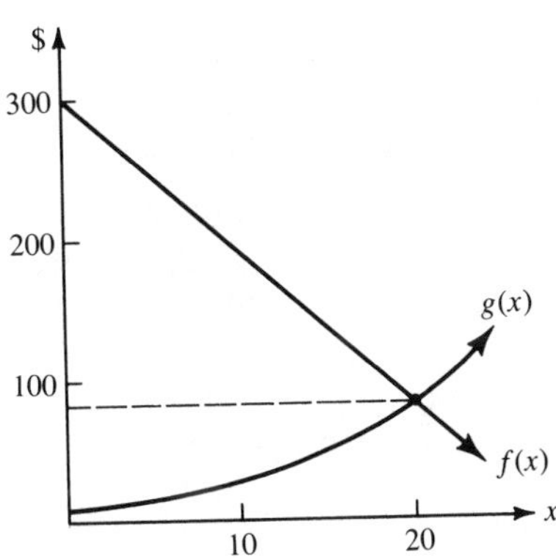

(b) \$90; 20 units
(c) \$2000
(d) \$1066.67

25. (a)

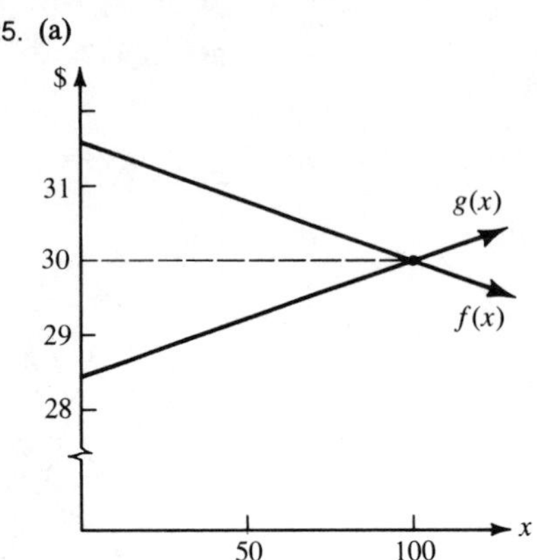

(b) \$30; 100 units
(c) \$81.85
(d) \$84.95

27. (a)

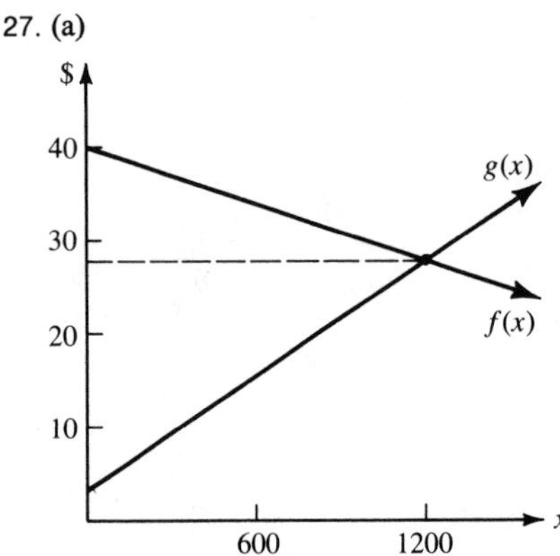

(b) \$28; 1200 units
(c) \$7200
(d) \$14,400

CHAPTER 8

Section 8.2

1. $\frac{1}{15}(3x+6)^5+C$
3. $\frac{1}{7}(x+2)^7+C$
5. $\frac{1}{5}(x^3+2)^5+C$
7. $-\frac{5}{3}(2-3t^2)^{1/2}+C$
9. $\frac{1}{2}\ln|y^2+4y+5|+C$
11. $\frac{1}{2}(z^4+2)^{1/2}+C$
13. $\frac{1}{2}e^{x^2}+C$
15. $-18(9-t)^{1/2}+\frac{2}{3}(9-t)^{3/2}+C$
17. -4
19. $\frac{61}{3}$
21. $9\frac{1}{3}$
23. $\frac{1}{2}\left(e^3-\frac{1}{e}\right)\doteq 9.859$
25. $C(x)=\frac{1}{3}(x^2+16)^{3/2}+228.67$; \$270.34
27. $P(x)=\frac{1}{6}(x^4+3600)^{3/2}-61{,}500$; \$202,836; \$368,400

Section 8.3

1. $\frac{1}{7}x(x+2)^7-\frac{1}{56}(x+2)^8+C$
3. $\frac{1}{3}xe^{3x}-\frac{1}{9}e^{3x}+C$
5. $\frac{1}{5}x^2(x+1)^5-\frac{1}{15}x(x+1)^6+\frac{1}{105}(x+1)^7+C$
7. $\frac{1}{3}x^3\ln x-\frac{1}{9}x^3+C$
9. $\frac{1}{9}(x+2)^2(x+5)^9-\frac{1}{45}(x+2)(x+5)^{10}+\frac{1}{495}(x+5)^{11}+C$
11. $\frac{1}{4}x^2e^{2x^2}-\frac{1}{8}e^{2x^2}+C$
13. $\frac{3}{4}x(x+5)^{4/3}-\frac{9}{28}(x+5)^{7/3}+C$
15. $2x(x+4)^{1/2}-\frac{4}{3}(x+4)^{3/2}+C$
17. $4\frac{3}{10}$
19. $\frac{3}{4}(5e^4-1)\doteq 204$
21. $P(x)=\dfrac{(x-1)(x-5)^3}{3}-\dfrac{(x-5)^4}{12}-\dfrac{175}{12}$; \$308.33; \$3,728.25

Section 8.4

1. $(5+x)-5\ln|5+x|+C$
3. $\frac{1}{320}(3+8x)^5-\frac{3}{256}(3+8x)^4+C$
5. $\ln|x+\sqrt{x^2-16}\,|+C$
7. $-\frac{1}{6}e^{-6x}+C$
9. $\dfrac{1}{2\ln 3}3^{2x}+C$

11. $3x\ln(2x)-3x+C$
13. $\frac{1}{3}(x^2+4)^{2/3}+C$
15. 0.6363
17. $x(\ln 3x)^2-2x\ln(3x)+2x+C$

Section 8.5

1. 2.7500 (exact 2.6667)
3. 1.6290 (exact 1.6094)
5. 2.6667
7. 1.6108
9. (a) 1.1038 (b) 1.1051
11. 250,400 square meters

Section 8.6

1. Divergent
3. Convergent; $\frac{1}{2}$
5. Convergent; $\frac{1}{3}$
7. Convergent; 0
9. Convergent; $\dfrac{1}{2e}$
11. 1
13. 1
15. $p>0$

Section 8.7

1. Yes
3. No (area not equal to 1)
5. Yes
7. No $[f(x)<0]$
9. No (area not equal to 1)
11. Yes
13. $k=\frac{1}{2}$; $E(X)=1$
15. $k=3$; $E(X)=1\frac{1}{2}$
17. $k=\frac{1}{8}$; $E(X)=1\frac{1}{3}$
19. $k=\frac{2}{33}$; $E(X)=\frac{39}{22}$
21. $k=\frac{3}{4}$; $E(X)=1$
23. (a) $\frac{1}{4}$ (b) $\frac{1}{4}$ (c) $\frac{1}{2}$ (d) $\frac{1}{2}$
25. (a) $\frac{1}{2}$ (b) $\frac{1}{4}$ (c) 1 (d) $\frac{1}{4}$
27. (a) $\frac{7}{65}$ (b) $\frac{56}{65}$ (c) $\frac{2}{5}$ (d) $\frac{7}{65}$
29. (a) $\frac{11}{16}$ (b) 1 (c) $\frac{189}{256}$ (d) 1
31. 0.0256; 0.5904; $E(T)=8$ hr
33. 0.104; 0.648; $E(T)=2\frac{1}{2}$ min
35. $\frac{56}{225}$; $\frac{40}{225}$; $E(T)=5$ weeks
37. $k=60$; .3; $\frac{9}{115}$

CHAPTER 9

Section 9.2

1. (a) $3x^2+3xy+x+3y+3y^2$ (b) $x^2+5xy+3x-3y-y^2$
   (c) $2x^4-2x^2+2x^3y+6xy+15xy^2+x^2y^2+7xy^3+3y^3+2y^4$
   (d) $\dfrac{2x^2+4xy+2x+y^2}{x^2-xy+3y-x+2y^2}$
3. (a) All real ordered pairs (x,y); $f(2,-1)=17$; $f(-1,2)=11$
   (b) All real ordered triples (x_1,x_2,x_3)
   $f(2,2,1)=0$, $f(1,1,2)=7$
   (c) All real ordered pairs (r,s) such that $r\geqslant 0$ and $s\geqslant 2$ or $r<0$ and $s<2$
   $f(2,3)=\sqrt{2}$ $f(3,2)=0$
   (d) All real ordered pairs (u,v) such that $3u\neq 2v$
   $g(1,1)=6$, $g(1,4)=\frac{-21}{5}$
   (e) All real ordered triples (x,y,z) such that $y^2\geqslant x^2$
   $h(1,2,3)=3+\sqrt{5}$ $h(4,5,6)=6+\sqrt{11}$
5. (a) $x+y+z=8$

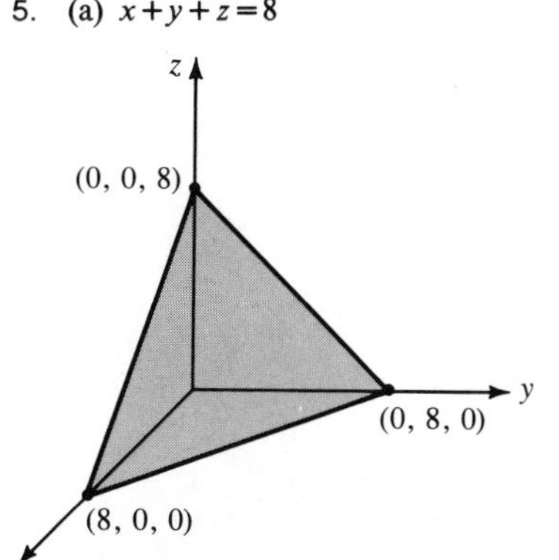

(b)

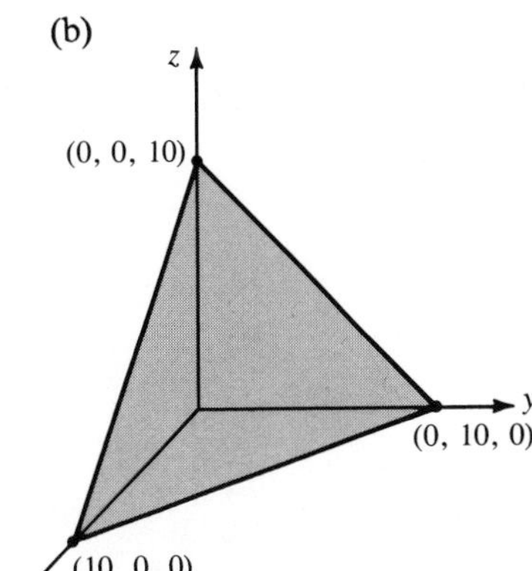

(c) $2x+4y+3z=12$

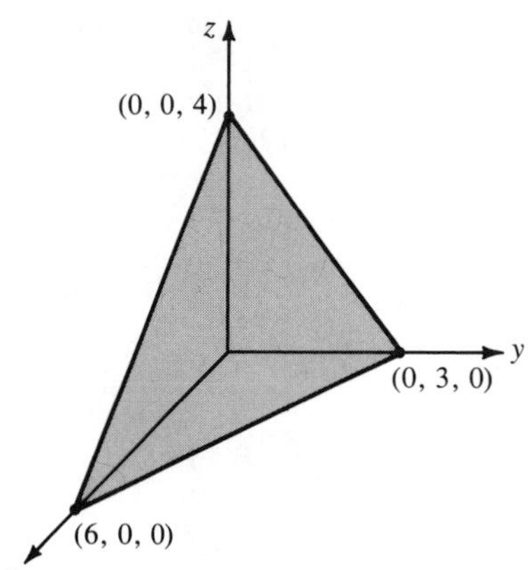

(d)

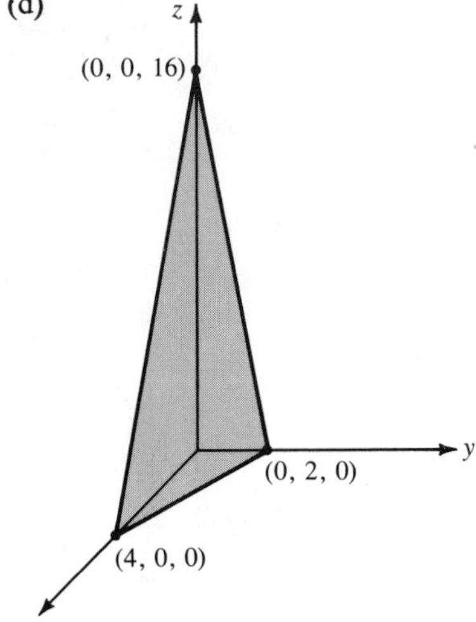

(e) $x+y=8$

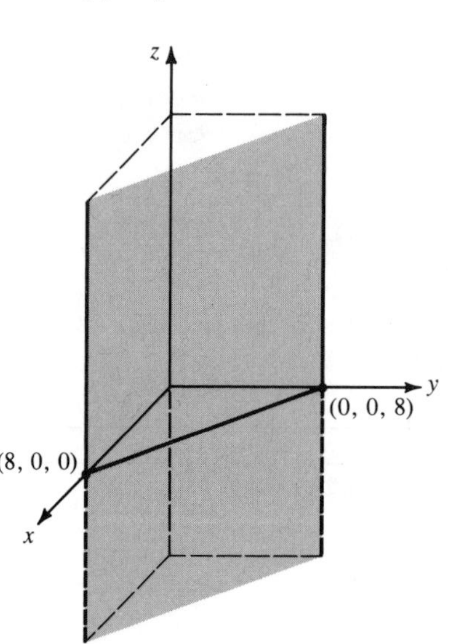

(f)

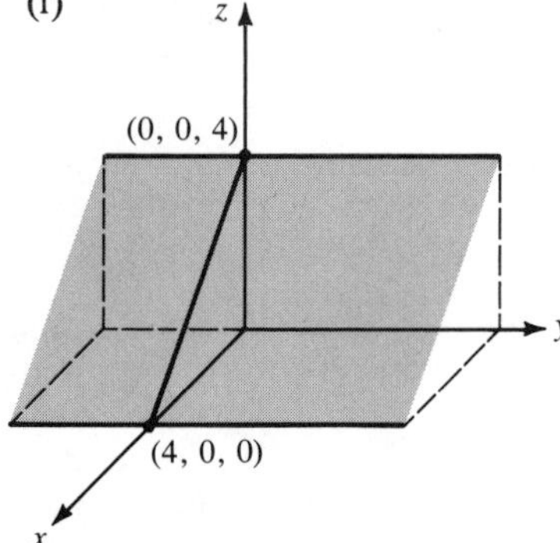

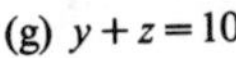
(g) $y+z=10$

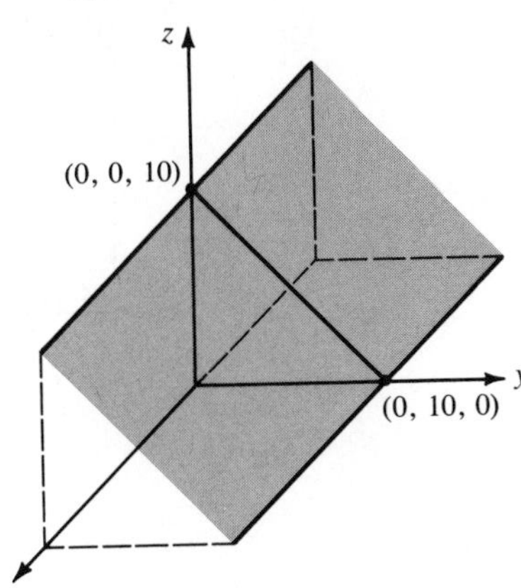

(h)

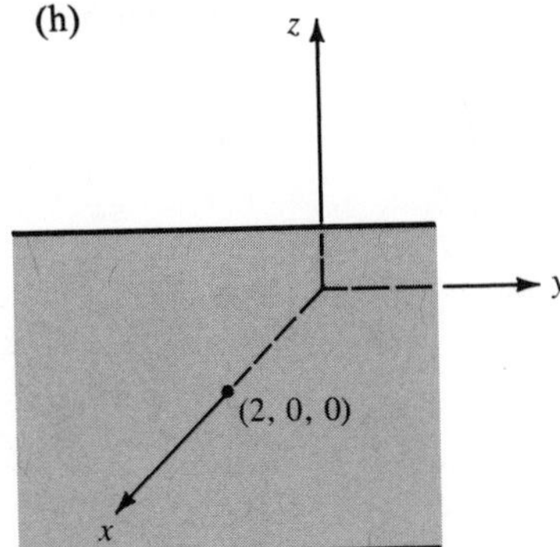

(i) $y=10$

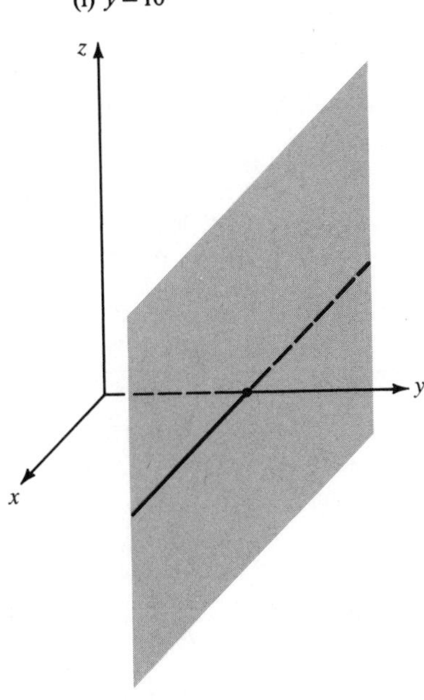

(j)

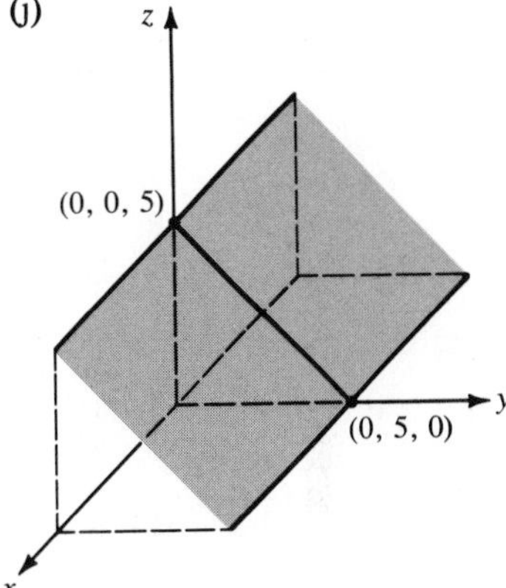

7. (a) $15x+12y=33{,}000$

(b)

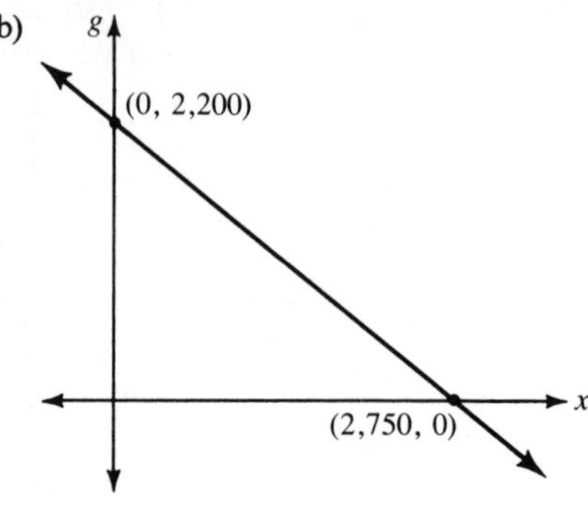

(c) 2,750 scientific models or 2,200 financial models

Section 9.3

1. $\dfrac{\partial f}{\partial x}=0$; $\dfrac{\partial f}{\partial y}=3$
3. $\partial f/\partial x=2$; $\partial f/\partial y=0$
5. $\partial f/\partial x=2x+2y$; $\partial f/\partial y=2x$
7. $\partial f/\partial x=6y+2x-4$; $\partial f/\partial y=2y+6x+3$
9. $\partial f/\partial x=2xye^{x^2y}$; $\partial f/\partial y=x^2e^{x^2y}$
11. $\partial f/\partial x=(3x^2+2xy)^2(18x+6y)$; $\partial f/\partial y=6x(3x^2+2xy)^2$
13. $\partial f/\partial x=2x4^{x^2-y^2}\ln 4$; $\partial f/\partial y=-2y4^{x^2-y^2}\ln 4$
15. $\partial f/\partial x=y/z$; $\partial f/\partial y=x/z$; $\partial f/\partial z=-xy/z^2$
19. $\partial f/\partial x=2x$; $\partial f/\partial y=6y$; $\partial^2 f/\partial x^2=2$; $\partial^2 f/\partial y^2=6$; $\partial^2 f/\partial x\,\partial y=0$
21. $\partial f/\partial x=y+4x$; $\partial f/\partial y=x-6y$; $\partial^2 f/\partial x^2=4$; $\partial^2 f/\partial y^2=-6$; $\partial^2 f/\partial x\,\partial y=1$
23. $\partial f/\partial x=2+3/y$; $\partial f/\partial y=-4-3x/y^2$; $\partial^2 f/\partial x^2=0$; $\partial^2 f/\partial y^2=6x/y^3$; $\partial^2 f/\partial x\,\partial y=-3/y^2$
25. $\partial f/\partial x=4x+ye^x$; $\partial f/\partial y=-1+e^x$; $\partial^2 f/\partial x^2=4+ye^x$; $\partial^2 f/\partial y^2=0$; $\partial^2 f/\partial x\,\partial y=e^x$
27. $\partial f/\partial x=y/x+4y^3$; $\partial f/\partial y=\ln x+12xy^2$; $\partial^2 f/\partial x^2=-y/x^2$; $\partial^2 f/\partial y^2=24xy$ $\partial^2 f/\partial x\,\partial y=1/x+12y^2$
29. (a) 8 (b) 12
31. (a) 17 (b) 0
33. (a) $N(x,y)=x^2+y^3-4xy$ (b) $2x-4y$ (c) $3y^2-4x$ (d) $-8{,}51$ (e) 2, $6y$, $-4,-4$

Section 9.4

1. (0,2) minimum
3. (0,0) saddle point
5. $(\frac{10}{3},-\frac{2}{3})$ minimum
7. (0,0) saddle point; (1,1) minimum
9. (0,0) saddle point
11. $(\frac{1}{9},-\frac{1}{5})$ saddle point; $(\frac{1}{9},1)$ maximum; $(1,-\frac{1}{5})$ minimum; (1,1) saddle point
13. (a) (0,0) saddle point; (32,16) minimum (b) 32 skilled; 16 unskilled (c) \$69,520
15. (a) $P(x,y)=-9x^2+84x+4xy+8y-y^2$ (b) (10,24) maximum (c) \$10,000 inventory value; 24,000 sq ft of floor space (d) \$516,000
17. (a) (0,0) saddle point; (.66,.55) minimum (b) \$.66 material; \$.55 labor (c) \$2.34
19. (a) (28.15,20.17) maximum (b) 20.17 boxes; 28.15 hr (c) \$292.02 (d) \$.058

Section 9.5

1. $(4,\frac{17}{2})$ minimum; $\lambda=-\frac{47}{2}$
3. $(2,-20)$ minimum; $\lambda=-40$
5. (3,5) minimum; $\lambda=-7$
7. Test fails
9. (10,10) minimum; $\lambda=-90$

11. 4,850 machines; 48.5 employees; \$2,352.25 profit
13. \$5 material; \$15 price; 200 demand

Section 9.6
1. 67.5 3. $260\frac{11}{12}$ 5. 10 7. $35\frac{2}{3}$ 9. $\frac{3}{20}$
11. $1\frac{1}{6}$ 13. $5\frac{1}{3}$ 15. $\frac{1}{3}$ 17. $37\frac{1}{3}$ cubic units 19. $27\frac{85}{162}$ cubic units

APPENDIX

Table A.1 Common Logarithms of Numbers

N	0	1	2	3	4	5	6	7	8	9
10	0000	0043	0086	0128	0170	0212	0253	0294	0334	0374
11	0414	0453	0492	0531	0569	0607	0645	0682	0719	0755
12	0792	0828	0864	0899	0934	0969	1004	1038	1072	1106
13	1139	1173	1206	1239	1271	1303	1335	1367	1399	1430
14	1461	1492	1523	1553	1584	1614	1644	1673	1703	1732
15	1761	1790	1818	1847	1875	1903	1931	1959	1987	2014
16	2041	2068	2095	2122	2148	2175	2201	2227	2253	2279
17	2304	2330	2355	2380	2405	2430	2455	2480	2504	2529
18	2553	2577	2601	2625	2648	2672	2695	2718	2742	2765
19	2788	2810	2833	2856	2878	2900	2923	2945	2967	2989
20	3010	3032	3054	3075	3096	3118	3139	3160	3181	3201
21	3222	3243	3263	3284	3304	3324	3345	3365	3385	3404
22	3424	3444	3464	3483	3502	3522	3541	3560	3579	3598
23	3617	3636	3655	3674	3692	3711	3729	3747	3766	3784
24	3802	3820	3838	3856	3874	3892	3909	3927	3945	3962
25	3979	3997	4014	4031	4048	4065	4082	4099	4116	4133
26	4150	4166	4183	4200	4216	4232	4249	4265	4281	4298
27	4314	4330	4346	4362	4378	4393	4409	4425	4440	4456
28	4472	4487	4502	4518	4533	4548	4564	4579	4594	4609
29	4624	4639	4654	4669	4683	4698	4713	4728	4742	4757
30	4771	4786	4800	4814	4829	4843	4857	4871	4886	4900
31	4914	4928	4942	4955	4969	4983	4997	5011	5024	5038
32	5051	5065	5079	5092	5105	5119	5132	5145	5159	5172
33	5185	5198	5211	5224	5237	5250	5263	5276	5289	5302
34	5315	5328	5340	5353	5366	5378	5391	5403	5416	5428
35	5441	5453	5465	5478	5490	5502	5514	5527	5539	5551
36	5563	5575	5587	5599	5611	5623	5635	5647	5658	5670
37	5682	5694	5705	5717	5729	5740	5752	5763	5775	5786
38	5798	5809	5821	5832	5843	5855	5866	5877	5888	5899
39	5911	5922	5933	5944	5955	5966	5977	5988	5999	6010
40	6021	6031	6042	6053	6064	6075	6085	6096	6107	6117
41	6128	6138	6149	6160	6170	6180	6191	6201	6212	6222
42	6232	6243	6253	6263	6274	6284	6294	6304	6314	6325
43	6335	6345	6355	6365	6375	6385	6395	6405	6415	6425
44	6435	6444	6454	6464	6474	6484	6493	6503	6513	6522
45	6532	6542	6551	6561	6571	6580	6590	6599	6609	6618
46	6628	6637	6646	6656	6665	6675	6684	6693	6702	6712
47	6721	6730	6739	6749	6758	6767	6776	6785	6794	6803
48	6812	6821	6830	6839	6848	6857	6866	6875	6884	6893
49	6902	6911	6920	6928	6937	6946	6955	6964	6972	6981
50	6990	6998	7007	7016	7024	7033	7042	7050	7059	7067
51	7076	7084	7093	7101	7110	7118	7126	7135	7143	7152
52	7160	7168	7177	7185	7193	7202	7210	7218	7226	7235
53	7243	7251	7259	7267	7275	7284	7292	7300	7308	7316
54	7324	7332	7340	7348	7356	7364	7372	7380	7388	7396
55	7404	7412	7419	7427	7435	7443	7451	7459	7466	7474
56	7482	7490	7497	7505	7513	7520	7528	7536	7543	7551
57	7559	7566	7574	7582	7589	7597	7604	7612	7619	7627
58	7634	7642	7649	7657	7664	7672	7679	7686	7694	7701
59	7709	7716	7723	7731	7738	7745	7752	7760	7767	7774
60	7782	7789	7796	7803	7810	7818	7825	7832	7839	7846
61	7853	7860	7868	7875	7882	7889	7896	7903	7910	7917
62	7924	7931	7938	7945	7952	7959	7966	7973	7980	7987
63	7993	8000	8007	8014	8021	8028	8035	8041	8048	8055
64	8062	8069	8075	8082	8089	8096	8102	8109	8116	8122
65	8129	8136	8142	8149	8156	8162	8169	8176	8182	8189
66	8195	8202	8209	8215	8222	8228	8235	8241	8248	8254
67	8261	8267	8274	8280	8287	8293	8299	8306	8312	8319
68	8325	8331	8338	8344	8351	8357	8363	8370	8376	8382
69	8388	8395	8401	8407	8414	8420	8426	8432	8439	8445
70	8451	8457	8463	8470	8476	8482	8488	8494	8500	8506
71	8513	8519	8525	8531	8537	8543	8549	8555	8561	8567
72	8573	8579	8585	8591	8597	8603	8609	8615	8621	8627
73	8633	8639	8645	8651	8657	8663	8669	8675	8681	8686
74	8692	8698	8704	8710	8716	8722	8727	8733	8739	8745
75	8751	8756	8762	8768	8774	8779	8785	8791	8797	8802
76	8808	8814	8820	8825	8831	8837	8842	8848	8854	8859
77	8865	8871	8876	8882	8887	8893	8899	8904	8910	8915
78	8921	8927	8932	8938	8943	8949	8954	8960	8965	8971
79	8976	8982	8987	8993	8998	9004	9009	9015	9020	9025
80	9031	9036	9042	9047	9053	9058	9063	9069	9074	9079
81	9085	9090	9096	9101	9106	9112	9117	9122	9128	9133
82	9138	9143	9149	9154	9159	9165	9170	9175	9180	9186
83	9191	9196	9201	9206	9212	9217	9222	9227	9232	9238
84	9243	9248	9253	9258	9263	9269	9274	9279	9284	9289
85	9294	9299	9304	9309	9315	9320	9325	9330	9335	9340
86	9345	9350	9355	9360	9365	9370	9375	9380	9385	9390
87	9395	9400	9405	9410	9415	9420	9425	9430	9435	9440
88	9445	9450	9455	9460	9465	9469	9474	9479	9484	9489
89	9494	9499	9504	9509	9513	9518	9523	9528	9533	9538
90	9542	9547	9552	9557	9562	9566	9571	9576	9581	9586
91	9590	9595	9600	9605	9609	9614	9619	9624	9628	9633
92	9638	9643	9647	9652	9657	9661	9666	9671	9675	9680
93	9685	9689	9694	9699	9703	9708	9713	9717	9722	9727
94	9731	9736	9741	9745	9750	9754	9759	9763	9768	9773
95	9777	9782	9786	9791	9795	9800	9805	9809	9814	9818
96	9823	9827	9832	9836	9841	9845	9850	9854	9859	9863
97	9868	9872	9877	9881	9886	9890	9894	9899	9903	9908
98	9912	9917	9921	9926	9930	9934	9939	9943	9948	9952
99	9956	9961	9965	9969	9974	9978	9983	9987	9991	9996
N	0	1	2	3	4	5	6	7	8	9

Source: This table is reproduced from *Handbook of Probability and Statistics with Tables* by R. S. Burington and D. C. May, second edition.

Table A.2 Natural, or Naperian, Logarithms

To find the natural logarithm of a number which is 1/10, 1/100, 1/1000, etc. of a number whose logarithm is given, subtract from the given logarithm $\log_e 10$, $2 \log_e 10$, $3 \log_e 10$, etc.

To find the natural logarithm of a number which is 10, 100, 1000, etc. times a number whose logarithm is given, add to the given logarithm $\log_e 10$, $2 \log_e 10$, $3 \log_e 10$, etc.

$\log_e 10 = 2.30259$	$6 \log_e 10 = 13.81551$
$2 \log_e 10 = 4.60517$	$7 \log_e 10 = 16.11810$
$3 \log_e 10 = 6.90776$	$8 \log_e 10 = 18.42068$
$4 \log_e 10 = 9.21034$	$9 \log_e 10 = 20.72327$
$5 \log_e 10 = 11.51293$	$10 \log_e 10 = 23.02585$

N	0	1	2	3	4	5	6	7	8	9
1.0	0.00000	.00995	.01980	.02956	.03922	.04879	.05827	.06766	.07696	.08618
.1	.09531	.10436	.11333	.12222	.13103	.13976	.14842	.15700	.16551	.17395
.2	.18232	.19062	.19885	.20701	.21511	.22314	.23111	.23902	.24686	.25464
.3	.26236	.27003	.27763	.28518	.29267	.30010	.30748	.31481	.32208	.32930
.4	.33647	.34359	.35066	.35767	.36464	.37156	.37844	.38526	.39204	.39878
.5	.40547	.41211	.41871	.42527	.43178	.43825	.44469	.45108	.45742	.46373
.6	.47000	.47623	.48243	.48858	.49470	.50078	.50682	.51282	.51879	.52473
.7	.53063	.53649	.54232	.54812	.55389	.55962	.56531	.57098	.57661	.58222
.8	.58779	.59333	.59884	.60432	.60977	.61519	.62058	.62594	.63127	.63658
.9	.64185	.64710	.65233	.65752	.66269	.66783	.67294	.67803	.68310	.68813
2.0	0.69315	.69813	.70310	.70804	.71295	.71784	.72271	.72755	.73237	.73716
.1	.74194	.74669	.75142	.75612	.76081	.76547	.77011	.77473	.77932	.78390
.2	.78846	.79299	.79751	.80200	.80648	.81093	.81536	.81978	.82418	.82855
.3	.83291	.83725	.84157	.84587	.85015	.85442	.85866	.86289	.86710	.87129
.4	.87547	.87963	.88377	.88789	.89200	.89609	.90016	.90422	.90826	.91228
.5	.91629	.92028	.92426	.92822	.93216	.93609	.94001	.94391	.94779	.95166
.6	.95551	.95935	.96317	.96698	.97078	.97456	.97833	.98208	.98582	.98954
.7	.99325	.99695	*.00063	*.00430	*.00796	*.01160	*.01523	*.01885	*.02245	*.02604
.8	1.02962	.03318	.03674	.04028	.04380	.04732	.05082	.05431	.05779	.06126
.9	.06471	.06815	.07158	.07500	.07841	.08181	.08519	.08856	.09192	.09527
3.0	1.09861	.10194	.10526	.10856	.11186	.11514	.11841	.12168	.12493	.12817
.1	.13140	.13462	.13783	.14103	.14422	.14740	.15057	.15373	.15688	.16002
.2	.16315	.16627	.16938	.17248	.17557	.17865	.18173	.18479	.18784	.19089
.3	.19392	.19695	.19996	.20297	.20597	.20896	.21194	.21491	.21788	.22083
.4	.22378	.22671	.22964	.23256	.23547	.23837	.24127	.24415	.24703	.24990
.5	.25276	.25562	.25846	.26130	.26413	.26695	.26976	.27257	.27536	.27815
.6	.28093	.28371	.28647	.28923	.29198	.29473	.29746	.30019	.30291	.30563

Source: This table is reproduced from S. Selby, *Standard Mathematical Tables*, 20th edition (Cleveland, Ohio: The Chemical Rubber Company, 1972). Used by permission.

Table A.2 (cont.)

N	0	1	2	3	4	5	6	7	8	9
.7	.30833	.31103	.31372	.31641	.31909	.32176	.32442	.32708	.32972	.33237
.8	.33500	.33763	.34025	.34286	.34547	.34807	.35067	.35325	.35584	.35841
.9	.36098	.36354	.36609	.36864	.37118	.37372	.37624	.37877	.38128	.38379
4.0	1.38629	.38879	.39128	.39377	.39624	.39872	.40118	.40364	.40610	.40854
.1	.41099	.41342	.41585	.41828	.42070	.42311	.42552	.42792	.43031	.43270
.2	.43508	.43746	.43984	.44220	.44456	.44692	.44927	.45161	.45395	.45629
.3	.45862	.46094	.46326	.46557	.46787	.47018	.47247	.47476	.47705	.47933
.4	.48160	.48387	.48614	.48840	.49065	.49290	.49515	.49739	.49962	.50185
.5	.50408	.50630	.50851	.51072	.51293	.51513	.51732	.51951	.52170	.52388
.6	.52606	.52823	.53039	.53256	.53471	.53687	.53902	.54116	.54330	.54543
.7	.54756	.54969	.55181	.55393	.55604	.55814	.56025	.56235	.56444	.56653
.8	.56862	.57070	.57277	.57485	.57691	.57898	.58104	.58309	.58515	.58719
.9	.58924	.59127	.59331	.59534	.59737	.59939	.60141	.60342	.60543	.60744
5.0	1.60944	.61144	.61343	.61542	.61741	.61939	.62137	.62334	.62531	.62728
.1	.62924	.63120	.63315	.63511	.63705	.63900	.64094	.64287	.64481	.64673
.2	.64866	.65058	.65250	.65441	.65632	.65823	.66013	.66203	.66393	.66582
.3	.66771	.66959	.67147	.67335	.67523	.67710	.67896	.68083	.68269	.68455
.4	.68640	.68825	.69010	.69194	.69378	.69562	.69745	.69928	.70111	.70293
.5	.70475	.70656	.70838	.71019	.71199	.71380	.71560	.71740	.71919	.72098
.6	.72277	.72455	.72633	.72811	.72988	.73166	.73342	.73519	.73695	.73871
.7	.74047	.74222	.74397	.74572	.74746	.74920	.75094	.75267	.75440	.75613
.8	.75786	.75958	.76130	.76302	.76473	.76644	.76815	.76985	.77156	.77326
.9	.77495	.77665	.77834	.78002	.78171	.78339	.78507	.78675	.78842	.79009
6.0	1.79176	.79342	.79509	.79675	.79840	.80006	.80171	.80336	.80500	.80665
.1	.80829	.80993	.81156	.81319	.81482	.81645	.81808	.81970	.82132	.82294
.2	.82455	.82616	.82777	.82938	.83098	.83258	.83418	.83578	.83737	.83896
.3	.84055	.84214	.84372	.84530	.84688	.84845	.85003	.85160	.85317	.85473
.4	.85630	.85786	.85942	.86097	.86253	.86408	.86563	.86718	.86872	.87026
.5	.87180	.87334	.87487	.87641	.87794	.87947	.88099	.88251	.88403	.88555
.6	.88707	.88858	.89010	.89160	.89311	.89462	.89612	.89762	.89912	.90061
.7	.90211	.90360	.90509	.90658	.90806	.90954	.91102	.91250	.91398	.91545
.8	.91692	.91839	.91986	.92132	.92279	.92425	.92571	.92716	.92862	.93007
.9	.93152	.93297	.93442	.93586	.93730	.93874	.94018	.94162	.94305	.94448
7.0	1.94591	.94734	.94876	.95019	.95161	.95303	.95445	.95586	.95727	.95869
.1	.96009	.96150	.96291	.96431	.96571	.96711	.96851	.96991	.97130	.97269
.2	.97408	.97547	.97685	.97824	.97962	.98100	.98238	.98376	.98513	.98650
.3	.98787	.98924	.99061	.99198	.99334	.99470	.99606	.99742	.99877	*.00013
.4	2.00148	.00283	.00418	.00553	.00687	.00821	.00956	.01089	.01223	.01357
.5	.01490	.01624	.01757	.01890	.02022	.02155	.02287	.02419	.02551	.02683
.6	.02815	.02946	.03078	.03209	.03340	.03471	.03601	.03732	.03862	.03992
.7	.04122	.04252	.04381	.04511	.04640	.04769	.04898	.05027	.05156	.05284
.8	.05412	.05540	.05668	.05796	.05924	.06051	.06179	.06306	.06433	.06560

Table A.2 (cont.)

N	0	1	2	3	4	5	6	7	8	9
.9	.06686	.06813	.06939	.07065	.07191	.07317	.07443	.07568	.07694	.07819
8.0	2.07944	.08069	.08194	.08318	.08443	.08567	.08691	.08815	.08939	.09063
.1	.09186	.09310	.09433	.09556	.09679	.09802	.09924	.10047	.10169	.10291
.2	.10413	.10535	.10657	.10779	.10900	.11021	.11142	.11263	.11384	.11505
.3	.11626	.11746	.11866	.11986	.12106	.12226	.12346	.12465	.12585	.12704
.4	.12823	.12942	.13061	.13180	.13298	.13417	.13535	.13653	.13771	.13889
.5	.14007	.14124	.14242	.14359	.14476	.14593	.14710	.14827	.14943	.15060
.6	.15176	.15292	.15409	.15524	.15640	.15756	.15871	.15987	.16102	.16217
.7	.16332	.16447	.16562	.16677	.16791	.16905	.17020	.17134	.17248	.17361
.8	.17475	.17589	.17702	.17816	.17929	.18042	.18155	.18267	.18380	.18493
.9	.18605	.18717	.18830	.18942	.19054	.19165	.19277	.19389	.19500	.19611
9.0	2.19722	.19834	.19944	.20055	.20166	.20276	.20387	.20497	.20607	.20717
.1	.20827	.20937	.21047	.21157	.21266	.21375	.21485	.21594	.21703	.21812
.2	.21920	.22029	.22138	.22246	.22354	.22462	.22570	.22678	.22786	.22894
.3	.23001	.23109	.23216	.23324	.23431	.23538	.23645	.23751	.23858	.23965
.4	.24071	.24177	.24284	.24390	.24496	.24601	.24707	.24813	.24918	.25024
.5	.25129	.25234	.25339	.25444	.25549	.25654	.25759	.25863	.25968	.26072
.6	.26176	.26280	.26384	.26488	.26592	.26696	.26799	.26903	.27006	.27109
.7	.27213	.27316	.27419	.27521	.27624	.27727	.27829	.27932	.28034	.28136
.8	.28238	.28340	.28442	.28544	.28646	.28747	.28849	.28950	.29051	.29152
.9	.29253	.29354	.29455	.29556	.29657	.29757	.29858	.29958	.30058	.30158

Table A.3 Exponential Functions

x	e^x	e^{-x}	x	e^x	e^{-x}	x	e^x	e^{-x}
0.00	1.0000	1.000000	0.40	1.4918	0.670320	0.80	2.2255	0.449329
0.01	1.0101	0.990050	0.41	1.5068	.663650	0.81	2.2479	.444858
0.02	1.0202	.980199	0.42	1.5220	.657047	0.82	2.2705	.440432
0.03	1.0305	.970446	0.43	1.5373	.650509	0.83	2.2933	.436049
0.04	1.0408	.960789	0.44	1.5527	.644036	0.84	2.3164	.431711
0.05	1.0513	0.951229	0.45	1.5683	0.637628	0.85	2.3396	0.427415
0.06	1.0618	.941765	0.46	1.5841	.631284	0.86	2.3632	.423162
0.07	1.0725	.932394	0.47	1.6000	.625002	0.87	2.3869	.418952
0.08	1.0833	.923116	0.48	1.6161	.618783	0.88	2.4109	.414783
0.09	1.0942	.913931	0.49	1.6323	.612626	0.89	2.4351	.410656
0.10	1.1052	0.904837	0.50	1.6487	0.606531	0.90	2.4596	0.406570
0.11	1.1163	.895834	0.51	1.6653	.600496	0.91	2.4843	.402524
0.12	1.1275	.886920	0.52	1.6820	.594521	0.92	2.5093	.398519
0.13	1.1388	.878095	0.53	1.6989	.588605	0.93	2.5345	.394554
0.14	1.1503	.869358	0.54	1.7160	.582748	0.94	2.5600	.390628
0.15	1.1618	0.860708	0.55	1.7333	0.576950	0.95	2.5857	0.386741
0.16	1.1735	.852144	0.56	1.7507	.571209	0.96	2.6117	.382893
0.17	1.1853	.843665	0.57	1.7683	.565525	0.97	2.6379	.379083
0.18	1.1972	.835270	0.58	1.7860	.559898	0.98	2.6645	.375311
0.19	1.2092	.826959	0.59	1.8040	.554327	0.99	2.6912	.371577
0.20	1.2214	0.818731	0.60	1.8221	0.548812	1.00	2.7183	0.367879
0.21	1.2337	.810584	0.61	1.8404	.543351	1.01	2.7456	.364219
0.22	1.2461	.802519	0.62	1.8589	.537944	1.02	2.7732	.360595
0.23	1.2586	.794534	0.63	1.8776	.532592	1.03	2.8011	.357007
0.24	1.2712	.786628	0.64	1.8965	.527292	1.04	2.8292	.353455
0.25	1.2840	0.778801	0.65	1.9155	0.522046	1.05	2.8577	0.349938
0.26	1.2969	.771052	0.66	1.9348	.516851	1.06	2.8864	.346456
0.27	1.3100	.763379	0.67	1.9542	.511709	1.07	2.9154	.343009
0.28	1.3231	.755784	0.68	1.9739	.506617	1.08	2.9447	.339596
0.29	1.3364	.748264	0.69	1.9937	.501576	1.09	2.9743	.336216
0.30	1.3499	0.740818	0.70	2.0138	0.496585	1.10	3.0042	0.332871
0.31	1.3634	.733447	0.71	2.0340	.491644	1.11	3.0344	.329559
0.32	1.3771	.726149	0.72	2.0544	.486752	1.12	3.0649	.326280
0.33	1.3910	.718924	0.73	2.0751	.481909	1.13	3.0957	.323033
0.34	1.4049	.711770	0.74	2.0959	.477114	1.14	3.1268	.319819
0.35	1.4191	0.704688	0.75	2.1170	0.472367	1.15	3.1582	0.316637
0.36	1.4333	.697676	0.76	2.1383	.467666	1.16	3.1899	.313486
0.37	1.4477	.690734	0.77	2.1598	.463013	1.17	3.2220	.310367
0.38	1.4623	.683861	0.78	2.1815	.458406	1.18	3.2544	.307279
0.39	1.4770	.677057	0.79	2.2034	.453845	1.19	3.2871	.304221

Table A.3 (cont.)

x	e^x	e^{-x}	x	e^x	e^{-x}	x	e^x	e^{-x}
1.20	3.3201	0.301194	1.60	4.9530	0.201897	2.00	7.3891	0.135335
1.21	3.3535	.298197	1.61	5.0028	.199888	2.01	7.4633	.133989
1.22	3.3872	.295230	1.62	5.0531	.197899	2.02	7.5383	.132655
1.23	3.4212	.292293	1.63	5.1039	.195930	2.03	7.6141	.131336
1.24	3.4556	.289384	1.64	5.1552	.193980	2.04	7.6906	.130029
1.25	3.4903	0.286505	1.65	5.2070	0.192050	2.05	7.7679	0.128735
1.26	3.5254	.283654	1.66	5.2593	.190139	2.06	7.8460	.127454
1.27	3.5609	.280832	1.67	5.3122	.188247	2.07	7.9248	.126186
1.28	3.5966	.278037	1.68	5.3656	.186374	2.08	8.0045	.124930
1.29	3.6328	.275271	1.69	5.4195	.184520	2.09	8.0849	.123687
1.30	3.6693	0.272532	1.70	5.4739	0.182684	2.10	8.1662	0.122456
1.31	3.7062	.269820	1.71	5.5290	.180866	2.11	8.2482	.121238
1.32	3.7434	.267135	1.72	5.5845	.179066	2.12	8.3311	.120032
1.33	3.7810	.264477	1.73	5.6407	.177284	2.13	8.4149	.118837
1.34	3.8190	.261846	1.74	5.6973	.175520	2.14	8.4994	.117655
1.35	3.8574	0.259240	1.75	5.7546	0.173774	2.15	8.5849	0.116484
1.36	3.8962	.256661	1.76	5.8124	.172045	2.16	8.6711	.115325
1.37	3.9354	.254107	1.77	5.8709	.170333	2.17	8.7583	.114178
1.38	3.9749	.251579	1.78	5.9299	.168638	2.18	8.8463	.113042
1.39	4.0149	.249075	1.79	5.9895	.166960	2.19	8.9352	.111917
1.40	4.0552	0.246597	1.80	6.0496	0.165299	2.20	9.0250	0.110803
1.41	4.0960	.244143	1.81	6.1104	.163654	2.21	9.1157	.109701
1.42	4.1371	.241714	1.82	6.1719	.162026	2.22	9.2073	.108609
1.43	4.1787	.239309	1.83	6.2339	.160414	2.23	9.2999	.107528
1.44	4.2207	.236928	1.84	6.2965	.158817	2.24	9.3933	.106459
1.45	4.2631	0.234570	1.85	6.3598	0.157237	2.25	9.4877	0.105399
1.46	4.3060	.232236	1.86	6.4237	.155673	2.26	9.5831	.104350
1.47	4.3492	.229925	1.87	6.4883	.154124	2.27	9.6794	.103312
1.48	4.3929	.227638	1.88	6.5535	.152590	2.28	9.7767	.102284
1.49	4.4371	.225373	1.89	6.6194	.151072	2.29	9.8749	.101266
1.50	4.4817	0.223130	1.90	6.6859	0.149569	2.30	9.9742	0.100259
1.51	4.5267	.220910	1.91	6.7531	.148080	2.31	10.074	.099261
1.52	4.5722	.218712	1.92	6.8210	.146607	2.32	10.176	.098274
1.53	4.6182	.216536	1.93	6.8895	.145148	2.33	10.278	.097296
1.54	4.6646	.214381	1.94	6.9588	.143704	2.34	10.381	.096328
1.55	4.7115	0.212248	1.95	7.0287	0.142274	2.35	10.486	0.095369
1.56	4.7588	.210136	1.96	7.0993	.140858	2.36	10.591	.094420
1.57	4.8066	.208045	1.97	7.1707	.139457	2.37	10.697	.093481
1.58	4.8550	.205975	1.98	7.2427	.138069	2.38	10.805	.092551
1.59	4.9037	.203926	1.99	7.3155	.136695	2.39	10.913	.091630

Table A.3 (cont.)

x	e^x	e^{-x}	x	e^x	e^{-x}	x	e^x	e^{-x}
2.40	11.023	0.090718	2.80	16.445	0.060810	3.20	24.533	0.040762
2.41	11.134	.089815	2.81	16.610	.060205	3.21	24.779	.040357
2.42	11.246	.088922	2.82	16.777	.059606	3.22	25.028	.039955
2.43	11.359	.088037	2.83	16.945	.059013	3.23	25.280	.039557
2.44	11.473	.087161	2.84	17.116	.058426	3.24	25.534	.039164
2.45	11.588	0.086294	2.85	17.288	0.057844	3.25	25.790	0.038774
2.46	11.705	.085435	2.86	17.462	.057269	3.26	26.050	.038388
2.47	11.822	.084585	2.87	17.637	.056699	3.27	26.311	.038006
2.48	11.941	.083743	2.88	17.814	.056135	3.28	26.576	.037628
2.49	12.061	.082910	2.89	17.993	.055576	3.29	26.843	.037254
2.50	12.182	0.082085	2.90	18.174	0.055023	3.30	27.113	0.036883
2.51	12.305	.081268	2.91	18.357	.054476	3.31	27.385	.036516
2.52	12.429	.080460	2.92	18.541	.053934	3.32	27.660	.036153
2.53	12.554	.079659	2.93	18.728	.053397	3.33	27.938	.035793
2.54	12.680	.078866	2.94	18.916	.052866	3.34	28.219	.035437
2.55	12.807	0.078082	2.95	19.106	0.052340	3.35	28.503	0.035084
2.56	12.936	.077305	2.96	19.298	.051819	3.36	28.789	.034735
2.57	13.066	.076536	2.97	19.492	.051303	3.37	29.079	.034390
2.58	13.197	.075774	2.98	19.688	.050793	3.38	29.371	.034047
2.59	13.330	.075020	2.99	19.886	.050287	3.39	29.666	.033709
2.60	13.464	0.074274	3.00	20.086	0.049787	3.40	29.964	0.033373
2.61	13.599	.073535	3.01	20.287	.049292	3.41	30.265	.033041
2.62	13.736	.072803	3.02	20.491	.048801	3.42	30.569	.032712
2.63	13.874	.072078	3.03	20.697	.048316	3.43	30.877	.032387
2.64	14.013	.071361	3.04	20.905	.047835	3.44	31.187	.032065
2.65	14.154	0.070651	3.05	21.115	0.047359	3.45	31.500	0.031746
2.66	14.296	.069948	3.06	21.328	.046888	3.46	31.817	.031430
2.67	14.440	.069252	3.07	21.542	.046421	3.47	32.137	.031117
2.68	14.585	.068563	3.08	21.758	.045959	3.48	32.460	.030807
2.69	14.732	.067881	3.09	21.977	.045502	3.49	32.786	.030501
2.70	14.880	0.067206	3.10	22.198	0.045049	3.50	33.115	0.030197
2.71	15.029	.066537	3.11	22.421	.044601	3.51	33.448	.029897
2.72	15.180	.065875	3.12	22.646	.044157	3.52	33.784	.029599
2.73	15.333	.065219	3.13	22.874	.043718	3.53	34.124	.029305
2.74	15.487	.064570	3.14	23.104	.043283	3.54	34.467	.029013
2.75	15.643	0.063928	3.15	23.336	0.042852	3.55	34.813	0.028725
2.76	15.800	.063292	3.16	23.571	.042426	3.56	35.163	.028439
2.77	15.959	.062662	3.17	23.807	.042004	3.57	35.517	.028156
2.78	16.119	.062039	3.18	24.047	.041586	3.58	35.874	.027876
2.79	16.281	.061421	3.19	24.288	.041172	3.59	36.234	.027598

Table A.3 (cont.)

x	e^x	e^{-x}	x	e^x	e^{-x}	x	e^x	e^{-x}
3.60	36.598	0.027324	4.00	54.598	0.018316	4.40	81.451	0.012277
3.61	36.966	.027052	4.01	55.147	.018133	4.41	82.269	.012155
3.62	37.338	.026783	4.02	55.701	.017953	4.42	83.096	.012034
3.63	37.713	.026516	4.03	56.261	.017774	4.43	83.931	.011914
3.64	38.092	.026252	4.04	56.826	.017597	4.44	84.775	.011796
3.65	38.475	0.025991	4.05	57.397	0.017422	4.45	85.627	0.011679
3.66	38.861	.025733	4.06	57.974	.017249	4.46	86.488	.011562
3.67	39.252	.025476	4.07	58.557	.017077	4.47	87.357	.011447
3.68	39.646	.025223	4.08	59.145	.016907	4.48	88.235	.011333
3.69	40.045	.024972	4.09	59.740	.016739	4.49	89.121	.011221
3.70	40.447	0.024724	4.10	60.340	0.016573	4.50	90.017	0.011109
3.71	40.854	.024478	4.11	60.947	.016408	4.51	90.922	.010998
3.72	41.264	.024234	4.12	61.559	.016245	4.52	91.836	.010889
3.73	41.679	.023993	4.13	62.178	.016083	4.53	92.759	.010781
3.74	42.098	.023754	4.14	62.803	.015923	4.54	93.691	.010673
3.75	42.521	0.023518	4.15	63.434	0.015764	4.55	94.632	0.010567
3.76	42.948	.023284	4.16	64.072	.015608	4.56	95.583	.010462
3.77	43.380	.023052	4.17	64.715	.015452	4.57	96.544	.010358
3.78	43.816	.022823	4.18	65.366	.015299	4.58	97.514	.010255
3.79	44.256	.022596	4.19	66.023	.015146	4.59	98.494	.010153
3.80	44.701	0.022371	4.20	66.686	0.014996	4.60	99.484	0.010052
3.81	45.150	.022148	4.21	67.357	.014846	4.61	100.48	.009952
3.82	45.604	.021928	4.22	68.033	.014699	4.62	101.49	.009853
3.83	46.063	.021710	4.23	68.717	.014552	4.63	102.51	.009755
3.84	46.525	.021494	4.24	69.408	.014408	4.64	103.54	.009658
3.85	46.993	0.021280	4.25	70.105	0.014264	4.65	104.58	0.009562
3.86	47.465	.021068	4.26	70.810	.014122	4.66	105.64	.009466
3.87	47.942	.020858	4.27	71.522	.013982	4.67	106.70	.009372
3.88	48.424	.020651	4.28	72.240	.013843	4.68	107.77	.009279
3.89	48.911	.020445	4.29	72.966	.013705	4.69	108.85	.009187
3.90	49.402	0.020242	4.30	73.700	0.013569	4.70	109.95	0.009095
3.91	49.899	.020041	4.31	74.440	.013434	4.71	111.05	.009005
3.92	50.400	.019841	4.32	75.189	.013300	4.72	112.17	.008915
3.93	50.907	.019644	4.33	75.944	.013168	4.73	113.30	.008826
3.94	51.419	.019448	4.34	76.708	.013037	4.74	114.43	.008739
3.95	51.935	0.019255	4.35	77.478	0.012907	4.75	115.58	0.008652
3.96	52.457	.019063	4.36	78.257	.012778	4.76	116.75	.008566
3.97	52.985	.018873	4.37	79.044	.012651	4.77	117.92	.008480
3.98	53.517	.018686	4.38	79.838	.012525	4.78	119.10	.008396
3.99	54.055	.018500	4.39	80.640	.012401	4.79	120.30	.008312

Table A.3 (cont.)

x	e^x	e^{-x}
4.80	121.51	0.008230
4.81	122.73	.008148
4.82	123.97	.008067
4.83	125.21	.007987
4.84	126.47	.007907
4.85	127.74	0.007828
4.86	129.02	.007750
4.87	130.32	.007673
4.88	131.63	.007597
4.89	132.95	.007521
4.90	134.29	0.007447
4.91	135.64	.007372
4.92	137.00	.007299
4.93	138.38	.007227
4.94	139.77	.007155
4.95	141.17	0.007083
4.96	142.59	.007013
4.97	144.03	.006943
4.98	145.47	.006874
4.99	146.94	.006806
5.00	148.41	0.006738

Values beyond the range of this table may be obtained by

$$e^{x+y} = e^x \cdot e^y$$

Table A.4 Single-Payment Compound Amount Factor

Conversion Periods	*Rate r*							
n	0.01	0.02	0.03	0.04	0.05	0.06	0.07	0.08
1	1.010	1.020	1.030	1.040	1.050	1.060	1.070	1.080
2	1.020	1.040	1.061	1.082	1.103	1.123	1.145	1.166
3	1.030	1.061	1.093	1.125	1.158	1.191	1.225	1.260
4	1.041	1.082	1.126	1.170	1.216	1.262	1.311	1.360
5	1.051	1.104	1.159	1.217	1.276	1.338	1.402	1.469
6	1.061	1.126	1.194	1.265	1.340	1.419	1.500	1.587
7	1.072	1.148	1.230	1.316	1.407	1.504	1.605	1.714
8	1.083	1.171	1.267	1.369	1.477	1.594	1.718	1.851
9	1.094	1.195	1.305	1.423	1.551	1.689	1.838	1.999
10	1.105	1.218	1.344	1.480	1.629	1.791	1.967	2.159
11	1.116	1.243	1.384	1.540	1.710	1.898	2.104	2.332
12	1.127	1.268	1.426	1.601	1.790	2.012	2.252	2.518
13	1.138	1.294	1.469	1.666	1.886	2.133	2.410	2.720
14	1.149	1.319	1.513	1.731	1.980	2.261	2.579	2.937
15	1.161	1.345	1.558	1.801	2.079	2.397	2.759	3.172
16	1.172	1.372	1.605	1.873	2.183	2.540	2.952	3.426
17	1.184	1.400	1.653	1.948	2.292	2.693	3.159	3.700
18	1.196	1.428	1.702	2.026	2.407	2.854	3.380	3.996
19	1.208	1.457	1.754	2.107	2.527	3.026	3.617	4.316
20	1.220	1.486	1.806	2.191	2.653	3.207	3.870	4.661
21	1.232	1.516	1.860	2.279	2.786	3.400	4.140	5.034
22	1.245	1.546	1.916	2.370	2.925	3.604	4.430	5.437
23	1.257	1.577	1.974	2.465	3.072	3.813	4.741	5.871
24	1.270	1.608	2.033	2.563	3.225	4.049	5.072	6.341

Source: This table is abridged from *Handbook of Financial Mathematics* by Justin H. Moore. Copyright © 1929 by Prentice-Hall, Inc.

Table A.4 (cont.)

Conversion Periods	*Rate r*							
n	0.01	0.02	0.03	0.04	0.05	0.06	0.07	0.08
25	1.282	1.641	2.094	2.666	3.386	4.291	5.427	6.848
26	1.295	1.673	2.157	2.773	3.556	4.549	5.807	7.396
27	1.308	1.707	2.221	2.883	3.733	4.822	6.214	7.988
28	1.321	1.741	2.288	2.999	3.920	5.111	6.649	8.627
29	1.335	1.776	2.357	3.119	4.116	5.418	7.114	9.317
30	1.347	1.811	2.427	3.243	4.322	5.743	7.612	10.063
31	1.361	1.848	2.500	3.373	4.538	6.088	8.145	10.868
32	1.375	1.885	2.575	3.508	4.765	6.453	8.715	11.737
33	1.389	1.922	2.652	3.648	5.003	6.840	9.325	12.676
34	1.402	1.961	2.732	3.794	5.253	7.251	9.979	13.690
35	1.417	2.000	2.814	3.946	5.516	7.686	10.677	14.785
36	1.430	2.040	2.898	4.104	5.792	8.147	11.424	15.968
37	1.445	2.080	2.985	4.268	6.081	8.636	12.224	17.246
38	1.460	2.122	3.075	4.439	6.385	9,154	13.079	18.625
39	1.474	2.165	3.167	4.616	6.705	9.703	13.995	20.115
40	1.489	2.208	3.262	4.801	7.040	10.286	14.974	21.724
41	1.503	2.252	3.360	4.993	7.391	10.902	16.023	23.462
42	1.519	2.297	3.461	5.193	7.761	11.557	17.144	25.339
43	1.534	2.343	3.565	5.400	8.150	12.250	18.344	27.367
44	1.549	2.390	3.671	5.617	8.557	12.985	19.628	29.556
45	1.565	2.438	3.781	5.841	8.985	13.765	21.002	31.920
46	1.580	2.487	3.895	6.075	9.434	14.590	22.473	34.474
47	1.596	2.536	4.011	6.318	9.905	15.466	24.046	37.232
48	1.612	2.587	4.132	6.571	10.401	16.394	25.729	40.210
49	1.628	2.639	4.256	6.833	10.921	17.378	27.530	43.427
50	1.645	2.691	4.384	7.107	11.467	18.420	29.457	46.902

INDEX